Fundamentals of Structural Mechanics, Dynamics, and Stability

Fundamentals of Structural Mechanics, Dynamics, and Stability

A.I. Rusakov

CRC Press
Taylor & Francis Group
Boca Raton London New York

CRC Press is an imprint of the
Taylor & Francis Group, an **informa** business

First edition published [2021]
by CRC Press
6000 Broken Sound Parkway NW, Suite 300, Boca Raton, FL 33487-2742

and by CRC Press
2 Park Square, Milton Park, Abingdon, Oxon, OX14 4RN

Library of Congress Cataloging-in-Publication Data

Names: Rusakov, A. I. (Alexander Ivanovich), 1955- author.
Title: Fundamentals of structural mechanics, dynamics, and stability / A.I. Rusakov.
Description: First edition. | Boca Raton : CRC Press, 2021. | Includes index.
Identifiers: LCCN 2020026403 (print) | LCCN 2020026404 (ebook) | ISBN 9781498770422 (hardback) | ISBN 9780429155291 (ebook)
Subjects: LCSH: Structural analysis (Engineering)--Textbooks. | Structural dynamics--Textbooks. | Structural stability--Textbooks.
Classification: LCC TA645 .R87 2021 (print) | LCC TA645 (ebook) | DDC 624.1/7--dc23
LC record available at https://lccn.loc.gov/2020026403
LC ebook record available at https://lccn.loc.gov/2020026404

ISBN: 978-1-4987-7042-2 (hbk)
ISBN: 978-0-429-15529-1 (ebk)

Typeset in Times
by Deanta Global Publishing Services, Chennai, India

Contents

SECTION IV Hinged Beams

SECTION V Statically Determinate Trusses

SECTION VI Energy Methods in Deflection Analysis

SECTION VII Force Method

SECTION VIII *Displacement Method and Mixed Method*

SECTION IX *Plastic Behavior of Structures*

SECTION XII Dynamics of Elastic Systems

About the Author

Alexander Ivanovich Rusakov has been engaged in scientific and teaching work in academic and design institutions for over 35 years, continues his scientific research in construction design, and publishes scientific papers and textbooks. Dr. Rusakov was born in 1955 in the city of Rostov-on-Don (RF), and graduated from Rostov State University (Rostov-on-Don, RF) with a degree in physics. From 1984 to 2001, he taught at Rostov Military Institute of Rocket Forces, where he defended his candidate thesis in 1988 and then his doctoral thesis in 1999. From 2001 to 2006, he held the chair in structural mechanics at Rostov State University of Transport Communication. From 2006 to 2013, Dr. Rusakov worked in design organizations of Rostov-on-Don, holding engineering positions where he continued scientific investigation in construction design and educational activity with engineering personnel. His latest position is chief designer at South Regional Scientific Research and Design Institute of City Planning (Rostov-on-Don). Dr. Rusakov was the top lecturer for a number of mathematical and technical courses as well as courses in mechanics, including mechanics of materials, structural mechanics, and theory of elasticity. He is the sole author of two books, *Strength of Materials* and *Structural Mechanics*. Some studies and research results by Dr. Rusakov are presented at the site http://www.rusakov.donpac.ru/index1.htm.

Introduction

This textbook is intended for teaching students in civil engineering specialties. The book is structured as a lecture course, which should be convenient for the instructor and understandable for the student. Every chapter of the book represents an expanded lecture. The author assumes that the instructor will finalize suggested chapters in order to shorten the presentation in specified boundaries of contact hours, whereas the student will find a way to read every proposed chapter thoroughly.

The composition and consistency of topics in the presented course is usual for structural analysis being studied at a bachelor level. The presented items, however, address a broader scope of this science, namely:

- The theory of displacements in elastic systems, which should precede studying the methods of analysis of statically indeterminate structures, is given in a wider extent than usually is given by lectors;
- The theory of analysis of ductile bar systems is given along with the proof of all principal theorems and some provisions of this theory which are usually formulated without proof;
- The finite element method is presented not only in the frames of the direct stiffness method but also in relation to planar bodies;
- In conclusion of the course, two topics are given which could constitute a separate discipline: "Stability and Dynamics of Structures."

The presented book was conceived by the author while reading courses in mechanics of deformed bodies in Rostov State University of Transport Communication during the 2000–2006 period. Already at that time, I tried to construct my lectures as the continuation of courses in mathematics: everything which should be taken for granted I stated as starting postulates; I proved all or almost all which could be proved – often by formulating assertions as theorems. This mathematical style is continued in the given book because students like it. It should not be thought that students dislike the dryness of the lecture exposition. It is much more important for them that the teacher's arguments are clear and persuasive. It is also important for students to see how abstract mathematical provisions suddenly acquire the brightness of life when they give new conclusive inferences wanted directly in the work of an engineer – that is, when a civil engineering student is aware that he couldn't design a building without this math.

So, this book is designed in a mathematical style, is abundant in theorems with proofs, and thus is longer and more detailed than usual lecture courses. Before shortening exposition, the instructor should have a look into future topics. Example: Chapter 17 contains the extension of the theory of displacements in elastic systems; much attention is paid to substantiation of positive definiteness of a flexibility matrix, though the concept of positive definiteness is not required in the subsequent articulation of the displacement method. But this property of a flexibility matrix will be needed in setting forth the dynamics of structures with multiple degrees of freedom. Having removed this issue from the lecture based on Chapter 17, the instructor is bound to return to it in the lecture based on Chapter 31.

As noted above, the following topics could be expanded into special courses: the finite element method, the theory of stability, and the dynamics of structures. I believe that shortened exposition of these courses in the given book is appropriate, because a bachelor's degree holder must have the perception of them. Students enrolled in master's degree programs as well as students specializing in the design of unique and responsible structures (e.g., bridges in complicated environments or hazardous industry facilities) should get more extensive training in these disciplines than is supposed in my chapters. Nevertheless, this doesn't mean that the three topics are treated in a layman's manner "little about a lot of things" in this book. Rather, I select the most interesting (maybe highly complex) issues from a lot of possible ones and try to explore them in detail. Such is the case, for example, of the displacement method for analysis of structure stability presented in Chapters 28 and 29, the exposition of which has replaced other well-known methods – for example, the Rayleigh-Ritz method.

The usual duration of an academic lecture is two hours. While devising the course, the author was never rigorously held by these time frames: some chapters, being set forth completely, may take up to four hours. In the case of excessively long chapters, I sought to select such a sequence of statement that the chapter might be painlessly reduced by finalizing it at one of the items. Naturally, the shortening of the chapter implies the vision of the whole topic by the instructor.

Many chapters have supplements wherein there are proofs of statements used in the text of the chapter, nontrivial examples, or clarifications of theoretical provisions. The supplements are mandatory for study under the direction of the instructor.

The basic ideas of structural analysis can be set forth in an illustrative form for plane systems and then generalized for space systems. In this book, a major object of research is the planar bar system. The problems of analysis of planar systems are supplemented by short analysis of spatial systems.

While working upon the book, the author has been taking into account the curriculum requirements of US universities. Whenever the chapter material necessitated addressing building regulations, the building codes of the US were employed.

Definitions and new notions are emphasized in italic font. Most important assertions are emphasized in bold.

Alexander Rusakov
Rostov-on-Don, Russia

Section I

Basic Concepts

This section contains the principal terms of structural mechanics with which the book operates – among them there are definitions of design model; geometrically stable, unstable, and instantaneously unstable systems. Principally, the plane systems are addressed in order to give afterwards the generalization of obtained results to space systems. The classification of connections between system members is presented, and geometrically stable systems are categorized. The concepts of static determinacy and indeterminacy are introduced, and the criteria of systems' geometrical stability, static determinacy and indeterminacy are substantiated.

1 Kinematic Analysis Basics in Structural Mechanics

1.1 BASIC TERMS: GEOMETRICAL STABILITY AND INSTABILITY OF SYSTEMS; DISCS, CONSTRAINTS, DEGREES OF FREEDOM

Objects of Structural Mechanics are constructions which represent assembly of constructive elements (bars, shells, plates, and massive bodies) connected with each other in a single entity. Engineering investigation of a construction is pursued on the basis of its design model. *The design model is a simplified representation of construction with an account of basic data which define the object's behavior under loading.*

Figure 1.1 shows the transition from real construction (a bridge) to its complete design model, where real bars are replaced by idealized smooth bars (with no notches, brackets, and other peculiarities), and the next transition to a simplified design model where the bar joints are supposed to be hinged.

The constructions and their design models differ in the elements composing them. Among other things under consideration, there are bar systems composed by bars; thin-walled systems composed by plates and shells; and massive systems composed by bodies approximately the same size in three dimensions.

Further, we mainly take under consideration plane bar systems, which have centroidal principal axes of the cross-sections of bars in the same plane. We call a plane bar system or system of plane bodies having the same middle plane a plane mechanical system. Plane systems are often incorporated into space systems. Investigation of plane systems is a basis for analysis of space systems.

Internally geometrically stable systems are systems of solid bodies – system elements – connected with each other, where the shape alteration is allowed only with elements' deformation.

Internally geometrically unstable (internally substatic) systems are systems of solid bodies connected with each other, where the finite relative displacements of bodies are possible without their deformation.

Internally instantaneously unstable systems are systems of solid bodies connected with each other, where the infinitesimal relative displacements of bodies are possible without their deformation.

In all these definitions, a system is assumed as being separated from surrounding bodies, i.e., the system's elements interact only among themselves. The word "internally" in these definitions may be omitted for systems which initially are not in contact with surrounding bodies.

In the last definition, we imply that whereas deformation arises from displacement, it has the second or higher infinitesimal order (Karnovsky and Lebed 2010).

The instance of a stable system is a hinged triangle; the instance of a substatic system is a plane hinged quadrilateral; and the instance of an instantaneously unstable system is two bars situated in line and a massive body (the earth), connected by hinges (Figure 1.2).

We shall term a plane system where the mutual arrangement of points doesn't change under the action of external forces (or the changes are negligible in a problem under consideration) as a *disc*. Those parts of systems where there are no noticeable mutual displacements we shall consider as discs.

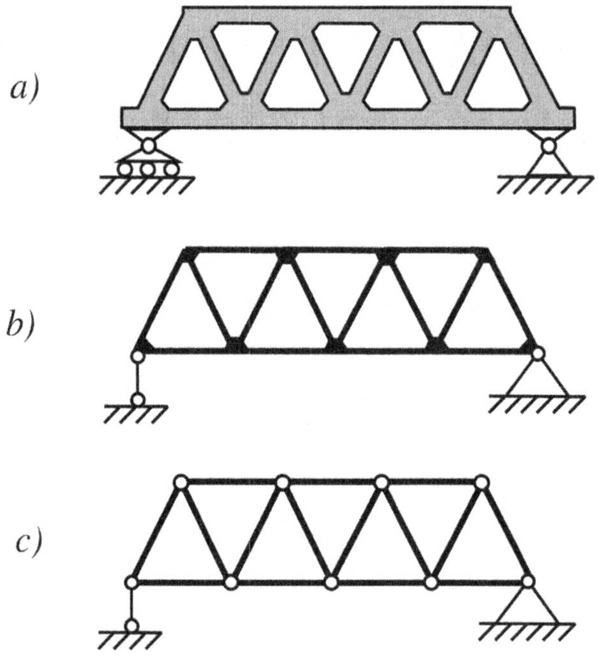

FIGURE 1.1 Transition from real construction to design model.

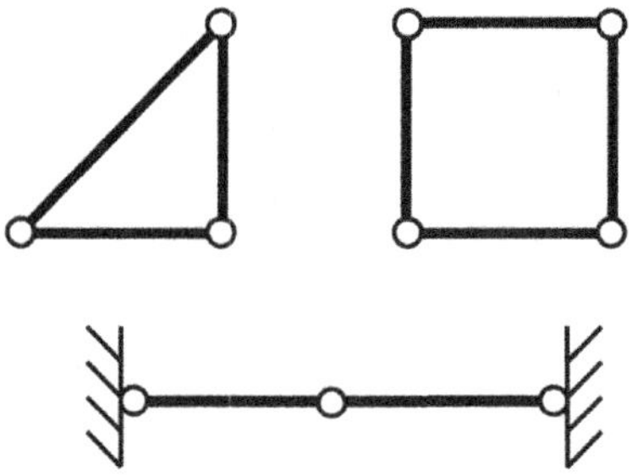

FIGURE 1.2 Examples of stable, substatic, and instantaneously unstable systems.

If external forces, applied to a system, and displacements of a system are considered in a disc-referenced coordinate system, then we call this disc the earth. The earth is not considered as belonging to a system unless otherwise arranged.

We call the ability of a system to move relative to the earth with no deformation of system elements "freedom of a system." We call the ability of a system to change shape by displacements of elements taken for rigid bodies "instability."

We call any parameter, belonging to the totality of independent parameters necessary to specify the system location relative to the earth, the system's degree of freedom (DOF). On specifying the system location, all system elements are considered non-deformable, that is, discs. The number of system's DOFs F^* serves as a measure for freedom of a system.

We call the number of independent parameters, specifying the system location relative to one of its discs taken for motionless, the system's degree of instability I. Quantity I is a measure of the system's instability.

For an isolated disc in the plane, we have $F^* = 3$, because such a disc can get two translational displacements along coordinate axes and one rotational displacement around any point.

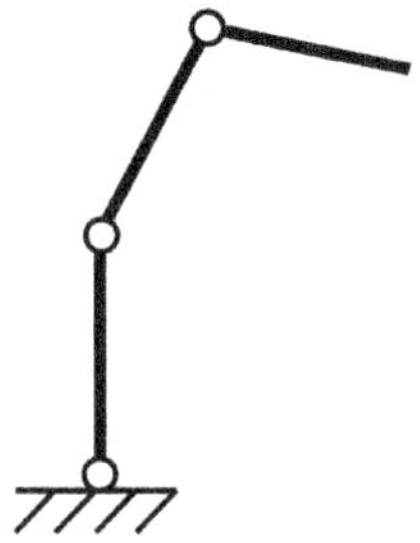

FIGURE 1.3 System of three members.

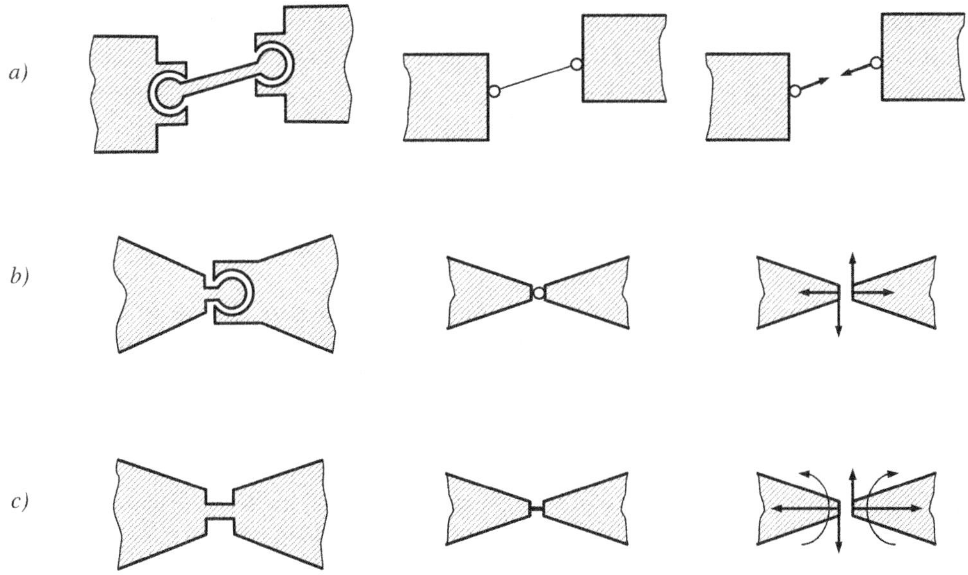

FIGURE 1.4 Connections of principle types and their reactions (left to right: outline, symbolic representation, reactions produced by connection): (a) Connection of the first type – link; (b) Connection of the second type – hinge; (c) Connection of the third type – rigid joint.

If the earth is included in the system composition, then $F^* = I$. If a system is not connected with the earth, then $F = I + 3$. In the example in Figure 1.3, $F^* = 3$ but $I = 2$.

Discs are connected with each other by specific connections. In plane systems, two discs can be connected by three principle types of connections as follows:

- *Connection of the first type* is a link, that is, a rigid rod with cylindrical hinges at the ends (Figure 1.4a). This connection can be called *a simple rod*. This one opposes translational displacement of one disc relative to the other and deletes one degree of freedom of two connected discs. The connection develops reactions acting up on the linked discs. These reaction forces are exerted along the axis line of the rod and have opposite directions.
- *Connection of the second type* is a cylindrical hinge, or so-called pin joint (Figure 1.4b). This one opposes any translational displacements of one disc relative to the other. The connection deletes two degrees of freedom of two joined discs. Possibility of external force exertion upon the connection is let in. If there is no external force applied to the connection, the reaction forces are equal, opposite, and act along the line crossing the hinge. Usually we decompose each reaction for this joint in two vector addends along given directions.

- *Connection of the third type* is a rigid joint, sometimes also called a fixed joint (Figure 1.4c). This one doesn't allow translational or rotational displacements of one disc relative to the other. It deletes three degrees of freedom and joins two discs in one. Possibility of exertion of external concentrated force and external force moment upon the connection is let in. A rigid joint develops the combination of reactions upon each assembled disc, consisting of a reaction force of any direction (total reaction force) and moment of forces. If there is no external loading to the connection, the reaction forces are equal, opposite, and act along the line crossing the joint; the reaction moments are equal and opposite in the sense of rotation.

Connections of principle types can take constructive realization different from the kind shown in Figure 1.4. For example, connection of the first type can be realized by a roller squeezed between plane surfaces of discs, and connection of the second type is formed by a sharp edge of one disc pressed to another (Figure 1.5a and b). The type of connection is defined not by the construction, but by the DOFs deleted with connection and the reactions produced. A hinge joint can be composed by two links, shown in Figure 1.4a, with a common hinge (Figure 1.5c). A rigid joint can be composed by three links (Figure 1.5d).

Sometimes, connections of the second and third types are considered not as devices but as separation surfaces between discs (Figure 1.6a). In this lecture course, we interpret the connections of the second or third types as devices jointing two discs, which **is a small size in comparison to discs** and can take up the external forces (Figure 1.6b).

Connections are also called constraints because they delete a system's degrees of freedom.

Three listed types of constraints in case of connecting with the earth prove to be hinged movable support, hinged immovable support, and rigid support, respectively.

The principal connections considered above join together two discs and are called simple connections. If a hinge or rigid joint binds up more than two discs (Figure 1.7), then such connections are called multiple. **Multiple connections can be considered as a few adjacent simple connections, and they replace as many simple hinges or rigid joints, as many discs are bound by them without one.** Like in the case of principle connections, a multiple hinge can be subjected to concentrated load, and multiple rigid joints can be subjected to concentrated load and external force moment.

Three or more discs can be connected not only by multiple hinged or rigid joints but also by mixed connection. Mixed connection is both hinged and rigid connections consolidated in single unit at small distances from one another (Figure 1.8).

Let us consider the constraints' character in stable systems composed of two or three discs with a minimum number of simple connections.

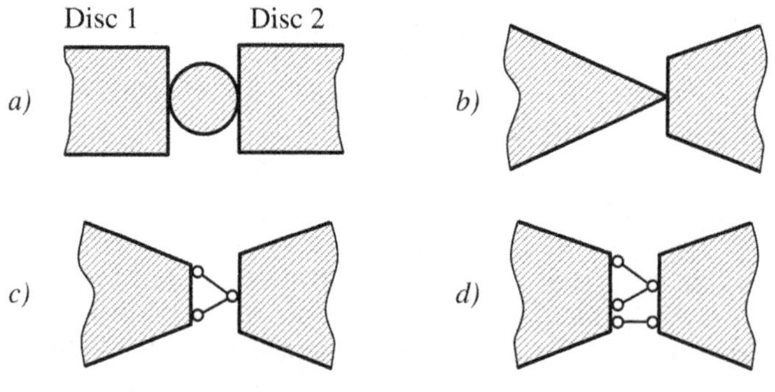

FIGURE 1.5 Possible constructions of connections: (a) Connection of the first type; (b) and (c) Connections of the second type; (d) Connection of the third type.

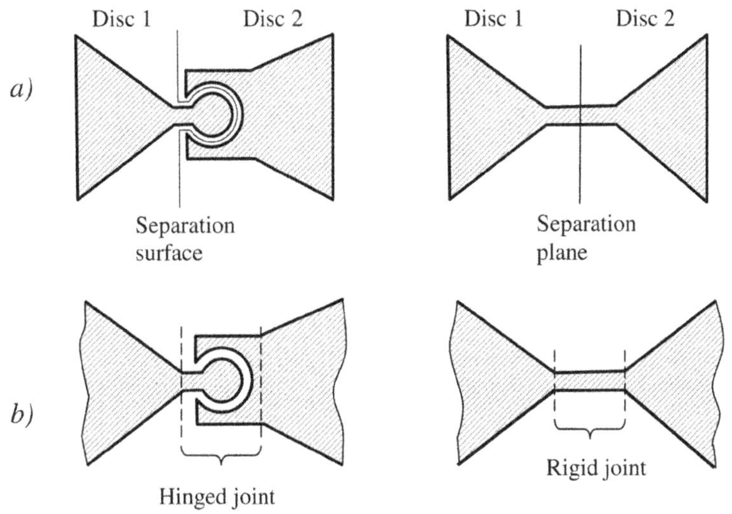

FIGURE 1.6 Interpretations of connections: (a) Connection as separation surface; (b) Connection as joint device.

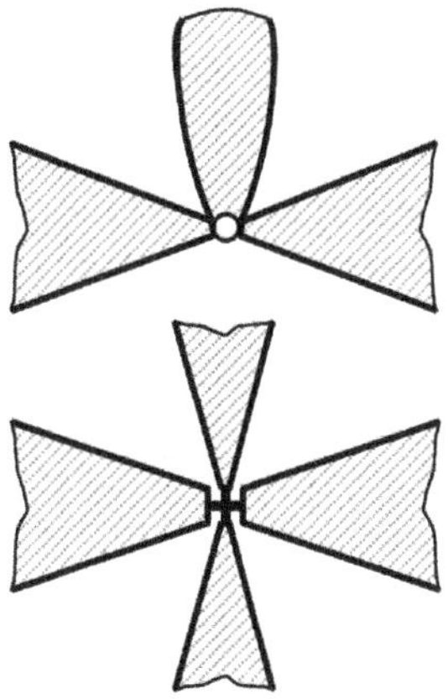

FIGURE 1.7 Multiple connections.

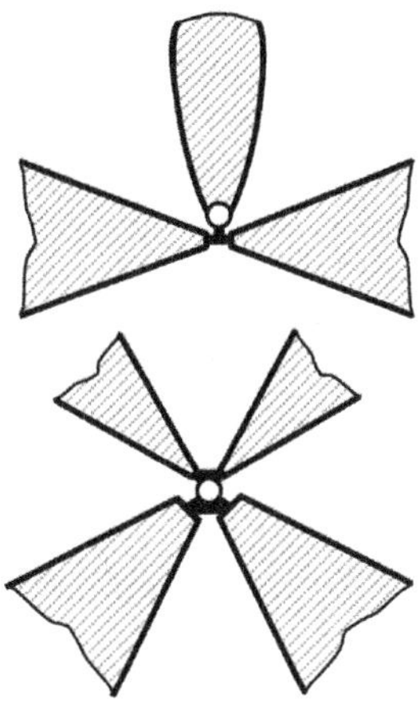

FIGURE 1.8 Mixed connections.

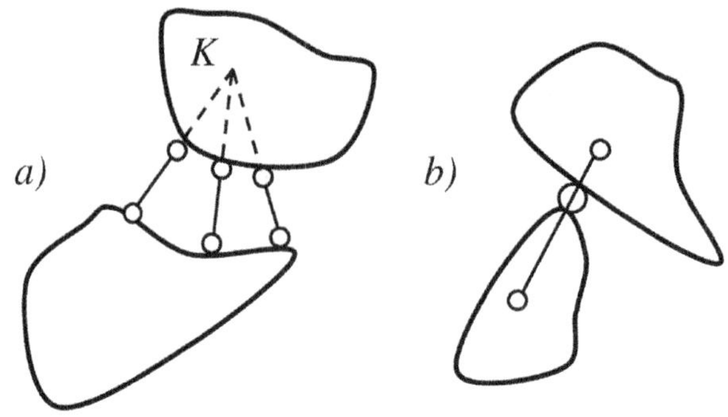

FIGURE 1.9 Instantaneously unstable systems.

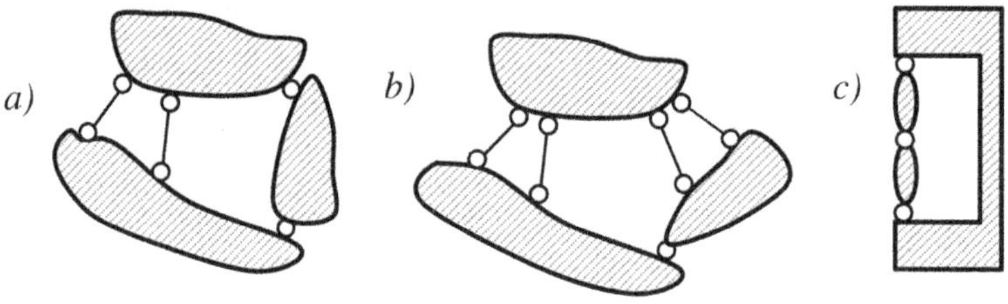

FIGURE 1.10 Three-disc systems.

1. **Two-disc systems**. We get stable systems of two discs with a minimum number of connections if we link the discs either by three simple rods, or by the rod and a hinge, or by a rigid joint. Removal of any connection results in instability of a disc couple. In the case of three links joining together discs, it is sufficient for stability that these links don't intersect at one point and are not collateral. If three or more links intersect in a common point, then we get an instantaneously unstable system, where the links don't oppose infinitesimal mutual rotation of two discs around instantaneous pivot-point K formed by intersection of links (Figure 1.9a). In case of a hinge and a link joining discs, it is sufficient for stability that the line of link doesn't intersect the hinge. Otherwise, the system happens also to be instantaneously unstable as it is shown in Figure 1.9b.
2. **Three-disc systems**. We can assemble a stable system of three discs by linking a stable two-disc system to the third disc using the rules described above. As well, we can point out special cases of forming a three-disc system when removal of but one connection makes a system unstable. Namely: three discs can be associated in pairs when each association is a hinge or couple of links (examples in Figure 1.10a, b). The stability condition of such a system is the requirement that each hinge and each intersection point of links joining together two discs – all three points – don't lie in a straight line. The matter is that intersection point of the link couple is the instantaneous pivot-point of one disc relative to other, and if instantaneous pivot-points of three disc couples happen to lie in a straight line, then there is the possibility of infinitesimal rotation in the three-disc system, when one pivot-point moves along the normal to the line connecting two other pivot-points. In other words, the system becomes instantaneously unstable like in Figure 1.10c.

Stable systems at any number of discs which are formed by the rules of two-disc or three-disc association are called simple systems.

The connections are distinguished not only by the reaction forces they develop and by the displacements they allow, but also by their role in stability and determinacy of a given system. By the functional feature every constraint in a system can be:

- Required, if its removal makes a system unstable;
- Redundant, if its removal leaves a system stable;
- False, if it allows infinitesimal mutual displacements of discs bound through it.

False connections exist in instantaneously unstable systems only.

For any disc of a system, we can set up three equations of statics where the unknowns are the reactions of connections: two equilibrium conditions for forces, applied to a disc, and one condition for moments of forces. To set up such equations, we imaginarily cut the disc off from a system, and at the places of removed connections apply unknown reaction forces. In the same way, any connection of a system can be mentally cut off, and the equations of statics for the one be set up. (In the next item of the chapter we shall set up these equations for simple connections.) The combination of equations of statics for all discs and connections composes the system's static equilibrium equation set. For a system in equilibrium, the totality of all constraint reactions obeys this equation set.

The constraint reactions can be statically determinate or indeterminate. A reaction is determinate if determined from the equation set of static equilibrium uniquely, and indeterminate if determined from the set ambiguously. It is known that **the reactions of required links are statically determinate, whereas the reactions of redundant constraints are statically indeterminate.**

We briefly clarify this property. If the equation set of statics had an ambiguous solution for reaction of the required link, then the nonzero reaction of this link would obey the named equation set for the unloaded system, but this is impossible. On the contrary, in a redundant link, we can artificially induce the reaction force of any wanted value with system stability being saved, i.e., this reaction for given loads is undeterminable.

The reaction of every false link in two-disc systems in Figure 1.9 is statically indeterminate, because in such a link, an arbitrary artificial reaction can be induced, which is counterbalanced by reaction forces developed by the adjacent connections (connection). However, there may be cases wherein the reaction of the false link appears to be statically determinate.

Now then, a system can be internally stable, unstable, and instantaneously unstable. These three possible attributes of a system we refer to as *attributes of stability*. While considering them, we assume that the system's connections with surrounding bodies have been removed.

A system that has connections with the earth is called attached. A system that has no connections with the earth is called free.

For attached systems three properties of external stability are considered along with the stability attributes – namely, external stability, instability, and instantaneous instability. *Three possible properties of external stability are the stability attributes of an attached system in which the earth has been included among its discs.* Example: an externally stable system is an attached system, which is stable in totality with the earth. Externally stable systems are shortly called structures. The properties of external stability are also called properties of stability relative to the earth.

All said before about assembling the stable systems from two or three discs refers to systems stable relative to the earth, if the earth is assumed as one of the discs.

The properties of external stability are to be introduced besides the attributes of stability, because the earth usually is not assumed to belong to a system, whereas in construction engineering, stability does matter for a system taken in totality with the earth. Further, the stability, instability, and instantaneous instability of attached systems are assumed by default as the properties of external stability.

1.2 NUMBER OF CONSTRAINTS AS CRITERION OF SYSTEM STABILITY AND DETERMINACY

From the mechanics of materials, we know that forces applied to construction are divided into external and internal. External forces, in their turn, are divided into loads and support reactions.

If internal forces, acting in a stable mechanical system in equilibrium, are determined at arbitrary loads and non-deformed state of the system elements from equations of statics for the whole system or its separate parts, then such a system is called statically determinate or isostatic.

If internal forces, acting in a stable mechanical system in equilibrium, for some load assemblage are not determined from equations of statics, then such a system is called statically indeterminate.

Usually in the course of static determinacy analysis, it is supposed that a system is comprised of statically determinate discs, i.e., internal forces in any disc are uniquely determined by constraint reactions for given loads. That is why **static determinacy of a system of discs is generally thought to be reduced to the determinacy of constraint reactions from equations of statics.**

Let us consider the equations of statics for the principle types of connections. Let constraint A join together discs 1 and 2 (Figure 1.11). Then:

- From the equilibrium conditions of connection of the first type, we obtain that its reactions lie in the centerline of the link and are oppositely directed and equal in magnitude (Figure 1.11a):

$$R_2 = R_1; \tag{1.1}$$

- For the connection of the second type being under load $\boldsymbol{P}$ we have the vectorial equation of force equilibrium (Figure 1.11b):

$$-\boldsymbol{R}_1 + (-\boldsymbol{R}_2) + \boldsymbol{P} = 0,$$

  where $\boldsymbol{R}_i$ is the vector of reaction applied to the i-th disc and sign "–" points to the force action from the disc. Thus:

$$\boldsymbol{R}_2 = -\boldsymbol{R}_1 + \boldsymbol{P}; \tag{1.2}$$

  (The latter vectorial equality is represented as two scalar equalities for corresponding components of reactions.)

- For the connection of the third type under action of load $\boldsymbol{P}$ and external moment M, the equation of force equilibrium coincides with the preceding equation and gives a solution (1.2).

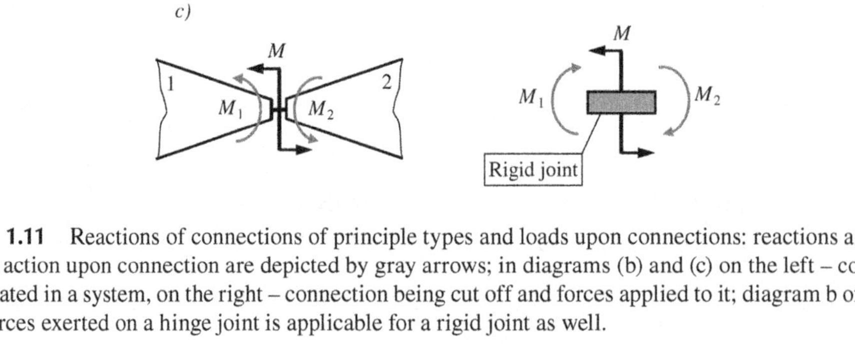

FIGURE 1.11 Reactions of connections of principle types and loads upon connections: reactions and forces of discs' action upon connection are depicted by gray arrows; in diagrams (b) and (c) on the left – connection incorporated in a system, on the right – connection being cut off and forces applied to it; diagram b of concentrated forces exerted on a hinge joint is applicable for a rigid joint as well.

For directions of force moment action, taken in accordance with Figure 1.11c, the equation of equilibrium of reactive moments M_1 and M_2 takes the form:

$$M_2 = -M_1 + M. \tag{1.3}$$

We see that in all these cases, the equilibrium equations for connection are able to represent the reactions applied to the second disc through reactions applied to first disc. Reactions, acting upon one of the discs, prove to be independent variables in these equations, and their number is equal to 1 for connection of the first type; 2 for connection of the second type; and 3 for connection of the third type.

Let the discs of a mechanical system be joined only by simple connections. In the statics equation set, written for all discs and constraints, we make the conversions as follows: for each connection we express reactions applied to one of discs by reactions applied to another disc in accordance with formulas (1.1)–(1.3), then make substitution of these reactions into equations of statics for discs. In such compacted form, the equation set of statics keeps only independent reactions of each connection and only equilibrium equations for discs. Further, we shall consider a compacted equation set of statics and speak only of independent reactions of constraints.

We consider the system which comprises D discs connected by simple connections in a number of L links, H hinges, and R rigid joints. Initially, we shall consider an attached system which links to the earth with simple rods in the number of L_{sup}.

Let us set up the relations expressing conditions of instability for such a system.

The total number of constraint reactions is defined by the value

$$N_{con} = L + 2H + 3R + L_{sup}. \tag{1.4}$$

If we represent connections of the second and third types as equivalent linkages by connections of the first type, then the recent formula specifies a total number of links. This number we call the *number of constraints (connections) in a system.*

Discs of a system with no connections have $3D$ DOFs. The number of DOFs which can be deleted by constraints is not greater than the number of constraints (1.4). So we get for the number of system DOFs:

$$F^* \geq 3D - (L + 2H + 3R + L_{sup}). \tag{1.5}$$

If a system is stable relative to the earth, then $F^* = 0$, and we come to *the necessary condition of stability for an attached system*:

$$3D \leq L + 2H + 3R + L_{sup}. \tag{1.6}$$

This condition is not sufficient because the connections may be situated with excess for some group of discs, but not to bind another group of discs in stable assembly. The question of stability or instantaneous instability of a system is solved by kinematic analysis.

If a system is stable and inequality (1.6) appears to be the equality, then we have a minimum possible number of constraints in a stable system. So, for a stable system under the condition:

$$3D = L + 2H + 3R + L_{sup} \tag{1.7}$$

– all constraints are required and the system is statically determinate.

Let us introduce *the degree of static indeterminacy* of a bar system as the difference between the number of unknown forces specifying the stress state of a system and the number of independent statics equations. The number of forces at the joints is the right-hand side of condition (1.6), and these forces specify the stress state of bars, whereas the number of statics equations is equal to the left-hand side of condition (1.6). As a result, we get the expression for the degree of indeterminacy of the attached bar system:

$$n = L + 2H + 3R + L_{sup} - 3D. \tag{1.8}$$

This number appears to be the number of redundant constraints in a system.

Relations (1.5)–(1.8) can be modified so they would enable us to make conclusions about instability and determinacy of free systems. In a free system, the total number of constraint reactions is defined by the value:

$$N_{con} = L + 2H + 3R. \tag{1.9}$$

This value is called the *number of connections between a system's discs* (*or number of constraints of the system's discs*), because it determines the total number of simple rods composing connections of all types.

With respect to any system's disc being supposed unmovable, the remaining portion of discs with no constraints has $3(D-1)$ DOFs. The number of DOFs which can be deleted by constraints is not greater than the number of constraints (1.9). Thus we get for the degree of instability:

$$I \geq 3(D-1) - (L + 2H + 3R). \tag{1.10}$$

If a system is stable then $I = 0$, and we get *the necessary condition of stability for a free system*:

$$3(D-1) \leq L + 2H + 3R. \tag{1.11}$$

For a stable system under the condition:

$$3(D-1) = L + 2H + 3R \tag{1.12}$$

– all constraints are required and system is statically determinate.

The left-hand side of relation (1.12) defines the number of statics equations for a free stable system. Note that, in comparison to an attached system, the number of independent statics equations decreases by 3. The point is that assemblage of loads exerted on a free plane system must satisfy three equilibrium conditions. These conditions make interdependent the equations of statics for discs and must be excluded from the complete set of these equations.

The number of internal forces in the joints is the right member of condition (1.12). The degree of indeterminacy of a bar system in this case takes the form:

$$n = L + 2H + 3R - 3(D-1). \tag{1.13}$$

SUPPLEMENT TO CHAPTER

Kinematic Analysis of Plane Systems by Examples

The purpose of kinematic analysis is to determine the kind of system: stable, unstable, or instantaneously unstable. After making a conclusion about the system type, it may be necessary to accomplish additional kinematic analysis to suggest additional constraints which make the system stable or to remove redundant constraints and attain static determinacy of the system.

In Figure 1S.1, we have a two-disc system where condition (1.6) takes the form:

$$3 \cdot 2 \leq 2 + 2 \cdot 1 + 4.$$

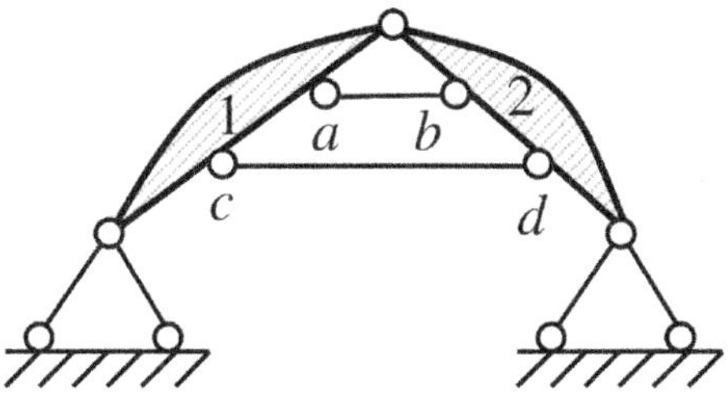

FIGURE 1S.1 Two-disc stable system.

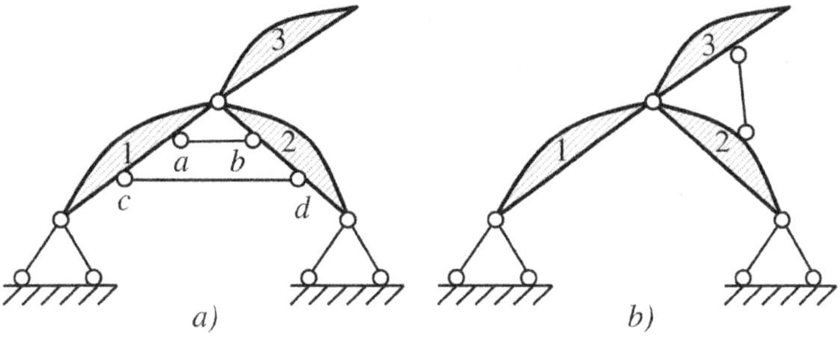

FIGURE 1S.2 Transition from unstable to stable system.

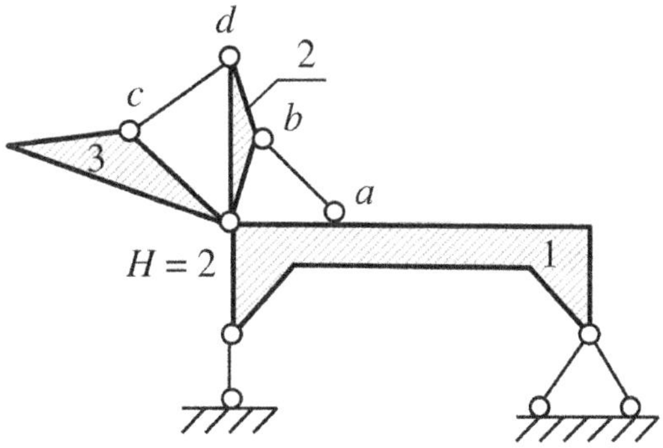

FIGURE 1S.3 Three-disc stable system.

The system may be stable. Remove two rods *ab* and *cd*, take the earth into account as a disc, and we come to a simple three-disc system. Thus, the system is stable and includes two redundant links. We make sure that, actually, there are two redundant links, by subtracting the number of possible DOFs from the total number of constraints in accordance with formula (1.8).

In Figure 1S.2a, we have a three-disc system built from the former system by hinged attachment of a third disc. The system is unstable though it has a redundant constraint:

$$n = (2 + 2 \cdot 2 + 4) - 3 \cdot 3 = 1.$$

We can come to a stable statically determinate system by removal of one link and replacement of another as shown in Figure 1S.2b.

The system in Figure 1S.3 is obtained by consecutive attachment of discs in an unchangeable way using a minimum number of links. The system is stable and statically determinate what is confirmed by equality (1.7):

$$3 \cdot 3 = 2 + 2 \cdot 2 + 3.$$

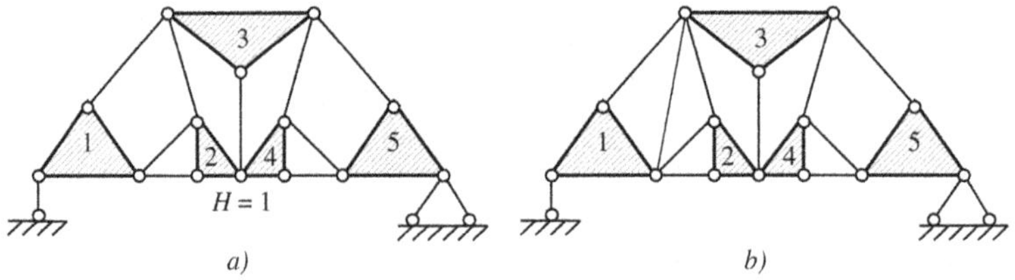

FIGURE 1S.4 Transition from unstable to stable simple system.

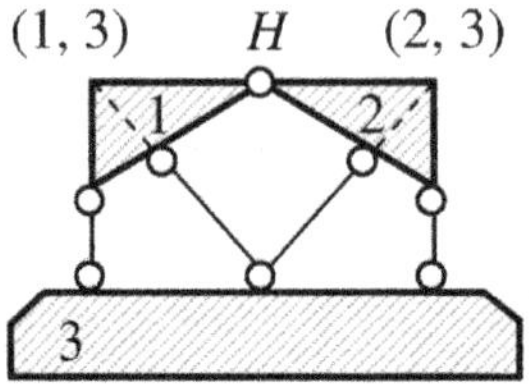

FIGURE 1S.5 Internally unstable system.

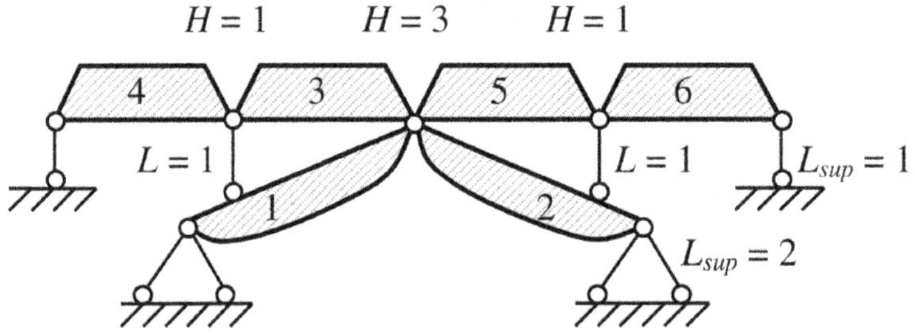

FIGURE 1S.6 Externally stable simple system.

The more interesting example is in Figure 1S.4a. The system is unstable, has five discs, nine rods, one hinge, and three support constraints. According to formula (1.5), this system has no less than one DOF:

$$F^* \geq 3 \cdot 5 - (9 + 2 \cdot 1 + 3 \cdot 0 + 3) = 1.$$

It is possible to establish deficient connections by assembling system elements into aggregative discs using the rules of joining noted above. We find out that discs No. 1, 2, 3, do not make up a stable three-disc system because of the lack of one link. If we enter this link into the system as it's depicted in Figure 1S.4b, then the further two remaining discs are joined in an invariable manner, and we get a statically determinate stable system.

The three-disc system in Figure 1S.5 is instantaneously unstable; the number of connections between disc couples corresponds to the rule of stable system building, but instantaneous pivot-points of discs (1, 3) and (2, 3) lie in a straight line with the hinge H.

Let us check the stability of the system in Figure 1S.6 by use of criterion (1.6):

$$3 \cdot 6 = 2 + 2 \cdot 5 + 3 \cdot 0 + 6.$$

The system may be stable. If we build this system by sequentially joining the discs to the earth in the order defined by their numbers, then the rules of a simple system's building are satisfied (the earth would be considered here as an additional disc).

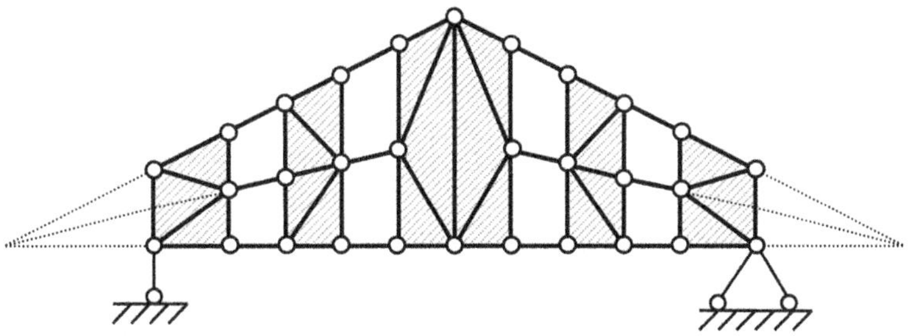

FIGURE 1S.7 Instantaneously unstable bar system.

The bar system in Figure 1S.7 is instantaneously unstable. The stable discs in this system are being marked with hatching, and it is shown that connections between these discs form instantaneous pivot-points. The reader is recommended to verify the fulfillment of condition (1.7) for this system.

Section II

Influence Lines

This section begins with the classification of bar systems. Then the section sets out the theory of influence lines: the basic terms are introduced; general properties of influence lines are studied and the peculiarities of a system response to the movable load series are established; the design position of load series on the load line is analyzed. Further, the section presents substantiation of design model for analysis of the response from the main beam of the bridge due to the action of vehicle. In conclusion, the section gives the specification of live load exerted upon highways and railroad bridges.

2 Movable Loads on a Beam

2.1 CLASSIFICATION OF BAR SYSTEMS

Plane bar systems are classified as follows.

We distinguish internally geometrically stable, unstable, and instantaneously unstable systems by number and placement of connection. Attached systems are distinguished as externally geometrically stable, unstable, and instantaneously unstable systems. Stable systems are separated into statically determinate – systems with required constraints only, and statically indeterminate – systems with redundant constraints. Geometrically stable attached bar systems are sometimes termed as framed structures.

By succession of assemblage, we distinguish simple and complex bar systems. Simple systems are formed by the rules of two-disc or three-disc association. The others are complex.

Framed structures are separated into thrusted and thrustless by the direction of their support reactions to vertical loading. In thrustless structures, vertical loading produces vertical support reactions only, whereas in thrusted structures vertical loading produces oblique reactions.

By constructive peculiarities and internal forces, we distinguish:

- *Hinged chains* – unstable systems consisting of tensed bars or discs, connected in series by hinges;
- *Beams* – thrustless structures formed by straight bars connected at the ends, positioned in a straight line, and working in bending. The example of a beam of three bars is shown in Figure 6.1;
- *Frames* – stable systems of bars connected together by rigid joints or by hinges and working in bending with tension-compression (Figure 2.1). Vertical frame members are referred to as *columns* (being compressed) and *hangers* (being tensed); horizontal frame members are referred to as *girders*;
- *Hinged trusses* – stable systems formed by straight bars joined together with hinges (Figure 7.1). Members in the outline of trusses, except vertical ones, are referred to as *chord members*. They constitute the top and bottom chords of the truss. The members inside the outline make up the web, or the lacing, of the truss. The regions between adjacent joints of the chord are called *panels*; sometimes the same term is used for chord members;
- *Trusses* – bar systems, where replacement of all rigid joints by hinges makes them into hinged trusses. The members of a truss under joint loading do work in tension-compression with additional bending produced by rigidity of joints;
- *Arches* – thrusted structures consisting of curved bars with convexity in the direction opposite to load. Arches work in compression with bending;
- *Combined structures* – combinations of the systems considered above assembled for composite action (Figure 2.2).

Statically determinate beams are classified as follows: *simple* ones – beams on two supports at the ends; *overhang ones* – beams on two supports with cantilevering end (or both ends); *cantilever ones* – beams fixed in one end; and *hinged ones* – structures assembled from two or more beams connected in series by hinges.

Bar systems can be completed by cables and cable subsystems. There are some peculiarities in the design of structures with cables, but, on the whole, cable members can be considered as tensed bars with hinged joints.

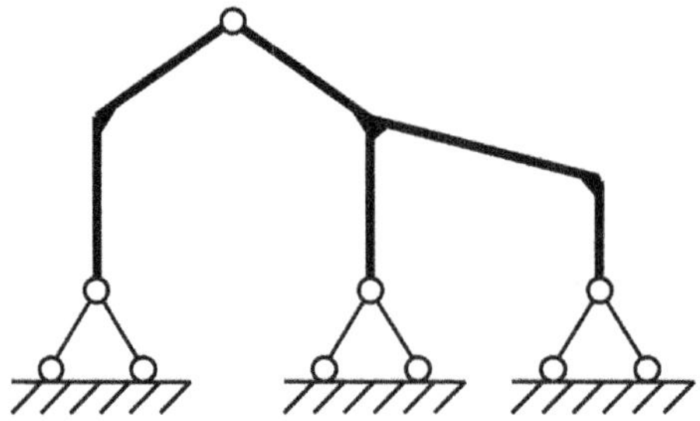

FIGURE 2.1 Example of frame.

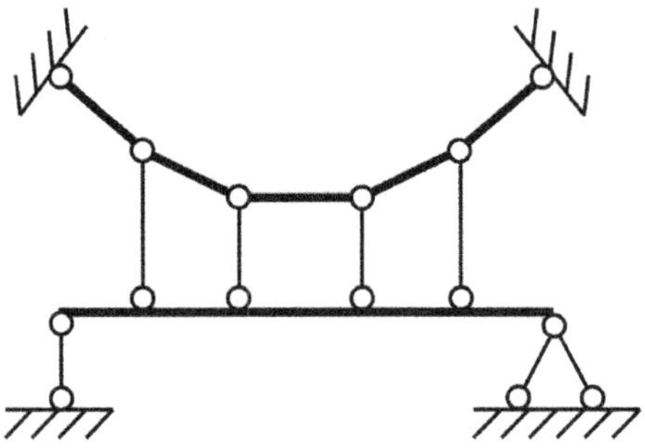

FIGURE 2.2 Combination of beam and chain.

2.2 INFLUENCE LINES AND PARTICULAR CASE OF BEAMS; PROPERTIES OF INFLUENCE LINES

Many a system is to be analyzed for the action of a vertical movable load, i.e., for an assemblage of vertical forces which changes location on a structure. For example, we can take the bridges where the vehicles move or lifting cranes including a load trolley with hoisting gear. Further, we confine ourselves to a plane series of vertical loads on a plane elastic structure. Elastic structures consist of elastic bodies bound by connections of a specific kind, which are called perfect connections. Connections of principle types studied in Chapter 1 belong to perfect ones. The concepts of perfect connection and elastic structure will be studied in Chapter 11. The important property of elastic structures is that they obey the superposition principle. This principle was studied in the mechanics of materials and states that stresses, strains, and displacements caused by an assemblage of loads exerted upon a structure are represented as the sum of stresses, strains, and displacements caused by each of loads acting separately.

We call the line of live loads motion the load line. In analysis of structures, we assume the composition of moving loads and their mutual position as given and investigate the action of a movable load series upon some member of a structure due to movement along the load line. *The position of the load, for which the internal forces in the member is most unfavorable, is referred to as the design position.* To determine the design position of a load and make further analysis of the construction member, the influence functions and lines are intended.

We call the dependence of given characteristic of the stress-strain state (SSS) of a structure versus the coordinate of concentrated unit load on the load line with no other loads being applied the *influence function* for this characteristic. *The graph of an influence function is referred to as an influence line.* Influence lines are constructed for reactions of connections, for internal forces at cross-sections of members, and deflections of these sections.

Influence lines are plotted by rules; we put the abscissa axis (termed as a basic line) athwart the load direction on the design diagram and point out the ordinate along the line of load action. Other rules (hatching, marking the signs and values) are the same as for the construction of an internal force diagram.

There are two approaches in the construction of influence lines for internal forces and reactions of connections, which are referred to as static and kinematic methods of construction.

In the case of statically determinate systems, the static method of construction consists in composing the equations of static equilibrium, which include the sought force and unit load's coordinate. After solving these equations, we obtain the wanted force as the function of coordinate, i.e., required influence function which is represented by an influence line. For instance, in Figure 2.3 the simple beam is depicted with the live load P specified by coordinate z. The support reactions are determined elementarily from statics equations in the form:

$$R_A = P\left(1-\frac{z}{l}\right); \quad R_B = P\frac{z}{l}.$$

Given $P = 1$ these relations are the influence functions, which define corresponding influence lines.

For the same beam, let us construct the influence line for a bending moment at section I (Figure 2.4). The bending moment is equal to either the sum of moments of right forces, provided that positive sense of rotation is counterclockwise with respect to the centroid of the section, or the sum of moments of left forces, provided that positive sense of rotation is clockwise. If the load is situated to the right of section I, then to the left there acts only reaction R_A producing the moment

$$M_{\text{I}} = R_A a.$$

Therefore, one can obtain the right portion of the influence line by multiplying the ordinates of the influence line for reaction R_A by distance a. If the load is situated to the left of the section under consideration, then it is convenient to add up the moments of the right forces. Accordingly, the left portion of the wanted influence line is found out by multiplying the ordinates of the influence line for R_B by b. So, depending on the load's position, we have *left and right branches* making up the influence line for the bending moment.

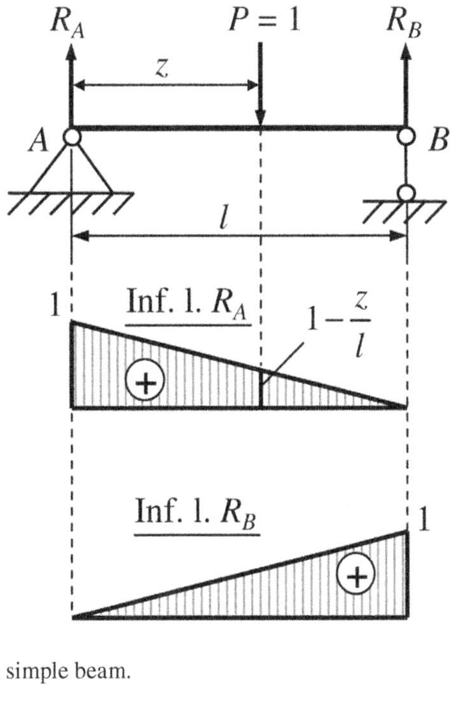

FIGURE 2.3 Live load upon simple beam.

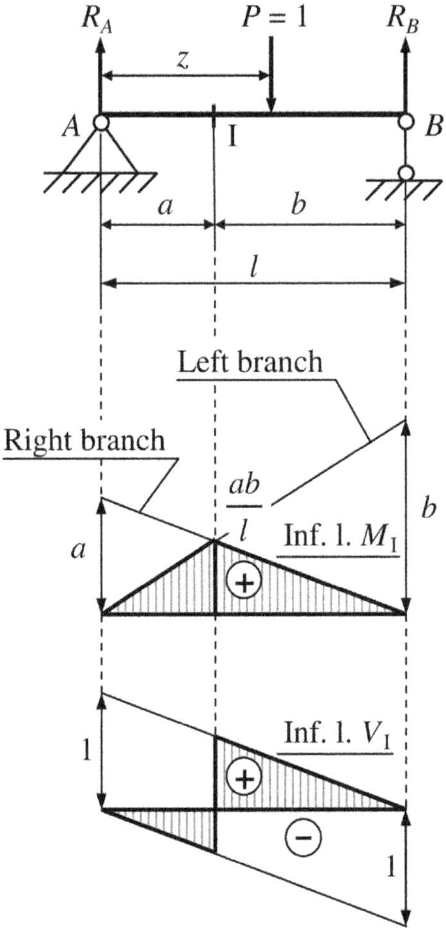

FIGURE 2.4 Influence line construction for bending moment and shear force.

The influence line for shear force is also made up of the left and right branches, according to the load's location on the left or right of the section. It is easy to see that the right branch coincides with the influence line for R_A, whereas the left branch is the influence line for quantity $-R_B$ (Figure 2.4).

The kinematic method of influence line construction will be studied in Chapters 6 and 16.

Further, the sought characteristic of the SSS, which complies with the principle of superposition, we call *response of the structure to the applied load*. Let us consider the most important properties of influence lines for any response of the structure.

Due to the principle of superposition, the assemblage of loads P_i, applied at different points on the load line, causes the response which equals a sum of responses from each load:

$$S = \sum P_i f(z_i), \tag{2.1}$$

where f is the influence function of response, z_i is the coordinate of the i-th load. It follows that, in particular, the uniformly distributed load w applied in the given segment $[a, b]$ of load line causes the response:

$$S = \int_a^b wf(z)dz = w\omega, \tag{2.2}$$

where $\omega = \int_a^b f(z)dz$ is the quantity numerically equal to the area under influence line in the region of load action. This quantity we refer to as the area of the influence line.

Thus, *property 1 of influence line*: Response of structure due to the action of load assemblage is determined as a sum of the products of loads and influence line ordinates under loads. The response due to a uniformly distributed load is determined as the product of load intensity and the area of influence line.

Property 2 of influence line: To determine the response due to action of vertical loads' series applied within the bounds of the straight portion of the influence line, it is allowed to substitute the resultant of this series for this series.

Let us prove this property. Take the load series of forces P_i with the resultant force R applied at the point with coordinate $\bar{z}$. This point is determined in the load line from the equality for moments of the load series and the resultant:

$$R\bar{z} = \sum P_i z_i. \tag{2.3}$$

The formulated property 2 is expressed by the equality:

$$\sum P_i f(z_i) = Rf(\bar{z}), \tag{2.4}$$

where, as usual, f is the notation of the influence function. We know that in the region of the loads' action, the influence line is specified by linear function:

$$f(z) = kz + c, \quad z = z_1, \; \ldots \; , z_i, \; \ldots \tag{2.5}$$

The reader is recommended to verify identity (2.4) by means of substitutions (2.3) and (2.5).

Let us consider the beam, upon which the load is transmitted via the row of other beams situated above (Figure 2.5). If transmission of load occurs with the help of intermediate elements supporting the ends of upper beams, then such a construction is called a *beam with indirect load application*. The bottom beam is referred to as the main beam or girder, the top beams are referred to as secondary beams or stringers, the girder's points taking up a load are referred to as panel points, and the portion between the neighboring panel points is referred to as a panel. The secondary beams are considered simple.

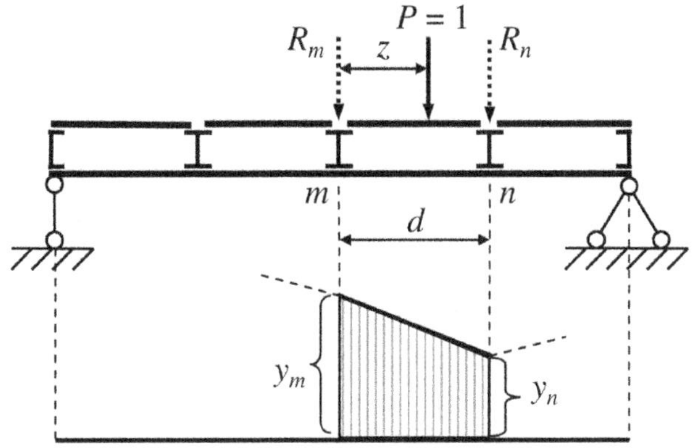

FIGURE 2.5 Beam with indirect load application and influence line within the panel.

Property 3 of influence line: Upon condition of indirect load application, the influence line for the response of a girder is continuous and is represented by the segment of a straight line within the panel extension.

We prove this property. Let the influence line's ordinates of response S at panel points m and n be equal, respectively, y_m and y_n (Figure 2.5). In order to prove property 3, it is sufficient to be convinced that the influence line within panel m–n is the segment connecting the boundary points having these ordinates. For the load's position between panel points, there are only two concentrated loads acting on the girder and being applied at these points. Denote them as R_m and R_n. Response S is produced in the girder; thus it is determined uniquely by loads applied to the girder. Load R_m could be caused by the same weight placed on the load line at the gap between the stringers, and thus the effect of this load is determined through the influence line and equals $R_m y_m$. The effect of load R_n is obtained as $R_n y_n$. In accordance with the superposition principle, we get for the response under consideration:

$$S = R_m y_m + R_n y_n. \tag{2.6}$$

But forces R_m and R_n are, at the same time, the reactions of stringer supports over the panel, i.e.:

$$R_m = 1\frac{d-z}{d} = \frac{d-z}{d}; \quad R_n = 1\frac{z}{d} = \frac{z}{d}. \tag{2.7}$$

After substitution of these relations into (2.6), we get the wanted property.

2.3 ANALYSIS OF THE STRESS-STRAIN STATE OF GIRDERS BY USING A BEAM WITH INDIRECT LOAD APPLICATION

A beam with an indirect load application often appears to be a convenient model for the analysis of the structure's SSS. Here, we demonstrate the transition to this model in the analysis of bridge girders. We consider a one-spanned highway bridge, where the framework consists of two girders with bearings at their ends, the series of floor beams which rest upon girders, and the slabs of the deck being supported by floor beams (Figure 2.6). The slabs are supplied with a covering, making up a traffic area and a sidewalk. Below, we formalize the problem of girders' SSS analysis under the assumption of elasticity of bridge framework. This assumption enables us to find out the SSS characteristics independently for each type of loading, and then, by using the superposition principle,

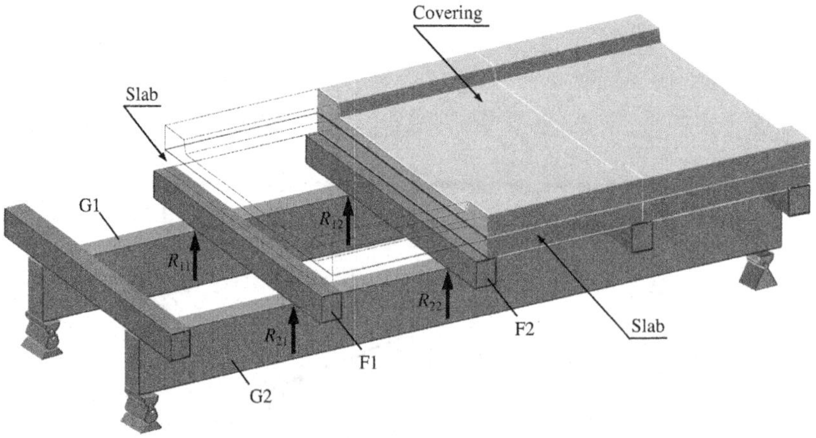

FIGURE 2.6 Diagram of a bridge with the deck mounted on floor beams (the selected slab and its covering are shown by the skeleton; the slab at the left is conditionally not shown).

to establish the actual state of construction by totalizing the components of SSS over all types of loads. Among different types of loads which may act simultaneously (being caused by own weight of constructions, wind, snow, temperature deformations, etc.), we are interested in the live load only, i.e., the load from vehicles and pedestrians on the bridge surface.

We assume the live load to be vertical. It is transmitted by slabs of deck upon the floor beams. We assume the slabs are smooth, transmitting the load at the stripes of contact with floor beams. The areas where slabs and floor beams rest on underlying members are the constraints where the reactions arise (the bands of contact of slabs with floor beams may be taken for closely spaced point-like connections). The *first assumption* of the SSS analysis is to neglect the moments and horizontal components of reactions developed by connections. In other words, we assume the verticality of constraint reactions according to the live load direction. This assumption enables us to construct the model for the analysis of girders without specifying the connection type. On the condition of rationally designed structure, this assumption doesn't produce noticeable errors in analysis.

Let us take under consideration one of the bridge slabs and denote the floor beams below it as F1 and F2. We denote the girders as G1 and G2 and their connections with floor beams as "11," "12," "21," "22," where the first digit is the girder's number, second digit is the number of floor beam below the selected slab. Denote reactions of girders as R_{11}, R_{12}, R_{21}, R_{22} (Figure 2.6). The *second assumption* of the SSS analysis states: loads applied upon any slab produce the reactions of only those connections between the main and floor beams, wherein floor beams support the given slab; other connections between beams do not develop reactions due to these loads. For example, while loads are exerted upon the slab between beams F1 and F2, there may be but four nonzero reactions R_{ij} which are shown in Figure 2.6. According to this, the loads upon girders equal to these reactions with opposite signs define the SSS of girders due to forces applied on the covering of the selected slab.

Let it be required to investigate some response in girder G1, given that a three-axle vehicle consisting of a two-axle tractor and semitrailer moves on the bridge (Figure 2.7a). The line of motion along the lane of the bridge in the middle of the axle track of the vehicle and the concentrated loads

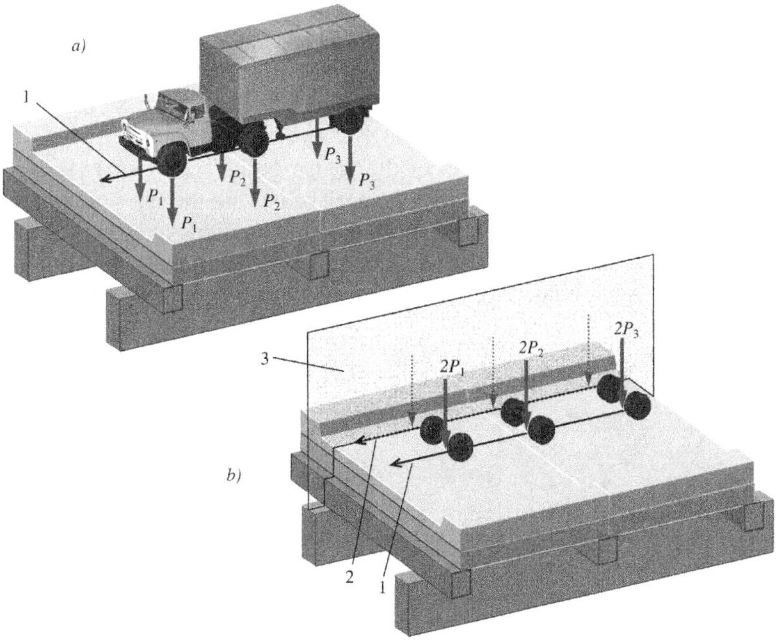

FIGURE 2.7 Action of the load from a vehicle upon the deck: (a) Real loads from the wheels; (b) Conversion of the load to the plane system of forces. 1 – line of vehicle's motion (the same is the load line); 2 – shifted load line; 3 – plane of loading for a girder.

from wheels: P_1 for front axle; P_2 for middle axle; P_3 for the rear axle, are given. The line of motion is positioned between the vertical symmetry planes of girders (i.e., between their planes of loading). The *third simplification* of the SSS analysis will be the next: we replace the loads from wheels on each axle by their resultant applied on the line of motion. We obtain the live system of three forces, for which the line of vehicle's motion is the load line (Figure 2.7b). This replacement, as a rule, doesn't make noticeable errors in analysis. The substantiation for this replacement is set out in the supplement to the chapter, and, in founding this, we employ the supposition that the length of a slab (dimension along the framework) is much greater than the axle track.

The fourth step of the problem's simplification is to transfer the load line into the plane of loading of the girder under consideration, i.e., we assume that the plane system of loads acts on the slab right over the girder (Figure 2.7b). The result of simplification is an increase of loads upon the main beam, and, therefore, we assume that analysis with the new design scheme will secure the strength capacity for the designed structure.

The fifth and last step of the problem's simplification will be to neglect the reactions of girder G2 in calculating the reactions of girder G1. This assumption enables us to determine reactions R_{11}, R_{12} from statics equations in the case of the action of vertical loads upon "outlined" slab in Figure 2.6. The design diagram for the determination of these reactions is shown in Figure 2.8, where F is the resultant of loads upon the slab. The required formulas differ from (2.7) only in denotations and have the form:

$$R_{11} = F\frac{d-z}{d}; \quad R_{12} = F\frac{z}{d}.$$

This fifth step involves some new error into the analysis because, though it is easy to obtain due to assumptions 1–4 that the reactions of girder G2 are zero in total ($R_{21} + R_{22} = 0$), but, if girder G1 has been deformed, it is impossible in a general case that $R_{21} = R_{22} = 0$. Really, let us take the slabs and the floor beams for absolute rigid. As a result of loading upon a rigid slab, the latter can rotate only as an entity – therefore, if the girder G2 is not deformed, then the pivot axis is positioned on beam G2 and both sites of connections 11 and 12 have to get the same displacement. We come to the absurd conclusion that **all** connections of girder G1 with floor beams are displaced in the same way due to load upon the selected slab. Thus, the last step of simplification may induce the noticeable distortions in the results of the analysis.

Owing to simplifications of the problem of the bridge girder's SSS analysis, we come to the design diagram depicted in Figure 2.5, where secondary beams are the slabs of the deck, intermediate elements are the floor beams of the framework, and the fictitious unit load should be replaced by the series of three movable loads. The obtained design scheme enables us to get an approximate solution to the problem. There is need for discussion about the usefulness of this scheme.

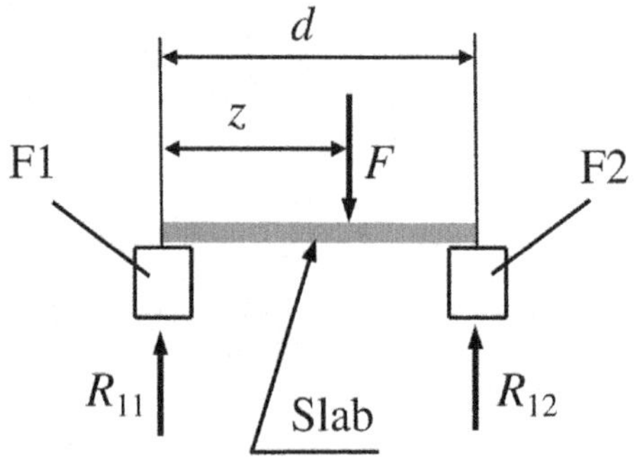

FIGURE 2.8 Slab on floor beams

The designing of complicated technical objects includes two stages: the stage of developing the object's structure and preliminary evaluation of its technical characteristics, and the stage of verification and adjustment of original engineering solutions. Analysis at the first stage is referred to as evaluative. At the second stage, it is called checking or implementation analysis. The model of a beam with an indirect application of load may be employed in evaluative calculations, where quick estimation of several engineering solutions is wanted, but in checking the structural calculations of a bridge, the theory of influence lines is applicable with valuable restrictions, because it is developed for plane structures. In implementation analysis, we have to consider the motion of the load assemblage on the three-dimensional framework, and for different hazardous positions of this assemblage, be assured that the SSS of the constructive member under analysis meets the requirements to its working capacity. In this analysis the reduction of a space system of loads to a plane system doesn't make significant simplification of the problem.

SUPPLEMENT TO CHAPTER

Reduction of Loads from Vehicle on Bridge to Plane System of Forces

Below we give the substantiation of the replacement of loads from each wheel couple of a vehicle by resultant force for the purpose of structural calculations of a bridge girder (see Figures 2.6 and 2.7). We shall consider the location of resultant of reaction forces developed by one of floor beam – for instance, beam F2 – due to the action of the slab in three cases of positioning the couple of wheels on the slab of the deck (Figure 2S.1):

1) The vehicle axle is near opposite beam F1;
2) The vehicle axle is in the middle between beams F1 and F2;
3) The vehicle axle is near the beam under consideration F2.

If in these three cases the resultant of reaction forces of floor beam doesn't considerably change by value and location due to replacement of two loads $\frac{F}{2}$ from the couple of wheels by one load F in the middle of the axle, then the action of each beam F1, F2 on a girder doesn't change either, and the proposed replacement is allowable. We take this as a reasonable foundation to accept such a replacement for other positions of wheel couple on the load line within the slab.

Further, we assume that the length of the slab is an order of magnitude greater than the axle track: $d \gg t$ (in practice it is sufficient that $d > 2.5t$). The resultant of reaction forces developed by a floor beam is shortly called the beam reaction.

In case 1, the replacement of two loads by one statically equivalent load in the course of reaction analysis for beam F2 is allowable in accordance with Saint-Venant's principle.*

To consider case 2, note at the beginning that three forces: the resultant F of the arbitrary system of vertical forces upon the slab and the reactions R_{F1} and R_{F2} produced by this system on the beams F1 and F2, respectively – cross the same straight line on the slab. This follows from the equilibrium equation for moments of forces (otherwise the slab would topple over). The named line is denoted as m for case 3 of the loading by wheel couple (Figure 2S.1). In case 2, this line is parallel to the traffic lane and $R_{F1} = R_{F2}$, because otherwise the symmetry of elasticity problem is violated. The same is obtained after the replacement of loads from a couple of wheels by an equivalent double load, and, therefore, such replacement is acceptable.

* Saint-Venant's principle: If any loading applied to a body within a small region is replaced by other statically equivalent loading applied within the same region, then changes of stress and strain produced at points in a body removed from the region of load application at a distance in order of magnitude greater than extension of this region will be of negligibly small value (Timoshenko and Goodier 1951).

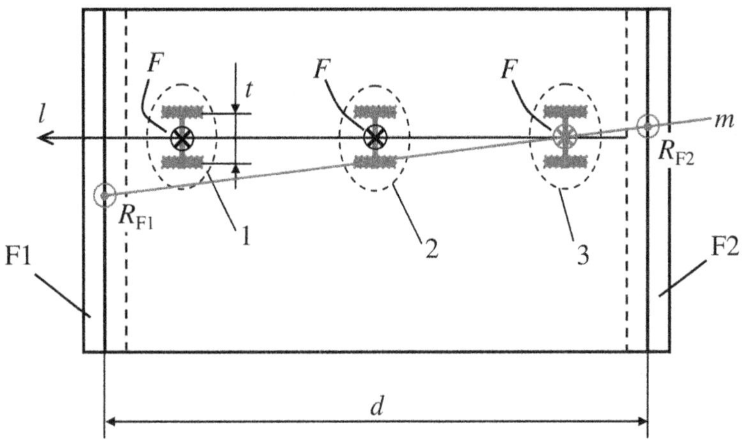

FIGURE 2S.1 Wheel couple on the slab of the bridge: l – line of vehicle's motion; 1, 2, 3 – possible positions of wheel couple; m – line of possible location of forces on the slab in case 3; R_{F1}, R_{F2} – resultants of reactions upon the slab in case 3; (resultants F, R_{F1}, R_{F2} are perpendicular to the slab and shown as the points).

In case 3 the value and location of reaction R_{F2} are uniquely defined by resultants of external forces upon the slab F and R_{F1}. All three forces lie in the same line, and reaction R_{F1} obeys case 1, i.e., it doesn't change due to replacement of two wheel loads $\frac{F}{2}$ with load F in the middle of the axle. Consequently, the same may be said about reaction R_{F2} – which was to be obtained.

3 Theoretical Basics of Calculations via Influence Lines

3.1 DETERMINATION OF DESIGN POSITION OF LIVE LOAD SERIES FOR GIVEN INFLUENCE LINE

In this item, we consider the case typical in engineering practice when an influence line consists of several segments of straight lines. Such lines are called polygonal. At the beginning, we define a discontinuous influence line more precisely. Assume that an influence line has a vertical portion with an origin at some vertex *B*, i.e., the influence line has a discontinuity at the point of a basic line with coordinate z_B(Figure 3.1). Then we complete a definition of the influence line by the limit:

$$f(z_B) = f(z_B + 0) \quad \text{or} \quad f(z_B) = f(z_B - 0).$$

We take as given that an influence line is defined for such vertices, depending on the problem under consideration: if we search for the maximum of response *S*, then we take the maximum of the two possible limits; otherwise, we take the minimum. Such a definition corresponds to the physical meaning of the problem: In practice, some small variations are always possible in configuration of load assemblage. Therefore, in the neighborhood of discontinuity, the more unfavorable portion of the influence line manifests itself. For definiteness, we assume below that the maximum of response *S* is searched for and, respectively, the influence function is defined at the break points through maximum:

$$f(z_B) = \max\big(f(z_B + 0), f(z_B - 0)\big). \tag{3.1}$$

We specify the position of a live load series by coordinate z of some point affixed to this series. We allow that the load series can move out of the load line. Beyond the load line, we extend the definition of influence line by equality: $f(z) = 0$. Therefore, the response function is defined over the all number axis, and when all loads of the series are positioned out of a structure, it holds $S(z) = 0$.

Let us prove that **in the case of a polygonal influence line, dependence of any structure's response *S* versus load series coordinate *z* is piecewise-linear, and every linear portion (excluding its boundary points) corresponds to possible positions of the series, when each load is located within the same segment of the influence line (having not reached the boundaries of the segment).** Figure 3.1 clarifies the assertion about the piecewise-linear form of the response function. The load series in this case consists of two loads with the value *P* for each; the distance between them is *c*. The position of the series is specified by the coordinate of the resultant, shown by a dotted line. The influence line comprises three segments and corresponds to the discontinuous function. The influence line is defined at the break point by its maximum value; in accordance with this, response *S* depends on the load series' coordinate as it is shown by the graph under the influence line. The graph consists of seven segments with three eliminated points designated by arrows.

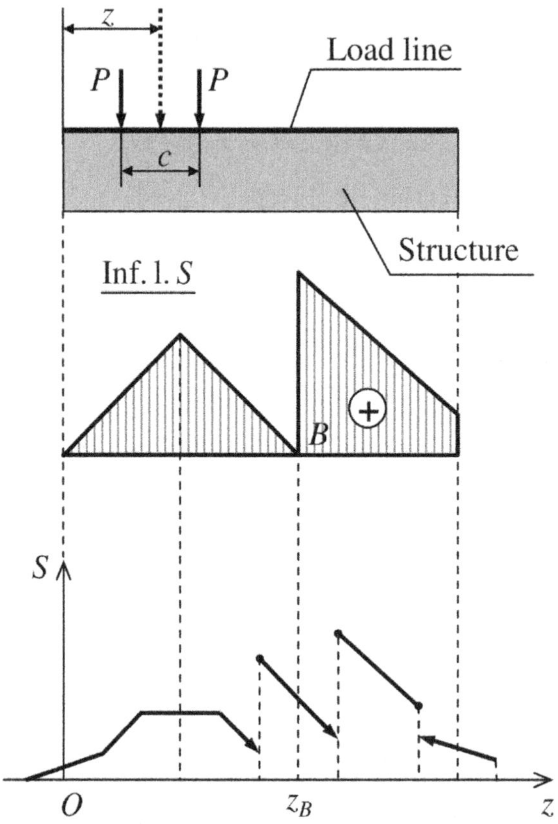

FIGURE 3.1 Response of structure to load series at given influence line.

We proceed to the proof. Let neither of loads of series be positioned on a vertical passing through any vertex of influence line. For each straight portion of the influence line, including one or more loads in its boundaries, we denote:

R_i – resultant of loads' totality positioned over the i-th portion;
h_i – ordinate of the i-th portion under resultant R_i;
tan α_i – derivative of influence line at the i-th portion (or, which is the same, the slope of the corresponding portion).

On small enough displacement Δz of a load series made along the load line, no load will go out from its portion. Therefore, the resultant for each portion will not change but only will be displaced by Δz. The ordinate of the resultant will get the increment:

$$\Delta h_i = \Delta z \tan a_i.$$

Before small displacement of load series, the response of structure equals:

$$S = \sum R_i h_i;$$

After displacement, the increment of the response will equal:

$$\Delta S = \sum R_i \, \Delta z \tan \alpha_i. \tag{3.2}$$

Thus, the increment of response is proportional to an increment of the coordinate of series while the loads don't trespass the bounds of their portions. The dependence of response versus coordinate appears to be linear in this specific case and piecewise-linear in general case.

Note that the piecewise-linear form of dependence $S = S(z)$ in the problem being considered allows for isolated points of its graph. This takes place, for instance, in the case of the influence line shown in Figure 3.2 for the system of two loads at distance c from each other. The reader is invited to supplement Figure 3.2 with dependence $S(z)$ in the same manner as it was done in Figure 3.1.

The next property of polygonal influence lines appears to be useful in searching the design position of the concentrated load series.

The design position of the live load series for the polygonal influence line implies the location of one or more loads over the vertices of this line.

Let us prove it. The design position of loads corresponds to the maximum of response S. Since the response function is piecewise-linear and the influence function has been defined at the break points by (3.1), it is sufficient to search for the maximum of dependence $S(z)$ amid the boundaries of linear portions, i.e., for cases of positioning a load over a vertex. Really, setting aside the issue of the existence of a maximum, let us search for the supremum of function $S(z)$. The latter may be found at the point between linear portions (the break-or-kink point) or be the limit value at the boundary of the linear portion. Thus, possible values of the sup are three values $S(z_B+0)$, $S(z_B-0)$, $S(z_B)$ for each break-or-kink point z_B. Note that at the point of discontinuity of function $S(z)$, the jump of this function caused by the argument's variance from the break point is possible only downward, because on getting the load off a vertex of the influence line, the jump of the value of the influence line is possible only downward. This means that both the left and right limits of searched function $S(z)$ at the break point can't be greater than the value of function $S(z)$ at this point. Therefore, the sup is the maximum of $S(z_B)$ values over all the break-or-kink points. We see that the maximum of function $S(z)$ is attained at its definitional domain and corresponds to the load's location over a vertex.

Possible discontinuities in dependence $S = S(z)$ result from discontinuity of the influence line. On the contrary, **if an influence line is continuous, then the dependence of response versus the coordinate of the load series is continuous as well.**

The load in a given load series, positioned over one of the vertices of polygonal influence line when the extreme value of response is reached, is referred to as the *critical load*.

The important particular case of a polygonal influence line is represented by a triangular line (Figure 3.3). We confine ourselves to the situation when the vertex of the triangle lies inside the load line, i.e. $a \neq 0$, $b \neq 0$. By virtue of this, the dependence of the response versus the coordinate of the load series is continuous. The sum of loads applied on either side of the vertex of a triangular influence line we call either the right or left total load (according to the position of the loads). If one of

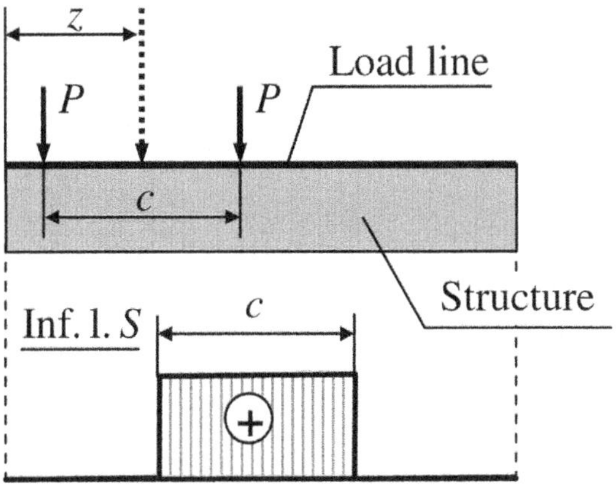

FIGURE 3.2 Case of structure's response when its graph has isolated point.

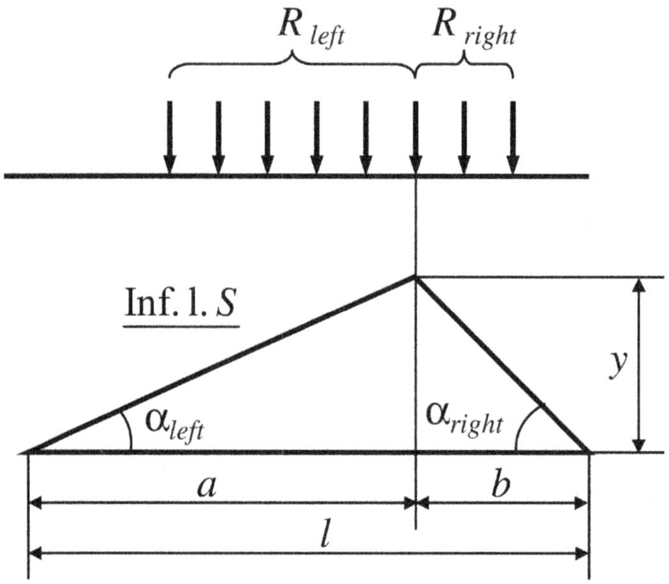

FIGURE 3.3 Triangular influence line with load positioned over its vertex.

the loads appears to be positioned over the vertex, we assume that its weight is distributed between right and left total loads. To determine the critical load, the next property of a triangular influence line may be employed.

For some load P_{cr} to be critical, it is sufficient that: (1) this load be positioned over the vertex of the triangular influence line; (2) the outermost loads of series be located inside load line; and (3) load P_{cr} can be partitioned between left and right total loads by the ratio:

$$\frac{R_{left}}{R_{right}} = \frac{a}{b}, \tag{3.3}$$

where a and b are the parameters of influence line's triangle; R_{left}, R_{right} are the magnitudes of left and right total loads, respectively.

Let condition (3.3) hold due to partitioning load P_{cr} into two nonzero components. Then, if we displace the system of loads to the left, condition (3.3) turns into inequality:

$$\frac{R_{left}}{R_{right}} > \frac{a}{b}, \tag{3.4}$$

Otherwise, if we displace this system to the right, then instead of (3.3) we get the inequality:

$$\frac{R_{left}}{R_{right}} < \frac{a}{b}. \tag{3.5}$$

For small enough displacements, when no load of series reaches the boundaries of the load line, all loads happen to be positioned inside the straight portions. Formula (3.2) for a shifted position of the load series takes the form (Figure 3.3):

$$\frac{\Delta S}{\Delta z} = R_{left} \tan \alpha_{left} - R_{right} \tan \alpha_{right} = \frac{R_{left} y}{a} - \frac{R_{right} y}{b} = \left(\frac{R_{left}}{a} - \frac{R_{right}}{b} \right) y.$$

Let $y > 0$. For the load series shifted to the left, inequality (3.4) enables us to find out that $\frac{\Delta S}{\Delta z} > 0$, and, in the same way, for the series shifted to the right, we establish that $\frac{\Delta S}{\Delta z} < 0$. The sign of derivative of the continuous function $S = S(z)$ changes as soon as load P_{cr} passes through the vertex; therefore, this load is critical.

The cases when condition (3.3) holds due to inclusion of load P_{cr} into the left or right total load as a whole may be considered in the same manner.

Note that **when features 1 and 2 of the property of triangular influence line work, feature 3 will be not only sufficient but also a necessary condition of the criticality for load P_{cr}**. The latter is proved after the described scheme by analysis of the sign of derivative $\frac{\Delta S}{\Delta z}$ under small biases of load series to the left and to the right.

3.2 BRIDGE DESIGN LIVE LOADS SPECIFIED BY DESIGN REGULATIONS

The philosophy of structural calculations is based on a probabilistic approach, i.e., while specifying the magnitudes of loads applied to a structure, we proceed from the idea that, with a probability of about one, there will not appear more dangerous loads during the design life of a structure. For specifying the live loads on a bridge deck, i.e., loads from vehicles and pedestrians, national building codes define some design situations, for which the loads produce most unfavorable effects. It is presumed that though the loads of design situations may be exceeded during the operating routine of structure, the probability of such exceedance is negligible. Further, we take under consideration only static loads which arise due to typical traffic, without hard braking or emergencies. (Static loads are considered as sustained and not inducing the fatigue of structural members.)

3.2.1 Highway Bridges

The specifications of loads acting on a bridge deck for analysis of bridge framework are standardized by the American Association of State Highway and Transportation Officials in the code referred to as *AASHTO LRFD Bridge Design Specifications* (2012). These specifications define the design situations for cases of traffic on a bridge along one lane only, along two lanes simultaneously, along three simultaneously, etc., according to the number of lanes. The sidewalks, if they are present, are taken for lanes with specific loads but will not be considered here. Possible design situations are as follows:

1. *The design situation of traffic on one lane.* In this case, we select among road users a *design vehicle* most dangerous for framework members and specify *the structure of the live load for this vehicle, namely: the wheel track, the distances between axles, and the loads from axles.* The structure of the load does not need all the axles of the design vehicle to be included; this may be, for example, the "axle tandem" – the couple of most loaded, closely spaced axles of a vehicle. All possible positions of the vehicle are allowed (with some limitations in distances from the roadway edge). Such a movable load constitutes the first component of transient live loading upon a bridge. The second component represents the remaining traffic in a homogeneous way and is specified by the *design uniformly distributed load* on the same lane. The design vehicle can be of two possible types, and for each type, one has to consider all possible positions of the movable load in combination with distributed load. Usually, it is assumed that there is only one vehicle on the lane, but there may be exceptions, when, for some constructions, two such vehicles are supposed on one lane simultaneously (Mertz 2012, pp. 7–8). Cases when two vehicles are located on the same lane are not under consideration here.

The account of a load's random variations about standard values is fulfilled due to multiplication of loads by the multiple presence factor m. In the case of traffic on a single lane, this factor is assigned of $m = 1.2$, i.e., it is granted that the probability of exceeding the standard values by the loads is sufficiently large.

2. *The design situation of traffic on two lanes.* In this case, the two-component loading introduced in Item 1 is situated on each lane. We take the multiple presence factor for loads on both lanes as $m = 1$, i.e., we neglect the possibility of simultaneous exceedance of standard loads on both lanes.

   The loading on each lane must be chosen as most unfavorable for the functioning of the constructive member under analysis. *Steel Bridge Design Handbook* (Mertz 2012), comprising the comments to AASHTO LRFD, allows the assignment of loading on the second lane to be the exact replica of loading on the first lane, i.e., when the loading on the second lane is obtained by lateral transference of loading on the first lane. But such simplification must be additionally grounded, because the designer remains in charge of it.
3. *The design situation of traffic on three lanes and the situations due to greater number of lanes.* In the case of three lanes, the two-component loading of Item 1 is situated on each lane and $m = 0.85$ is assumed. For a greater number of lanes, we also employ the loading of Item 1 on all lanes and $m = 0.65$. An engineer can simplify the analysis by reproducing the same loading on all lanes – if there are sufficient reasons for it.

The first component of design loading – the design vehicle introduced above in Item 1 – is either represented by the three-axle lorry consisting of a tractor and semitrailer (Figure 2.7) or specified by the axle tandem. The plan of wheels for a three-axle design vehicle is shown in Figure 3.4, on the left; the loads are denoted in the same way as in Figure 2.7. The wheelbase of a semitrailer varies within the limits of 14–30 ft to produce the most unfavorable force effect in the constructive member for which analysis is performed. The plan of wheels for the axle tandem is shown in Figure 3.4, on the right. The mutual location of loads and their magnitudes specified by AASHTO LRFD (2012) are given in the figure caption. The minimum distance d of any wheel from the face of the curb is 1 or 2 ft according to the member under consideration.

The second component of design loading – the design distributed load – is specified as 0.64 klf with uniform distribution along the lane. The region of its action must include the load from the design vehicle. This region is stripe-shaped and its extension and location must produce the most unfavorable forces in the constructive member. The width of exertion for this load is 10 ft or is equal to the width of the lane, if the lane is no wider than 10 ft.

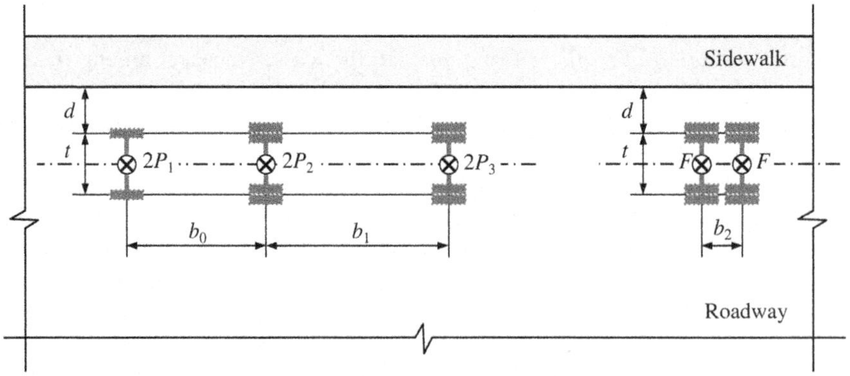

FIGURE 3.4 Design vehicles on roadway and loads from them: on the left two-axle tractor with semitrailer; on the right axle tandem. Geometrical requirements: $t = 6$ ft – wheel track; $b_0 = 14$ ft – wheelbase of tractor; $b_1 = 14 \div 30$ ft – wheelbase of semitrailer; $b_2 = 4$ ft – axle spacing of tandem; $d \geq 1$ ft – distance to the curb. Requirements to axle loads: $2P_1 = 8$ kip; $2P_2 = 2P_3 = 32$ kip; $F = 25$ kip.

Specifications (AASHTO LRFD 2012) comprise allowances which enable to take into account the dynamic loads. This is done by multiplying the first component of the live static load by the corresponding allowance. In most cases, the static loads from design vehicles have to be increased by 33% to account for the dynamics of interaction between structure and traffic flow.

3.2.2 Railroad Bridges

The specifications of loads upon a deck for the design of railroad bridges' frameworks are standardized by American Railway Engineering and Maintenance of Way Association in vol. 2 of the code referred to as *Manual for Railway Engineering* (2015). This manual defines the design situations during movement of a train on one track of the deck and generalizes them for cases of train movement on two tracks simultaneously or on three and four tracks simultaneously. Before designing, one has to specify the requirements of the railroad loads on the bridge. These requirements are defined by the Cooper rating of loading, which specifies the design train on the deck. Further, we consider the generally used rating of E80, where the letter means the type of loading, i.e., mutual location and ratios of loads on the track, and the number means the magnitude of the greatest concentrated load. Possible design situations for a bridge deck with one or two tracks are as follows:

1. *The design situation of railroad traffic on one track.* The *design train* moves on a given track of the bridge and comprises two components of loading. The first component of design loading is made up of two hitched locomotives and specified as a live system of concentrated loads of definite structure; the second component of design loading represents the train of cars following the locomotive and is specified by the uniformly distributed load on the track. The Cooper load is clarified in Figure 3.5, where the concentrated loads from axles of locomotives and the distributed load per unit of track's length are depicted. The extension of a car train must produce the most unfavorable forces in the constructive member.

   Note: The silhouettes of steam locomotives in Figure 3.5 are not by chance: American engineer Theodore Cooper proposed this type of two-component loading in 1894. Cooper's standard was E10, where number 10 means the load in kilopounds from each of four driving axles of real steam locomotives. In modern bridge engineering, we deal with conventional steam locomotives, which produce a load of 80 kip by each driving axle (and other characteristics changed proportionally).
2. *The design situation of railroad traffic on two tracks.* In this situation, the two-component loading introduced in Item 1 is positioned on each track. The location of trains and their extensions must meet the requirement of greatest forces in constructive member.

Just as for the highway bridges, it is feasible to account for dynamic loads by applying allowances to axle static loads. Empiric formulas for these allowances are given in *Manual for Railway Engineering*, Vol. 2 (2015).

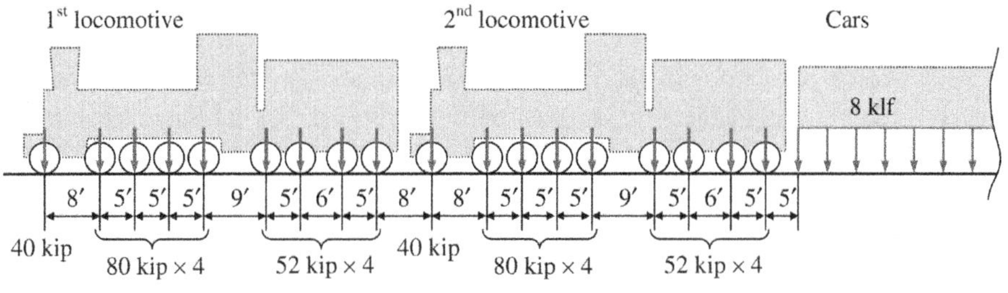

FIGURE 3.5 Design train on track: mutual location and magnitudes of loads.

Note in conclusion that technique of decomposing the live loading into two components, one of which is the system of concentrated loads for description of the heaviest vehicle and the other is the uniformly distributed load for description of the homogeneous traffic, is accepted in worldwide design routine, see, for example, the code of the European Union (*Eurocode 1: Actions on Structures – Part 2* 2003).

Section III

Three-Hinged Arches

This section sets out the theory of the three-hinged arch calculation: the conceptual framework is introduced, including the concept of reference beam; the design formulas are derived for support reactions and internal forces acting in an arch; the concept of arch rational axis is introduced and the rule to determine the one is justified; the peculiarities of arch influence lines are established, and the methods are suggested for their construction.

4 Arches
General Info and Diagrams of Internal Forces

4.1 CONCEPT OF ARCHES AND CALCULATION OF SUPPORT REACTIONS FOR THREE-HINGED ARCHES

Arches are thrusted structures consisting of curvilinear bars with convexity in a direction opposite to the load. Further, loads are taken as vertical and directed downward and arches as convex upward. Arches can be statically determinate and indeterminate, and static determinacy of an arch depends on the number of hinges in its construction. Statically indeterminate arches can be unhinged, one-, and two-hinged. From among statically determinate arches, we shall further consider three-hinged arches which comprise two curved bars connected by an intermediate hinge and two hinged immovable supports.

The shape of an arch is specified by its axial line and cross-sections of bars with centroids in the axial line. We assume that an arch's axial line lies in the vertical plane, which simultaneously is the plane of loads. The shape of cross-sections is allowed to be variable along the axial line, but the principal centroidal axis of every section lies in the arch plane.

The upper point of an arch axis is referred to as the *crown* or *key*. In a three-hinged arch, the intermediate hinge C is referred to as the *crown* or *key hinge* and usually is located at this point (Figure 4.1). Massive masonry or concrete support under the arch is referred to as an *abutment* (in design diagrams an abutment is usually depicted as the earth); a slanting surface of abutment where the arch is supported is referred to as the *skewback*; support hinges of arches are also referred to as *skewback hinges*; the portion of the arch from abutment to crown is referred to as a *haunch* – further, we employ this term for a member of a three-hinged arch; straight line AB between support hinges is referred to as the *support line*; distance l between arch supports in horizontal projection is referred to as the *span*. For arches being studied in mechanics, the arch axis doesn't project beyond the span boundaries.

The vertical drawn from the crown to the support line is referred to as the *arch rise*. Also, the term "arch rise" is used for the height of elevation f of the crown hinge over the support line and distance f' from the median hinge to the support line (Figure 4.1). Further, we employ this term for distance f' and elevation height f of the crown hinge we call the *arch vertical rise.*

Due to the convexity of the arch axis in a direction opposite to the vertical load, the supports of an arch develop inclined reactions; the horizontal component of these reactions is directed inside the span. This is the character of arched structures: the forces of the haunches' action upon supports have not only vertical but also horizontal components – the latter aim to move the supports apart. On the other hand, at cross-sections of arches, bending moments are essentially lesser in comparison to the beams of the same span, and therefore arched constructions are easier and cheaper than beamed ones. In order to eliminate horizontal reactions, some arches are supplied by tie bar – a bar tying the haunches nigh their abutments.

Arches are employed in bridge superstructures and as the rafters for sheds and roofs of large extensions. The usage of arches in bridge superstructures enables the enlargement of the spans to several hundred meters. For instance, the arch bridge in the Hamm district of Dusseldorf depicted in

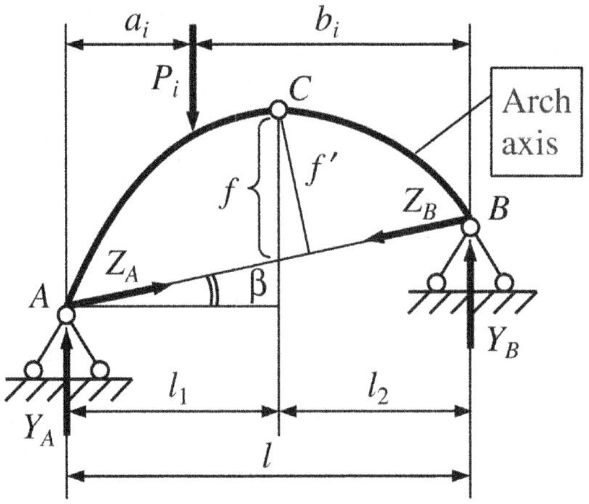

FIGURE 4.1 Three-hinged arch.

FIGURE 4.2 Railroad bridge over Rhine River in Dusseldorf. Photo by Johann H. Addicks. (License CC-by-ND. Source: https://commons.wikimedia.org/wiki/File:Hammer_Brücke_Düsseldorf_seitlich.jpg.)

Figure 4.2 has a span of 250 m. The arches may be situated either above the deck or below it; the latter is characteristic of viaducts. For example, the construction in Figure 2.6 might be supplemented with two three-hinged arches; each would support one of the girders through the row of columns and be situated below the beam in its plane of loading. This solution enables the enlargement of the bridge span with more economical cross-sections of girders. The subsystem "girder–arch" is shown in Figure 4.3. In particular, the possible constructions of the arch's hinges are represented in this figure. The hinges must not only allow for the small angular displacement of haunches, but prevent the collapse of the arch from accidental and seismic loads. For this purpose, the construction of hinges provides anti-seismic ties.

By means of statics equations, one can determine the internal forces acting upon cross-sections of the three-hinged arch. We confine ourselves to concentrated loads. Denote the distances from the line of action of each load P_i to support hinges as a_i and b_i, respectively. We represent every support reaction as the vector sum of two components: vertical Y and sloped Z directed along the support line (Figure 4.1). Let us introduce the reference beam for a given arch as the simple beam of the same span, subjected to the same assemblage of loads (Figure 4.4). We show below that the support reactions and internal forces at the sections of an arch can be obtained by using the support reactions and diagrams of internal forces for the reference beam.

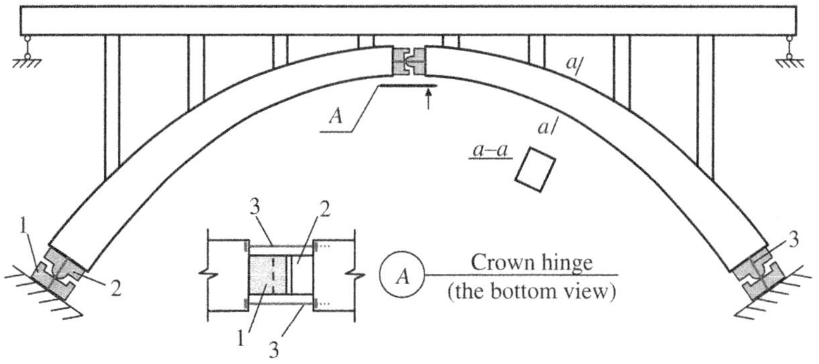

FIGURE 4.3 Subsystem "girder–arch" in bridge superstructure. Crown and skewback hinges of the arch are of the same construction and comprise the members: 1 – foot bearing; 2 – foot; 3 – anti-seismic bolt.

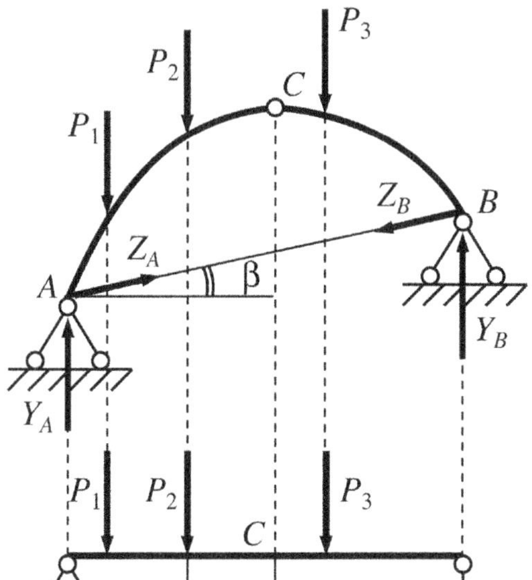

FIGURE 4.4 Three-hinged arch and substitute beam.

In the beginning, we determine the support reactions. The equations of equilibrium of moments for the arch with respect to pivots *A* and *B* have the next forms, respectively:

$$Y_B l - \sum P_i a_i = 0; \tag{4.1}$$

$$Y_A l - \sum P_i b_i = 0 \tag{4.2}$$

(sloped reactions don't produce the moment relative to support hinges). The equations of moment equilibrium for the beam are just the same; consequently, **vertical components of reactions of support hinges appear to be equal to support reactions of the reference beam**. The vertical force completing the sloped reaction to the support reaction is called the *beam reaction force*. In contrast to vertical components, sloped reactions are specific only to an arch; therefore they are called *arch reaction forces*.

Horizontal components of reactions in thrusted structures are referred to as *thrusts*. For the thrusts of the arch under consideration, it holds:

$$H_A = Z_A \cos\beta; \quad H_B = Z_B \cos\beta \tag{4.3}$$

(the positive sense of thrusts is taken inward the span).

From the condition of external forces' equilibrium in horizontal projection, by taking into account the verticality of loads, we establish that the **left and right thrusts equal each other**:

$$H_A = H_B = H. \tag{4.4}$$

Accordingly:

$$Z_A = Z_B = Z. \tag{4.5}$$

To determine the left sloped reaction, we employ the equilibrium equation for moments of forces acting on the left haunch relative to the crown hinge. This equation may be written in the form:

$$Y_A l_1 - \sum_{left} M_C(P_i) - Z_A f' = 0, \tag{4.6}$$

where $M_C(P_i)$ is the moment of force P_i relative to the crown hinge with counterclockwise rotation taken as positive; the sum is taken over all the loads on the left haunch. Formula (4.6) enables us to use the bending moment diagram for the reference beam while calculating the sloped reactions. For vertical load P_i it holds:

$$M_C(P_i) = P_i(l_1 - a_i). \tag{4.7}$$

Accordingly, the bending moment at cross-section C of the simple beam equals (Figure 4.4):

$$M_C^0 = Y_A l_1 - \sum_{left} M_C(P_i). \tag{4.8}$$

Out of this, under vertical loads, we get:

$$Z = \frac{M_C^0}{f'}, \tag{4.9}$$

and from (4.3), after substitution $f' = f \cos\beta$, we obtain the thrust:

$$H = \frac{M_C^0}{f}. \tag{4.10}$$

The reader is recommended to obtain formula (4.9) from an equilibrium condition for moments of forces acting upon the right haunch of the arch.

The bending moment produced by loads acting vertically downward is positive at cross-sections of simple beams, and thus, in accordance with formula (4.10), **horizontal reactions of supports of the three-hinged arch are directed inside the span**.

The complete vertical reactions are established with account of sloped reactions which make the contribution $Z \sin\beta = H \tan\beta$:

$$Y_A^* = Y_A + H \tan\beta; \; Y_B^* = Y_B - H \tan\beta. \tag{4.11}$$

We emphasize that both reactions of support hinge Y and Y^* are vertical, but the first one is determined in oblique coordinates and the second one in rectangular coordinates.

4.2 ANALYTICAL CALCULATION OF INTERNAL FORCES AND THE RATIONAL AXIS OF A THREE-HINGED ARCH

In mechanics of materials, we have been considering internal forces acting upon the cross-section of a member as follows:

- Axial (or longitudinal) force, vector $\boldsymbol{N}$ of which is directed perpendicularly to the section;
- Shear force, vector $\boldsymbol{V}$ of which lies in the section;
- Bending moment, which is the vector component of internal moment $\boldsymbol{M}$ with respect to the centroid of the section, lying in the section and produced by normal stresses;
- Torque moment, which is the vector component of internal moment $\boldsymbol{M}$ with respect to the centroid of the section, normal to the section and produced by shear stresses.

If a bar is subjected to the action of a plane system of forces passing through the bar's axial line, then the torque is nil, shear force lies in the plane of loads, and the bending moment is orthogonal to this plane. Therefore, in this case, the lines of action for axial and shear forces as well as the direction of the bending moment are defined unambiguously, and we have three independent forces defined by scalar quantities at the cross-section.

Note: The specifying of internal moments relative to the centroid of a cross-section is accepted in the so-called basic theory of bending, which is restrictedly applicable to the bending of members with nonsymmetric cross-sections and may not be applicable to thin-walled nonsymmetric sections at all. For nonsymmetric sections, the improved theory of bending has been developed, which demands calculation of a torque moment relative to the especial point of the section called the center of bending. Further on, we employ the basic theory of bending, having in mind that **for a bar with constant cross-section symmetric relative to the plane of loads**, this theory doesn't involve errors into the SSS analysis of member, and for nonsymmetric sections it gives an approximate estimation of the SSS.

To investigate internal forces upon cross-sections of an arch, we introduce the arch-bound coordinate system (CS) Axy in the plane which comprises the axial line of the arch. Place origin A on the left support, direct axis Ax inside the span horizontally, and direct axis Ay upward vertically. The bound CS enables us to specify the equation of the arch axial line and the location of the arbitrary cross-section. We are reminded that internal forces are the forces and moments evoked by the action of any part of a structure upon the cross-section bordering this part from the remaining one. This part is assumed as the source of external forces and mentally is cut off from the structure, i.e., removed from consideration. We accept the part of the arch which doesn't comprise "initial" support A, i.e., the part to the right of the section, as removed. We determine internal forces in the sliding CS $Oxyz$, the origin of which is placed at the centroid of the section, axis Oz is directed to the cut-off part along the normal to section, and axis Oy is directed to the center of curvature of the arch axis.

Internal forces are introduced as the components of the shear force vector $V \equiv V_y$, axial force vector $N \equiv N_z$, and the complete moment vector $M \equiv M_x$. The CSs and positive directions of internal forces arisen due to action of the right part of the arch upon the left part are shown in Figure 4.5a and c. In complicated plane structures, the next sign convention is recommended to determine the shear force: **shear force is positive if it rotates the object (the portion of the member) clockwise**. The definition of shear force proposed for an arch corresponds to the sign convention.

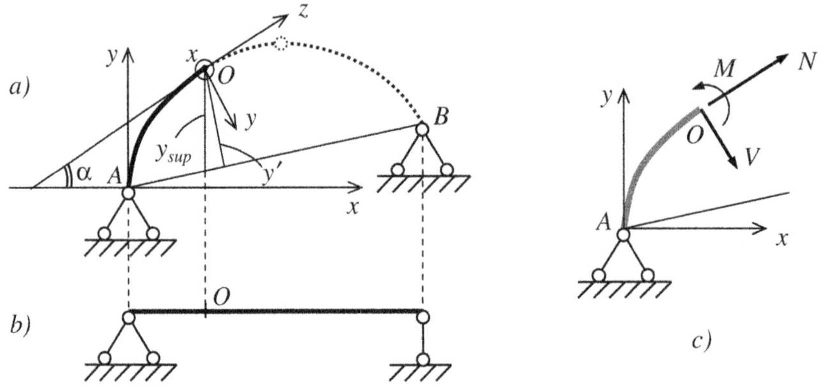

FIGURE 4.5 With regard to calculation of forces in cross-section of arch: (a) Coordinate systems; (b) Reference beam of the arch; (c) Positive directions of internal forces.

Further, we put the denotation of the section's center of gravity in quotes to denote the cross-section as a whole. We can determine bending moment M at any section "O" having noticed that this one counterpoises the sum of moments of external forces on either side of the cross-section, taken with respect to the centroid – for example, the sum of moments of the left forces. This sum is obtained in the same way as the total moment of forces on the left of the crown hinge in equation (4.6). In the given case it holds:

$$M = Y_A x - \sum_{left} M_O(P_i) - Z_A y', \tag{4.12}$$

where x is the coordinate of section "O"; $M_O(P_i)$ is the moment of force P_i relative to point O with counterclockwise positive sense; the summing is made over all loads to the left of section "O"; and y' is the distance from the centroid of the section to the support line. Just as in formula (4.8), the first two addends in (4.12) determine bending moment M^0 at section "O" of the reference beam. The last addend in (4.12) can be represented through thrust H and quantity y_{sup}, which is the elevation height of the cross-section over the support line (Figure 4.5a):

$$Zy' = Z\cos\beta \frac{y'}{\cos\beta} = Hy_{sup}.$$

Finally we get:

$$M = M^0 - Hy_{sup}. \tag{4.13}$$

Shear and axial forces at section "O" are established from the condition of equilibrium of forces for the left part of the arch (Figure 4.6a). The total vertical component of left forces relative to section "O" with positive direction "upward" equals:

$$P_O^{left} = Y_A - \sum_{left} P_i + H\tan\beta. \tag{4.14}$$

The total horizontal component of left forces with positive direction "to the right" equals thrust H.

To make convenient the composing of statics equations, external and internal forces being in equilibrium are shown in Figure 4.6b, with a common origin at point O. The equilibrium condition for forces in projection onto axis Oy is as follows:

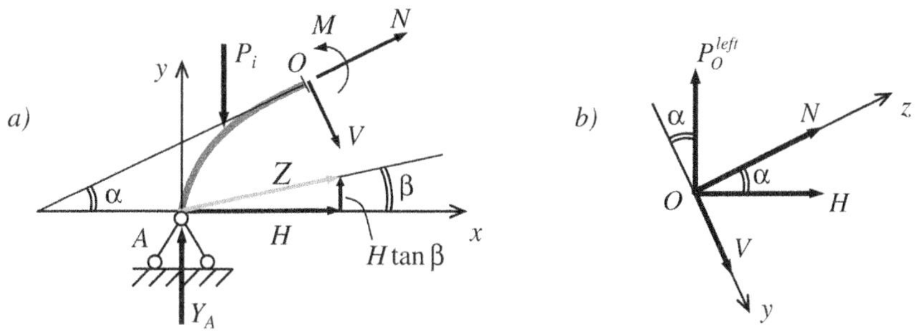

FIGURE 4.6 Diagrams of forces applied to the left part of the arch.

$$V + H\sin\alpha - P_O^{left}\cos\alpha = 0, \tag{4.15}$$

where α is an angle of the axial line slope. Note that two first addends in formula (4.14) are the sum of left forces with respect to section "O" in the reference beam. Denoting the shear force at this section of the beam as V^0, we can write down:

$$V^0 = Y_A - \sum_{left} P_i. \tag{4.16}$$

After substitution of force P_O^{left} from (4.14) into formula (4.15) and replacement according to (4.16), we obtain the expression for shear force at any section of an arch:

$$V = \left(V^0 + H\tan\beta\right)\cos\alpha - H\sin\alpha. \tag{4.17}$$

The equilibrium condition for forces depicted in Figure 4.6b, in projection onto axis Oz, has the form:

$$P_O^{left}\sin\alpha + H\cos\alpha + N = 0. \tag{4.18}$$

Hence:

$$N = -(V^0 + H\tan\beta)\sin\alpha - H\cos\alpha. \tag{4.19}$$

Thus, for support reactions and internal forces at the cross-section, we have obtained relations (4.10), (4.11), (4.13), (4.17), and (4.19). We see that internal forces in a hinged arch are determined through the forces in its reference beam.

For an arch with supports at the same level, it is handy to represent formulas (4.17) and (4.19) in the form:

$$V = \left(V^0 - H\frac{dy_{sup}}{dx}\right)\cos\alpha; \quad N = -\left(V^0\frac{dy_{sup}}{dx} + H\right)\cos\alpha, \tag{4.20}$$

where $\frac{dy_{sup}}{dx}$ is the derivative of the function specifying the axial line of an arch.

Formula (4.13) leads us to the conclusion that bending moments acting on cross-sections of an arch can be decreased in comparison to bending moments in the beam under the same loading if we choose the curvature of the arch in a proper way. It turns out that these moments can be zeroed due to the next theorem.

Theorem about Rational Axis of Arch

In order that bending moment in a three-hinged arch being subjected to vertical loading equals zero at all cross-sections, it is necessary and sufficient that vertical ordinates of all points of the arch axis, determined from the support line, be proportional to corresponding ordinates of beam diagram M^0.

Proof

Let $y_{sup}(x) = kM^0(x)$ with some k for all coordinates x. Then $f = y_{sup}(x_C) = kM_C^0$. Substitute thrust (4.10) into (4.13) and thus obtain:

$$M = M^0 - Hy_{sup} = M^0 - \frac{M_C^0}{f} y_{sup} = M^0 - \frac{M_C^0}{kM_C^0} kM^0 = 0,$$

wherefrom the sufficiency follows. The necessity of the theorem's affirmation follows from formula (4.13).

The axis of an arch is referred to as rational for a given load if bending moments do not arise at all cross-sections of the arch. For an arch with supports at the same level and a rational axis, the shear forces at sections also do not arise. The reader is recommended to verify the latter by using the first formula (4.20).

Now then, we can diminish bending moments in an arch a good deal, even to zero, owing to appropriate shaping of the arch axis. At the same time, in contrast to the reference beam, in an arch there arise axial forces (4.19) that are compressive, as a rule. For extreme normal stresses in the cross-section of a member, it is known that due to compression with bending:

$$\sigma_{\substack{\max \\ \min}} = \frac{N}{A} \pm \frac{M}{S},$$

where A is the area of cross-section and S is the section modulus. If a beam covers a given span, then both tension and compression stresses arise in its sections, whereas only compression stresses arise in an arch of rational axis and horizontal support line. The tensile strength of brittle materials (stone, concrete) is one or more orders of magnitude less than compressive strength, and the usage of beams made of uniform brittle materials is not reasonable. At the same time, arches made of brittle materials withstand essential loads because they work in compression.

Relative to the magnitude of compressive forces in an arch, note that usually they are of the order of magnitude of the thrust, but in correspondence with formula (4.10) the thrust can be made arbitrarily large due to sufficiently small rise of arch. Therefore, we have to choose the rise of the arch large within reasonable limits.

The equation of the axial line of a symmetric parabolic arch can be represented in the form (Figure 4.7):

$$y = \frac{4f}{l^2} x(l - x). \tag{4.21}$$

The axis of this arch happens to be rational if the arch is subjected to the load uniformly distributed along its span, for the diagram of beam moments in this case is also of parabolic form symmetrical relative to the rise of the arch.

The diagrams of internal forces in an arch are constructed either on its axial line or on the horizontal under the arch. Figure 4.7 shows examples of diagrams constructed on axial and horizontal basic lines. While constructing the diagrams, one has to determine the internal forces at characteristic cross-sections of the arch (where there are kinks, jumps, and extrema of internal force profiles)

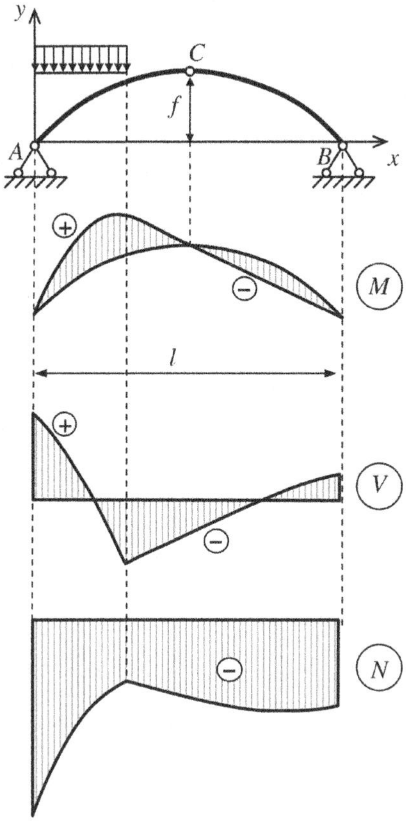

FIGURE 4.7 Diagrams of internal forces: M – diagram of the moment on the axial line of the arch; V and N – diagrams of shear and axial forces on the horizontal below the arch.

FIGURE 4.8 Theodor Heuss Bridge over the Rhine River (connects cities of Mainz and Wiesbaden). Photo by Heidas (http://heiko-dassow.de/). (License CC-by-SA. Source: https://commons.wikimedia.org/wiki/File:Mainz-Theodor-Heuss-Bruecke-2005-05-16a.jpg.)

and, if necessary, at some intermediate cross-sections. To determine forces V and N by formulas (4.20), the values of $\frac{dy_{sup}}{dx}$ and $\cos\alpha$ are wanted, which, in the case of a parabolic arch, are calculated by the formulas:

$$\frac{dy_{sup}}{dx} = \frac{4f}{l^2}(l-2x); \tag{4.22}$$

$$\cos\alpha = \frac{1}{\sqrt{1+\tan^2\alpha}} = \frac{1}{\sqrt{1+\left(\frac{dy_{sup}}{dx}\right)^2}}. \tag{4.23}$$

In conclusion, we give an example of a dainty match of architectural ideas and structural solutions. Figure 4.8 presents a photo of the arch bridge over the Rhine River in Germany. The two-hinged arches strengthen the superstructure from below. Massive piers withstand horizontal loads from the arches. This bridge has a maximum span of 103 m and was erected in 1885.

5 Three-Hinged Arches under Live Static Load

5.1 ANALYTICAL TECHNIQUE OF INFLUENCE LINE CONSTRUCTION AND NIL POINT METHOD

Influence lines for an arch are constructed as the dependences of a response versus the horizontal coordinate x of the unit load acting upon an arch; the horizontal below the plot of an arch is taken as the basic line.

The *influence line for the thrust* is constructed on the basis of the formula obtained before:

$$H = \frac{M_C^0}{f}. \tag{5.1}$$

Due to (5.1), the influence line for the thrust is the triangle with a vertex under the crown hinge (Figure 5.1); the ordinate of the vertex is:

$$\bar{H}_{\max} = \frac{l_1 l_2}{lf}. \tag{5.2}$$

An *influence line for a bending moment* at any section "O" is constructed on the basis of formula (4.13):

$$M = M^0 - Hy_{sup}. \tag{5.3}$$

Two addends on the right side are represented with two influence lines, for which the subtraction is easy to do by overlaying these lines in the same diagram. The ordinates of the difference can be initially constructed as vertical segments between two influence lines; finally, one draws them on the horizontal (Figure 5.1). We see that **the influence line for a bending moment is a piecewise-linear continuous line of three portions; the ordinates of this line prove to be null at the supports.**

By the technique described, the lines for H and M are constructed for an arch of general form (Figure 4.1). Further, we confine ourselves to the case of an arch with supports at the same level.

An *influence line for shear force* at any section "O" is constructed on the basis of formula (4.17):

$$V = V^0 \cos\alpha - H \sin\alpha. \tag{5.4}$$

Figure 5.2 demonstrates the result of subtraction of constituents on the right side.

An *influence line for axial force* is determined by the expression:

$$N = -(V^0 \sin\alpha + H \cos\alpha). \tag{5.5}$$

Figure 5.2 demonstrates the result of summation of constituents on the right side.

We see that **influence lines for V and N are piecewise-linear lines consisting of three portions, with the discontinuity below the section and the kink below the crown hinge; the portions adjacent to the section are parallel; at the supports the ordinates of these lines prove to be null.**

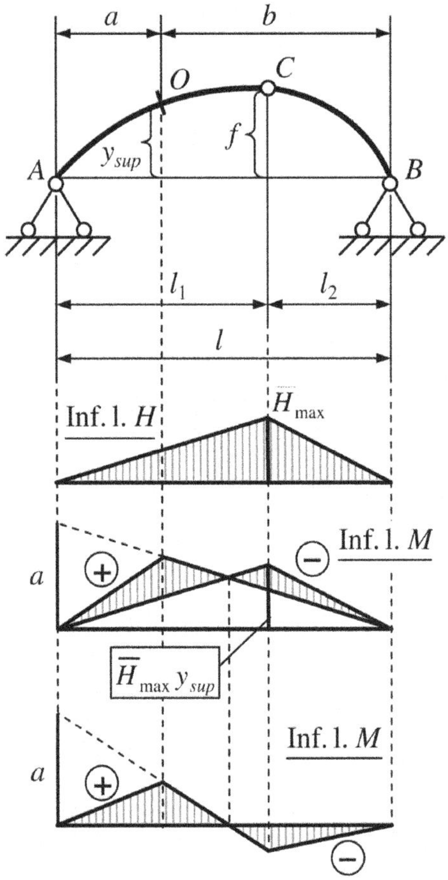

FIGURE 5.1 Construction of influence lines for thrust and internal moment.

Formulas (5.3)–(5.5) enable us to calculate the ordinates of influence lines at the characteristic points of the basic line (under the section and the crown hinge), and after that, these lines can be drawn with observation of the scale. Besides, there is the nil point method, which secures a prompt way of drawing these lines without calculations, by means of graphical constructions only.

Influence lines for forces *M*, *V*, and *N* exerted upon the cross-section of an arch are made up of three branches: left, right, and median. The median branch corresponds to the position of the load at the arch's portion *OC*, i.e., between the section and the crown hinge. We call the point of intersection of a median branch with a basic line the nil point of the influence line. Having determined the nil point and employed the noted properties of the influence line, one can construct the wanted line without scaling. Moreover, by using the known ordinate of the median line's extension to the vertical below one of the supports (Figures 5.1 and 5.2), one can also specify the scale. The nil point is found graphically on the plot of the arch.

The nil point of the bending moment is located under point *E* of the intersection of two straight lines, one of which runs from the support hinge of a haunch comprising the section to the centroid of the section while another runs from the opposite support hinge to the crown hinge (lines *AO* and *BC* in Figure 5.3). Indeed, let the section is selected, for instance, at the left haunch. Then point *E* is situated over portion *OC* of the left haunch, and, for load positioned under point *E*, external forces act upon the right haunch through the hinges. Accordingly, the reaction of the right support is directed to the crown hinge (along line *BC*). Besides this reaction and unit load, a third force is exerted on the structure – it is the reaction of the left support and equilibrium of the arch taken as a

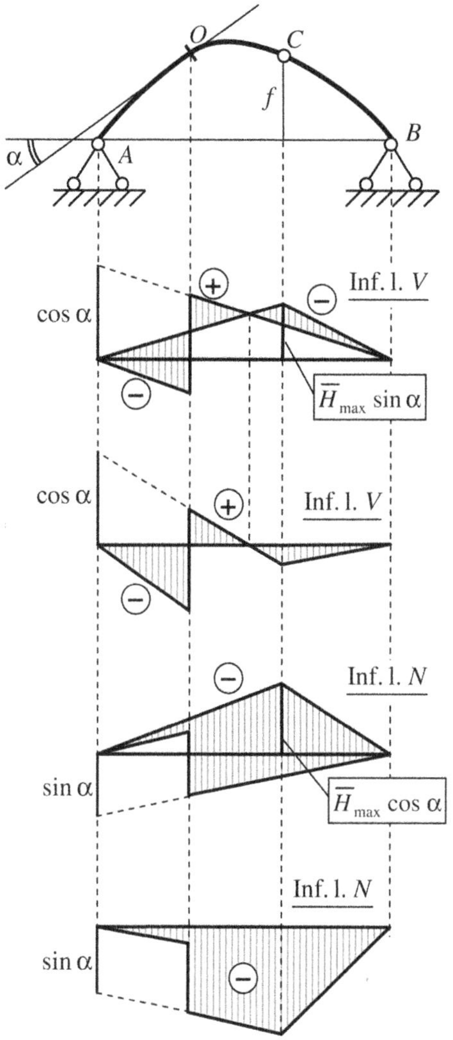

FIGURE 5.2 Construction of influence lines for internal forces.

whole implies intersection of these forces at a single point. To fulfill this condition, the reaction of the left support must be directed along line AO. Thus, this reaction crosses the centroid of the section under consideration and doesn't produce a bending moment at the section (Figure 5.3). We see that when the load is located under point E, the nil point of the influence line is reached. Note that nil point e for a bending moment is always situated below the portion OC.

To substantiate the rules of finding the nil point of shear and axial forces, we have to consider the dependence of the wanted force versus the position of load $P = 1$ at the auxiliary cantilever fixed at portion OC. Such cantilevers are depicted in Figures 5.4 and 5.5. We assume that the load's position on the cantilevers is uniquely determined by coordinate x. In the beginning, we shall get convinced that **the influence line for the force at section "O" caused by a unit live load positioned at the auxiliary cantilever coincides with the median branch of the sought influence line**. Due to this property, we may search the nil point of the median branch as the nil point of the influence line for the sought force in the substitute arch with a load line formed by cantilevers.

The substantiation of the assertion is as follows. If the load is located on the auxiliary cantilever, the force at the section "O" depends only on the support reaction upon the haunch comprising the

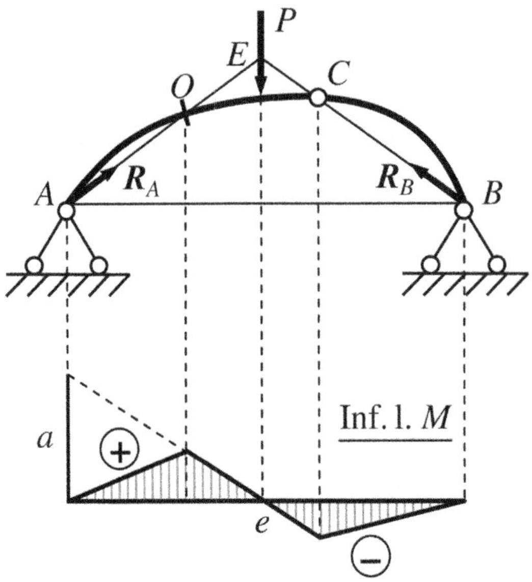

FIGURE 5.3 Construction of influence line for moment M by nil point method.

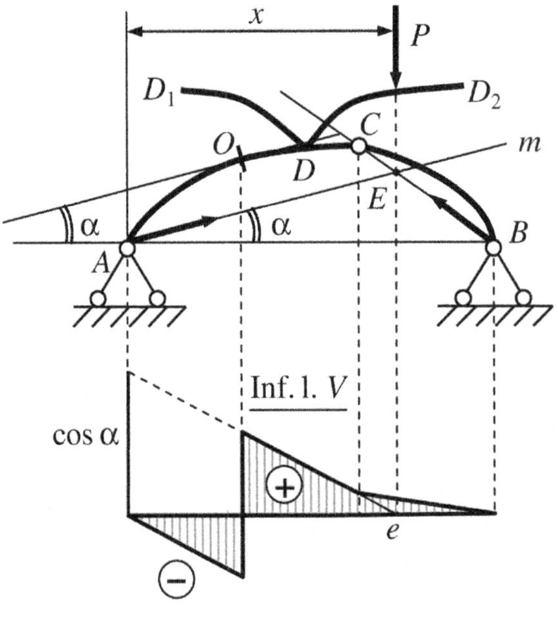

FIGURE 5.4 Construction of influence line for force V by nil point method.

section. For definiteness, we assume that the section is located to the left of hinge C. Then the force is defined by the beam component of left support reaction Y_A and thrust H. By using the calculation technique from Chapter 4, it is easy to show that reactions Y_A and H depend linearly on coordinate x of the load located on the cantilever. We can also establish that if the load acts on the arch's portion OC directly, the dependencies of reactions versus coordinate x remain the same, and these reactions define the sought internal force as before. So, the force at section "O" depends linearly on x if the load is located on the cantilever, and this dependence doesn't change due to the load's transference to portion OC.

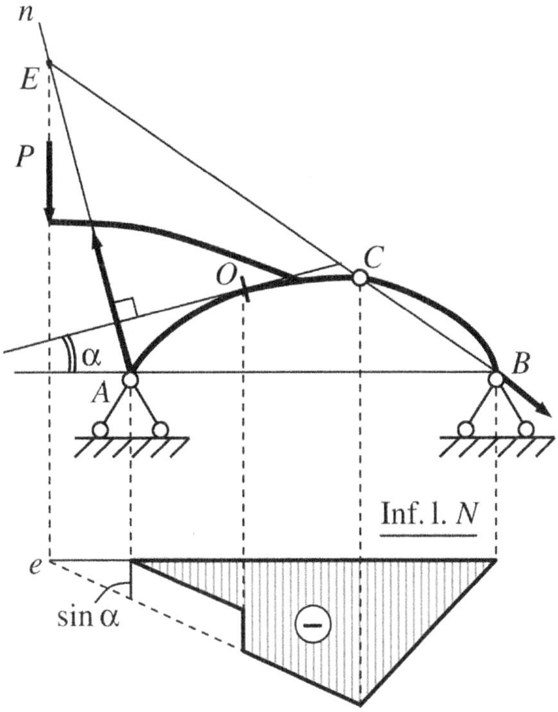

FIGURE 5.5 Construction of influence line for force N by nil point method.

The nil point of shear force is found according to the next rule. From the support of the haunch containing section "O," we draw a straight line parallel to the arch axial line at section "O" (in Figure 5.4 it is line Am). From the opposite support, we draw another line through the crown hinge (in Figure 5.4 it is line BC). Point E of intersection of these lines is positioned over the sought nil point e.

Let us show that the point obtained in this way is the nil point for the influence line of shear force in case of the unit load's movement on auxiliary cantilevers D_1D and D_2D (Figure 5.4). At the arch portion AO, a single external force is exerted – the sloped support reaction, which is counterpoised by internal forces acting upon the cross-section. Evidently, this reaction doesn't produce the shear force if it doesn't have a shear component. To this effect, the support reaction upon the haunch comprising the section must be directed in parallel to the normal to this section, i.e., along line Am. In the case of a load applied along the vertical passing through point E, this requirement is fulfilled: to comply with the condition of equilibrium of three external forces, the reaction of support A has to be directed to point E.

Transition to the substitute arch supplied with cantilevers is necessary when the median segment of the wanted influence line doesn't reach the abscissa axis. In this case, the nil point is obtained by extending the median segment and the movement of the load on the cantilever outside arch portion OC makes this extension. If the nil point belongs to the median segment, then one can find this point from the condition of zero internal force due to the load's position at portion OC, and the substitute arch isn't required.

The nil point of axial force is found according to the next rule. From the support of the haunch containing section "O," we draw a straight line perpendicular to the arch axial line at section "O" (in Figure 5.5 it is line An). From the opposite support, we draw another line through the crown hinge (in Figure 5.5 it is line BC). Point E of intersection of these lines is positioned over the sought nil point e. This rule follows the requirement that when the load acts on the auxiliary cantilever, the support reaction upon the haunch containing the section must be directed in parallel to the plane of the section (Figure 5.5).

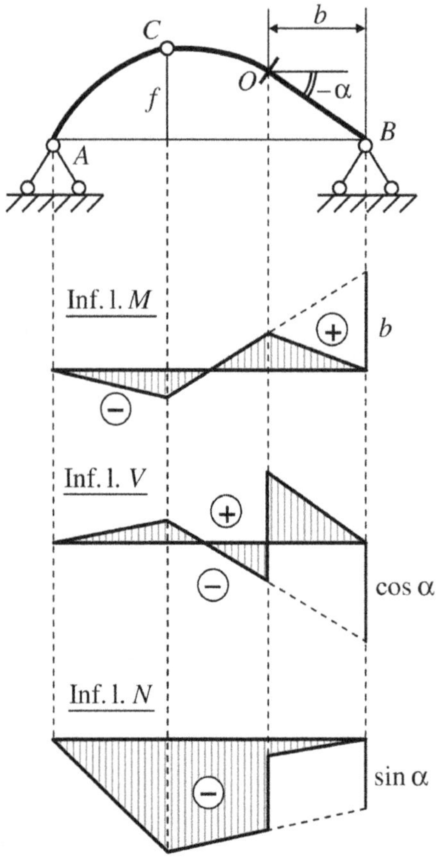

FIGURE 5.6 Influence lines when the section is positioned to the right of crown hinge.

Up to here, the influence lines were constructed for section "O" located to the left of the crown hinge. In Figure 5.6, the influence lines for internal forces are presented for the section located to the right of the crown hinge.

5.2 KERN MOMENTS AND NORMAL STRESSES

Below we assume that the cross-section of an arch doesn't vary along the axial line.

Known internal forces enable one to determine the stresses in the section of an arch. In calculating stresses, it is convenient to use the concept of kern limits of a section. The definition of kern limits is based on the notion of the kern of the section being studied in mechanics of materials. The kern (or core) of the section is a zone around the centroid of the section within which action of axial compressive force will not cause tensile normal stresses in the section (Timoshenko 1948, p. 235).

Kern limits are introduced for a bar (in general case, curvilinear) which is loaded by the plane system of forces and has the symmetry plane coinciding with the plane of loads. The kern limits of the section (also termed as top and bottom core points of the section) are the outermost points of the kern, lying in the plane of loads. For any cross-section with principal centroidal axis Oy lying in the plane of loads, we denote (Figure 5.7):

E_1 and E_2 – the top and bottom extreme points, respectively;*

* We call the point of the section most remote from the neutral line in either half-plane the extreme point of the section in bending with tension-compression. On either side of neutral line there is at least one extreme point, where the tensile or compressive stress reaches its largest value.

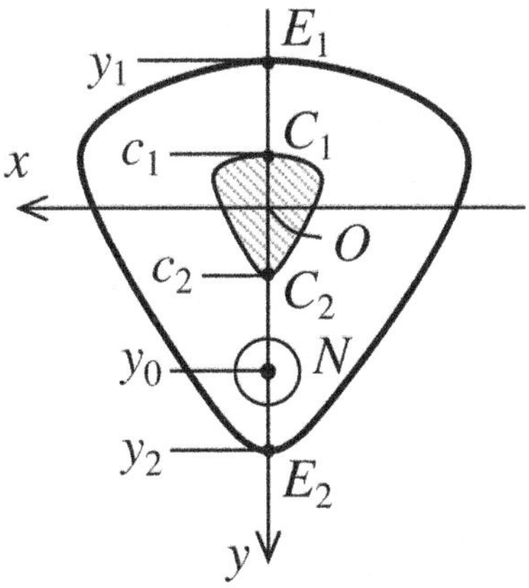

FIGURE 5.7 Resultant of normal internal forces and kern of section (hatched): The case of $N > 0$ is shown; $M = y_0 N$.

y_i – vertical coordinates of extreme points (i.e., $y_i = OE_i$ – taking the sign into account);
$S_i = I_x / |y_i|$ – the section moduli for top and bottom fibers;
C_1 and C_2 – the top and bottom kern limits, respectively;
c_i – vertical coordinates of kern limits (i.e., $c_i = OC_i$ – taking the sign into account).

We can construct the influence line for the largest normal stresses by using the influence lines for M and N and the following relation to transform these lines:

$$\sigma_{1,2} = \frac{N}{A} \mp \frac{M}{S_{1,2}}. \tag{5.6}$$

Here σ_1 is the stress at the top fiber crossing the section, and σ_2 is the one at the bottom fiber. (These stresses are referred to as *edge stresses* in the arch section.)

Nevertheless, there is no need to sum up influence lines for the two addends in (5.6) if we introduce the total moments of elementary internal forces* in the section relative to the kern points of section. *The total moments of normal internal forces in a section, taken relative to the kern limits are referred to as kern (core) moments.* We remind readers that the bending moment is produced by the totality of internal forces normal to a section with respect to the centroid of the section. Therefore, kern moments differ from bending moments only in choosing the pole for calculation.

Let us prove the next *theorem about kern moments*, which fulfills both straight and curved members.

Theorem

If external forces lie in the symmetry plane of a member, the total moment of internal normal forces taken relative to the kern limit determines the stress at the opposite extreme point by the formula of largest stresses due to bending:

$$\sigma_2 = \frac{M_c^{(1)}}{S_2}; \quad \sigma_1 = -\frac{M_c^{(2)}}{S_1}, \tag{5.7}$$

where $M_c^{(1)}$ and $M_c^{(2)}$ are the kern moments relative to the top and bottom kern limits, respectively.

* Elementary internal force is the force exerted upon elementary (infinitesimal) area of cross-section.

The proof is based on the property of kern limits as follows: **if axial force is applied to the kern limit of a section, then the normal stress equals zero at the extreme point lying opposite** (Karnovsky 2012, p. 87). In the case under consideration, the stress at any point of the section is determined by the formula:

$$\sigma = \frac{N}{A} + \frac{My}{I_x}, \tag{5.8}$$

where y is the coordinate of the chosen point. We denote as y_0 the coordinate of a point where resultant N of elementary internal forces normal to the section is applied. The relation between bending moment $M = Ny_0$ and kern moment can be written as

$$M_c^{(1)} = N(y_0 - c_1) = M - Nc_1.$$

Represent now (5.8) in the form:

$$\sigma = \frac{N}{A} + \frac{(M_c^{(1)} + Nc_1)y}{I_x} = \left(\frac{N}{A} + \frac{Nc_1}{I_x}y\right) + \frac{M_c^{(1)}}{I_x}y.$$

The augend in parentheses is the normal stress for the case when axial force is applied at the top kern limit. Due to the property of kern limits, if $y = y_2$, this constituent turns into null. The first equality (5.7) has been proved. The second one is proved likewise.

According to the theorem of kern moments, the maximum stress, for example, at the bottom fiber of an arch is proportional to single internal force – the kern moment relative to the top core point. The technique of influence line construction for this moment doesn't differ from the technique developed for moment *M*. The only feature to remember is that while bending moment *M* equals the sum of moments of external forces at the one side of the section, taken with respect to the centroid, the kern moment equals the same sum with respect to the kern limit. The influence lines for bending and kern moments are shown in Figure 5.8. The quantities x_O and x_{C_1} are the horizontal coordinates of corresponding points of section; e_O and e_{C_1} are the nil points.

Also, Figure 5.8 presents the influence line for largest stress σ_2, obtained by totalizing two influence lines by (5.6). This line has discontinuity below the section. The discrepancy of influence lines for $M_c^{(1)}$ and σ_2 is explained by the difference in load lines: $M_c^{(1)}$ is involved with the movement of load $P = 1$ on the line of top core points, whereas σ_2 is involved with the load's movement on the arch axial line. Note that when the load is placed in the vicinity of section "*O*," formulas (5.6)–(5.7) may be inapplicable for stress analysis, because deformation hypotheses of bending theory become inapplicable for such loading.

But if we even proceed from the deformation hypotheses, we find out that *all* influence lines shown in Figure 5.8 differ from "genuine" lines in the vicinity of section "*O*." The matter is that load can't be applied physically to the points inside the member – the line of its movement would be taken with good reason on the upper edge of the arch (Figure 5.9). Due to such choice of the load line, formula (5.3) may produce an error when the load is positioned in the vicinity of the section. This occurs because of the discrepancy between the left (right) forces' assemblage in an arch and the similar assemblage in the reference beam: On an arch and on the reference beam, the load with the same horizontal coordinate can be positioned at the different sides of section (Figure 5.9). Let, for example, bending moments are calculated by totalizing the moments of left forces according to formula (4.12). Since the force *P* in Figure 5.9 appears to be right for the arch but left for the reference beam, the transition from formula (4.12) to formula (4.13) happens to be incorrect because in such a transition there appears the error equal to the moment of force *P* relative to centroid *O*. A similar situation takes place with the axial force being determined by formula (5.5).

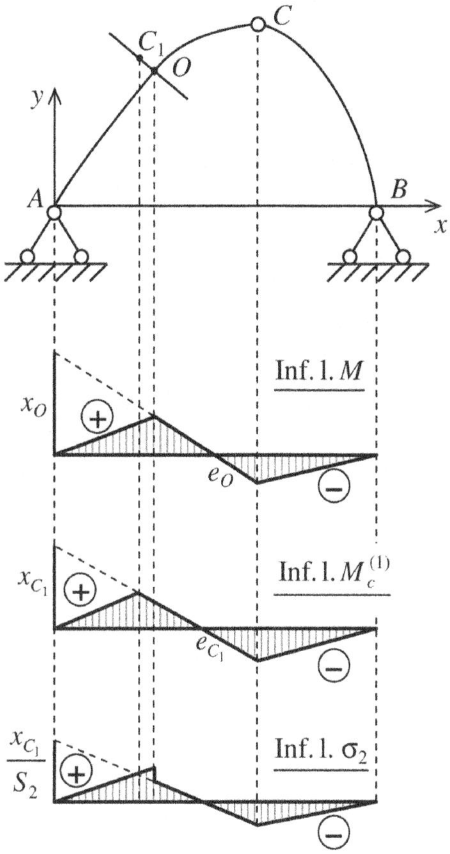

FIGURE 5.8 Influence lines for internal moments and edge stress.

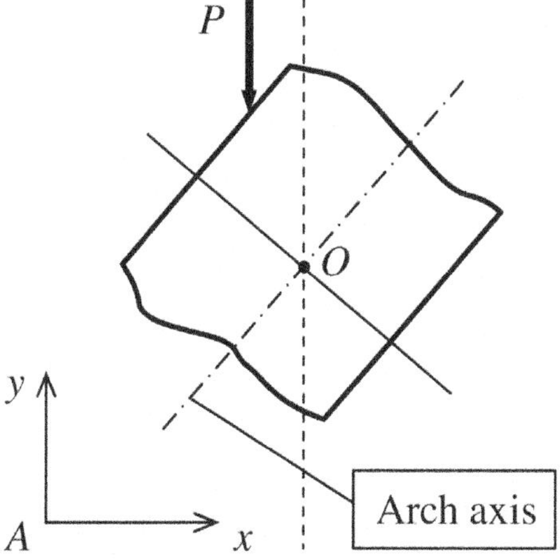

FIGURE 5.9 The case when the load is positioned to the right of section "O" on the arch and to the left of a similar section on the reference beam.

Remark

The main results of the stated items (including formulas (5.6)–(5.8)) are applicable for an arch with slight variability of the cross-section along the axial line. The comment about necessary minuteness of this variability can be found, in (Hibbeler 2011, p. 287). In structural calculations, it is usually sufficient that at every point on the outline of an arbitrary cross-section of an arch, the dihedral angle between the tangent plane to the surface of the arch and cross-section be in the range of 85–95°.

Section IV

Hinged Beams

This section contains the definition of a hinged beam as a statically determinate compound beam and presents the techniques of its kinematic analysis and calculation of internal forces. The static and kinematic methods of influence line construction are under consideration. Kinematic method, also known as the Müller-Breslau principle, is substantiated by using the principle of virtual work.

6 Hinged Beams

6.1 CONCEPT OF A HINGED BEAM AND ITS KINEMATIC ANALYSIS; INTERACTION SCHEME

The *hinged beam* (or *multi-span cantilever beam*) is a statically determinate system assembled by sequentially positioned straight bars connected at the ends by hinges and working in bending. The members of a hinged beam are connected by simple hinges and called simple beams. An interval between two neighboring supports is called a span.

Hinged beams are also called *Gerber beams* in honor of German engineer Heinrich Gerber, who patented the hinged beam in 1866 and employed this constructive solution in bridge building. The theory of hinged beams was developed by H. Gerber and Russian engineer G.S. Semikolenov during the period 1863–1871.

Hinged beams are applied to subsystems of structures, and they are used as the model of some structures. Unthrusted hinged trusses under vertical loads may be simulated by the hinged beam. The reference beam for such a truss is constructed by replacement of its discs with simple beams (Figure 6.1). The results of analysis of a hinged beam are used further in analysis of its prototype – the hinged truss. In particular, the support reactions of both systems are the same.

When comparing a hinged truss with its reference hinged beam, oftentimes one calls the truss an open-web structure while the beam is called a solid-web structure.

Hinged beams are assumed to be horizontally situated and are designed to support vertical loads and force couples. To eliminate longitudinal forces in members, the hinged movable supports are positioned vertically; besides these supports, a hinged beam must include one hinged immovable support or rigid support. Vertical loads produce vertical support reactions (and maybe the reaction moment in a rigid support).

Kinematic analysis of a hinged beam is made using the usual technique. The necessary condition of stability and static determinacy (1.7) is represented in the form:

$$H+3=L_{sup}. \tag{6.1}$$

This we could expect: three equations of statics for the system as a whole can be supplemented by conditions of equality to zero for the total moment of external forces exerted upon the system's part at one side of the connecting hinge, relative to this hinge. After solving the set of $H + 3$ equations of statics for the support reactions of the same number, we can determine the internal forces in the members.

The stability analysis for a hinged beam is made by consecutive attachment of its members to the earth in a stable fashion. Nevertheless, sometimes the rules of two-disc or three-disc association can't be applied, and then the stability relative to the earth may be verified by removal of connections between members. Having removed one of the hinges, one considers possible displacements of disjoined ends. If constraints allow only horizontal displacements for one of the ends but vertical displacements for another, then returning the hinge back into the system would make any displacements of reconnected members impossible, i.e., the system constitutes a solid disc together with the earth. An example of a system which doesn't comply with the rules of two-disc or three-disc association is shown in Figure 6.2a. The permitted displacements of disjoined subsystems are shown in Figure 6.2c. The mechanism in Figure 6.2c has two DOFs, which are deleted by connecting the ends at point *C*. The reader is recommended to point out in the diagram the parameters specifying the location of this mechanism.

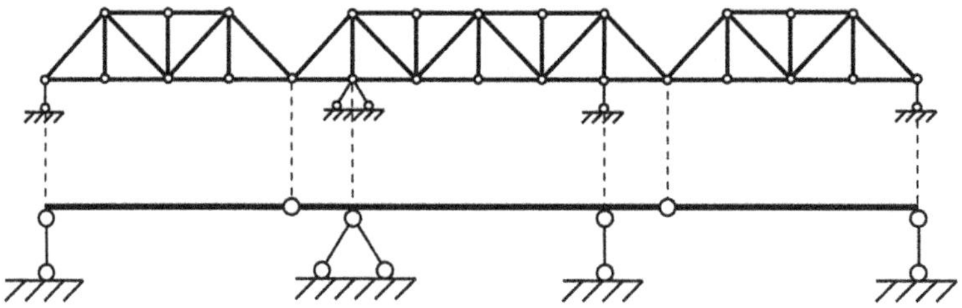

FIGURE 6.1 Hinged truss and its reference beam.

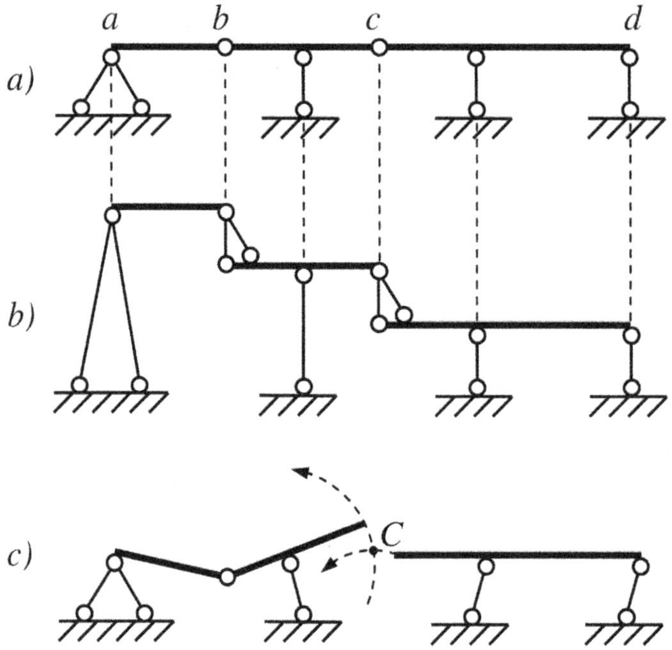

FIGURE 6.2 Hinged beam (a), its interaction scheme (b), and displacements of members after removal of one hinge (c).

The following rule of distribution of simple hinges connecting members of a hinged beam is the necessary condition of its static determinacy: **each span of a hinged beam may contain no more than two hinges; spans with two hinges must alternate with spans without hinges.**

Analysis of a hinged beam is based on an *interaction scheme*, by which the problem of analysis is reduced to consecutive calculation of separate members as statically determinate beams. In this scheme, the members are depicted on different levels depending on the sequence of analysis; on that account, this scheme is also referred to as a *multilevel diagram.*

The interaction scheme is a multilevel representation of a hinged beam, wherein loads upon members of the lower level do not influence internal forces in the members of the upper level, and, conversely, internal forces in the members of the lower level depend on loads applied to an adjacent member of the upper level. Among two adjacent members of the interaction scheme, the lower member is called the main, or primary, simple beam, and the upper one is termed the suspended, or secondary, simple beam.

An example of a hinged beam with possible variants of its multilevel diagram is shown in Figure 6.3. The constraint reactions arising between members of different levels are depicted in diagram b of Figure 6.3. These reactions can be developed by the connections represented in diagrams

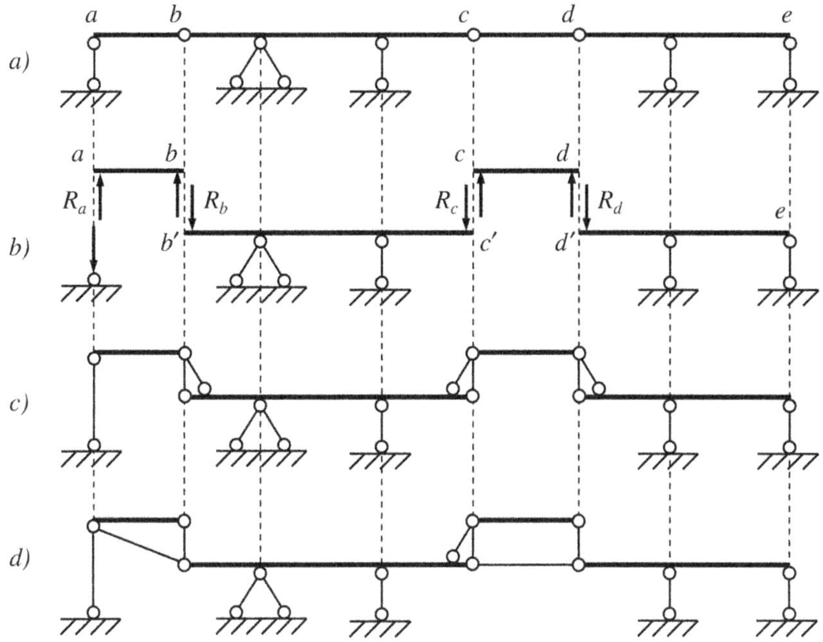

FIGURE 6.3 Hinged beam and possible representations of its interaction scheme.

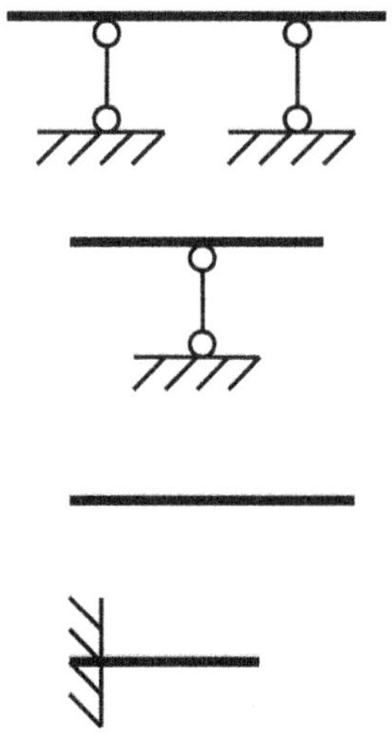

FIGURE 6.4 Possible elements of hinged beam.

c and *d* (Figure 6.3). The two latter schemes are equivalent in a static sense, but diagram c is more typical and common. Another example of a multilevel diagram is shown in Figure 6.2b.

An interaction scheme has the important *property of step-like shape*: **adjacent simple beams are situated at different levels of the multilevel diagram.**

To prove this property, let us consider possible displacements of the ends of adjacent members after removal of the connecting hinge. We establish, at the beginning, that one and only one of the disjoined ends can move in a vertical direction (with pivoting of the member).

A hinged beam is assembled by beams and mechanisms, of which several are presented in Figure 6.4. While connecting possible elements with hinges, we can discover that the free end of the system is either immovable, or allows horizontal or vertical displacement only, or both noted displacements. Let us disconnect two members in the hinged beam by removal of the hinge. The obtained system acquires two DOFs, and these DOFs can't be specified by horizontal displacements of both disconnected ends. Really, in the latter case, the connecting hinge doesn't oppose coherent horizontal displacements of both subsystems, i.e., the return of the hinge wouldn't restore the system stability. Just the same, two DOFs can't be specified by vertical displacements of both ends. So, only one of the disconnected ends can move vertically.

Now we prove that **if a member among two adjacent simple beams can pivot after removal of the connecting hinge, then this member is secondary.** From this assertion, we get the property of the step-like shape of an interaction scheme as the corollary. This assertion has significance in itself, because it can be employed as a kinematic feature of the secondary member in a hinged beam.

We assume there are no loads at that part of the system which contains the member with the ability to pivot after disconnection. Then the shear reaction is not developed by a connecting hinge, for the shear force being applied to the vertically-movable end would evoke the movement of the whole mechanism. But this means that a member with the ability to pivot (after disconnection) remains unstressed (before disconnection) with no regard to loads upon the adjacent member. In the same way, we can substantiate that loads upon the member with the ability to pivot produce nonzero reactions in the connecting hinge and, therefore, nonzero internal forces in the adjacent member. Thus, the property of step-like shape has been proved.

Owing to the property of step-like shape of an interaction scheme, we can calculate a hinged beam as the totality of statically determinate beams studied before. To accomplish that, one has to begin the calculation of constraint reactions and internal forces from the simple beams of the top level, descending step-by-step on the diagram.

To construct the interaction scheme, we may compare adjacent members in pairs by the kinematic feature. Besides the kinematic feature, we can use the static feature of the secondary member: **the member of the top level has exactly two unknown constraint reactions.** Member *ab* in Figure 6.3 has one unknown support reaction and one reaction of the right connection; therefore, this one is located at the top level with respect to the adjacent member. Simple beam *cd* is also at the top level because it has two unknown reactions – of left and right connections, respectively. In counting the number of reactions, we don't take into account horizontal ones (they are absent in hinged beams).

6.2 STATIC METHOD OF INFLUENCE LINE CONSTRUCTION

In construction of influence lines for internal forces and constraint reactions in hinged beams, static method is implemented through following rule.

Let an influence line be constructed for internal force or constraint reaction acting in a given member. Construction is to be started with location of the unit load on the given member. The portion of an influence line below any other member is a segment. The construction proceeds by joining up the portions under adjacent members. Ascending upon a member of the upper level means continuous transition to the new segment of the influence line. The nil point of this new segment corresponds to the positioning of the load over the support or at the next hinge (if the support is absent). Descending on a member of the lower level means "zeroing" of the influence line's part, i.e., equating the ordinates of all remained portions to zero.

Figure 6.5 shows examples of influence line construction for the internal forces at given cross-section I of member *CD*. The suggested rule may be clearly substantiated by using these examples. After transition of the unit load onto member *DE*, the force in neighboring beam *CD* is defined only by the reaction of hinge *D*. But this reaction depends linearly on the coordinate of the load positioned on member *DE*; thus the portion of the influence line below *DE* is the segment of a straight line. The intersection of this segment with the abscissa axis corresponds to the zero reaction of connection *D*, and the latter occurs only when the load is positioned over support *H*.

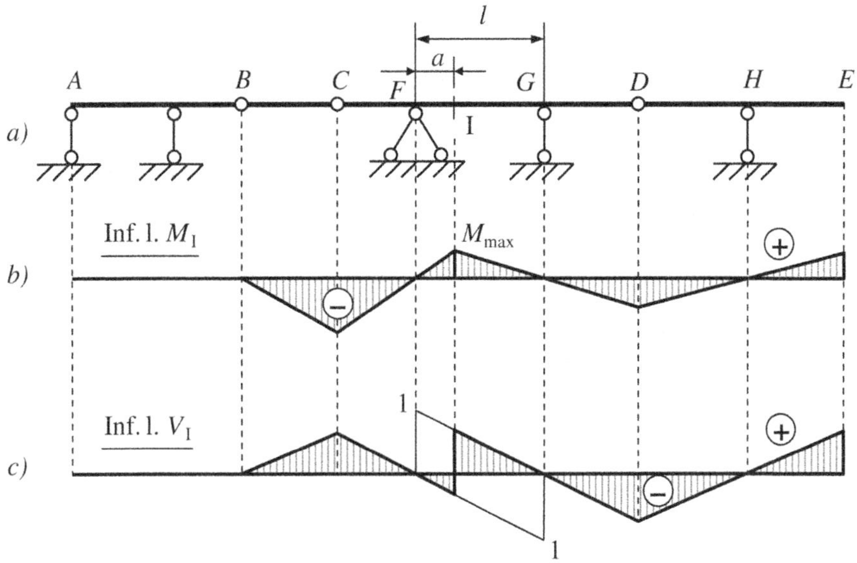

FIGURE 6.5 Examples of an influence line construction by the static method; inf. l. for bending moment M_I is constructed from the point with ordinate $M_{max} = \frac{a(l-a)}{l}$; inf. l. for shear force V_I is constructed from the points with unit ordinates

Similar reasoning is fair for the portion of the influence line under member *BC*. In this case, the reaction of hinge *C* becomes zero for the load positioning at hinge *B*.

Simple beam *AB* is located under member *BC* in the multilevel diagram; therefore, the load upon beam *AB* can't produce forces in member *BC* as well as in the next member *CD*.

We may change the positioning of supports in the system under consideration so that members *AB* and *CD* would be situated at the upper level relative to the member *BC* of the bottom level. Then the forces in member *BC* would not produce the forces in *CD*, and, as before, the influence line portion below *AB* would be zeroed.

6.3 KINEMATIC METHOD OF INFLUENCE LINE CONSTRUCTION

The kinematic method for construction of influence lines in its brief formulation is known as the Müller-Breslau principle. This method is founded on the principle of virtual work being studied in theoretical mechanics. The principle of virtual work in its modern formulation was developed by Swiss mathematician Johann Bernoulli in 1717. The method for construction of influence lines based on this principle was suggested by German engineer and professor of bridge design Heinrich Müller-Breslau in 1886.

There exist several formulations of the principle of virtual work which differ by a level of generality and objects under consideration. In this course, the simplest formulation is given, in which the geometrically unstable systems of rigid bodies are considered (we confine ourselves to plane systems).

Virtual displacements are infinitesimal displacements permitted by constraints in a system. Infinitesimal displacements are displacements small enough that, in calculation of required physical values, the errors of analysis may be neglected. The total work done by an assemblage of forces exerted on a system as it moves through a set of virtual displacements is called *virtual work*.

For plane systems being studied in structural mechanics, the principle of virtual work states: **if an unstable system of discs with connections of the main types is in equilibrium, then the virtual work done by external forces is zero.**

Through the use of this principle, we can construct influence lines for any internal force or the constraint reaction S in a statically determinate bar system. **For this purpose, we transform the statically determinate system into a mechanism with a single DOF by removal of the connection at the site of action of the sought force S. To secure equilibrium of the mechanism, we exert one or two additional external forces (moments) X equal to the sought force at this site.**

The connection here is considered in the abstract, as restriction on displacements in a system. Removal of a connection in a bar system means the transformation of the system in a fashion that depends on the response under investigation S. If this response is the reaction developed by connection of the first type, then transformation is really reduced to removal of this connection from a system, and forces X are exerted at the sites of joining the simple rod to members instead of reactions developed by connection. If the connection supports a member, then only one force is applied to the system (Figure 6.6a). But if the connection under investigation joins two members, then two forces X are required for equilibrium of the mechanism. If response S is the internal force acting upon a cross-section of a member, then as removal of the connection, we imply the insertion of a kinematic joint at the site of the cross-section, which deletes this response at the neighboring sections. A kinematic joint deleting a bending moment is shown in Figure 6.7a; kinematic joint deleting shear force is shown in Figure 6.8a. In the same diagrams, the additional forces (moments) X, which are equal and directed oppositely, are depicted. One or two external forces (moments) ensuring equilibrium of a system with a removed connection can be called generalized force. The presented examples for investigation of response S in hinged beams are plentiful. The transition to a system with a removed connection is treated more minutely in Chapter 12.

To construct an influence line, we exert the unit load on the load line of the single-DOF mechanism and set up the virtual work equation. This equation has the simple form:

$$1 \cdot \Delta_P + X \cdot \Delta_X = 0, \tag{6.2}$$

where Δ_P and Δ_X are the virtual displacements; namely, Δ_P is the vertical displacement of the point of unit load action, and Δ_X is mutual displacement of system members at the site of a removed internal connection or displacement of a member relative to the earth at the site of the removed support connection. Quantity Δ_X is also called *displacement in the direction of the removed connection.* This quantity may be not only translational displacement but also the mutual pivot angle for two parts of the mechanism. It is important to specify this displacement such that work of an additional generalized force has the form $X \cdot \Delta_X$.

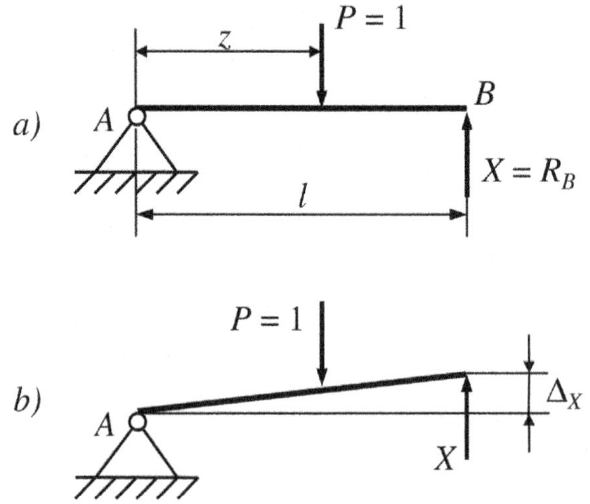

FIGURE 6.6 Removal of the right support link of a simple beam.

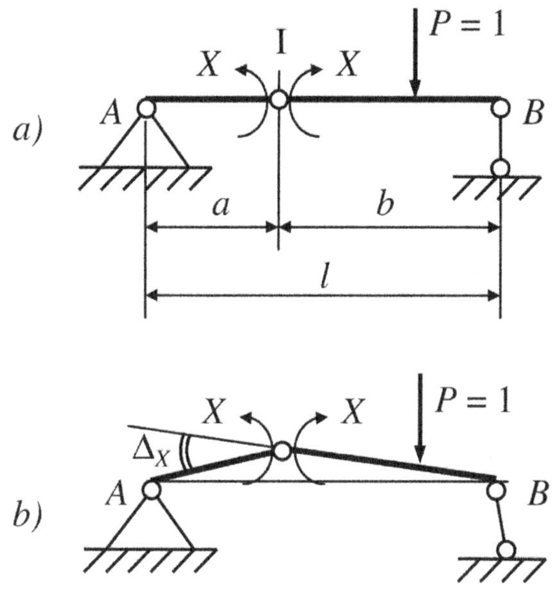

FIGURE 6.7 Removal of constraint producing a bending moment at section I; $X = M_I$.

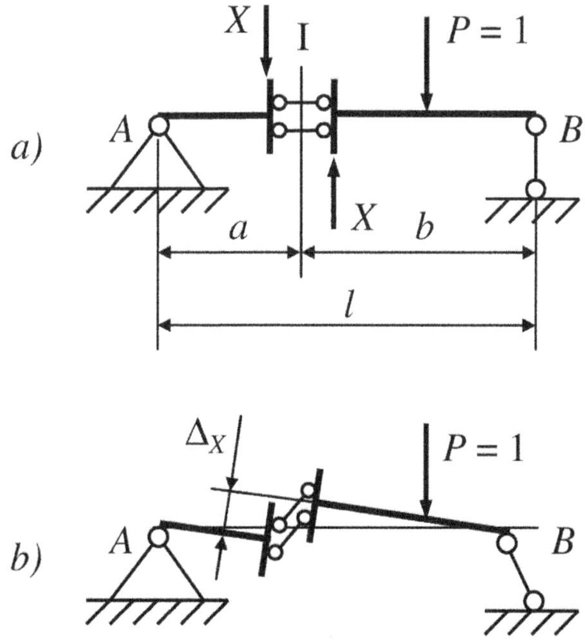

FIGURE 6.8 Removal of constraint producing shear force at section I; $X = V_I$.

Finally, the wanted response S is obtained in the form:

$$X = -\frac{\Delta_P}{\Delta_X}. \tag{6.3}$$

Further, we call two graphs of functions proportional if they are constructed relative to the same abscissa axis and the ratio of their functions is constant. We call vertical displacement Δ_P "deflection of the load line." The positive direction of deflection coincides with the positive direction of the unit load, that is, downward. In accordance with the formula (6.3), **the influence line of response $S = X$ is proportional to the diagram of load line deflections due to virtual displacement in the system with a removed connection at the location and in the direction of this connection**. According to the kinematic method, the problem of influence line construction is solved by construction of the diagram of load line deflections and its next normalizing. The infinitesimal displacements in (6.3) should be read as small enough displacements.

Let $\Delta_X > 0$. According to (6.3), to obtain the sought influence line, we have to divide the ordinates of the diagram of load line deflections by displacement Δ_X and change the signs of ordinates. The latter we can fulfill by interchanging signs "+" and "–" on the constructed diagram, without changing the diagram itself. In equation (6.2), we took the downward direction for deflections as positive according to the direction of the moveable load. Thus, the interchange of signs on the normalized diagram of deflections transforms it into the wanted influence line with the upward direction as positive.

The above may be stated as the *Müller-Breslau principle*:

The influence line for a force (or moment) response of a system is proportional to the diagram of load line deflections in the system obtained by removing the constraint producing this response and by giving virtual displacement at the location and in the direction of removed constraint. This diagram can be converted into an influence line via dividing ordinates by named virtual displacement and interchanging the ordinates' signs.

Also, we cite the traditional formulation of Müller-Breslau principle, which, being briefer, contains the inaccuracy fraught with errors in application of the principle:

The influence line for a force (or moment) response of a system can be constructed as the diagram of load line deflections in the system obtained by removing the constraint corresponding to this response and by giving a unit displacement at the location and in the direction of removed constraint. (The positive values of deflections are assumed downward; the positive values on an influence line are reckoned upward from the basic line.)

The inaccuracy of this statement lies in ambiguous interpretation of unit displacement, because it is not stated as infinitesimal.

In the presented formulations of the Müller-Breslau principle, it is not stated that the system under consideration is statically determinate – in this chapter it is implied. But further, in Chapter 16, we establish that this principle is applicable in the construction of influence lines for statically indeterminate systems.

Below, three simple examples of influence line construction by the kinematic method are presented. We confine ourselves to the construction of *qualitative influence lines*, i.e., graphs proportional to the sought influence lines with a positive proportionality factor. (To normalize the graphs in these examples is not a laborious problem.)

1. The influence line for response R_B of the beam in Figure 2.3 may be obtained by the removal of the right support and displacement of the right end of the beam (Figure 6.6b). The reaction of support is taken with the sign "+" if it is directed upward. Accordingly, the virtual displacement is to be directed upward, and the same direction becomes positive for the deflection of the load line after the interchange of signs. In the same manner, the influence line for response R_A is constructed if we interpret the left support as the assemblage of horizontal and vertical links and remove the vertical link.
2. The influence line for bending moment M_I in the beam in Figure 2.4 may be obtained by substitution of the hinge for a rigid joint at the section's zone. External moment X is applied to both ends adjoining the hinge and secures the equilibrium of the mechanism if it equals the bending moment (Figure 6.7a). Displacement Δ_X in this case is a sufficiently small angle of the mutual pivot of the beam's parts (see Figure 6.7b). By giving positive virtual displacement to such a single-DOF mechanism, we obtain the diagram of load line deflections of the triangle shape with the vertex at the hinge.

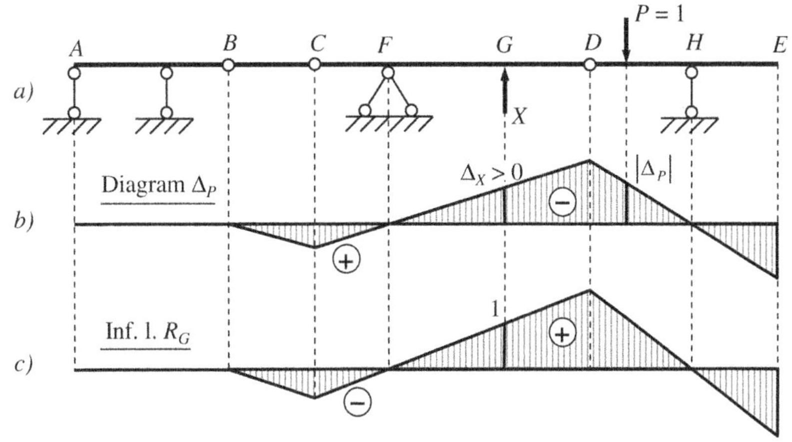

FIGURE 6.9 Construction of an influence line for the support reaction at point G of a hinged beam (see Figure 6.5) by the kinematic method: (a) System with removed constraint; (b) Diagram of vertical displacements of the system in diagram a; (c) Sought influence line.

Through this example, one can see the following shortcoming of the second formulation of the Müller-Breslau principle: in accordance with the principle, we have to turn the parts of the beam separated by hinges so that the mutual pivot would be that of the angle $\Delta_X = 1$ rad. Then the angle between branches of the obtained influence line must be equal to $\pi - 1 = 180° - 57.3°$, which is impossible. In the square coordinate grid, the least value of this angle equals $180° - 53.13°$ and is reached when $a = b$.

3. The influence line for shear force V_1 in the beam in Figure 2.4 may be obtained by the use of a kinematic joint made up of two connections of the first type (Figure 6.8a). Such connection deletes the shear force in the neighborhood of the section, and for equilibrium, we have to exert external shear force X to both cut-up ends of a beam. Small displacement of this mechanism is shown in Figure 6.8b. The displaced load line defines the influence line for shear force (compare with Figure 2.4).

Employment of the kinematic method in the case of hinged beams doesn't differ from the case of simple beams considered above. Let us construct, for example, the influence line for the reaction of support G of the beam in Figure 6.5. Removing this support and displacing the obtained mechanism in the direction of the removed constraint enables one to get the diagram of displacements of the beam's axial line represented in Figure 6.9. Through virtual displacement of the point of removed support along the vertical upward, the work of external force X is positive, thus, $\Delta_X > 0$. According to the kinematic method, the diagram of displacements differs from the sought influence line in the scale only if we replace the ordinates' signs by the opposites. The necessary scaling may be established, if we notice that for load $P = 1$ applied at the point G, the displacements on the right side of (6.3) are equal and opposite in sign, and therefore, the wanted influence function at this point $f(z_G) = 1$.

Section V

Statically Determinate Trusses

This section begins with the definition and classification of trusses, and then exposes the theory and the methods of analysis of statically determinate trusses. Three chapters of the section are devoted to the plane trusses' analysis and the fourth and final chapter generalizes obtained results for space trusses. The composing of the equation set of statics for a plane truss is given; the obtained equation system is further employed for substantiation of the null load method for investigation of stability and static determinacy of trusses. While considering the calculation of internal forces in statically determinate trusses, the section presents the methods of moment point, of projections, and of joints, as well as the method of substitute members. The properties of influence lines for trusses are established, which enable the construction of influence lines through calculation of internal forces. The section explains the way of influence line construction for complex truss through the method of substitute members coupled with the kinematic method. The last chapter gives the classification of connections in space systems, considers the peculiarities of kinematic analysis for space trusses, and develops the technique of transformation of statics equation set for calculation of space systems. It is also shown in the chapter how the methods of plane truss analysis are generalized for calculation of space trusses. The method of decomposition of space truss into plane trusses is substantiated. The peculiarities of reticulated trusses and Schwedler trusses are studied.

7 Preliminaries on Trusses

INTRODUCTION FROM LINEAR ALGEBRA

Let us consider the linear equation set:

$$\begin{aligned} &A_{11}x_1 + A_{12}x_2 + \ldots + A_{1n}x_n = b_1; \\ &A_{21}x_1 + A_{22}x_2 + \ldots + A_{2n}x_n = b_2; \\ &\ldots\ldots\ldots\ldots\ldots\ldots\ldots \\ &A_{n1}x_1 + A_{n2}x_2 + \ldots + A_{nn}x_n = b_n. \end{aligned} \tag{7I.1}$$

This set is characterized by the square matrix:

$$\mathbf{A} = \begin{pmatrix} A_{11} & A_{12} & \ldots & A_{1n} \\ A_{21} & A_{22} & \ldots & A_{2n} \\ \ldots & \ldots & \ldots & \ldots \\ A_{n1} & A_{n2} & \ldots & A_{nn} \end{pmatrix}, \tag{7I.2}$$

which is referred to as the coefficient matrix of the linear equation set.

We call a matrix consisting of a single column (single row) a column vector (row vector), or sometimes just a vector. Let us introduce the vector of unknowns and the vector of absolute terms of the set (7I.1):

$$\mathbf{x} = \begin{pmatrix} x_1 \\ x_2 \\ \vdots \\ x_n \end{pmatrix}; \quad \mathbf{b} = \begin{pmatrix} b_1 \\ b_2 \\ \vdots \\ b_n \end{pmatrix}.$$

It is convenient to consider set (7I.1) as the vector equation:

$$\mathbf{Ax} = \mathbf{b}. \tag{7I.3}$$

A square matrix with a zero determinant is referred to as a singular matrix; otherwise a square matrix is called nonsingular. The determinant of matrix $\mathbf{A}$ is denoted as det $\mathbf{A}$. The existence and uniqueness of the solution of a linear equation set is defined by whether the coefficient matrix is singular or not. The solution of the linear equation set in vector form (7I.3) exists and is unique on the obligatory condition: det $\mathbf{A} \neq 0$. It follows that for an absolute term vector $\mathbf{b} = 0$ and upon the condition det $\mathbf{A} \neq 0$, the solution of equation (7I.3) is only and trivial: $\mathbf{x} = 0$. If det $\mathbf{A} = 0$, the solution may not exist, but if it exists, then there exists an infinite set of other solutions. Thus, for absolute term vector $\mathbf{b} = 0$ upon condition det $\mathbf{A} = 0$, one can find the nontrivial solution $\mathbf{x} \neq 0$.

7.1 CLASSIFICATION OF TRUSSES

In Chapter 2, we defined hinged trusses as stable systems formed by straight bars joined together with hinges. We remind the reader that the members forming the outline of a truss, except the vertical ones, constitute the top and bottom chords of the truss; the members inside the outline make up the web, or the lacing of the truss; and the regions between adjacent joints of each chord are called panels (Figure 7.1). The height between chords of a truss is referred to as depth. The inclined members of lacing are referred to as diagonals; the vertical ones are referred to as posts, hangers, or verticals.

We distinguish trusses by the outline as having either parallel chords or polygonal chords (Figure 7.1).

Trusses differ by web type as follows:

- Trusses with triangular lacing, in which the diagonals make up the continuous zigzag (Figure 7.2a);
- N-trusses, in which the web makes up the continuous zigzag with alternation of diagonals and verticals (Figure 7.2b);
- K-trusses, in which every diagonal runs from a chord to the middle of the post (Figure 7.2c);
- Lattice trusses with rhombic webs (Figure 7.2d);
- Lattice trusses with double triangular lacing, which is composed of two triangular lacings (Figure 7.2e), and with multi-triangular lacing, which has a larger number of such lacings (Figure 7.2f).

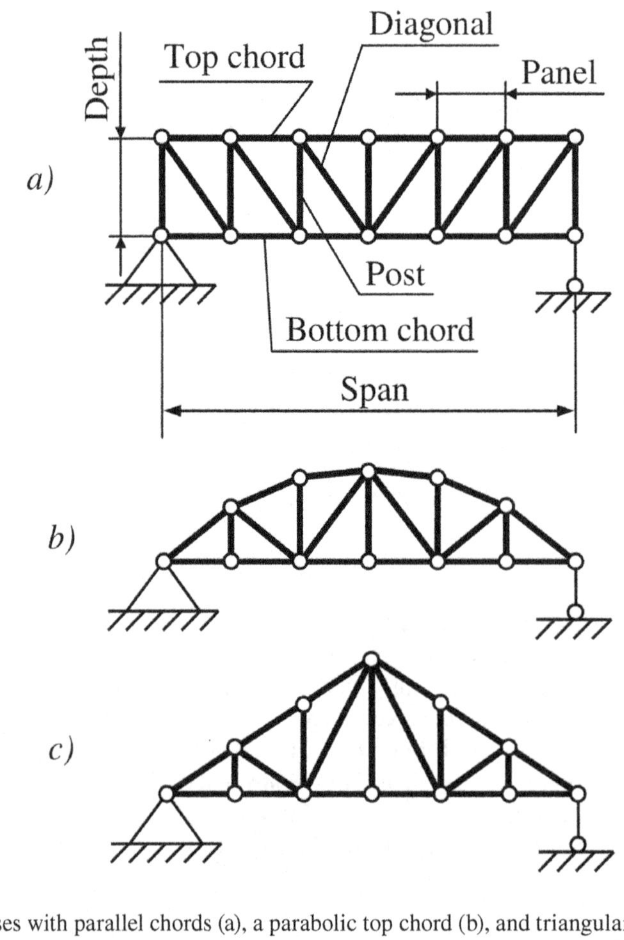

FIGURE 7.1 Trusses with parallel chords (a), a parabolic top chord (b), and triangular outline (c).

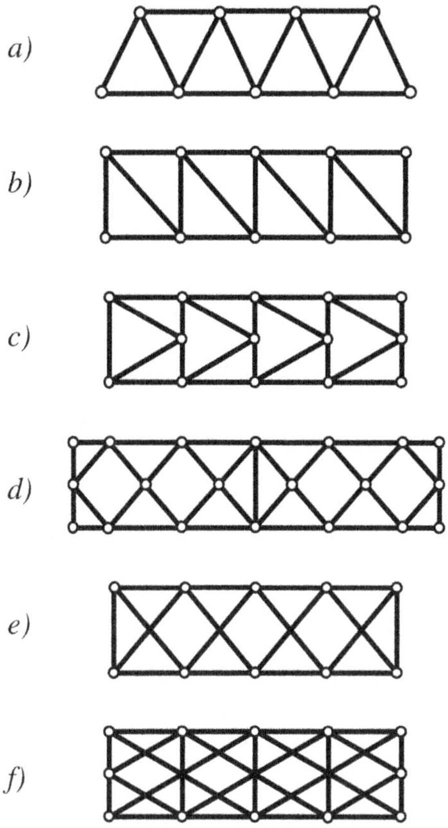

FIGURE 7.2 Trusses with webs of different types.

Plane internally stable trusses we refer to as *truss girders* if they are horizontally located and support vertical loads, being indispensable constructive members of a structure. Truss girders may be classified by location of their supports and direction of support reactions as follows:

- Simply supported truss girders, which are supported at the outermost hinges (Figure 7.3a);
- Cantilever truss girders with simply supported spans, which have one or two overhanging ends extending out of the span (Figure 7.3b and c);
- Cantilever truss girders fixed at one end only (Figure 7.3d);
- Arched truss girders (Figure 7.3e).

Simply supported and cantilever truss girders are thrustless structures. Cantilever truss girders supported at only one end are taken for thrustless as well, because the total horizontal reaction of the supporting wall produced by vertical loads is equal to zero for these trusses. Arched truss girders are thrusted structures.

By the sequence of construction, we distinguish the following types of trusses:

- *A simple truss* is a truss constructed from a hinged triangle through consecutive linking of every new hinge by two bars which don't lie in a straight line. Construction of such a truss meets the rule of three-disc association and complies with the definition of a simple system of discs from Chapter 1. The trusses in Figure 7.1 are simple;

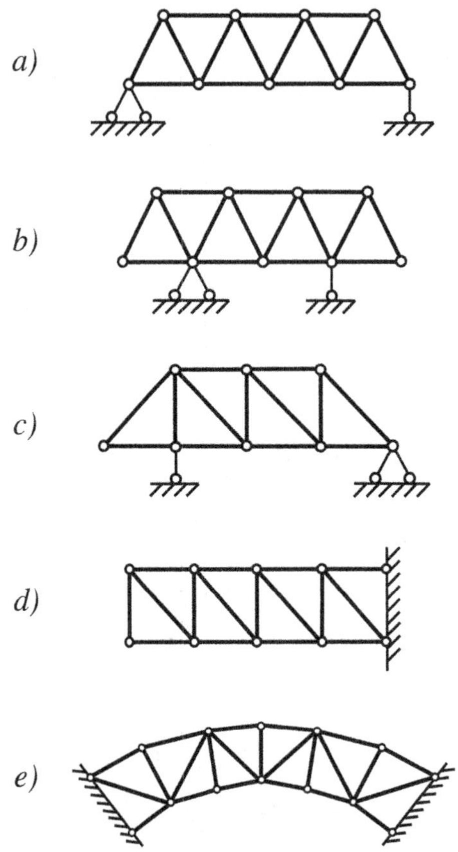

FIGURE 7.3 Different conditions of the resting of trusses.

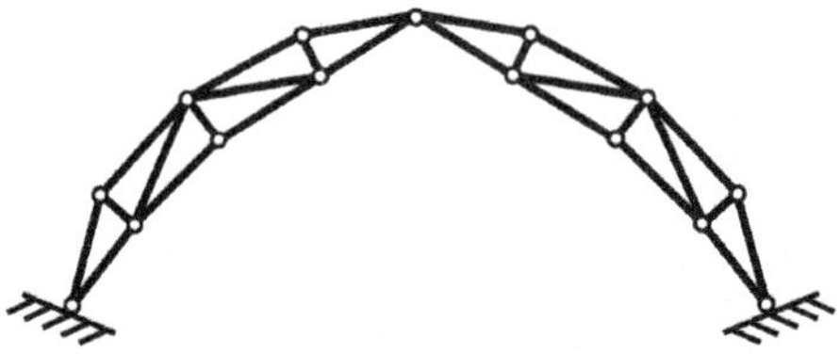

FIGURE 7.4 Example of compound truss.

- *A compound truss* is a simple multi-disc system, where the discs are simple trusses and may be the earth. The connections between simple trusses are made through their hinges. The example of an arch-shaped compound truss is shown in Figure 7.4. In the given case, two simple trusses and the earth are connected by the rule of three-disc association with the least number of constraints;
- *A complex truss* is a truss which does not appertain to simple nor compound trusses.

As every simple system of discs, simple and compound trusses are statically determinate.

The purpose of a truss can also serve as the classification feature; for example, we distinguish roof trusses, tower trusses, crane trusses, and bridge trusses.

Many trusses have names which distinguish them by a combination of characteristics, usually by the purpose, the shape of chords, and the web type. The varieties of roof trusses are shown in Figure 7.5 (*Trussed Rafter: Technical Manual* 2004). Some of the most familiar bridge trusses are

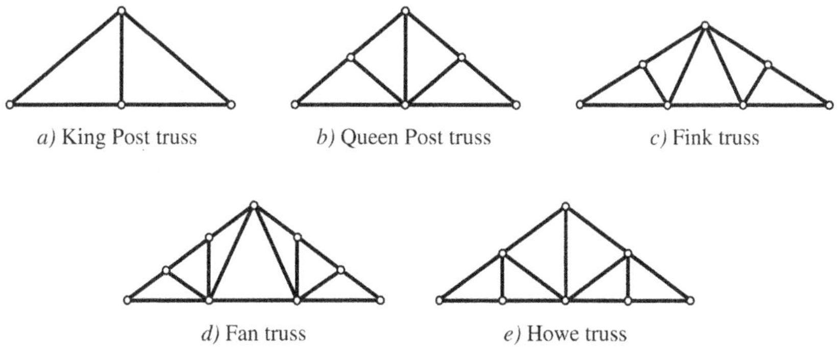

FIGURE 7.5 Roof trusses: (a) King Post truss; (b) Queen Post truss; (c) Fink truss; (d) Fan truss; (e) Howe truss.

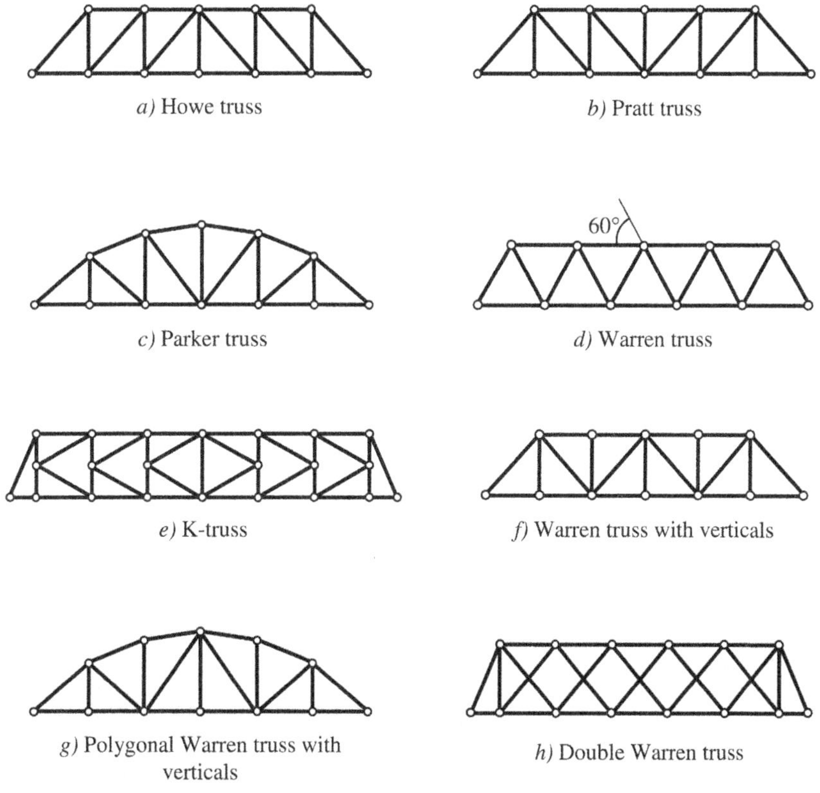

FIGURE 7.6 Bridge trusses: (a) Howe truss; (b) Pratt truss; (c) Parker truss; (d) Warren truss; (e) K-truss; (f) Warren truss with verticals; (g) Polygonal Warren truss with verticals; (h) Double Warren truss.

presented in Figure 7.6. Among these, trusses *a*, *b*, *c*, and *d* are referred to by the names of bridge engineers who patented them in the 19th century. Bridge trusses comprising triangular lacing are referred to as Warren trusses regardless of the real deviser (Figure 7.6f, g, and h).

In truss bridges, the superstructure includes two plane main trusses located on either side of the deck. We distinguish main trusses as follows, according to the location of the deck and lateral bracing: deck trusses with the traffic level on top of the superstructure; through trusses with the traffic level through the superstructure and the top chords being cross-braced; and pony trusses with the traffic level through the superstructure and the top chords unconnected to each other (Figure 7.7).

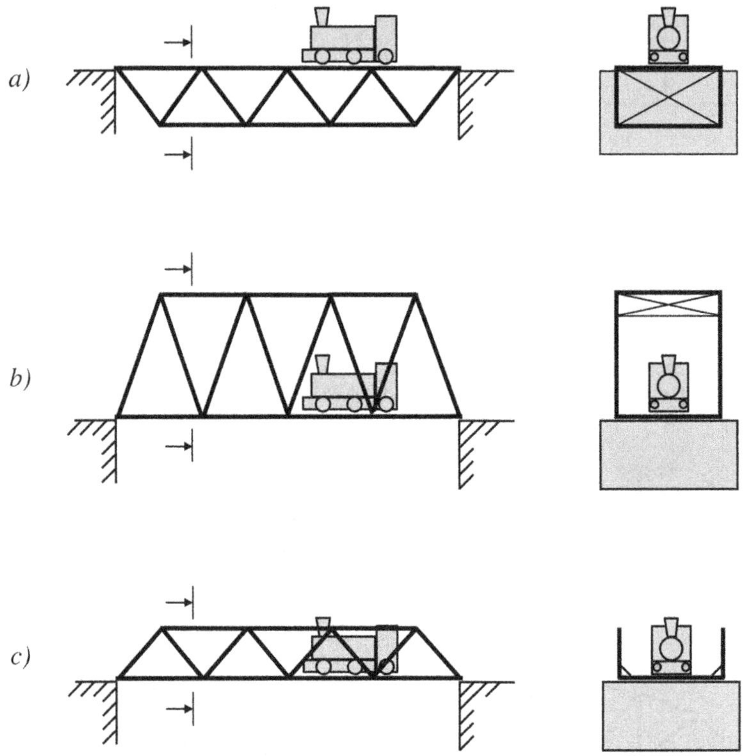

FIGURE 7.7 Classification of bridge trusses according to location of deck and lateral bracing: (a) Deck trusses; (b) Through trusses – top chords are cross-braced; (c) Pony trusses – top chords are not cross-braced.

FIGURE 7.8 The Steel Bridge in Portland, Oregon. Photo by Cacophony – own work. (License CC BY 3.0. Source: https://commons.wikimedia.org/w/index.php?curid=3576136.)

The bridge of through configuration (Figure 7.7b) may have a second deck erected over the upper lateral bracing; an example of such a bridge is shown in Figure 7.8.

Main trusses are combined into a superstructure by a system of lateral beams, struts, and bracings (Figure 7.9). In steel bridges, the main lateral force members are the floor beams, which brace up the truss girders and withstand the live load. The stringers (longitudinal beams) are mounted upon the floor beams, and a deck is erected upon the stringers. The stringers may connect the floor beams at the same level, as is depicted in Figure 7.9, or be laid on them. Usually two stringers suffice in a steel railway single track bridge.

The typical bridge of through configuration is shown in Figure 7.10. In this example, the members of the main trusses are connected by rigid joints. The joints of bridge trusses, as a rule, are

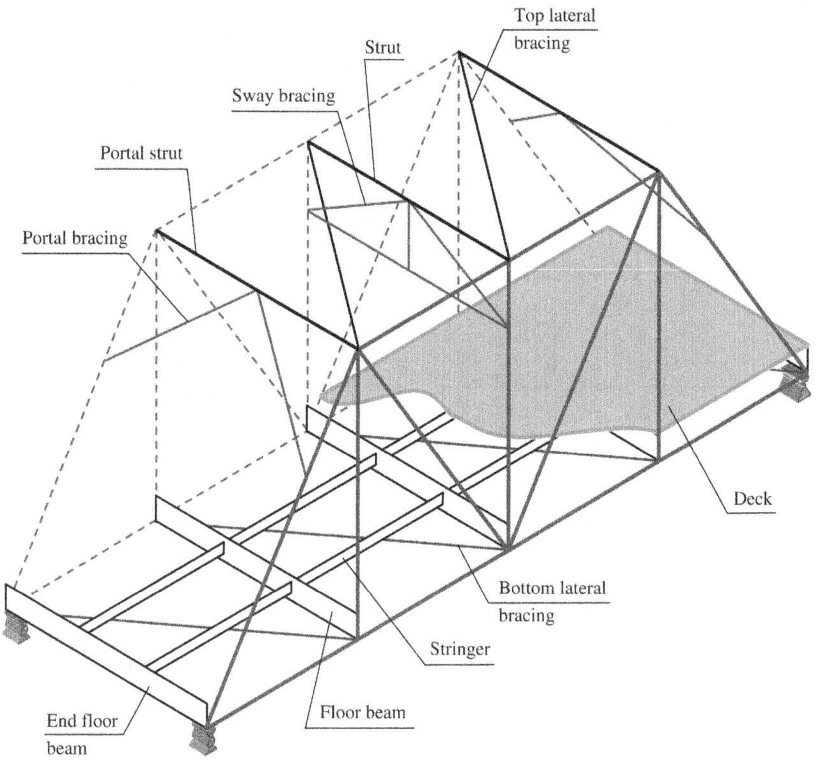

FIGURE 7.9 Simplified scheme of steel railway single track bridge.

FIGURE 7.10 Railway bridge over the Mura river in Mursko Središće, Croatia. Photo by Silverije. (License CC-by-SA. Source: https://commons.wikimedia.org/wiki/File:Željeznički_most,_Mursko_Središće_(Croatia).1.jpg.)

rigid, but due to the large span, they may be implemented hinged. The rigid joints comprise one or more plates, to which the members are connected by welding, riveting, or bolting. One can see that a joint is noticeably bigger than the cross-sections of connected bars. In rigid joints, the stress concentration arises, which is to be accounted for in their design. Rigid joints are substituted by hinges in design diagrams of trusses.

7.2 STABILITY AND DETERMINACY OF TRUSSES

Let us rewrite the necessary condition of stability (1.6) for a system attached to the earth:

$$3D \leq L + 2H + 3R + L_{sup}. \tag{7.1}$$

This condition can be represented for a truss as the relation between the number of joints J and the number of bars B. We have $D = B$; $L = R = 0$. Each bar terminates in a hinge at either end. If there are as many simple hinges at a joint as there are bars connected by it, then the equality $H = 2B$ will take place. But the simple hinges are one less at each joint, thus $H = 2B - J$. Replacing D with B, we come to the condition of stability for the attached truss:

$$3B \leq 2(2B - J) + L_{sup},$$

where L_{sup} is the number of support connections. After simplification, we get:

$$B + L_{sup} \geq 2J. \tag{7.2}$$

The necessary condition of stability for a system of the general type with the least number of connections is obtained by the pass to equality in (7.1). This equality is also the necessary condition of static determinacy. The necessary condition of stability for a truss with the least number of connections can be obtained by the shift to equality in (7.2):

$$B + L_{sup} = 2J. \tag{7.3}$$

This equality is also the necessary condition for static determinacy of a truss.

In the case of a free truss, the necessary condition of stability (1.11) takes the form:

$$B \geq 2J - 3; \tag{7.4}$$

and the condition of static determinacy (1.12) is of the form:

$$B = 2J - 3. \tag{7.5}$$

Kinematic analysis of a truss is to be initiated from verification of the condition of static determinacy (7.3) or (7.5), and then, if this condition is fulfilled, trying the possibility of constructing a simple or compound truss is recommended. If the types of simple or compound trusses aren't confirmed, kinematic analysis may be accomplished by the method of null load. Having in mind the grounding of this method, we shall obtain the statics equation set for attached trusses of the general type.

We assume that a truss is subjected to joint loads only. This means that among internal forces in the members of the truss, only axial forces may be nonzero. Really, each member interacts with the surroundings only through the joints, and if the forces at the ends of the member are directed off the axis, then an external moment would arise. This moment can't be balanced by bending moments at the ends, for bending moments are deleted by hinges.

We take as given that support connections are simple rods, and they are included into the system's composition. The total number of members in a system supplemented with support connections is

$n = B + L_{sup}$. If the system is statically determinate, then, according to (7.3), $n = 2J$. The complete statics equation set for any structure is composed of statics equations for each disc and each joint. In the case of a hinged truss under the joint loads, the statics equation for the j-th bar has the trivial form:

$$N_j \equiv N_j^1 = N_j^2, \tag{7.6}$$

where N_j^1, N_j^2 are the reactions of adjoining hinged joints, and "≡" means "equal by definition." Reactions N_j^i coincide, in our case, with the axial force exerted on the j-th member. We use denotation N_j, which was introduced in equation (7.6), for the axial force in a member of the truss; if the member is the support link, then N_j is the reaction of this link.

We represent the equilibrium conditions for the i-th joint in the form:

$$\begin{cases} \sum_{j=1}^{n} I_{ij} N_j \cos\alpha_j + P_{ix} = 0; \\ \sum_{j=1}^{n} I_{ij} N_j \sin\alpha_j + P_{iy} = 0. \end{cases} \tag{7.7}$$

Here α_j is the angle between the j-th member and coordinate axis x. This angle is taken as the angle of the member's pivot about one of the ends (the center of rotation) from its initial position when another end is directed along the x-axis (Figure 7.11). The positive sense of rotation is, as usual, counterclockwise. We refer to the quantity I_{ij} as a truss structure indicator. It takes the values:

$$I_{ij} = \begin{cases} 0, & \text{if } j\text{-th member is not connected with } i\text{-th joint;} \\ 1, & \text{if, in counting the angle } \alpha_j\text{, the center of member's} \\ & \text{rotation coincides with } i\text{-th joint;} \\ -1, & \text{if the center of } j\text{-th member's rotation is positioned} \\ & \text{at the end opposite to } i\text{-th joint.} \end{cases}$$

P_{ix}, P_{iy} are the components of the vector load exerted on the i-th joint (Figure 7.11).

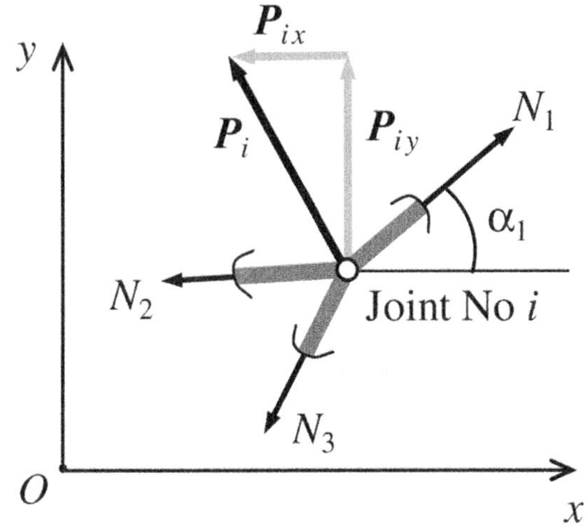

FIGURE 7.11 Forces acting upon a joint.

For any statically determinate truss, we can set up and solve the set of n linear equations (7.7), but this way of truss calculation may appear excessively laborious. There are more simple methods of truss analysis, which are treated in the next chapter.

Let us consider a truss attached to the earth and having the least number of connections when stability is possible, i.e., under the condition (7.3). The *method of null load* has been developed for analysis of stability and static determinacy of such a truss. Internal forces and reactions of constraints, which act in a mechanical system in the absence of loads, are referred to as *initial forces and reactions*. **The method of null load lies either in determining the totality of nonzero initial forces and support reactions that prove instability (instantaneous instability) of a hinged bar system with the least number of connections, or in substantiating that these forces and reactions equal zero, which proves the stability and static determinacy of such a system.** We accept that if the statics equation set for any mechanical system under some loading has a solution, then the corresponding state of equilibrium is possible (i.e., really arises under some conditions). The method of null load is substantiated by the next theorem.

Theorem of Null Load

An attached hinged bar system with the least number of connections and loads exerted on the joints is unstable or instantaneously unstable if its equilibrium is possible with nonzero initial forces or support reactions. The same system is stable and statically determinate if the initial forces and support reactions in it are necessarily zero.

Proof

When discussing the instability of a system below, we mean the instantaneous instability as well. We represent the statics equation set for a truss in the form:

$$\mathbf{AN} + \mathbf{P} = 0, \tag{7.8}$$

where $\mathbf{N}$ is the vector of unknown forces and support reactions; $\mathbf{P}$ is the vector of joint loads; and $\mathbf{A}$ is the coefficient matrix, the rows of which are made up from the coefficients of equilibrium equations for joints (7.7). In other words, we introduce the denotations:

$$\mathbf{N} = \begin{pmatrix} N_1 \\ \vdots \\ N_{2J} \end{pmatrix}; \quad \mathbf{P} = \begin{pmatrix} P_{1x} \\ P_{1y} \\ \vdots \\ P_{Jx} \\ P_{Jy} \end{pmatrix}; \quad \mathbf{A} = \begin{pmatrix} I_{11}\cos\alpha_1 & \cdots & I_{1n}\cos\alpha_n \\ I_{11}\sin\alpha_1 & \cdots & I_{1n}\sin\alpha_n \\ \dots & \dots & \dots \\ I_{J1}\cos\alpha_1 & \cdots & I_{Jn}\cos\alpha_n \\ I_{J1}\sin\alpha_1 & \cdots & I_{Jn}\sin\alpha_n \end{pmatrix}. \tag{7.9}$$

A hinged system complying with the theorem's conditions has a square matrix $\mathbf{A}$.

To prove the theorem, let us show, at the beginning, that the necessary and sufficient condition of a system's instability is the singularity of its matrix $\mathbf{A}$, i.e., the equality: $\det \mathbf{A} = 0$. Really, an unstable system can't keep equilibrium under some loads. For these loads, the equation set (7.8) is incompatible, and this implies a zero determinant of the coefficient matrix. Conversely, if the coefficient matrix of the statics equation set is singular, this set is incompatible for some loads, which means unbalance of the system and its instability.

Now let nonzero initial forces or support reactions be possible in the system. This means that the statics equation set (7.8) has a nontrivial solution subject to zero vector of absolute terms $-\mathbf{P}$, and the latter is possible only if $\det \mathbf{A} = 0$, i.e., when the system is unstable. On the contrary, if initial forces and reactions in a system are necessarily zero, then the statics equation set on zero loads doesn't have a nontrivial solution. Thus, det **A** $\neq 0$, and the system is stable.

The theorem is proved.

The truss in Figure 7.12 complies with the necessary condition of static determinacy (7.3). Let us employ the method of null load to make a conclusion about the static determinacy of this system. Assume that there are no loads upon the truss and at least one of members 8 or 9 is stressed. Dissect the truss by means of section I–I and consider the equilibrium conditions for the part over this section. Absent loads, the resultant of forces in members 8 and 9 must be opposite in direction to the resultant in members 6 and 2. The line of the resultant of forces 8 and 9 passes through point O_3; the line of the resultant of forces 6 and 2 passes through point O_1. Therefore, both resultants act along line O_1O_3. Now, dissect the truss by means of section II–II (Figure 7.12). It is easy to see that for equilibrium of the part of the structure over section II–II, it is necessary that the resultant of forces 8 and 9 acts along line O_2O_3. This is possible only if points O_1, O_2, and O_3 lie in the same straight line. But the structure in Figure 7.12 doesn't comply with this requirement; that is why, in the absence of loads, internal forces in members 8 and 9 are zeroth; ⇒ the same is true of members 6 and 2, and also 5 and 3; ⇒ all the members are not stressed; ⇒ the given truss is statically determinate.

An example of a similar truss is presented in Figure 7.13a, where the outline members make up the regular hexagon. In this case, line $O_1O_3O_2$ is the straight line, and there is reason to try to determine nonzero initial forces in members. Assume that identical internal forces are induced in members 8 and 9 – for instance, the tensile forces of magnitude N – and the support links do not develop reactions. For equilibrium of the part of the system over section I–I, the compressive forces of the same magnitude must act in members 6 and 2 (Figure 7.14a). For equilibrium of each joint a and b, the compressive force of magnitude $N_1 = 2N\cos 60^\circ = N$ must act in member 1. In the same manner, we establish that balance requirements for joints e and d are satisfied if members 5, 3, and 4 are compressed by force N (Figure 7.14b). Finally, by cutting out joints f and c and setting up for each of them the equilibrium equation for horizontal forces, we get the tensile force in member 7 of magnitude $N_7 = 2N\cos 60^\circ = N$ (Figure 7.14c). So, all joints of the system are in equilibrium if the

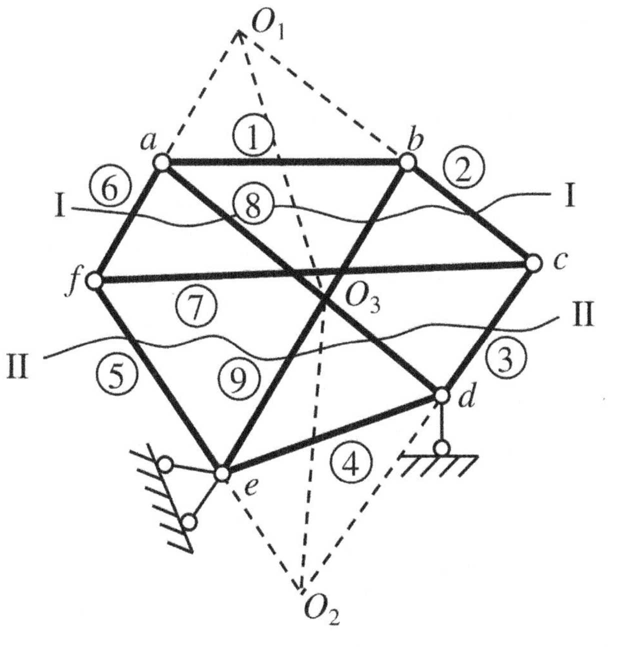

FIGURE 7.12 Stable statically determinate truss.

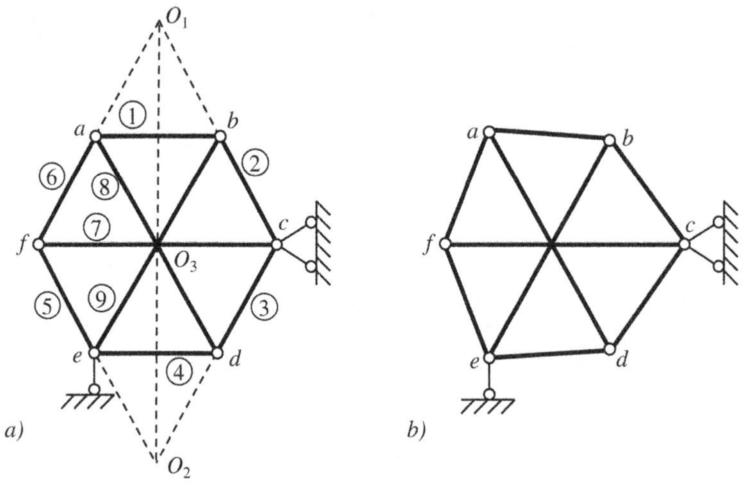

FIGURE 7.13 Instantaneously unstable truss: (a) undeformed shape; (b) deformed shape.

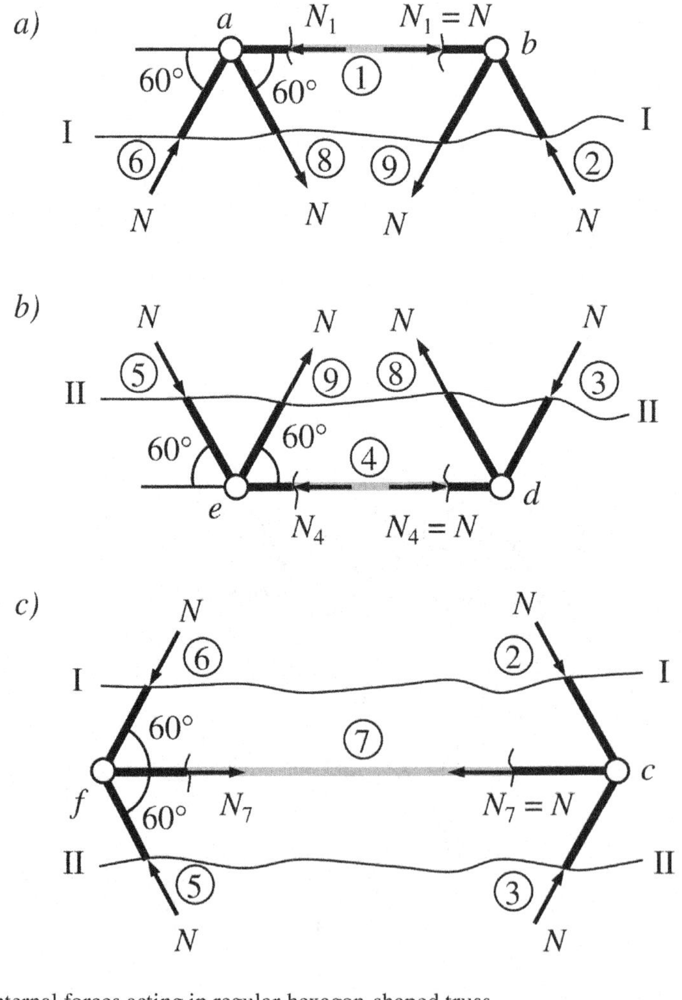

FIGURE 7.14 Internal forces acting in regular-hexagon-shaped truss.

radial members are tensed and the outline members are compressed by the same force N, whereas the supports do not work.

In projective geometry, a hexagon inscribed in a circle is referred to as a Pascal hexagon. We shall use this term for the regular-hexagon-shaped construction in Figure 7.13a. From this analysis, we conclude that a Pascal hexagon with three support links is not a geometrically stable system. It is known that this system is instantaneously unstable. An example of straining in this construction is shown in Figure 7.13b. The straining happens mainly due to the gaps in the hinges and their deformations.

7.3 CONVERSION OF A LOAD EXERTED UPON A MEMBER OF A STATICALLY DETERMINATE TRUSS TO ITS JOINTS

The methods of statically determinate truss analysis are based on the assumption of loads' action upon joints, whereas, in practice of design, some members of a truss take up the load directly instead of through joints. Nevertheless, there exists the rule of conversion of a member load to the joints of a truss, which enables us to apply common methods of truss calculation in the case of intermediate loads (i.e., loads applied to a member directly). To substantiate this rule, we require the *superposition principle for statically determinate systems of discs*: **any constraint reaction produced by a totality of loads in an attached statically determinate system can be obtained as the sum of reactions of this constraint produced by separate loads of the totality**. In this formulation, the principle is always fulfilled and follows from statics equations. Indeed: The statics equation set for an arbitrary system of discs is represented in the form (7.8), where vector **N** is the totality of constraint reactions; vector **P** is the totality of resultant vectors and total moments of loads exerted on each disc and each joint; and **A** is the matrix of the system's structure. In an attached statically determinate system, matrix **A** is square and has a nonzero determinant. Owing to this, the solution of equation set (7.8) is unique, and the sum of solutions corresponds to the sum of given load vectors **P**. The latter affirmation constitutes the mathematical substance of the superposition principle.

The rule of conversion of intermediate load to the joints reads as follows:

If any given member of an attached statically determinate truss bears a concentrated load of arbitrary direction, then for calculation of internal forces in other members, this load may be replaced by a lateral joint loads equal and opposite in direction to reactions of supports of a simple beam, and the longitudinal component of a load applied to any joint of the given member.

The proof of this assertion is as follows. Let load $\boldsymbol{P}$ of any direction be applied at point C of member AB, which belongs to a statically determinate truss (Figure 7.15), and there are no other loads upon the truss. Decompose the member load into longitudinal $\boldsymbol{P}_{long}$ and lateral $\boldsymbol{P}_{lat}$ components. Denote the reactions of joints A and B, produced by lateral load $\boldsymbol{P}_{lat}$, as $\boldsymbol{R}_A$ and $\boldsymbol{R}_B$. It is evident enough that reactions of joints to the lateral load coincide with support reactions of the simple beam. Let us supplement the system of given loads $\boldsymbol{P}_{long}$ and $\boldsymbol{P}_{lat}$ by two forces $\pm\boldsymbol{P}_{long}$ exerted on any joint

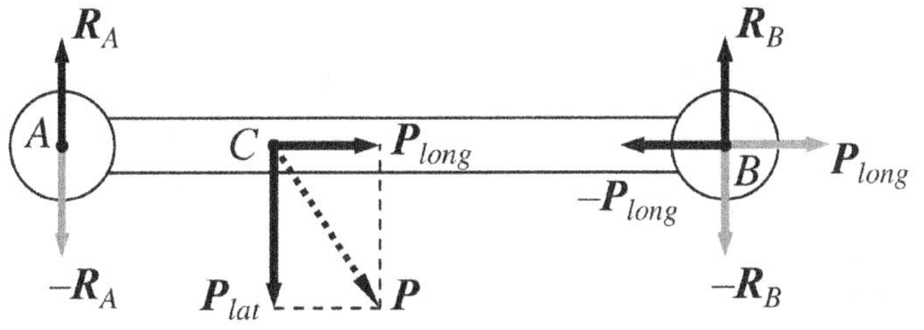

FIGURE 7.15 Representation of load P exerted on a given member by two systems of forces: gray arrows constitute the system of joint loads for calculation of the truss; black solid arrows constitute the system of balanced loads.

of member AB, for example, on B. At both joints A and B, we introduce two counterpoised loads, equal to reactions of these joints, namely: at the left joint $\pm\boldsymbol{R}_A$ and at the right joint $\pm\boldsymbol{R}_B$. Now the complete force assemblage consists of eight forces shown in Figure 7.15 by the black solid and gray arrows. The system of the next five loads is balanced among them: $\boldsymbol{P}_{lat}$, $\boldsymbol{R}_A$, $\boldsymbol{R}_B$, $\boldsymbol{P}_{long}$ at point C, and $-\boldsymbol{P}_{long}$ at point B. All these forces secure the equilibrium of member AB together with adjoining hinges if we take the internal forces in other members for zero. We see that a given truss will be in equilibrium under condition of zero reactions of all joints except A and B; and, because the solution of statics equations in the given problem is unique, the system of five enumerated loads may produce reactions only in joints A and B. On the strength of the superposition principle, reactions of all joints of the truss except A and B are defined by the three remaining loads from the complete assemblage, namely: $-\boldsymbol{R}_A$, $-\boldsymbol{R}_B$, and $\boldsymbol{P}_{long}$. These loads constitute the system of joint loads stated in the rule. In a truss, a joint reaction is the force in a member not subjected to intermediate load; thus we obtain that the forces in all members of the truss except AB are defined by the joint loads only. Obviously, this inference remains correct if intermediate load $\boldsymbol{P}$ acts together with joint loads specified initially for analysis of a truss.

The rule substantiated above doesn't resolve the problem of internal force calculation for the very member under a load. Forces in this member are determined on the basis of the superposition principle as the result of exertion of two composed systems of loads (and, maybe, other loads upon a truss). *The calculation rule for internal forces in the loaded member of a truss* reads as follows:

The internal forces in a member belonging to a statically determinate truss and bearing an intermediate load are established through the totality of external forces exerted upon this member, which consists of the given load and the reactions of adjoining hinges. In the absence of other loads, each of these reactions is obtained by summing up the reaction from the system of joint loads obtained by reducing the given load to joints, and the reaction from the system of balanced loads supplementing the system of joint loads to the given load.

In conclusion, note that the theorem about null load proved above permits generalizations. For instance, the next homonymous theorem is known for a system of discs:

If nonzero constraint reactions are possible on null load in an attached system of discs with the least number of connections required for stability, then the system is unstable (instantaneously unstable); if they are impossible, then the system is statically determinate.

This is proved in the same way as the similar theorem for trusses was proved but with the next particularity: vector equation (7.8) in the given case is the statics equation set for the totality of reactions of the disc connections, but not the internal forces in bars, as it was in the case of trusses.

8 Internal Forces in Statically Determinate Trusses

8.1 METHOD OF MOMENT POINT

In the first and second sections of this chapter, we are concerned with truss girders that have simply supported spans and bear vertical loads. When finding internal forces in such trusses, the equations of statics may be often set up in a simple form, and there are some methods to do it, based on the method of sections. We remind readers that loads are assumed to be exerted on joints of a truss unless otherwise arranged.

The *method of moment point* is to construct the section cutting a truss into two parts in such a manner that all axial lines of the cut members but one are concurrent at the same point. This point is referred to as the *moment point*. The equilibrium equation for moments of forces exerted on the part of the truss on one side of the section, with respect to the moment point, enables us to determine the force in the crossed member the axis of which doesn't run through this point. If the section cuts three members, the axes of which are not concurrent at the same point and not parallel, then one can find three moment points for the given section and obtain all forces in the members.

We denote the axial forces in members of the top chord (**over** the truss web) as O, the forces in members of the bottom chord (**under** the truss web) as U, the forces in diagonal members as D, and the forces in vertical members as V.* To determine forces O_{24}, D_{34}, U_{35} in the truss in Figure 8.1, let us draw section I–I and employ the moment point method. For moment point 4, we have the equilibrium equation:

$$\sum_{left\,4} M_4(P_i) + U_{35}h = 0,$$

where $M_4(P_i)$ is the moment of external force P_i relative to joint 4; the positive sense of moment is pointed out on the diagram (counterclockwise); and the sum is taken over all the external forces on the left of the section, including the support reaction. In the given case, one can sum up all the forces on the left of the vertical drawn from hinge 4 (this is marked under the summation symbol). This sum of moments can be calculated by means of the reference beam, which has the same extension, positioning of supports, and distribution of loads as the given truss girder. We accept that the cross-section of the reference beam is symmetrical relative to the plane of loading. We denote bending moments at the sections of the reference beam with superscript "0." Because the loads are presumed to be vertical, the moment of external force relative to any selected joint of the truss is the same as the moment of corresponding force relative to the beam's cross-section under this joint. Evidently, we have the bending moment acting upon section 4 of this beam as follows:

$$M_4^0 = -\sum_{left\,4} M_4(P_i),$$

whence we get the force in the member of the bottom chord:

$$U_{35} = \frac{M_4^0}{h}. \tag{8.1}$$

* These denotations are common in Europe and originated from German words *obere*, *untere*, *diagonal*, and *vertikal*.

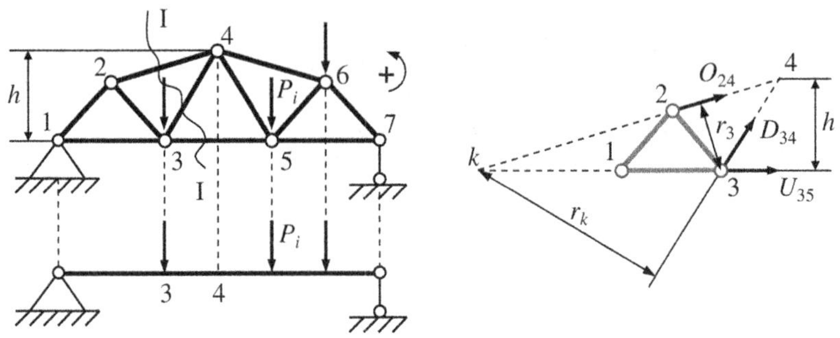

FIGURE 8.1 Truss with triangular lacing, its reference beam, and diagram of moment points' construction.

The force in the member of the top chord is obtained by summing up all moments of left forces exerted on the truss (including internal forces) relative to moment point 3. Subject to the action of vertical loads only, it is easy to obtain:

$$O_{24} = -\frac{M_3^0}{r_3}, \tag{8.2}$$

where the beam bending moment at section 3 is on the right side.

The force in the diagonal is determined by using the moment point k:

$$D_{34} = -\frac{1}{r_k}\sum_{left\ 4} M_k(P_i). \tag{8.3}$$

Summation is made, as before, over the left external forces. Here, we may also use internal forces at the sections of the reference beam, but it is necessary to introduce a new concept to do that.

For any beam, let us introduce the right-handed CS $O'xyz$ bound to this beam, with the origin O' positioned at the end of a beam, axis $O'z$ directed along its axial line to another end, and axis $O'y$ directed downward along the vertical (an example in Figure 8.2b). Let us consider the case of a plane loading that produces symmetrical bending of a beam, and introduce the total moment of internal forces at a given section "O" with respect to any pole k at the axial line of the beam. We shall determine the moment about pole k for internal forces exerted upon the part of the beam that comprises the origin of beam-bound CS, and specify it by the projection on the axis being normal to the plane of loading. If pole k belongs to section "O," then the internal moment under consideration is a bending moment; otherwise we call it an *internal moment with an offset pole.*

Figure 8.2a reproduces the diagram of the truss's loading in Figure 8.1 with representation of support reactions in the assemblage of external forces (the positive direction for forces is taken downward). Diagrams b and c in Figure 8.2 show the reference beam for the truss and loading the part of this beam at the left of the section under joint 4. The axis $O'y$ lies in the plane of loading; the internal moment at cross-section "O" is directed along axis $O'x$ and is taken in projection on this axis. We denote the bending moment and shear force exerted on section "O" as M_O and V_O, respectively, and the horizontal coordinates of section "O" and point k as z_O and z_k, respectively. By using the design diagrams in Figure 8.2b and c, it is easy to obtain for the moment with offset pole:

$$M_{kO} = M_O - V_O(z_O - z_k). \tag{8.4}$$

On the right-hand side, the first component is defined by the normal stresses in the section only; the second component is defined by the shear stresses only. Note that counterclockwise action of the moment on the left part of a beam is taken as positive.

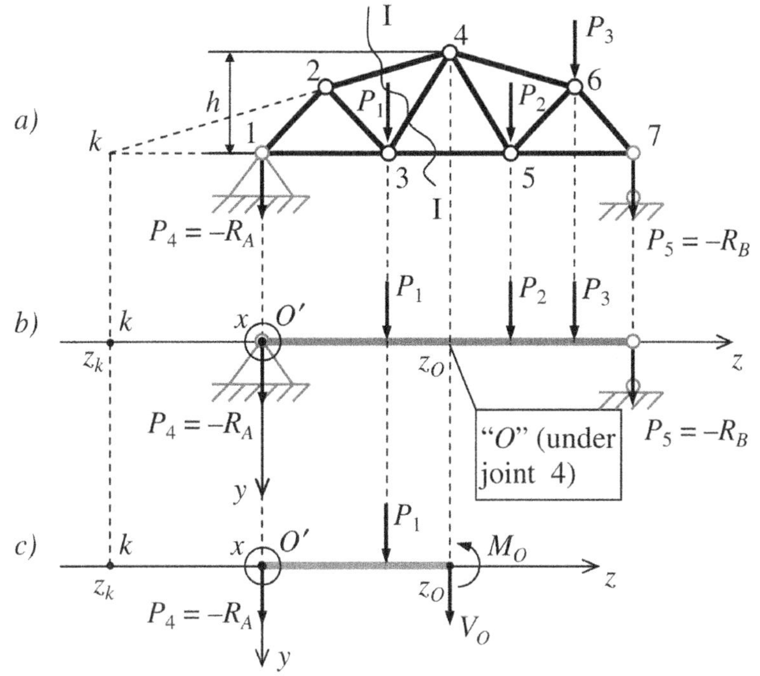

FIGURE 8.2 Transition from given truss to reference beam and diagram of loading for the left part of the beam. $O'xyz$ is the beam-bound CS.

The total moment of external forces exerted on the left part of the beam relative to section "O," about any point k of the axial line, when summed with the moment of internal forces acting upon this section, must result in zero. The diagrams of loading in Figure 8.2b and c clarify the condition of moment equilibrium:

$$\sum_{left\ O} M_k(P_i) + M_O - V_O(z_O - z_k) = 0.$$

As usual, the moment is taken with sign "+" if it is produced by the force acting counterclockwise relative to the pole. From the last equation, we obtain that moment of left external forces about any pole k equals the internal moment about the same pole with the opposite sign:

$$\sum_{left\ O} M_k(P_i) = -M_{kO}. \tag{8.5}$$

Let us choose point k for the reference beam under the moment point of the truss (Figure 8.2). It is permissible to calculate the moment of left forces by means of formula (8.3) with replacement of the truss by the beam. In the given case, we take into account external forces on the left of joint 4 and section "O" is taken under this joint. After substitution (8.5) and addition of superscript "0," we get:

$$D_{34} = \frac{1}{r_k} M_{kO}^0. \tag{8.6}$$

Note that the moment with the offset pole is not defined at the sections where concentrated loads are exerted. In this respect, it is similar to shear force since both these forces get a jump on passing through the section subjected to load. Therefore, if there is a load upon joint 4 in the example under consideration, we would have to bias the section "O" a little distance to the left or calculate the left shear force at section "O" of the reference beam.

The moment point method was proposed by German engineer August Ritter in 1861. Without the help of the beam analogy, the calculation of internal force N by this method is reduced to the formula:

$$N = \frac{\text{Moment of external forces}}{\text{Arm of internal force}}. \tag{8.7}$$

8.2 METHOD OF PROJECTIONS AND METHOD OF JOINTS; ZERO-FORCE MEMBERS

The method of projections lies in constructing the section that cuts a truss into two parts and setting up the equation of force balance for one of the parts in projection on some axis. The idea of this method is to choose the direction of this axis so that statics equation would have a single unknown force in order to determine this force. The method of projections performs well if the composed equation involves more than one internal force, but only one of them is unknown.

The most common implementation of the projection method is to construct the section cutting a number of members which are all parallel but one. Let us select the axis perpendicular to the parallel members. The equation of force balance in projection on this axis enables us to determine the force in the crossed member inclined to this axis.

Figure 8.3 depicts the Pratt truss for which the forces with vertical components can be calculated by the projection method. For instance, the equilibrium condition for the forces acting on the left of section I–I enables us to get the force in post 5–6:

$$V_{56} = -\sum_{left} P_{iy},$$

where P_{iy} is the projection of external force $\boldsymbol{P}_i$ on axis Ay. The last formula gives:

$$V_{56} = -V_5^0, \tag{8.8}$$

where the beam shear force at section 5 is on the right-hand side (if the load on joint 5 is taken for the left force and the load on joint 6 for the right one).

In the same way, we obtain for the section II–II:

$$D_{67} \sin\beta = \sum_{left\ 7} P_{iy},$$

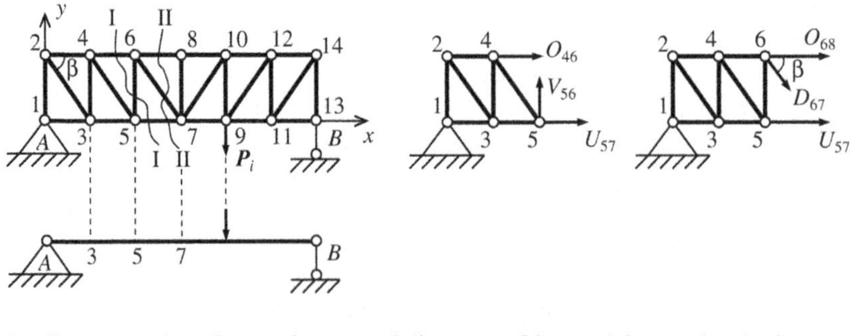

FIGURE 8.3 Pratt truss, its reference beam, and diagrams of internal forces. *Axy* is the truss-bound CS (coordinate plane comprises axial lines of members).

from whence:

$$D_{67} = \frac{V_{5-7}^0}{\sin\beta}, \tag{8.9}$$

where V_{5-7}^0 is the beam shear force at the portion between sections 5 and 7.

The *method of joints* is an important special case of the projection method, when the part of the truss under investigation is a single joint. The method assumes setting up the equations of force balance for the selected joint and finding unknown forces by solving these equations. The equations referred to have the form (7.7), but the method of joints is not reduced to setting up the complete system of such equations. The matter of the method is to choose the joint so that the equations of its equilibrium enable us to find unknown forces. To this effect, it is necessary that there be no more than two unknowns in the joint. For example, for a simple truss, we might select the joints in reverse order to how they make up the truss, sequentially employing this method. Each time, there will be no more than two unknown forces in the joint, which are determined from equations of equilibrium.

The calculation of forces in a truss is to be started from localizing the members with zero internal forces – in order that we might further ignore these members while setting up the statics equations. The two types of hinged connection of members are known, when there are no axial forces in one or two members:

- If a joint connects only two members not lying in a straight line, and is not subjected to the load, then forces in these members are zero. This case is presented in Figure 8.4a. The set of equilibrium equations obtained for joint A in projections on the coordinate axes Ax and Ay has the unique solution $N_1 = N_2 = 0$ subject to $0 < \alpha < 180°$;
- If three members are concurrent in a joint, two of them lie in a straight line, and the joint is not subjected to the load, then the force in the member directed at a slant to this line is zero. Really, the equilibrium equation for joint A in Figure 8.4b, made up in projection on coordinate axis Ay, has the form: $N_3 \sin\alpha = 0$. If $0 < \alpha < 180°$, then we get $N_3 = 0$.

The methods of projections and joints were elaborated by German engineers Johann W. Schwedler and Carl Culmann in the period 1851–1866. Also, in 1851 J. Schwedler substantiated the application of reference beams for analysis of trusses.

8.3 METHOD OF SUBSTITUTE MEMBERS

The method of substitute members, also called the Henneberg method,* permits us to determine the forces in a given attached statically determinate truss by substituting it with a truss that has the same location of joints but another positioning of members. By this method, one can calculate the complex truss by replacing it with a simple truss.

According to the method of substitute members, we introduce the replacement truss instead of a given attached statically determinate truss comprising n members by eliminating k members from the given system ($k < n$) and installation of k substitutive members in such manner that the number and location of joints don't change. Besides the given joint loads, k unknown generalized forces X_j, $j = \overline{1,k}$, are entered in the replacing truss; each generalized force X_j consists of two external forces of magnitude X_j, which are applied to the joints of the removed bar along its axial line in opposition to each other. We take $X_j > 0$ if these forces are directed towards each other.

The members of the replacing system are enumerated so that substitutive members are first in order. The method of substitute members assumes finding out such external forces X_j that the internal forces in the substitutive members turn into zero. If we withdraw the substitutive members

* German professor of mathematics and mechanics Lebrecht Henneberg suggested this method in 1886.

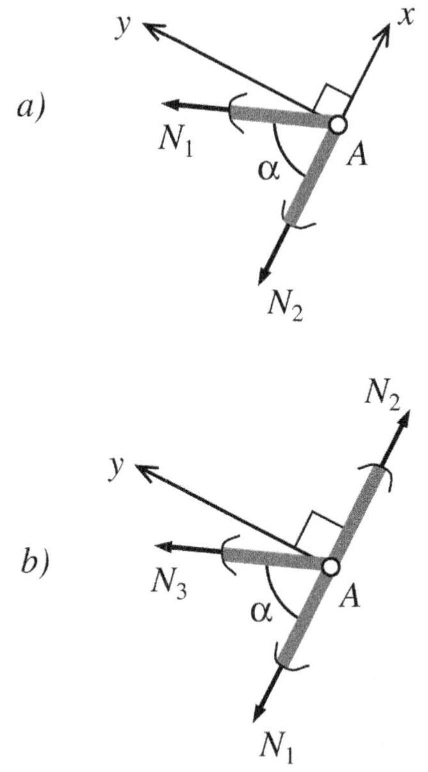

FIGURE 8.4 Joints with zero-force members.

subject to exertion of these forces X_j, then the balance of forces will be kept for all joints of this mechanism. After this withdrawal, in the remaining members, there act forces N_i, $i = \overline{k+1, n}$, obtained for the replacing truss. Next, let us bring back the members removed initially from the given truss and suppose that established generalized forces X_j are equal to the axial forces in these members and exerted by action of these members upon the joints. The equilibrium of joints will be kept, and, because equilibrium conditions for joints specify the internal forces unambiguously, our supposition is correct, and obtained totality of forces X_j, N_i really acts in the given truss. We may say that there is the equivalence of given and replacing systems on the condition of zero forces in substitutive members.

On the strength of the superposition principle, we have the relation for axial forces in all members of the replacing truss:

$$N_i = N_{iP} + \sum_{j=1}^{k} \bar{N}_{ij} X_j \quad i = \overline{1, n}, \tag{8.10}$$

where N_{iP} is the force produced by the given loads only, and $\bar{N}_{ij}$ is the force in the i-th member produced by external force $X_j = 1$. The conditions of equivalence for genuine and replacing systems are represented by the equation set as follows:

$$\sum_{j=1}^{k} \bar{N}_{ij} X_j + N_{iP} = 0 \quad i = \overline{1, k}. \tag{8.11}$$

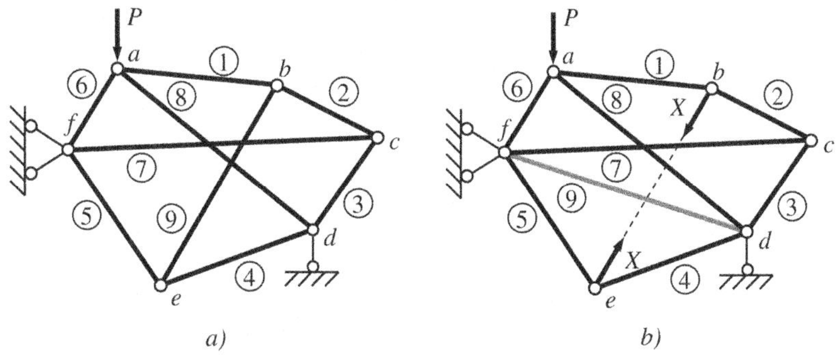

FIGURE 8.5 Given complex truss (a) and replacing simple truss (b).

These equations are termed canonical equations of the method of substitute members. While employing this method, it is required to calculate the quantities N_{iP} and $\overline{N}_{ij}$ at the beginning and then to solve the set of canonical equations – by doing this, the forces X_j acting in removed members are determined – and then we determine the forces in remained members by formula (8.10) at $i = \overline{k+1, n}$.

Figure 8.5a shows the truss for which the static determinacy has been established in Chapter 7 by the method of null load. The replacing truss, which was made by transposition of member 9, is depicted in Figure 8.5b. This truss is simple, and forces in its members are easily determined, for example, by the method of joints. The force in the substitutive bar is established by the formula:

$$N_9 = N_{9P} + \overline{N}_{9X}X, \tag{8.12}$$

where N_{9P} is the force from the load only, and $\overline{N}_{9X}$ is the internal force evoked by unit generalized force X. Out of that, we get the expression for the force in member 9 of the given truss:

$$X = -\frac{N_{9P}}{\overline{N}_{9X}}.$$

The forces in all other members of the original truss are calculated by the formula obtained for the members of the replacing truss:

$$N_i = N_{iP} + \overline{N}_{iX}X \quad i = \overline{1,8}.$$

The method of substitute members may be employed for stability verification of a hinged system which complies with the condition of static determinacy (7.3). While verifying stability, we have to pass from a given system to the replacing one, which is a statically determinate truss a fortiori and made by transposition of k members. After that, we calculate the determinant of the canonical equation set; if the one is nonzero then the given system is stable, otherwise it is unstable or instantaneously unstable. This verification procedure is grounded by the following theorem.

Theorem

In order for stability and static determinacy of an attached truss which may be statically determinate due to the number of constraints, it is necessary and sufficient that in the method of substitute members the set of canonical equations for this truss have a unique solution.

Proof

The uniqueness of the solution to the canonical equation set (8.11) means that the determinant of its coefficient matrix is nonzero:

$$\det(\bar{N}_{ij}) \neq 0. \tag{8.13}$$

To prove the assertion of the theorem, we initially prove that the set of canonical equations (8.11) in unknown forces X_j acting in removed members, being augmented by expressions (8.10) for the forces N_i, $i = \overline{k+1,n}$, in other members, is equivalent to the set of equilibrium equations for the joints of the original truss, composed in the same unknowns. We have shown that the totality of forces X_j, $j = \overline{1,k}$; N_i, $i = \overline{k+1,n}$, established through solving canonical equations (8.11), meets the equilibrium of the joints for the given truss. It is easily seen that the inverse is correct too. Really, if the totality of forces X_j, N_i is the solution of statics equations for the joints of a given system, then the mechanism made through replacement of k members by generalized forces X_j keeps balance, and insertion of substitutive members subject to zero internal forces doesn't break this balance. Therefore, these external forces X_j together with forces N_i secure zero forces in substitutive members of replacing truss, i.e., appear to be solutions of the equation set made up by (8.11) and equations (8.10) where $i = \overline{k+1,n}$.

Note further that in the case of zero given loads, the internal forces in all members of the substitute system are zeroth: $N_{iP} = 0$. Accordingly, the set of canonical equations (8.11) becomes homogeneous. Assume the given bar system is unstable (instantaneously unstable). Then, by virtue of the theorem about null load, there exists a nonzero totality of initial forces X_j, N_i which may act in its members. This totality must satisfy the homogeneous set of canonical equations (8.11), which may be only if its determinant is zero: $\det(\bar{N}_{ij}) = 0$. The sufficiency of stability condition (8.13) is proved. Inversely, if the determinant of system (8.11) is zero, then there exists a nonzero totality of generalized forces X_j in the replacing system under null load, which satisfies canonical equations, and, respectively, there exists a nonzero totality of forces in the members of a given system X_j, N_i subject to null load. But this is possible only on instability (instantaneous instability) of this system.

The theorem is proved.

The example of stability verification by the method of substitute members: if the truss in Figure 8.5a were the Pascal hexagon (Figure 7.13a), then the replacement of member 9 suggested above would result in equality $\bar{N}_{9X} = 0$. Thus according to expression (8.12), there would be ambiguity of reaction X developed by member 9 of the original truss in the absence of loads. This confirms the instantaneous instability of the Pascal hexagon.

8.4 METHOD OF LOAD CONVERSION FOR ANALYSIS OF TRUSSES WITH SECONDARY TRUSS MEMBERS

Sometimes, the problem of diminishing the weight of truss structures while keeping its bearing capacity may be resolved by means of secondary truss members (STMs). An STM is a bar subsystem entered into the truss fabric to reinforce it. We shall consider the STMs of statically determinate trusses, intended for the transference of loads from one set of joints to another.

In Chapter 7, we substantiated the rule of conversion of an intermediate load exerted on a member of a truss. The rule is to represent the load upon a member by two systems of forces; the first one is the balanced force system acting on the member, and the second one constitutes loads transmitted to the joints. The transference of loads with the help of STMs is implemented in the same way.

An STM contains the joints under external loads and the joints under the loads transmitted by the STM. The transference of loads is understood as the representation of loads acting on the STM by two systems of forces:

- The system of forces balanced in the boundaries of the STM, which includes the given loads and reactions of joints that take up the loads transmitted by the STM;
- The system of transferred loads opposite to reactions of joints noted above.

The analysis on the basis of loads' transference is used to determine internal forces by means of the superposition principle applied to statically determinate trusses with STMs.

The analysis assumes calculation of internal forces produced by the balanced system of external forces which act in the boundaries of each STM; calculation of internal forces produced by loads transferred through STMs; and totalization of the obtained forces in every bar. The balanced system of forces loads an STM only; the system of transferred loads stresses the truss as a whole, including the bars of STM.

A truss with STMs has a prototype – it is the truss of the same configuration but with no STMs. The prototype is convenient to establish the forces in a truss with STMs, produced by the loads transferred by STMs.

The transition from a truss with STMs to its prototype means the removal of some of the STM's members together with excessive hinges appeared due to the removal of members. On this removal, we shall keep the denotations of those hinges and members of the truss with STMs which remain in the prototype. The excessive hinges connect two members in the straight line; when the excessive hinge is removed, two bars are incorporated in one. The prototype of a truss with STMs is referred to as the *primary truss*. The primary truss can be mentally laid on the original truss with STMs so that homonymous hinges coincide. The possibility of transition to a primary truss by means of removing the bars and hinges from a truss with STMs is the first feature of a primary truss.

The prototype is correctly determined if under arbitrary loads exerted upon the hinges of truss with STMs, which are kept in the primary truss, the forces do not arise in the bars of STMs which are eliminated due to transition to the prototype. An absence of forces in the bars of STMs, which are removed for the transition to the prototype, under condition of the same loading of both trusses is the second feature of a primary truss. Owing to this feature, the forces in homonymous members of both trusses coincide for the same system of loads.

The reason for analysis on the basis of loads' transmission is that internal forces evoked by loads transferred through STMs are calculated for the prototype. We get two sets of forces to be summed up. One acts in STMs only; another is obtained for the prototype with account of loads' transmission. We replace one complicated calculation required for a truss with STMs according to the method of sections with two simple calculations. The method of analysis based on summing the internal forces from loads balanced in STMs and forces from transformed loads acting in the prototype of truss with STMs, is referred to as the *method of loads' conversion to a primary truss*.

The problem of loads' transference arises in the analysis of trusses with subdivided panels. The truss chord, the joints of which take up the working load directly, we shall call the *load chord*. These joints are connected with the floor beams in a bridge truss or the purlins in a roof truss. The subdivision of panels by means of STMs is used in trusses of different assignments, but we confine ourselves to main bridge trusses. For subdivision of panels, the intermediate joint is added in the middle of each panel of the load chord, and the STM is installed, which includes this joint and transmits the loads from the joint to the neighboring joints of one of the chords. This arrangement results in the decrease of loads upon the beams below the deck. Figure 8.6a and b, shows examples of panels' subdivision in the Pratt truss and the Warren truss with verticals. The Pratt truss with subdivided panels differs from the corresponding prototype in Figure 7.6b in installation of triangular STMs marked in Figure 8.6 with gray color. The internal hanger of the STM is stretched by the load exerted on the joint dividing the panel, and two diagonal bars of the triangle are compressed and transmit this load to adjacent joints of the chord. Figure 8.6b, shows a more complicated example of a truss with a top load chord. This truss contains the STMs of two types. Two STMs with the post inside the triangle redistribute the loads among the joints of the top chord, and four STMs with

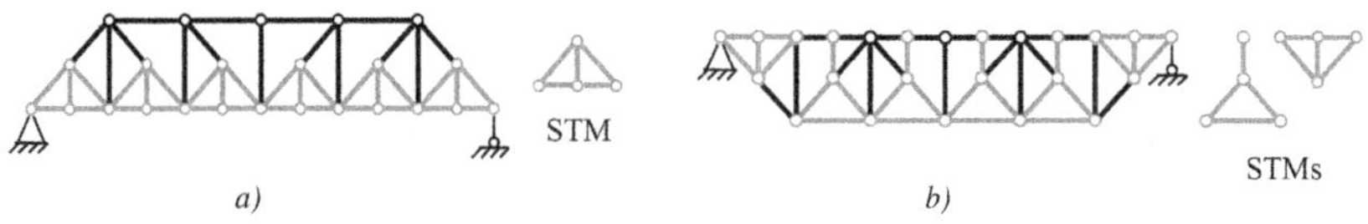

FIGURE 8.6 Trusses with subdivided panels: (a) Pratt truss; (b) Warren deck truss with verticals.

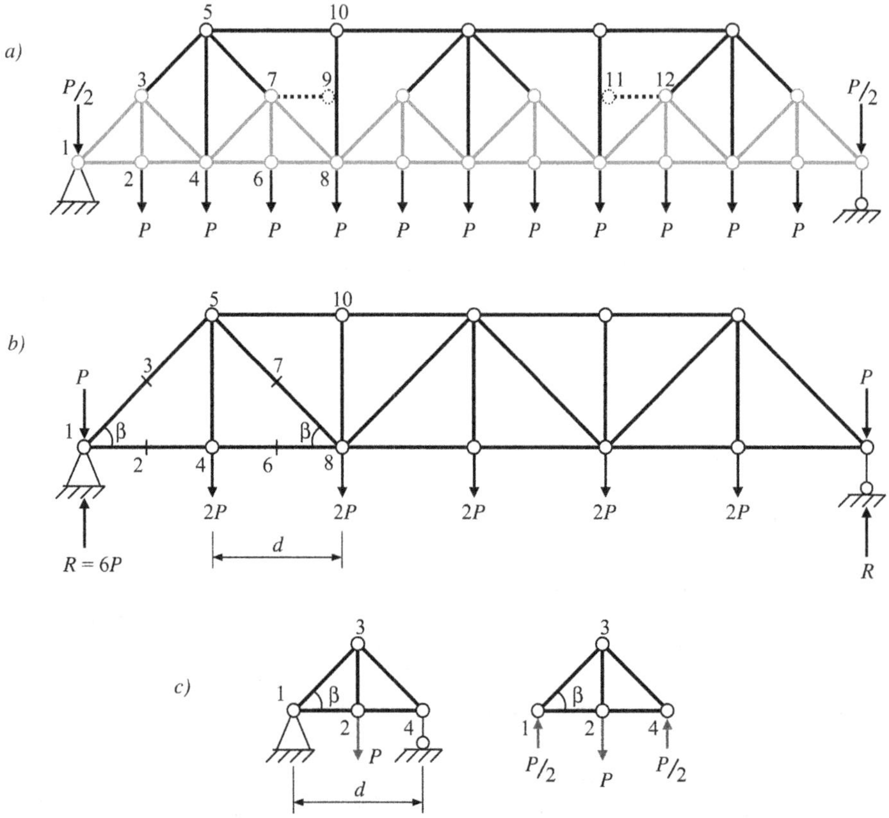

FIGURE 8.7 Warren through truss with verticals and subdivided panels (a), its prototype (b), and schemes of STM (c).

the post over triangle transmit the load from the top chord to the bottom chord. In both cases, the transmission of loads is done through the post and two diagonals of the STM.

The main trusses of the bridge in Figure 7.10 are Warren through trusses with verticals and subdivided panels, for which the design diagram is shown in Figure 8.7a. Using the example, we shall clarify the technique of analysis of trusses with STMs. The truss in the diagram is supplied with two horizontal members 7–9 and 11–12, which are used for raising the critical buckling force of the posts connected with them. Further, we don't take into account the work of these members within the truss. The STMs in the diagram are emphasized with gray. We shall consider the exertion of a uniformly distributed load on the deck of a bridge, of total magnitude $2P$ at the region in the boundaries of the panel. Accordingly, the floor beams transmit the load P upon internal joints of the load chord and the load $0.5P$ upon the outermost joints.

Figure 8.7b shows the primary truss; at two of its panels, the sections are marked where hinges were removed. The transmission of load on the support joints through the STM is shown in Figure 8.7c in the example of STM 1–3–4: on the left – the scheme of loading for calculation of joint reactions; on the right – the balanced system of forces in the STM. The scheme of loads' transmission for other STMs is precisely the same.

Let us verify the second feature of a primary truss in the example of STM 1–3–4 of given a truss. Hanger 2–3 and diagonal 3–4 of this STM are absent in the primary truss. Under the loads possible in the primary truss, the forces in these bars must be zero. Indeed, the hanger of the STM is not loaded because the force in it is equal to the vertical load upon hinge 2 which is absent in the primary truss. We have $V_{23} = 0$. On account of this, the equilibrium condition for joint 3 in projection on the normal to a straight line 1–3–5 gives $D_{34} = 0$. So, the feature of the prototype is fulfilled.

After the transference of loads by the STMs from the intermediate joint of each subdivided panel to its outermost joints, we come to the diagram of loading for the primary truss depicted in Figure 8.7b. We point out the forces in the members according to this diagram by superscript "*prim*." The balanced system of forces at every STM produces the forces in its members in correspondence with schemes in Figure 8.7c. We point out these forces with superscript "*sec*." According to the superposition principle, the virtual force in any member *i–j* of a given truss is obtained as follows:

$$N_{ij} = \begin{cases} N_{ij}^{prim} + N_{ij}^{sec}, & \text{if member } i-j \text{ belongs both to the primary truss and STM;} \\ N_{ij}^{prim}, & \text{if member } i-j \text{ belongs to the primary truss only;} \\ N_{ij}^{sec}, & \text{if member } i-j \text{ belongs to the STM only.} \end{cases} \qquad (8.14)$$

Let us determine, for instance, the forces in members in the region of subdivided panel 4–8 of the truss in Figure 8.7a. We mark the forces in the reference beam for a primary truss with superscript "0." Under the given extension of the subdivided panel and the slope of the diagonal member, the depth of the truss is $h = d\tan\beta$.

The equilibrium conditions for joint 1 in Figure 8.7c give the relations for internal forces in the STM as follows:

$$O_{13}^{sec} = -\frac{P}{2\sin\beta}; \quad U_{12}^{sec} = -O_{13}^{sec}\cos\beta = \frac{P}{2\tan\beta}.$$

Employing the methods of moment point and projections to the primary truss, we obtain:

$$U_{46} = U_{48}^{prim} + U_{46}^{sec} = \frac{M_5^0}{h} + U_{12}^{sec} = \frac{5Pd}{d\tan\beta} + \frac{P}{2\tan\beta} = \frac{11P}{2\tan\beta}; \quad U_{68} = U_{46};$$

$$D_{57} = D_{58}^{prim} = \frac{V_{5-8}^0}{\sin\beta} = \frac{3P}{\sin\beta};$$

$$D_{78} = D_{58}^{prim} + D_{78}^{sec} = \frac{3P}{\sin\beta} - \frac{P}{2\sin\beta} = \frac{5P}{2\sin\beta};$$

$$O_{5-10} = O_{5-10}^{prim} = -\frac{M_8^0}{h} = -\frac{5P\cdot 2d - 2P\cdot d}{d\tan\beta} = -\frac{8P}{\tan\beta}.$$

It is evident as well:

$$D_{47} = D_{47}^{sec} = -\frac{P}{2\sin\beta}; \quad V_{45} = V_{45}^{prim} = 2P; \quad V_{8-10} = 0.$$

One can obtain the same results with no conversion of loads exerted upon the given truss with STMs to its prototype, but the amount of calculation will essentially increase.

9 Influence Lines for Internal Forces and Support Reactions in Trusses

9.1 CONSTRUCTION OF INFLUENCE LINES FOR INTERNAL FORCES IN TRUSS GIRDERS USING THE STATIC METHOD

Let us accept that one of the chords of a statically determinate truss is subjected to loads of indirect application after the pattern described in Chapter 2 for a beam with indirect load application. Namely, we take as given that the chord is horizontal, and there is a series of secondary beams over it, forming the line of the live load's action. These beams rest on intermediate elements, which transmit the load to the joints of a chord. An example of an indirect load application upon the truss chord is shown in Figure 9.2. Usually, the series of secondary beams are not depicted but implied,* as in Figure 9.4. The chord taking up the movable load is referred to as the load chord. On the conditions arranged above for the live load action, the following *general property of influence lines for trusses is fulfilled*: **influence lines for internal forces in statically determinate trusses are continuous and consist of segments corresponding to the location of the load within a panel.** This affirmation is similar to property 3 of influence lines, stated in Chapter 2, and is proved in the same way.

In this first item of the chapter, we confine ourselves to the study of truss girders with simply supported spans (the girder may have cantilevers). The static method of construction of influence lines consists in using the relations obtained for a sought force from statics equations and applying the named general property for trusses. The static method may be actualized through all known methods of finding internal forces by means of statics equations, including the methods of moment point, projections, and joints. The methods of moment point and projections are most applicable, because constructing influence lines on their basis requires only a few rules.

Let force N acting in section I–I be determined by the moment point method, and this section crosses member m–n of the load chord (Figure 9.1). Pursuant to the moment point method, the influence line is constructed from three segments, or branches, namely, the left, right, and median branches. The right branch corresponds to the unit load's position on the right of cut panel m-n, the left branch is constructed for the unit load's position on the left of the cut panel, and the median branch corresponds to the load's position at this panel. Let us show how these branches are constructed.

Below, we shall distinguish the left and right parts of a truss cut into two parts by section I–I. The part on the left side of this section includes the left hinge of the dissected panel (hinge m). The part on the right side of the section includes the right hinge (n).

We consider the most common case when the supports of a truss are located on different sides of section I–I (i.e., belong to different parts of a truss dissected by this section). Introduce the truss-bound CS Axy, the plane of which comprises axial lines of members, position the origin at the left support hinge, and direct axis y upward along the vertical and axis x inside the span along the

* There is a reason not to show the series of secondary beams in diagrams: if we displace the concentrated load along the vertical from the secondary beam to the horizontal member of the load chord, then axial forces in all members of the truss won't change. Thus, in calculation of axial forces, we may accept that the live load moves at the load chord directly.

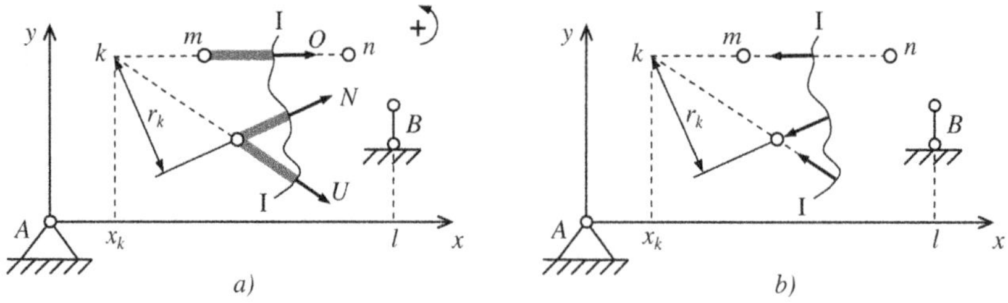

FIGURE 9.1 Example of internal forces' exertion in section I–I of a truss girder: (a) Forces *O*, *N*, *U* are exerted on the left part; (b) The same forces are exerted on the right part. *A* and *B* are supports of a truss (location of supports is permitted on different levels); *k* is the moment point for the determination of force *N*.

horizontal. From equilibrium condition for moments of forces exerted on the left part of the truss (Figure 9.1a), we obtain:

$$N = \mp \frac{1}{r_k} \sum_{left\ n} M_k(P_i), \tag{9.1}$$

where on the right-hand side, the summation is done for moments of left forces taken about moment point k. A moment is assumed to be positive if it acts counterclockwise; sign "–" on the right side corresponds to such positioning the arm r_k of an unknown force when positive force evokes a positive moment, otherwise the sign "+" is taken. In the case in point, force N defined by the right branch is caused by only one left force – support reaction R_A. From formula (9.1) we obtain:

$$N = \pm \frac{R_A x_k}{r_k}, \tag{9.2}$$

where x_k is the coordinate of the moment point. The reaction holds the linear relation to the coordinate of the unit load:

$$R_A = \frac{l - x}{l},$$

from whence it follows that right branch of the influence line is really a straight line. The ordinate of the right branch under support joint A equals:*

$$y_{right,A} = \pm \frac{x_k}{r_k}. \tag{9.3}$$

Relation (9.3), together with the requirement of the nil point's position for the right branch under the right support, specifies this branch.

Now we can consider the equilibrium condition for moments of right forces, which gives similar relations for construction of the left branch of the influence line:

$$N = \pm \frac{1}{r_k} \sum_{right\ m} M_k(P_i) \tag{9.4}$$

* The branch is implied as a segment or a straight line containing this segment (due to the context).

(sign "+" corresponds to the same position of arm r_k of an unknown force, i.e., positive force now evokes negative moment, Figure 9.1b);

$$N = \pm\frac{R_B(l - x_k)}{r_k}; \tag{9.5}$$

$$R_B = \frac{x}{l};$$

$$y_{left,B} = \pm\frac{l - x_k}{r_k}. \tag{9.6}$$

Linear dependence of force N versus coordinate x means that the left branch of the influence line is a straight line. Obviously, the nil point of this branch is located under the left support.

The left branch can be drawn with no regard for relation (9.6). After substitution of the reaction-versus-coordinate functions into (9.2) and (9.5), it is easily seen that left and right branches are concurrent under the moment point. Therefore, we can draw the left branch from the intersection of the right branch and the vertical downward from the moment point to the known nil point.

After drawing both the left and right branches, the median branch is drawn at the region between the verticals drawn from joints m and n of the dissected chord member. The median branch is uniquely defined by the requirement of continuity of the influence line.

Figure 9.2 presents the influence line for force D_{56}, constructed by using the stated technique. The influence line for force U_{57} is obtained in the same way, but here it is easier to utilize the influence line for the beam's bending moment because

$$U_{57} = \frac{M_6^0}{h_{67}\cos\alpha},$$

where M_6^0 is the bending moment in section 6 of the reference beam (verify this formula yourself).

Now, assume that the force under investigation N is determined by the method of projections, and all redundant forces are parallel to axis Ax (Figure 9.3). Just like in the previous problem, the influence line for N is specified by the left, right, and median branches. As before, we assume that the supports of a truss are located on different sides of section I–I. The analysis of this case doesn't substantially differ from the one previously done, though its form is simpler. By using the condition of left forces' equilibrium, we obtain the relations for the right branch, instead of (9.2) and (9.3), as follows:

$$N = -\frac{R_A}{\sin\beta}; \tag{9.7}$$

$$y_{right,A} = -\frac{1}{\sin\beta} \tag{9.8}$$

(angle β is shown in Figure 9.3; the positive reading for this angle is counterclockwise). For the left branch, one can set up the condition of right forces' equilibrium and obtain, instead of (9.5) and (9.6), the relations as follows:

$$N = \frac{R_B}{\sin\beta}; \tag{9.9}$$

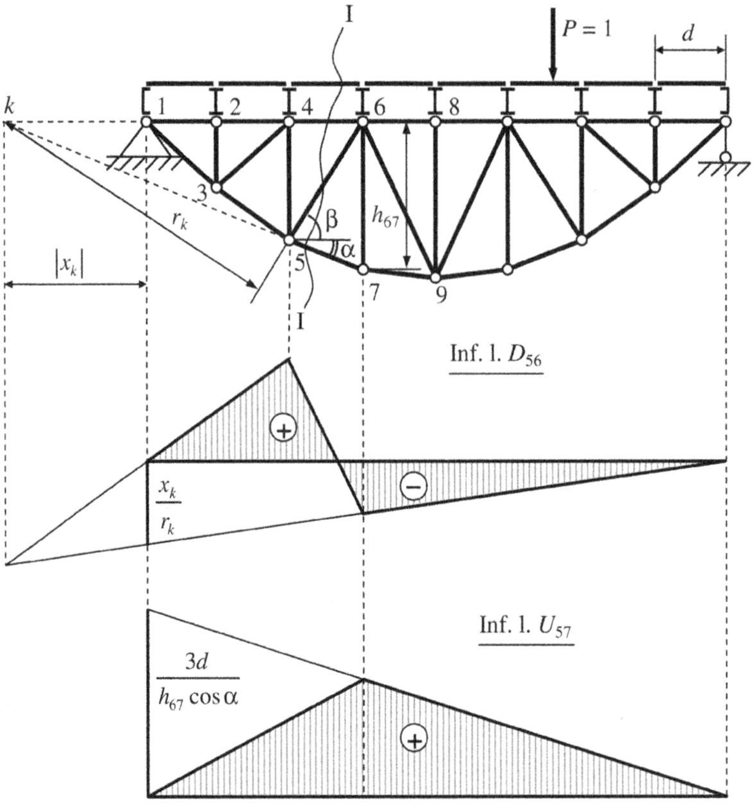

FIGURE 9.2 Construction of influence lines by moment point method.

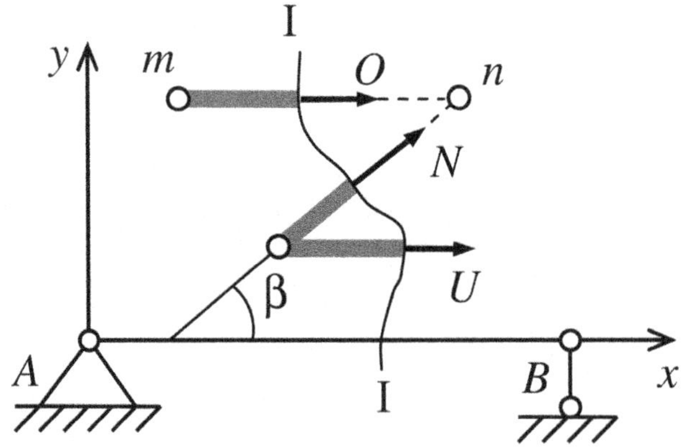

FIGURE 9.3 Example of internal forces' exertion in section I–I of a truss girder: forces O and U are parallel to the x-axis.

$$y_{left,B} = \frac{1}{\sin\beta}. \tag{9.10}$$

The left branch has the left nil point under the left support and the nil point of the right branch is under the right support. The noted nil points, together with ordinates (9.8) and (9.10), enable us to draw these branches as well as to establish their parallelism.

The consecution of construction of an influence line is the same for both the method of moment point and the method of projections; first, the left and right branches are drawn, using the known ordinates under supports and considering the mutual disposition of branches. Next, the median branch is drawn so that the influence line would be continuous. The instance of the influence line's construction by the method of projections is shown in Figure 9.4.

The summary for the case considered above (when supports belong to different parts of a truss dissected by section I–I, constructed for calculation of a wanted force) is the following:

1. **An influence line of force determined by the methods of moment point or projections consists of three segments (or branches): left, right, and median. The boundaries of the median branch are located under the joints of the load chord's panel which is dissected by the section constructed for calculation of the force.**
2. **For the force established by the moment point method, the left and right branches concur at the vertical drawn downward from the moment point. For the force established by the method of projections, these branches are parallel.**
3. **The influence line has nil points under supports.**

Coming up next, let us consider the case when supports of a truss are located on one side of section I–I – for example, on the left side – and force N is determined by the method of moment point (Figure 9.5). Then, given that the unit load is positioned on the right of joint n, there are two external forces exerted on the part of a truss at the left of section I–I – these are support reactions R_A and R_B. According to formula (9.1), we get the following expression for the wanted influence line at $x \geq x_n$:

$$N = \pm\frac{R_A x_k - R_B(l - x_k)}{r_k} = \pm\frac{l-x}{l}\cdot\frac{x_k}{r_k} \mp \frac{x}{l}\cdot\frac{l-x_k}{r_k}. \tag{9.11}$$

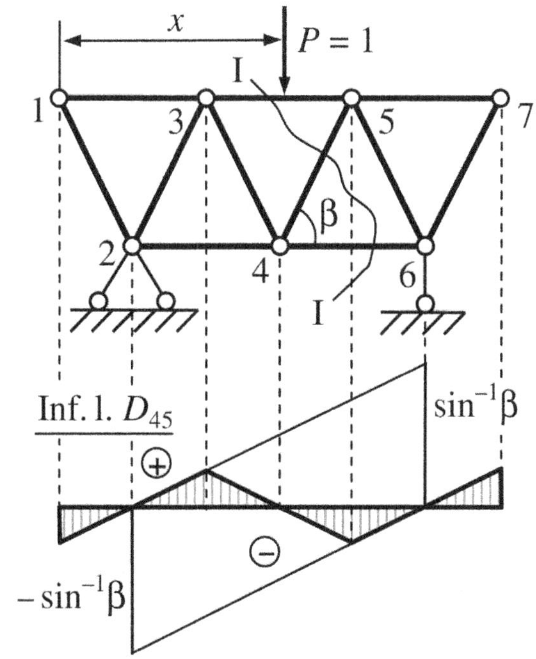

FIGURE 9.4 Construction of influence line by projection method.

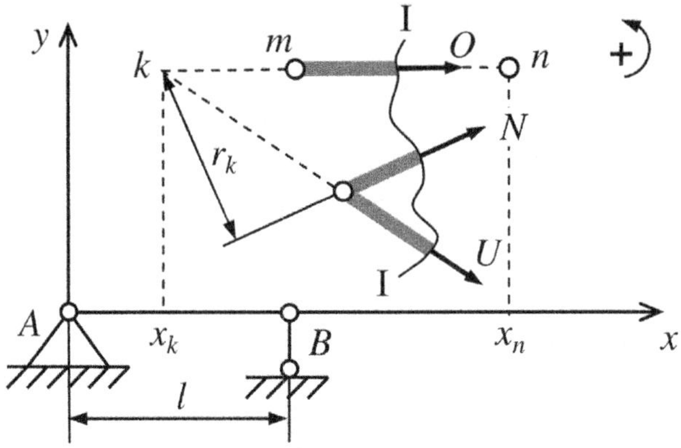

FIGURE 9.5 Example of internal forces' exertion in section I–I of a cantilever truss girder.

Linearity of the obtained function means that the right branch of the influence line is a straight line. The ordinates of this line under supports are obtained as follows:

$$y_{right,A} = \pm\frac{x_k}{r_k};\quad y_{right,B} = \mp\frac{l-x_k}{r_k}. \tag{9.12}$$

The nil point of the right branch proves to be in the same vertical with the moment point, because formula (9.11) gives $N=0$ at $x=x_k$. The left branch in the given case becomes zero, because there are no forces exerted upon the right part of a truss for positioning the load on the left of joint m. Thus, the left and right branches intersect under the moment point.

While considering the case when supports of a truss are located on the right side of section I–I, we come to similar inferences, but now the right branch is zeroed, whereas the ordinates of the left branch under supports have the form:

$$y_{left,A} = \mp\frac{x_k}{r_k};\quad y_{left,B} = \pm\frac{l-x_k}{r_k}. \tag{9.13}$$

If the force is determined by the method of projections, and supports of a truss are located on the left side of section I–I, then the equilibrium condition for the left part of the structure defines the right branch by the following relation, instead of (9.11):

$$N = -\frac{R_A+R_B}{\sin\beta} = -\frac{1}{\sin\beta}\left(\frac{l-x}{l}+\frac{x}{l}\right) = -\frac{1}{\sin\beta}. \tag{9.14}$$

Instead of (9.12), it holds:

$$y_{right,A} = y_{right,B} = -\frac{1}{\sin\beta}, \tag{9.15}$$

whereas the left branch becomes zero.

If the force is determined by the method of projections and supports of a truss are located on the right side of section I–I, then the zero branch is right-handed, while for the left branch it holds:

$$y_{left,A} = y_{left,B} = \frac{1}{\sin\beta}. \tag{9.16}$$

In the considered cases, when the projection method is employed, the left and right branches appear to be horizontal.

So, in the case when the section cutting a truss into two parts does not partition supports of the truss, inferences 1 and 2 stated above about the shape of influence line remain in force (in particular, the left and right branches concur under the moment point or parallel). Instead of inference 3, we infer that one of the branches of the influence line – left or right – proves to be zeroed, whereas for another branch, the formula for the ordinate under support $y_{right,A}$ or $y_{left,B}$ remains correct. When the unit load is positioned over the zeroed branch, the part of a truss being cut off from supports is not affected with loading. This conclusion suffices for construction of influence lines.

Figure 9.6 shows the instance of construction of an influence line by the moment point method. The moment point for force V_{56} coincides with joint 9 of the depicted deck truss. The zeroed branch is located on the left, because both supports are on the left of section I–I. The right branch is drawn by using the ordinate under left support (9.12) and the nil point under the moment point. The sought force produces a negative moment applied to the left part of the structure, so the ordinate is taken with sign "–." Figure 9.7 shows two more examples of influence lines in the case when section I–I is beyond the span boundaries. Diagram a in this figure reminds readers that the load is transmitted to the load chord through secondary beams. In diagram b, the left branch is zeroed, and the right one concurs with the left under the moment point, which coincides with support hinge 4. The ordinate of the right branch under the left support is again determined by formula (9.12). The influence line in diagram c is constructed by the method of projections. Here, the left branch is zeroed and the right one is parallel to the left and has the ordinate (9.15).

9.2 METHOD OF SUBSTITUTE MEMBERS FOR CONSTRUCTION OF INFLUENCE LINES AND UTILIZATION OF THE KINEMATIC METHOD

In the case of complex trusses, construction of influence lines may be done by the method of substitute members. Construction of influence lines based on this method is usually simple enough. We confine ourselves to installation of a single substitutive member, which we will take as the first in order among n members of a replacing truss; the force in the member removed from a truss we

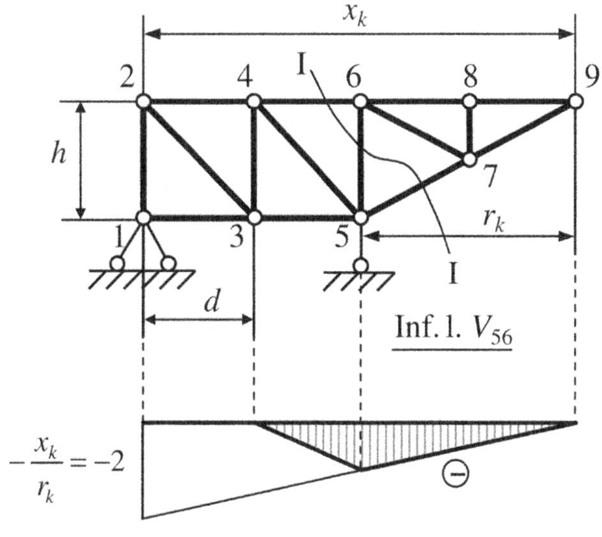

FIGURE 9.6 Construction of influence line by moment point method.

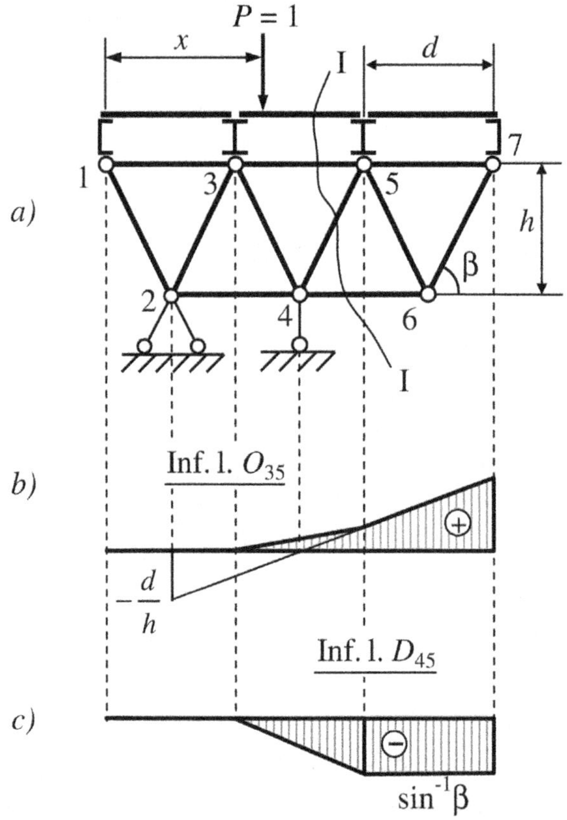

FIGURE 9.7 Construction of influence lines in the case of a cantilever truss girder.

denote as ***X***. For arbitrary loads upon a given truss, the forces in members of the replacing truss are determined by expression (8.10), which is, subject to replacement of one member, written in the form:

$$N_i = N_{iP} + \bar{N}_{iX} X \quad i = \overline{1,n}. \tag{9.17}$$

Here, as usual, N_{iP} is the force caused by given loads only; $\bar{N}_{iX}$ is the force in the i-th member produced by external force $X = 1$ only.

The force in the member removed from a system is obtained from the condition of zero force in the substitutive member:

$$X = -\frac{N_{1P}}{\bar{N}_{1X}}. \tag{9.18}$$

The forces in other members of the original truss are obtained by substitution of (9.18) into (9.17):

$$N_i = N_{iP} - \frac{\bar{N}_{iX}}{\bar{N}_{1X}} N_{1P} \quad i = \overline{2,n}. \tag{9.19}$$

Expressions (9.18) and (9.19) enable us to find the force in any selected member of a given truss, caused by the action of moveable load $P = 1$ upon the load chord. For certain positions of a load, one can obtain corresponding magnitudes of the force which will define the influence line. According to the general property of influence lines for trusses pointed out above, it is sufficient to set the load sequentially at all the joints and determine the forces produced by it. Moreover, kinks of the influence line arise at a small number of joints of the load chord, and, if these joints are known, it suffices to determine the ordinates only under these joints and get around complicated calculations.

To establish the general shape of the influence line for a statically determinate truss, we may employ the kinematic method of construction of influence lines (see Chapter 6). In accordance with this method, an influence line is obtained as the normalized diagram of load line deflections due to virtual displacement in the system after elimination of the member, for which the internal force is under investigation. The vertical displacements of the load chord may be accepted as deflections. The virtual displacement must ensure mutual displacement of the joints from which the member was taken out towards each other (according to directions of external forces applied to the joints in replacement of action of positive force in the removed member). In practice, this method is useful in finding the joints under which there are kinks of the influence line. When these joints are identified, the static method is involved for determination of the kink ordinates.

The qualitative influence line for force U_{57} in the truss in Figure 9.2 is easily established by the removal of member 5–7 and construction of the diagram of load chord deflections due to displacement of joints 5 and 7 towards each other. The removal of the member turns the truss into a two-disc system with hinged connection 6. On account of presumed displacement, the shape of the load chord is affected by the kink in point 6 that is represented by the influence line depicted below the truss. In the same way, by removing member 5–6, we may establish a qualitative influence line for force D_{56}. The mutual displacement of chord portions due to the displacement of joints 5 and 6 towards each other is displayed by an influence line constructed by static method in the same figure.

Example

Construct the influence lines for reaction R_5 and force N_{67} acting in the truss in Figure 9.8a. Take as given that $h = 0.5d$; $\angle 6\text{–}3\text{–}5 = 90^\circ$.

Transform the truss to the simple type. In order to do that, take out the middle support link and put in substitutive post 5–6 (Figure 9.8d). To secure the equivalence of given and replacing trusses, apply reaction X of eliminated connection on joint 5 so that the force in the post would be $N_{56} = 0$. The calculation formulas of the method of substitute members are as follows:

$$R_5 = X = -\frac{N_{56,P}}{\bar{N}_{56,X}}; \tag{9.20}$$

$$N_{67} = N_{67,P} - \frac{\bar{N}_{67,X}}{\bar{N}_{56,X}} N_{56,P}. \tag{9.21}$$

In order to determine the shape of the influence line for reaction R_5, let us eliminate connection of joint 5 with the earth and give the virtual displacement to this joint. The load chord takes the form of an isosceles triangle with the vertex at joint 6. This triangle is already shown in Figure 9.8b as the very influence line for force R_5. The ordinate of this line under joint 6 is determined further.

In order to get a qualitative influence line for force N_{67}, let us take out member 6–7 and investigate displacements of the top chord in the obtained mechanism due to virtual displacement of joints 6 and 7 towards each other. We see that vertical displacements of joints 2 and 10 are absent (neglecting the infinitesimals of the second order). Really, joint 2 moves only horizontally, because it belongs to originally vertical member 1–2 which is rotated about joint 1; joint 10 moves only horizontally because it belongs to the initially vertical hinged chain "9′–9–10" where hinge 9′ is fixed (Figure 9.8e). Zero deflections of the load chord at joints 2 and 10 means that nil points of sought

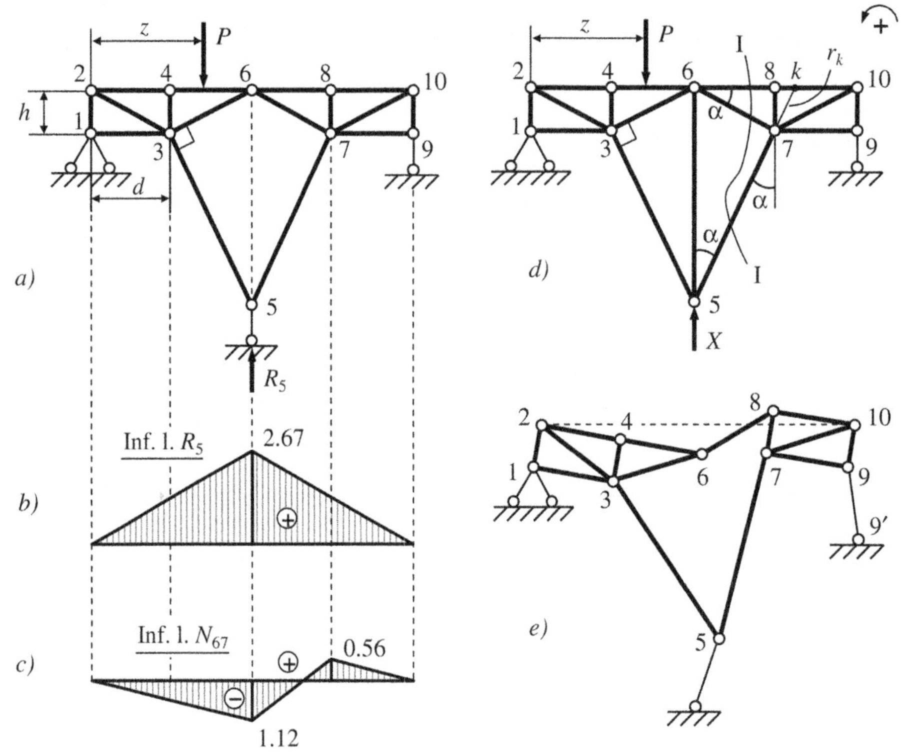

FIGURE 9.8 Given truss (a), influence lines for forces in given truss (b, c), replacing truss (d), and displacement of the system with eliminated member 6–7 (e).

influence line are located under these joints. Note also that portion 2–6 of the load chord remains straight during displacement of the mechanism, because the part of the truss below this portion makes up a disc. It follows that the influence line may be specified by two ordinates located under joints 6 and 8. The small real displacement of the system is shown in Figure 9.8e, and the corresponding influence line in Figure 9.8c. We further determine the stated ordinates by the method of substitute members.

The ordinate of the influence line for R_5 under joint 6 is the magnitude of force $X = X(2d)$ (i.e., the magnitude of force $R_5 = X$ at the coordinate of unit load $z = 2d$). One can establish this force by formula (9.20). To do that, let us first determine force $\bar{N}_{56,X}$. The equations of force balance for joint 5 have the form:

$$\bar{N}_{35,X} = \bar{N}_{57,X}; \quad \bar{N}_{56,X} + X + (\bar{N}_{35,X} + \bar{N}_{57,X})\cos\alpha = 0. \tag{9.22}$$

For the given ratio of the posts and panels, we have the diagonal's length $l_{67} = \frac{\sqrt{5}}{2}d$, from whence $\cos\alpha = \frac{d}{l_{67}} = \frac{2}{\sqrt{5}}$. Because $X = 1$, we obtain from equations (9.22):

$$\bar{N}_{56,X} = -1 - \bar{N}_{57,X}\frac{4}{\sqrt{5}}. \tag{9.23}$$

To find force $\bar{N}_{57,X}$, we employ the moment point method for section I–I (Figure 9.8d). The condition of moment equilibrium for any assemblage of vertical loads has the form:

$$\sum_{left\ 8} M_6(P_i) + N_{57} l_{67} = 0.$$

In this formula, pass to a beam bending moment by using the substitution:

$$M_6^0 = -\sum_{left\ 8} M_6(P_i).$$

We obtain:

$$N_{57} = \frac{2}{d\sqrt{5}} M_6^0. \tag{9.24}$$

Due to exertion of external force $X = 1$, the beam bending moment equals $\bar{M}_6^0 = -d$, and formulas (9.24) and (9.23) give:

$$\bar{N}_{57,X} = -\frac{2}{\sqrt{5}};$$

$$\bar{N}_{56,X} = -1 + \frac{8}{\sqrt{5} \cdot \sqrt{5}} = \frac{3}{5}. \tag{9.25}$$

The equations of force balance for joint 5 in the case of load $P = 1$ exerted on the load chord produce the equality analogous to (9.23):

$$N_{56,P} = -N_{57,P} \frac{4}{\sqrt{5}},$$

and we may rewrite this, involving the beam moment:

$$N_{56,P} = -\frac{8}{5d} M_6^0. \tag{9.26}$$

Let us place unit load at joint 6 (underneath it, the ordinate of influence line for R_5 is wanted). At that: $\bar{M}_6^0 = d$, so we obtain from (9.26):

$$N_{56,P} = -\frac{8}{5}.$$

From this, we have the ordinate of the influence line for R_5 under joint 6:

$$X(2d) = -\frac{N_{56,P}}{\bar{N}_{56,X}} = \frac{8}{3} \cong 2.67.$$

Next, we turn to finding the characteristic ordinates of the influence line for N_{67}. To employ formula (9.21), one needs the constant coefficients $\bar{N}_{56,X}$, $\bar{N}_{67,X}$. The first of them was obtained as (9.25). The second one is determined by the moment point method with the same section I–I. Let us set up the equation of moment equilibrium for right forces with respect to pole k (Figure 9.8d):

$$\sum_{right\ 6} M_k(P_i) - N_{67} r_k = 0.$$

This equation is correct for arbitrary vertical loads exerted upon the replacing truss. The arm of unknown force r_k is determined as follows:

$$r_k = l_{67} \tan\alpha = \frac{\sqrt{5}}{4} d.$$

Finally, we obtain:

$$N_{67} = \frac{4}{d\sqrt{5}} \sum_{right\ 6} M_k(P_i). \tag{9.27}$$

The distance between points 8 and k equals $l_{8k} = h\tan\alpha = 0.25d$; accordingly, the coordinate of the moment point equals $z_k = 3.25d$.

At the external force $X = 1$, the sum of moments of right forces in (9.27) is $-\frac{1}{2}\frac{3}{4}d = -\frac{3}{8}d$. Thus, for the sought "unit" force we have:

$$\bar{N}_{67,X} = -\frac{3}{2\sqrt{5}}.$$

Let us determine the ordinate of the influence line under joint 6. We set the unit load at this joint and get from (9.27) and (9.26):

$$N_{67,P} = \frac{3}{2\sqrt{5}}; \quad N_{56,P} = -\frac{8}{5}.$$

After that, we substitute calculated values into formula (9.21) and obtain:

$$N_{67}(2d) = -\frac{5}{2\sqrt{5}} \cong -1.12.$$

Let us determine the ordinate of the influence line under joint 8. We set the unit load at this joint and get:

$$N_{67,P} = \frac{13}{4\sqrt{5}}; \quad N_{56,P} = -\frac{4}{5}; \quad N_{67}(3d) = \frac{5}{4}\frac{1}{\sqrt{5}} \cong 0.559.$$

So, we can construct the influence line as it is depicted in Figure 9.8c.

This example is also interesting because if the height of posts is taken as $h = d$, then the given truss becomes instantaneously unstable. The latter is revealed in trying to establish wanted forces by formulas (9.20) and (9.21). We discover $\bar{N}_{56,X} = 0$; thus the wanted forces are not defined. For the given hinged system at $h = d$, there are loadings under which the equilibrium of undeformed system is impossible, and this is detected in division by zero in the stated formulas.

10 Space Statically Determinate Trusses

10.1 SPACE CONNECTIONS AND GEOMETRICAL STABILITY

Investigation of space systems is based on the same approaches which have been developed for plane systems. Space systems consist of bodies, i.e., geometrically stable objects with extension in three dimensions (sometimes, bodies are referred to as blocks). Every free body has six DOFs relative to another body. Connections between bodies delete DOFs. The following principal types of connections are distinguished in space systems:

A *link*, or *planar-moving ball connection*, or *simple rod*, is a rigid rod connecting two bodies by means of spherical hinges at the ends (Figure 10.1a). The connection deletes translational displacement of one connected body relative to the other (one degree of freedom), and develops reaction forces directed along the axial line of the rod.

A *ball and slot*, or *linear-moving ball connection*, is made up of two links with a common hinge at one of their ends (Figure 10.1b). This connection deletes translational displacements of one body relative to the other in the plane of two links (two DOFs) and develops a reaction force upon each body, determined by two components in this plane.

A *ball and socket*, or *immovable ball connection*, is assembled from three links not lying in the same plane, with a common hinge at one of their ends (Figure 10.1c). The connection deletes all translational displacements of the constrained body (three DOFs), and develops a reaction force on the body, determined by three components in the space.

Other types of connections between bodies are made by combining connections of the principal types. For instance, a rigid connection can be formed from three connections of the principal types positioned near enough to each other subject to right their arrangement (see comments to Figure 10.2 below).

Straight bars connected with other bodies of a space system by spherical hinges at their ends can freely rotate about their own axial lines, and this rotation doesn't affect the system's integrity. Usually, it is not taken into account as a degree of freedom. When we say bars of a system, we mean straight bars. In evaluation of the instability parameters of a space system, bars with hinges at their ends are represented by their axial lines. A free space body has six DOFs, whereas a free bar has five DOFs (if it is hinged, i.e., will be joined to a system by spherical hinges). In the course of composing the design diagram, the rigid joints at the ends of bars are usually replaced by spherical hinges – subject to the stability of the construction being saved.

In this chapter, hinged bar systems are primarily under consideration. Such a system, in the case of its stability, is a space hinged truss. Below, we establish the conditions of geometrical stability and static determinacy for an attached hinged bar system which comprises B members, L_{sup} support links, and J joints. During this analysis, we will refer to both members and support links as hinged bars. The location of a system is defined by the location of its joints being considered as material points. If all the system's hinged bars are cut off near the joints and eliminated from the system, then the totality of disconnected joints would have $3J$ degrees of freedom. Now, let us consecutively bring the removed bars back into the system. Every hinged bar will delete one DOF if it opposes mutual displacement of connected joints, or installation of bar will not change the number of DOFs if the distance between these two joints was fixed before by other hinged bars. Thus, hinged bars

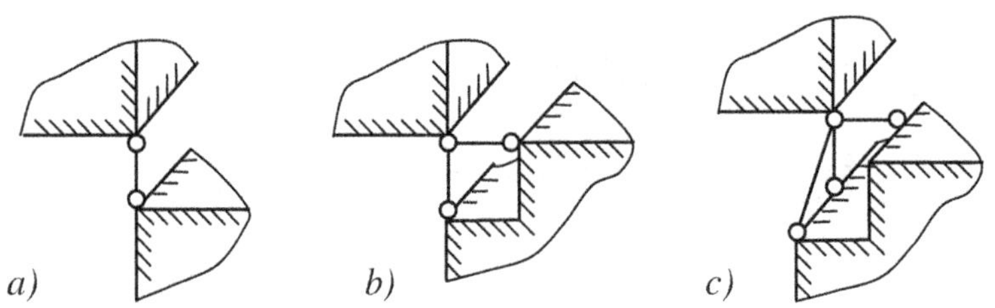

FIGURE 10.1 Types of space connections.

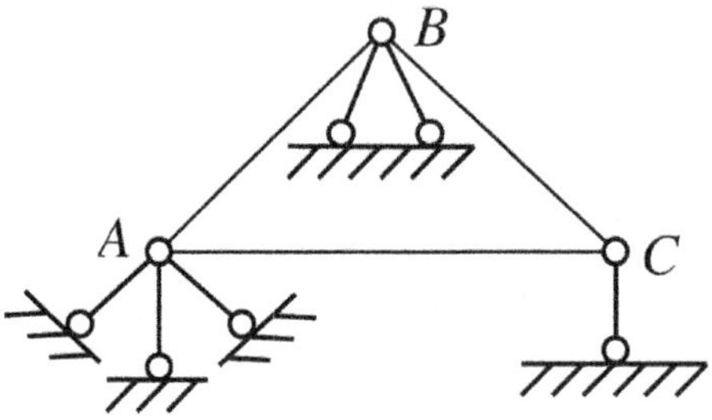

FIGURE 10.2 Example of stable joining a body to the earth.

may delete no more than $B + L_{sup}$ degrees of freedom, and the number of DOFs of a given system complies with inequality:

$$F^* \geq 3J - B - L_{sup}. \tag{10.1}$$

The necessary condition of stability for a space truss is obtained in the form:

$$B + L_{sup} \geq 3J. \tag{10.2}$$

Just as in the planar case, we can compose the statics equation set for a truss by using the conditions of force balance for every joint; in total, we have $3J$ equations of statics. The left side of the latter inequality represents the total number of internal forces securing the equilibrium of the system, i.e., the number of unknowns of the statics equation set. Requirement (10.2) means that the number of forces securing the equilibrium of a system can't be less than the number of statics equations.

Let J joints of a truss be connected by the least number of hinged bars, i.e., the following condition is fulfilled:

$$B + L_{sup} = 3J. \tag{10.3}$$

Each hinged bar of such a system is a required connection, i.e., its removal makes the system unstable. It was pointed out in Chapter 1 that reactions of the required hinged bars of a system are statically determinate. Thus equality (10.3) is the characteristic of static determinacy of a truss.

So far, we have considered attached trusses. The characteristic of static determinacy for a free space truss is expressed in the form:

$$B = 3J - 6. \tag{10.4}$$

As before, the right side of this relation represents the number of independent equations of statics.

A system has one of three attributes of stability: it can be internally stable, unstable, or instantaneously unstable. In the case of a plane bar system, internal stability is investigated in the coordinate system referenced to one of the members (or stable assemblage of members). In the case of a space hinged bar system, the axis of any member is the axis of its revolution, and therefore, a 3D coordinate system can't be bound to a separate member. The coordinate system for analysis of stability is to be specified by using three hinges; the plane of these hinges is taken as the coordinate plane, and the origin of coordinates is placed at one of them.

For an attached plane or space system, three properties of external stability are also introduced, which are defined as attributes of stability for a system taken in the aggregate with the earth. Investigation of a system's stability (i.e., kinematic analysis) may be done qualitatively, by investigation of the possibility of bodies' displacements relative to each other and the earth, or by analytical methods. Kinematic analysis is carried out in two stages; first a necessary condition of stability is verified, and then, either possibility of bodies' mutual displacements is considered in qualitative analysis, or analytical methods are employed for analysis of statics equations. The second stage of qualitative kinematic analysis usually is to determine the sequence of building the stable system. In kinematic analysis of plane systems, one could reveal stable subsystems consisting of two or three discs by identifying their distinguishing traits, but even in planar problems, it is not uncommon to fail in analysis of stability by consecutively joining up the discs. Analysis of space system assemblage is all the more complex.

Let us consider the stability conditions for a free system of two bodies. The number of degrees of freedom determined for a free system, with respect to one of its bodies, is referred to as *degree of instability* (see Chapter 1). Subject to an absence of connections, a system of two bodies has a degree of instability $I = 6$. Every link when "correctly" installed decreases the degree of instability by one. Therefore, the minimum number of constraints required for the stability of a two-body system equals six. On special arrangement of these constraints, there arises instability or instantaneous instability of the body couple (for instance, when all six simple rods are positioned in one plane). In order to verify the correct arrangement of these connections, one may try, at first, to establish the instantaneous pivot-point of one body relative to another, then to reveal the instantaneous pivot-axis, and then to establish impossibility of mutual rotation about this axis. An instantaneous pivot-point is the intersection point of three links not lying in the same plane; the instantaneous pivot-axis runs through this point and the intersection point of two more links not lying in the same plane with the instantaneous pivot-point. Finally, the sixth and last rod must ensure mutual immovability of the connected bodies. An example of such a stable joining is presented in Figure 10.2. In this example, the connections of the body with the earth are located at the vertices of triangle ABC, so that the ball and socket is located at point A, the ball and slot is located at point B in such fashion that the plane of its links doesn't pass through point A, and the simple rod is positioned at point C in such fashion that its axis doesn't intersect line AB. Kinematic analysis of the obtained system turns out to be simple: point A is the rotation center of the attached body, segment AB is immovable relative to the earth and, consequently, is the rotation axis of the body, and constraint C prohibits rotation around this axis.

Stability analysis of a two-body system may be employed for qualitative kinematic analysis of a truss if geometrically stable bar subsystems are disclosed in its fabric. Such subsystems can be considered as bodies at their further conjoining. A hinged bar triangle could serve as an example of a stable subsystem in a truss.

For space bar systems, there are two cases when analysis of stability usually doesn't cause difficulties. In the first case, the construction appertains to the type of *simple truss*. The space truss is referred to as simple if it contains an initial ("primary") hinged tetrahedron and is formed through consecutive linking of every new hinge by three bars which don't lie in the same plane. Figure 10.3, *a*, shows the simple truss made up by first joining hinge 5 to tetrahedron 1–2–3–4, and then by addition of hinge 6 to the system. Space simple truss is statically determinate, as well as a simple truss on the plane. In construction structures, a simple truss is usually included in a more complicated space truss. For example, bearing parts of tower trusses often comprise bar tetrahedrons (Figure 10.4).

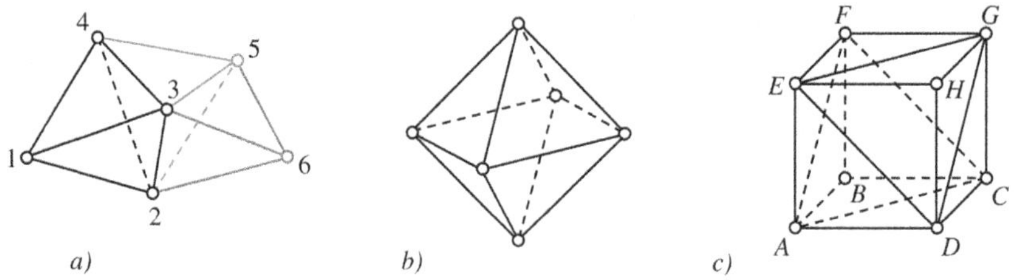

FIGURE 10.3 Examples of space trusses: (a) Simple truss; (b) Hinged octahedron – reticulated truss; (c) Cube-shaped truss comprising octahedral reticulated truss.

FIGURE 10.4 Gillerberg Observation Tower (North Rhine-Westphalia, Germany). On the left – overall view; on the right – bearing part of the tower. Photo by Uv22e (License CC-by-SA 3.0. Source: https://commons.wikimedia.org/w/index.php?curid=19352847).

In the second case, the truss appertains to the type of *reticulated truss*. Let us term a polyhedron with triangular facets a *simplicial polyhedron*. A hinged bar system is called reticulated if the axial lines of its members are the edges of a convex simplicial polyhedron. American mathematician Max Wilhelm Dehn proved the theorem about the rigidity of a convex simplicial polyhedron in 1916 (Pak 2006). It follows from this theorem that a reticulated bar system is stable.* Also, there is Euler's polyhedral formula derived in 1750, which establishes a relation between the number of vertices, edges, and facets of a convex polyhedron. According to this formula, the relation (10.4) between members (edges) and joints (vertices) is correct for reticulated trusses, i.e., a reticulated truss is statically determinate. The hinged octahedron is an example of a reticulated truss (Figure 10.3b). The British Museum's roof is erected on the truss of a reticulated structure with fixed margins (Figure 10.5). This truss is convex at the largest part of its extension, which ensures its bearing capacity.

The hinged cube in Figure 10.3c can be built through attachment of two joints to a reticulated truss by the rule of simple truss construction. Really, the cube comprises an irregular hinged octahedron enclosed between planes *AFC* and *EGD*.† The hinged triangle *AFC* of the octahedron can be supplemented to a hinged tetrahedron by joining vertex *B*; respectively, triangle *EGD* is supplemented to a tetrahedron by joining vertex *H*. These adjunctions result in construction of a stable system, the given hinged cube. In this example, we may see that stability of a reticulated truss is useful for qualitative kinematic analysis of trusses.

* In the literature, there is affirmation that geometrical stability of reticulated trusses follows from Cauchy's theorem about convex polyhedrons (1813). In reality, from Cauchy's theorem, there follows zeroth degree of instability of a reticulated truss, which doesn't exclude instantaneous instability.

† In order to see this octahedron, redraw the picture of the hinged cube on a sheet of paper and erase joints *B* and *H* together with the adjoining bars.

FIGURE 10.5 The Great Court of the British Museum, with tessellated roof designed by Foster + Partners. Photo by Andrew Dunn, http://www.andrewdunnphoto.com/ – Own work (License CC-by-SA 2.0. Source: https://commons.wikimedia.org/w/index.php?curid=439653).

If qualitative kinematic analysis of a space truss causes difficulties, one may employ any analytical method of stability analysis. Formerly, the method of null load and the method of substitute members have been substantiated as the instruments of analysis of stability and static determinacy for plane trusses with a minimum number of constraints. The same methods are applicable for conclusions about the stability of space trusses complying with condition (10.3). If a space truss does not relate to a simple one, then it is advisable to transform it into a simple truss by using the method of substitute members, and then a canonic equation set can be checked for uniqueness of the solution.

10.2 EQUATIONS OF STATICS IN SPACE

For a body belonging to a space system, the equilibrium conditions for forces and moments are written in the form:

$$\sum \boldsymbol{P}_i = 0; \quad \sum \boldsymbol{M}_A(\boldsymbol{P}_i) = 0, \tag{10.5}$$

where $\boldsymbol{P}_1, \boldsymbol{P}_2, \ldots$ are the external forces exerted upon the body, and $\boldsymbol{M}_A(\boldsymbol{P}_1), \boldsymbol{M}_A(\boldsymbol{P}_2), \ldots$ are the moments produced by these forces relative to some point A, respectively. The directions of constraint reactions in a stable system are known in advance (they coincide with the directions of the links forming these constraints). Taking this into account, we treat conditions (10.5) as the equations of statics in algebraic values of reactions acting on a body under consideration.

Six scalar equations (10.5) enable one to determine no more than six reactions of connections with surrounding bodies. If a body with six support links makes up an externally stable system, then all these connections are statically determinate, i.e., their reactions are established from equations (10.5). The original statics equation set (10.5) can be substituted by an equivalent equation set containing from three to six equations of axial moment equilibrium. One can choose, for instance, the totality of six axes in space, given definite requirements to their mutual location, and write down the statics equation set as follows:

$$\begin{gathered}\sum M_1(\boldsymbol{P}_i)=0;\ \sum M_2(\boldsymbol{P}_i)=0;\ \sum M_3(\boldsymbol{P}_i)=0;\\ \sum M_4(\boldsymbol{P}_i)=0;\ \sum M_5(\boldsymbol{P}_i)=0;\ \sum M_6(\boldsymbol{P}_i)=0,\end{gathered} \tag{10.6}$$

where $M_k(\boldsymbol{P}_i)$ is the moment of force $\boldsymbol{P}_i$ about the k-th axis. The axis is called a moment axis if the moments are determined with respect to this axis when setting up the equilibrium equation. The mutual location of six axes ensuring equivalence of equilibrium conditions (10.6) to the statics equation set for a solid body (10.5) has been studied in theoretical mechanics. In particular, this equivalence is reached when six moment axes make up the edges of a tetrahedron. Another case of equivalence of statics equation sets (10.5) and (10.6) takes place when three moment axes make up the sides of a triangle – the base of the triangular prism – and other three axes are parallel and coincide with the side edges of this prism. In the supplement to this chapter, there are proofs of these assertions.

Let us consider the problem of finding six support reactions for an attached body constituting a statically determinate system. In this problem, one can specify the totality of axes so that the system of six equations (10.6) will be decomposed into independent systems of two equations (Loitsianskii and Lurie 1940). Owing to this, the reactions often may be obtained in simple analytical form. In order to determine any reaction, the moment axis should be chosen in such a manner that lines of action of as many as possible other unknown reactions intersect this axis or run parallel to it. Then these other reactions do not affect the total axial moment and are not included in the equilibrium equation for axial moments. Let there be given axial lines 1, 2, 3, 4 of four links supporting a body. According to the position of these lines, the moment axes for determination of reactions developed by two remaining links are specified as follows:

Let lines 1 and 2 concur at point A. We term the plane containing lines 1 and 2 "plane 1" and introduce plane 2 passing through point A and line 3. Denote the trace of line 3 on plane 1 as B and the traces of line 4 on planes 1 and 2 as C and D, respectively (Figure 10.6a). Straight lines AD and BC can serve as axes for calculation of moments;

Let lines 1 and 2 concur at point A and lines 3 and 4 concur at point B. Then axial moments could be calculated relative to axes AB and CD, where CD is the intersection of the planes, in which given lines lie in pairs (Figure 10.6b).

Let us determine support reactions for the structure in Figure 10.7. The structure consists of a cylindrical body on six support rods and has axial symmetry of the third order. The views of the system in Figure 10.7 are presented in axonometric projection and in plan view. Force $\boldsymbol{P}$ is the resultant of the loads; N_1 and N_2 are wanted support reactions. In order to determine forces N_1 and N_2, we set up the equations of statics for the body being mentally cut off from supports. Taking into account the recommendations given before, we use the equilibrium condition for axial moments with respect to axis I–I. The corresponding equation can be obtained in the form:

$$Pa+(N_1+N_2)c\sin\alpha=0. \tag{10.7}$$

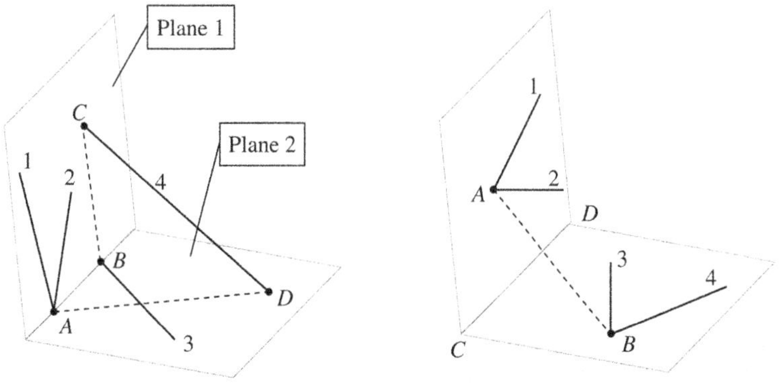

a) *b)*

FIGURE 10.6 Finding two moment axes for attached body when the other four axes are given.

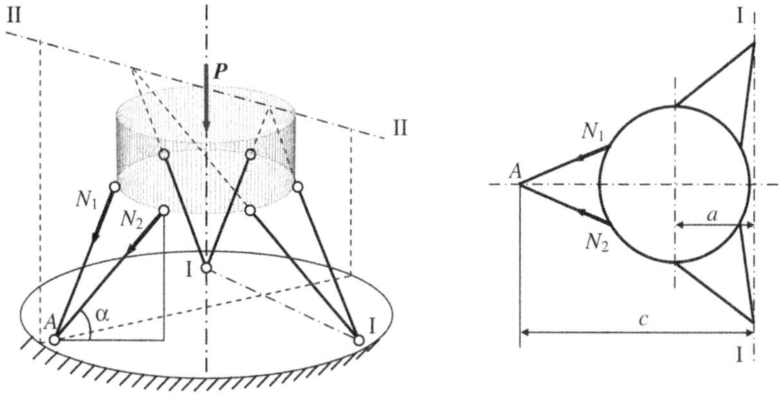

FIGURE 10.7 Specifying the moment axes for stable body with axial symmetry.

The first addend on the left-hand side is the total axial moment of loads. Let us make certain that the second addend is the total axial moment of forces $\boldsymbol{N}_1$ and $\boldsymbol{N}_2$ exerted on the body from the rods. Really, the moment of force will not change due to moving this force along its line of action. Let us transfer reactions $\boldsymbol{N}_1$ and $\boldsymbol{N}_2$ to point A and determine the axial moment of the total force $\boldsymbol{N}_1 + \boldsymbol{N}_2$. The horizontal component of this sum doesn't affect the moment of the total force about axis I–I. The vertical component is equal to $(N_1 + N_2)\sin\alpha$ and makes a contribution to the axial moment represented by the second addend.

Taking into account the symmetry of the design diagram, the solution of equation (10.7) is obtained in the form:

$$N_1 = N_2 = -\frac{Pa}{2c\sin\alpha}.$$

In the case of asymmetric loading, equation (10.7) must be modified as related to loads, and we have to supplement it with an equilibrium equation for moments about axis II–II (Figure 10.7, on the left). Note that in this case, two equations of moment equilibrium determine internal forces at any loads unambiguously, and under null load, these forces are zeroth. This means that the given system is statically determinate.

10.3 METHODS OF DETERMINATION OF INTERNAL FORCES IN SPACE TRUSSES

In this part of the chapter, we give a brief review of calculation methods for statically determinate space trusses and show that these methods are either based on methods developed for plane trusses or their generalizations.

There is the *method of decomposition of a space statically determinate truss into plane trusses*, which enables us to reduce the 3D problem of analysis to several plane problems. The method assumes the existence of plane statically determinate trusses in the fabric of a space one and requires decomposition of loads into components acting in the planes of such trusses. The method is to calculate plane trusses and implement superposition of internal forces: If the member belongs to several plane trusses simultaneously, then forces are to be summed. If it belongs only to a single truss, then the established force doesn't require conversion. The next assertion is the basis for the method.

Theorem about Decomposition of a Space Truss

If the loading on a space attached statically determinate truss consists of the forces lying in the same plane and exerted upon the joints of a plane truss incorporated into the space one and located in the plane of loads, then loading is taken up only by the members of the plane truss, whereas the forces in all other members of the space truss are equal to zero.

Remark

The theorem's wording implies that a plane truss includes support links; in calculation of this truss, its spherical hinges are replaced by cylindrical ones; and the plane truss is externally stable.

Proof

If the forces in members of a plane truss comply with the statics equations for the plane truss, and the forces in all other members are zero, the equilibrium of joints for the space truss is fulfilled. In statically determinate trusses, the statics equation set has unique a solution, hence, the named forces act in the space truss.

The theorem is proved.

Application of the method of decomposition of a statically determinate space truss into plane trusses is exemplified in the Schwedler truss. Johann Schwedler had suggested this structure in 1866 and implemented it in designing the roofs of Berlin gasholders. A Schwedler truss has the form of a cylinder or truncated dome and consists of several tiers of bar subsystems. Each tier includes an upper bar chord shaped like a regular polygon and rests upon the chord of the lower tier. The elementary bar subsystem, or cell, of the tier is an isosceles bar trapezoid with a diagonal (Figure 10.8). Nowadays, we may run into Schwedler trusses, for instance, in cylindrical frameworks of gasholders. The feature of Schwedler's constructive solution is presented in Figure 10.9. We can see that every bar trapezoid (in the given case, a rectangle) has two diagonal members constructed as a string and working in tension. According to the load direction, only one diagonal does work, whereas another "sags." Consequently, the diagonal members bearing the load at neighboring cells of the tier may have a common joint. The "sagged" diagonals are not included in the design scheme.

The convenient way of manual calculation of Schwedler trusses is to calculate every cell as a plane truss, tier by tier from top to bottom. Initially, let us consider a one-tiered Schwedler truss shaped like a regular triangular truncated pyramid (Figure 10.10a). In order to calculate this structure as the system of three plane trusses, the support links should be positioned as shown in Figure 10.10b: one of the rods is collinear with the meridional member (i.e., the leg of the trapezoid); two others

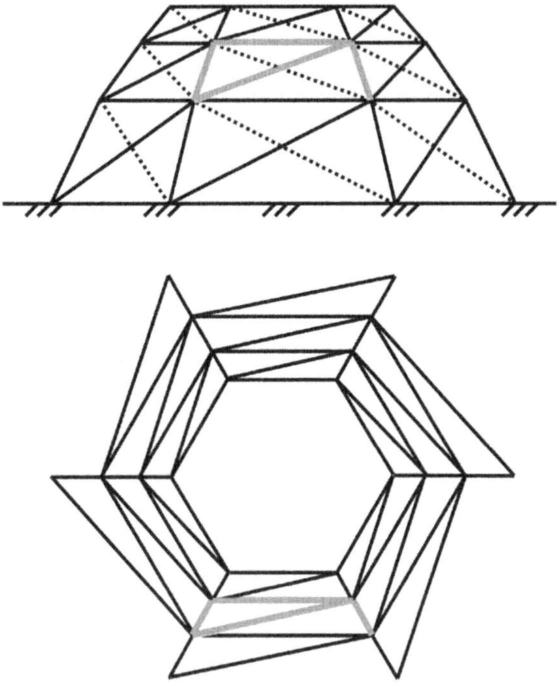

FIGURE 10.8 Three-tiered Schwedler cupola with hexagonal chords (one of the cells is marked in gray).

FIGURE 10.9 Gas holder at Cross Gates, Leeds, UK: On the left – overall view, on the right – view with the cap at its lowest. Photo on the left is own work by Mtaylor848 (License CC-by-SA 3.0. Source: https://commons.wikimedia.org/w/index.php?curid=8047808). Photo on the right by Mr Barndoor (License CC by 3.0. Source: https://commons.wikimedia.org/w/index.php?curid=14980822).

are horizontal and lie in the planes of the side facets of the pyramid. The load exerted on any joint of the given truss can be decomposed into three components lying in the planes of trusses constituting the given one. The internal forces are to be calculated for every plane truss; after that, for members that belong to two plane trusses simultaneously, the obtained forces must be totalized. Figure 10.10 shows the case of vertical load $\boldsymbol{P}$'s exertion upon one of the joints. This load can be decomposed into three components: horizontal $\boldsymbol{H}_1$ and $\boldsymbol{H}_2$, and tilted $\boldsymbol{T}$. The design diagram of loading on the front facet of the pyramid is shown in Figure 10.10b. The calculation of the left rear facet should be done for only the action of load $\boldsymbol{H}_1$.

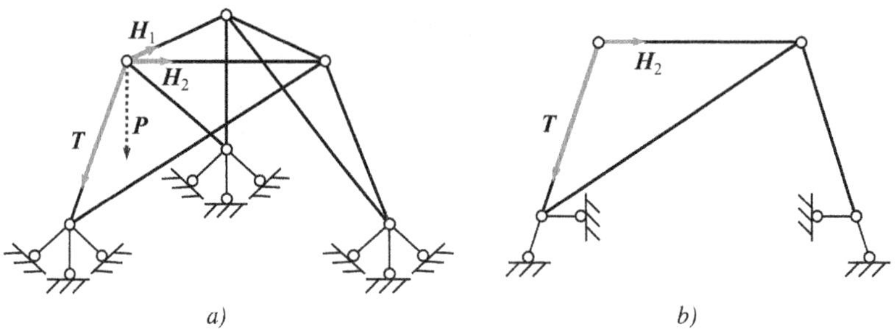

FIGURE 10.10 Space truss (a) made up of three plane trusses (b). The diagrams present the case of vertical load exertion on a joint and corresponding loading on the front plane truss.

When calculating the plane trusses in the given case, the space ball-and-socket supports are replaced by hinged immovable supports with two links for each. The support reactions for plane trusses should be calculated as reactions in horizontal and meridian links. In the transition to space truss, the forces in the meridian support link are to be summed. Note that there is no sense in determining the vertical support reaction for the plane truss, because, in reality, this truss is not vertical.

Now, let the truss in Figure 10.10 be the top tier of a multi-tiered truss, i.e., the chord of the lower tier is taken for the earth. In this case, the evaluation of forces is not changed, because, due to static determinacy of the top tier, the forces in its members do not depend on the strain state of the lower part of the truss. After calculation of the top tier, the obtained support reactions define the loads exerted upon the lower tier by the top one in accordance with Newton's third law. Therefore, it is possible to calculate tier by tier from top to bottom. This technique of analysis is also applicable for a truss with tiers with more than three cells.

To determine the forces in space trusses, there are methods based on the method of sections that are similar to the methods developed for plane trusses. Part of a statically determinate truss is separated by some surface. If the surface dissects six members exactly, then we may set up the equilibrium equations for the cut off part of the structure in unknown axial forces acting in these members and find the wanted forces by solving these equations. If each equilibrium equation is set up by totalizing the moments relative to some axis, then we have the *method of moments*. This method is realized on the basis of recommendations about choosing of axis, given in the previous item of the chapter and is analogous to the method of moment point. If each equation of balance is set up for projections of all internal and external forces on some axis, then we have the *method of projections* that is similar to the method of projections for planar problems. Often, the equation set in unknown forces includes the equations of both methods.

If an intersecting surface cuts off the joint of the truss, then we have the *method of joints*. This method assumes setting up and solving equations of force balance for cut-off joint. This method gives three equations for a joint and enables us to determine axial forces in dissected members if there are but three of them and they do not lie in the same plane. The method is also applicable if a joint is formed by more than three members, but the forces are unknown only in three of them. In the latter case, the members with known internal forces must be considered as the sources of the load upon the joint – in addition to the given loading on the truss. For determination of forces in three dissected members, it is sufficient to decompose the load exerted on the joint into components in the directions of these members. We denote the direction vectors of members forming the joint under consideration C as $\boldsymbol{n}_1$, $\boldsymbol{n}_2$, $\boldsymbol{n}_3$ (Figure 10.11a), and represent the decomposition of the load into these directions in the form:

$$\boldsymbol{P} = P_1\boldsymbol{n}_1 + P_2\boldsymbol{n}_2 + P_3\boldsymbol{n}_3.$$

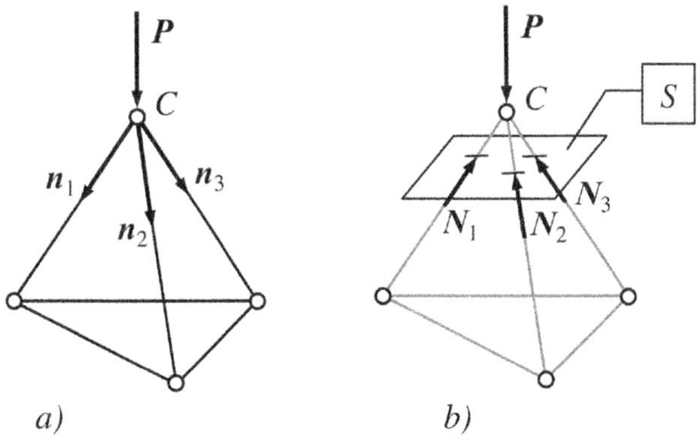

FIGURE 10.11 Direction vectors of members (a) and diagram of loading on joint (b). S is the dissecting plane; $\boldsymbol{N}_i = N_i\boldsymbol{n}_i$ are the vectors of internal forces.

Axial forces in the members are equal to the coefficients of this decomposition with a minus sign:

$$N_i = -P_i,\ i = 1,2,3. \tag{10.8}$$

Really, the conditions of balance for a cut-off joint are reduced to the vectorial equality:

$$\boldsymbol{P} + \boldsymbol{N} = 0, \tag{10.9}$$

where $\boldsymbol{N} = \Sigma N_i\boldsymbol{n}_i$ is the resultant vector of internal forces exerted upon the joint by the members (Figure 10.11b). From this, we obtain the requirement to axial forces:

$$\sum (P_i + N_i)\boldsymbol{n}_i = 0,$$

which, taking into account linear independence of vectors $\boldsymbol{n}_i$, leads to solution (10.8).

Just as in the case of plane trusses, the method of joints is effective subject to the right order of analysis of the joints. For instance, in the case of a simple truss, the joints should be taken in inverse order of building the truss.

For plane trusses, there are two types of joint geometry when the forces are zeroth in known members (see Chapter 8). For space trusses, there exist analogous types of hinged connections with zero force members:

1. If a joint of a space truss is formed by three members not lying in the same plane and is not subjected to a load, then forces in these members are zero. This affirmation follows from conditions of balance (10.8) due to $\boldsymbol{P} = 0$.
2. If all members forming a joint but one lie in the same plane, and load is not exerted upon the joint, then the force in the member directed at a slant to the plane of other members is equal to zero. Let, for example, three members lie in plane Axz and the fourth member make up an angle α with this plane (Figure 10.12). Then equation of force balance for joint A in projection on axis Ay has the form: $N_4 \sin\alpha = 0$, where N_4 is the axial force in the member slanted to the plane. If $0 < \alpha < 180^\circ$, then $N_4 = 0$.

The indications of zero force members not only enable simplification of the calculation of internal forces in a truss, but may be used for kinematic analysis. Let us return to analysis of a Schwedler truss. It was granted that this truss is stable. It is easily seen that the necessary condition of static

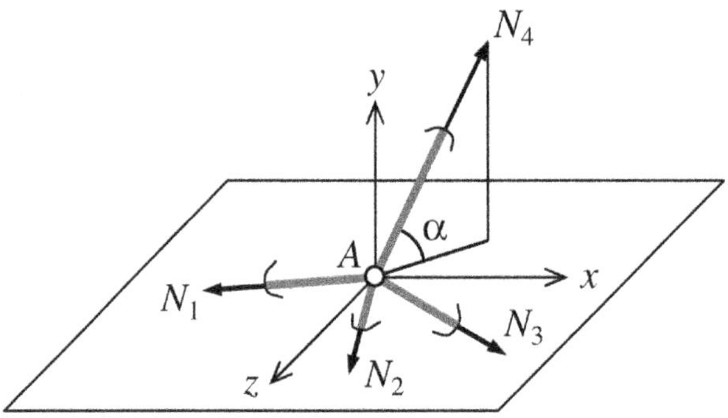

FIGURE 10.12 Joint with zero-force member.

determinacy (10.3) is fulfilled for this truss, and therefore, we may use the method of null load for verification of its stability. Let us establish the members with zero forces, provided there are no loads upon the truss in the simplest case shown in Figure 10.10a. The chord member on the right in the diagram of the truss is a zero-force member in the absence of loads, because this member is slanted to the plane of other members connected at the chord joint on the right (i.e., to the plane of the front cell). This inference is in effect for any chord member of the given truss due to axial symmetry of the third order. Having excluded the chord members from consideration, we see that every chord joint binds up only two members, and therefore the rule of finding zero force members in plane trusses is applicable to these joints. Thus, we come to the conclusion that all members of the truss are not stressed. We may come to the same conclusion if the number of cells in a one-tiered truss is greater than three. In the case of a multi-tiered truss, we sequentially analyze tier by tier from top to bottom, establishing each time that in every tier, all members are not stressed. The absence of initial forces in members means geometrical stability of a Schwedler truss.

Besides methods based on the method of sections and considered above, the *method of substitute members* is also used for calculation of space trusses; this one doesn't differ in substance from the homonymous method for plane trusses.

SUPPLEMENT TO CHAPTER

Representation of a Complete Set of Statics Equations for a Rigid Body in the Form of Equations of Axial Moment Equilibrium

At the beginning, we shall show that the statics equation set (10.5) in unknown forces exerted on a solid body is equivalent to the set of equations (10.6) of axial moment equilibrium if the moment axes are the edges of a tetrahedron (Figure 10S.1a). Note that if, for a given force system, the conditions of balance (10.5) are fulfilled, then the total moment of these forces with respect to the arbitrary axis is equal to zero. Therefore, it is sufficient to establish that the balance of forces corresponding to the complete set of statics equations for a rigid body (10.5) follows from the conditions of moment equilibrium (10.6). The proof is done in two steps:

Step 1: The proof of existence of the force system's resultant complying with conditions (10.6). Let us determine the total moment of a given force system relative to one of the vertices of the tetrahedron, for example, vertex A. It is known that the vector of a total moment of forces relative to a point determines the axial moment relative to the arbitrary axis running through this point as a scalar projection of the moment vector on this axis. It follows from conditions (10.6), that the projections of a moment's vector with respect to point A upon axes constituting the edges of the three-edged angle A equal zero. Since these axes concur at a point but do not lie in the same plane, this means that the moment vector is zeroth. The existence of such a point A, that the moment vector for

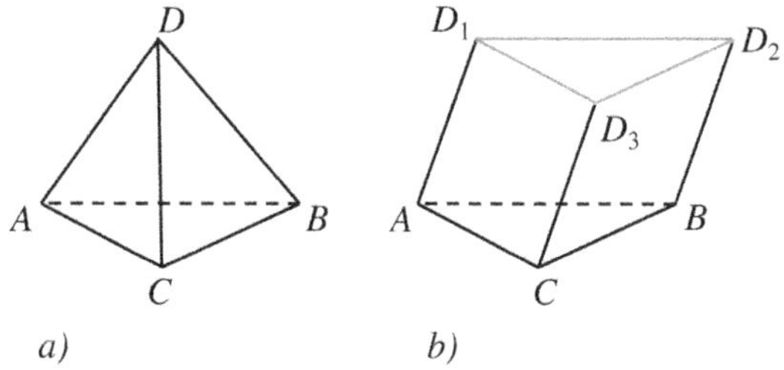

FIGURE 10S.1 Examples of six moment axes position for a body in equilibrium.

a force system relative to this point equals zero, means the existence of the force system's resultant which runs through point A.

Step 2: The proof of equality to zero for the resultant. Assume that the resultant of the force system is nonzero. Let us consider the conditions of moment equilibrium for the resultant force relative to the axes forming the edges of facet BCD (opposite to vertex A). In order that the moment of the resultant be zero relative to axis BC, it is necessary that the vector of the resultant lies in the same plane with this axis, i.e., belongs to facet ABC. From the condition of moment equilibrium relative to axes CD and BD, we establish that the vector of the resultant must belong to facets ACD and ABD. So, the nonzero vector must lie in all facets of the three-edged angle, which is impossible. Hence, the resultant of the given force system is zero.

The inference: the given force system is balanced, i.e., satisfies the complete statics equation set for a body (10.5).

Now, let the totality of moment axes in conditions (10.6) make up the edges of the prism's base ABC and its lateral edges AD_1, BD_2, CD_3 (Figure 10S.1b). We prove that relations of force balance (10.5) follow from conditions (10.6). The first step of the proof doesn't differ from the previous one; the force system has a resultant with the line of action running through vertex A. At the second step, we establish equality to zero for the resultant by negative proof: In order for its moment be zero relative to axis BC, it should belong to facet ABC. In order for its moment to be zero relative to axis CD_3, it should belong to facet AD_1D_3C. To belong to both these facets, the resultant should be directed along edge AC. But in this case, it doesn't lie in the same plane with axis BD_2, about which its moment is also assumed to be zero. Hence, the resultant is zero and the given force system is balanced. Thus, the conclusion: equation sets (10.5) and (10.6) are equivalent.

Section VI

Energy Methods in Deflection Analysis

This section is devoted to the theory of elastic systems. The superposition principle is stated, and the relationship between the strain energy of elastic system and internal forces in members is established. The section introduces the concepts of generalized force and generalized displacement; the latter ensures the generalization of the theory's results obtained further on. Betti's theorem of reciprocal works is proved and Maxwell's theorem of reciprocal displacements is deduced as a corollary. Mohr's formula for displacement in an elastic system is derived. Also, the section presents the analysis of constraint reactions in an elastic system: it is shown that internal forces at the cross-section of a member are represented as the reactions of connections; Rayleigh's theorems of reciprocal unit reactions and of reciprocal unit reaction and displacement are proved; the formula has been obtained for constraint reaction in an elastic system under given loads.

11 Foundations of Energy Approach in Displacement Analysis of Elastic Systems

11.1 PROPERTIES OF ELASTIC SYSTEMS; STRAIN ENERGY OF BAR SYSTEMS

In mechanics of materials, we have studied the generalized Hooke's law, which establishes a linear relation between stresses and strains in a body's point (Hibbeler 2011, p. 508). A system of interconnected bodies is referred to as linear elastic (or, briefly, elastic) if the generalized Hooke's law is fulfilled for every body of a system. Elastic systems have two important properties:

1. Internal forces acting in elastic systems are conservative. This means that for interaction forces between particles of a system, potential energy is defined. In order to specify the potential energy of a system unambiguously, one has to specify a zero state for which the energy is taken for zero. The potential energy of internal forces counted from an unstrained zero state of an elastic system is called strain energy. Admissibility to specify potential energy for an elastic system is defined by admissibility to specify this energy for each body of a system and by perfection of constraints. (Reactions of perfect constraints not subjected to loads do not perform the work through displacements in a system; the principal types of connections studied in Chapters 1 and 10 are perfect.) In the supplement to the present chapter, it is proved that an elastic (maybe unstable) system is a conservative system, and its potential energy is the sum of the strain energies of the bodies comprising it.
2. An externally stable elastic system complies with the superposition principle: **stresses, strains, and displacements in elastic structures, caused by the totality of loads, can be determined as the sum of stresses, strains, and displacements caused by each of loads acting separately**. The conditions of applicability of the formulated principle lie in the absence of support settlements and thermal impact upon a system; it is also assumed that displacements in a system are sufficiently small (or the loads causing the displacements are sufficiently small). The superposition principle is proved in the theory of elasticity.

Further, we consider elastic structures (i.e., externally stable systems). In the next chapter, we also study elastic – not necessarily stable – systems obtained by removal of some connections from a structure. In particular, we study the processes in elastic systems, i.e., the variation of their state in time. A process in any system of material points is referred to as quasistatic if it evolves slowly enough that the system can be considered to be in an equilibrium state at any moment in time. In conservative systems, the work of external forces through displacements in quasistatic process equals the change of the system's potential energy. We shall evaluate the strain energy as the work of external forces due to quasistatic – i.e., slow enough – loading upon a system.

EXAMPLE

Determine the strain energy of an externally stable elastic body affected by concentrated load P which produces the displacement of a load point in the direction of its action equal to Δ (Figure 11.1).

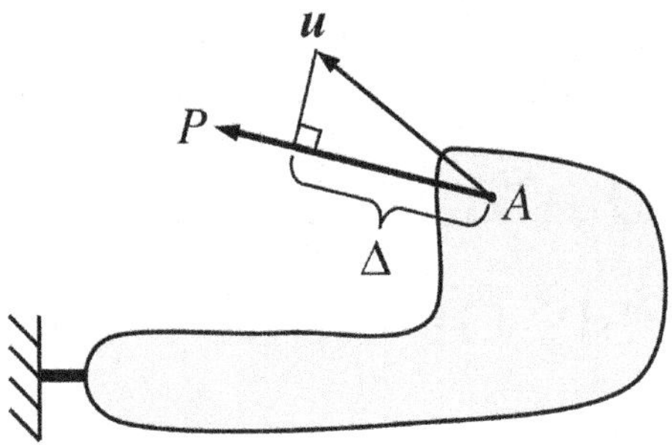

FIGURE 11.1 Displacement caused by force P: $\boldsymbol{u}$ is displacement vector; Δ is displacement in direction of force P's action.

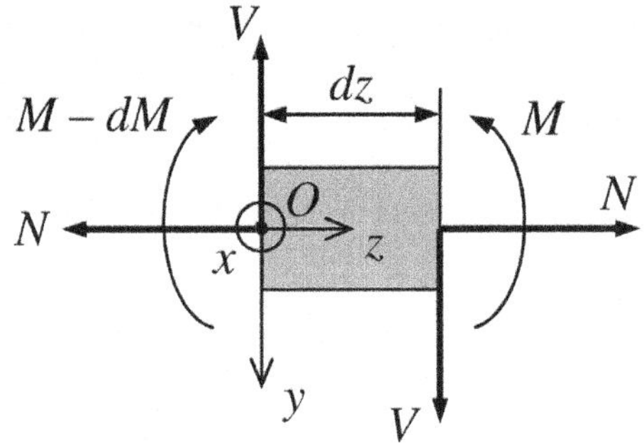

FIGURE 11.2 Internal forces action upon elementary portion of member.

The displacement in a given direction is the scalar projection of the displacement vector onto this direction. Figure 11.1 shows the displacement vector for point A of load exertion and projection of this vector on the load direction at this point. The operator implementing scalar projection upon vector $\boldsymbol{P}$ is denoted as $\mathrm{comp}_{\boldsymbol{P}}(\cdot)$. In the given case $\Delta = \mathrm{comp}_{\boldsymbol{P}}\boldsymbol{u}$. Owing to the superposition principle, the displacements of an elastic body are proportional to loads, i.e., $\Delta = \alpha P$, where α is the proportionality factor. On quasistatic increase of load P to a given magnitude, the work done through infinitesimal displacement $d\Delta$ amounts to $Pd\Delta = \alpha P dP$; the total work done through displacement of the load point equals:

$$W = \int P d\Delta = \int_0^P \alpha P dP = \frac{\alpha P^2}{2}. \tag{11.1}$$

Strain energy equals this work. By excluding coefficient α from the latter relation, we obtain:

$$U = W = \frac{1}{2} P\Delta. \tag{11.2}$$

Formula (11.2) expresses Clapeyron's theorem in the case of single load action on an elastic body: the strain energy of the elastic body is half of the work done by the **constant** external force through displacement of the load application point.

Let us consider a plane elastic bar system which has the plane of symmetry coinciding with a plane of loading. The members of this system work in bending with tension-compression, and three internal forces are exerted at any cross-section of members: M, N, and V. We shall determine the strain energy of the member's elementary portion of length dz, for which purpose we introduce sliding CS $Oxyz$ at the one of boundary sections and direct axis z inside the element (Figure 11.2).

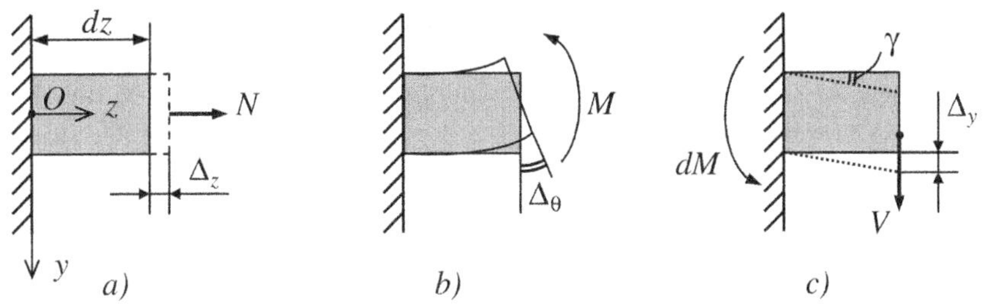

FIGURE 11.3 Deformations of member's elementary portion caused by internal forces acting separately.

This CS serves as the reference system, i.e., the left facet is accepted as immovable. We take as given that external loads are not exerted upon the element, and internal forces on boundary sections are taken for external. We shall determine potential energy as the work of external forces done during the element's deformation.

At the beginning, let us evaluate the work under assumption of independent action for each of the internal forces. In the presence of axial force N only, we have, in accordance with Hooke's law (Figure 11.3a):

$$\Delta_z = \frac{N}{EA}dz,$$

where EA is the axial rigidity of the member (E is Young's modulus; A is cross-sectional area). The latter formula can be written as follows:

$$\Delta_z = \alpha N; \;\; \alpha \equiv \frac{dz}{EA}.$$

The work due to the increase of the axial force by dN, done through corresponding displacement, equals $N d\Delta_z = \alpha N dN$. By calculating the work integral similarly to inference (11.1), we obtain:

$$dW_N = \int N d\Delta_z = \frac{N^2 dz}{2EA}. \tag{11.3}$$

In the presence of a bending moment only, the angle Δ_θ of the right section's pivoting motion relative to the left section (Figure 11.3b) is proportional to this moment. Really, the next equation is known for the radius of curvature of an elastic curve due to symmetrical bending (Hibbeler 2011, p. 572):

$$\frac{1}{\rho} = \frac{M}{EI},$$

where ρ is the radius of curvature, and EI is flexural rigidity (I is the principal moment of inertia of the cross-section). The angle of mutual rotation for two closely located sections can be presented in the form:

$$\Delta_\theta = \frac{dz}{\rho}.$$

Out of this, we obtain the relation of the mutual rotation angle and bending moment:

$$\Delta_\theta = \frac{M}{EI}dz = \alpha M; \;\; \alpha \equiv \frac{dz}{EI}.$$

It is known from theoretical mechanics that the work of a force system with zero resultant, exerted on a solid body, is the product of the total force moment and the angle of the body's rotation in the direction of moment action. Through the pivoting motion of the cross-section by the angle $d\Delta_\theta$, the moment M will perform the work:

$$d(dW_M) = Md\Delta_\theta.$$

The total work of bending deformation is calculated by analogy to (11.1):

$$dW_M = \int Md\Delta_\theta = \frac{M^2dz}{2EI}. \tag{11.4}$$

Now, we consider the action of the shear force only (Figure 11.3c). We simplistically assume that shear stress τ is the same at all points of the cross-section, i.e., it holds:

$$\tau = \frac{V}{A}.$$

Employing Hooke's law for shear, we determine the shear of the right section relative to the left one:

$$\Delta_y = \gamma dz = \frac{\tau}{G}dz,$$

where γ is the shear strain, and G is the shear modulus. This gives the proportional relation:

$$\Delta_y = \alpha V; \alpha \equiv \frac{dz}{GA}.$$

The product GA is called transversal rigidity. The total work of shear deformation is obtained through integration:

$$dW_V = \int Vd\Delta_y = \frac{V^2dz}{2GA}.$$

We can evaluate this work more precisely by taking into account the distinction of the actual distribution of shear stresses from the uniform distribution. We write down the final formula in the general form:

$$dW_V = \frac{V^2dz}{2GA}\eta, \tag{11.5}$$

where correction factor η depends on the shape of the cross-section.

Further, we notice that the work of force N, done through displacements caused by factors M and V equals zero, i.e., the total work of force N doesn't depend on the rest of the exerted internal forces. This also can be said about the work done by forces M and V. As a result, we obtain for the strain energy of the element under consideration:

$$dU = dW = dW_N + dW_M + dW_V.$$

The energy of the whole bar system is written as follows:

$$U = \sum\int_0^l \frac{M^2dz}{2EI} + \sum\int_0^l \frac{N^2dz}{2EA} + \sum\int_0^l \frac{V^2dz}{2GA}\eta, \tag{11.6}$$

where summation is made over all the members and integrals are taken over the entire length of every member. The obtained formula is applicable not only for straight bars but for bars with small curvature.

In the case of a space bar system, we obtain in similar fashion:

$$U = \sum \int_0^l \frac{M_z^2 dz}{2GJ} + \sum \int_0^l \frac{M_x^2 dz}{2EI_x} + \sum \int_0^l \frac{M_y^2 dz}{2EI_y} + \sum \int_0^l \frac{N^2 dz}{2EA} + \sum \int_0^l \frac{V_x^2 dz}{2GA} \eta_x + \sum \int_0^l \frac{V_y^2 dz}{2GA} \eta_y. \tag{11.7}$$

Here, J is the torsional constant of a member (Case, Chilver, and Ross 1999), and I_x and I_y are the principal centroidal moments of inertia of a cross-section.

11.2 GENERALIZED FORCE AND DISPLACEMENT; THEOREMS OF RECIPROCAL WORKS AND RECIPROCAL DISPLACEMENTS

Let us introduce the concepts of generalized force and generalized displacement, which are linked to some assemblage of forces acting in a mechanical system and are used in conjunction.

We say that a scalar or vector characteristic of a force system satisfies the superposition principle if this characteristic, taken for two systems of forces exerted upon an object conjointly, equals the sum of its values for these force systems exerted separately. Examples of such characteristics include the resultant vector of an arbitrary system of forces and the algebraic value of the total moment of any plane force system with a zero resultant.

The scalar characteristic P of a force system $\boldsymbol{P}_1, \boldsymbol{P}_2, \ldots, \boldsymbol{P}_n$, exerted at known points of the attached mechanical system in given directions, is referred to as *generalized force* for this assemblage of forces if it determines this force system uniquely and satisfies the superposition principle.

Let us specify the positive direction of each force under consideration $\boldsymbol{P}_i$ by unit vector $\boldsymbol{n}_i$, i.e., each force is defined in the form $\boldsymbol{P}_i = P_i\boldsymbol{n}_i$, where P_i is the algebraic value of the force ($P_i < 0$ is allowed). We introduce the denotations of displacements in the given mechanical system as follows: $\boldsymbol{u}_i$ is the displacement vector for the point of force $\boldsymbol{P}_i$ exertion, and Δ_i is the scalar projection of displacement vector $\boldsymbol{u}_i$ in the positive direction of the corresponding force, i.e., $\Delta_i = \text{comp}_{\boldsymbol{n}_i} \boldsymbol{u}_i$ (Figure 11.4). A linear function of displacements $\Delta_1, \Delta_2, \ldots, \Delta_n$ is referred to as *generalized displacement* Δ *for generalized force* P if the product $P\Delta$ determines the work done by the assemblage of forces $\boldsymbol{P}_1, \boldsymbol{P}_2, \ldots, \boldsymbol{P}_n$ when the system moves through infinitesimal displacements allowed by constraints in this system:

$$\sum P_i \Delta_i = P\Delta.$$

Displacement Δ is also referred to as *displacement in the direction of generalized force P's action*.

EXAMPLES

1. Let the force couple produce concentrated moment M_c at some cross-section of a beam. The actual displacements of a beam are accompanied by a pivoting motion by an angle θ (Figure 11.5). The work done by this force couple through actual displacements equals $M_c\theta$. In the given case, moment M_c could be called generalized force and angle θ generalized displacement for the force couple at the given cross-section.
2. Two opposite forces of magnitude X, acting in the replacing truss in Figure 8.5b along the line of member 9 removed from the original system, could be called generalized force if

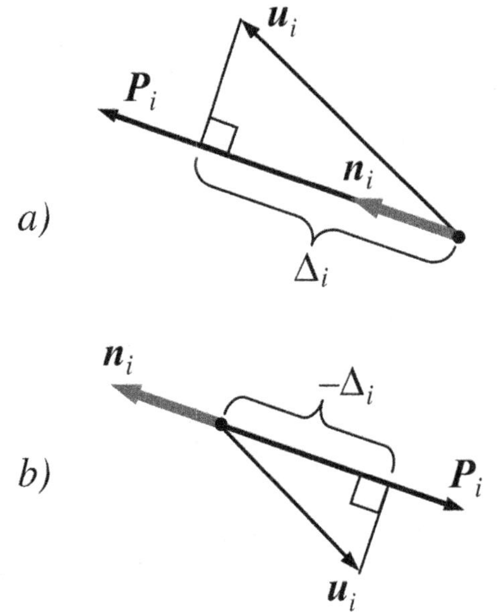

FIGURE 11.4 Instances of forces and displacements for the i-th point of a system: (a) Case $P_i > 0$; $\Delta_i > 0$; (b) Case $P_i < 0$; $\Delta_i < 0$.

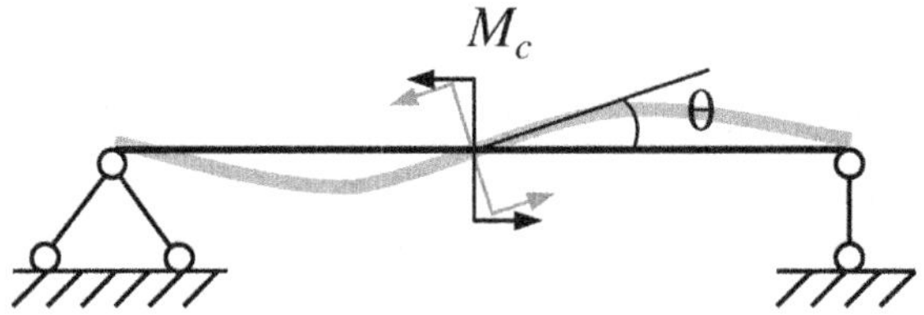

FIGURE 11.5 Example of generalized force M_c and corresponding generalized displacement θ.

the corresponding generalized displacement is measured by the diminution of the distance between the joints b and e. Really, introduce the displacement in the direction of force X as the difference of segment lengths:

$$\Delta_X = be - b_1 e_1,$$

where b and e are the points of joint location in an unstrained truss, and b_1 and e_1 are the same points after displacement. Then the work of forces X done through displacements of joints b and e is the product $X\Delta_X$.

3. For the mechanism in Figure 6.7, the work done by external moments X through angular displacements is equal to $X\Delta_X$. These moments could be called generalized force, and angle Δ_X of members' mutual rotation in the mechanism could be called generalized displacement.

The concepts of generalized force and generalized displacement provide an easy way to generalize the results obtained in the theory of elastic systems' displacements. These results will be further obtained for forces and displacements in their usual meaning, but the similar statement always happens to be right for generalized forces and displacements. Admissibility of expanding the result to generalized forces and displacements will be further stated with no proof.

Note that generalized forces and displacements employed in structural mechanics are not connected with generalized coordinates of a mechanical system and corresponding generalized forces being studied in theoretical mechanics.

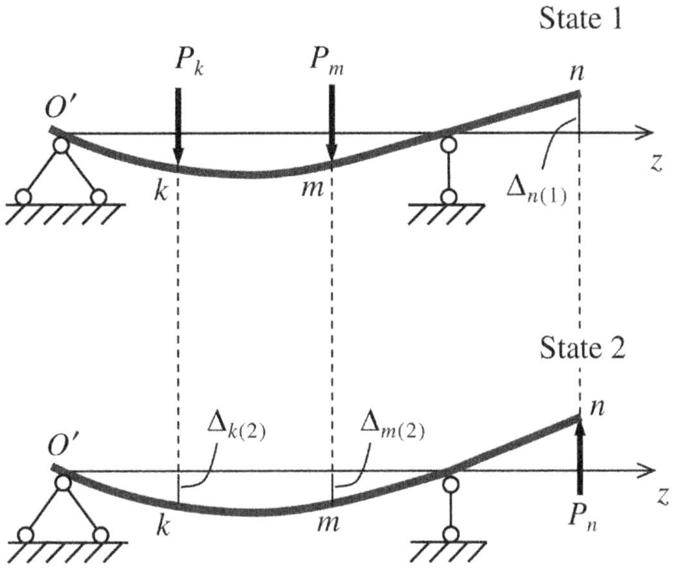

FIGURE 11.6 Two states of a beam under loads.

In 1872, Italian mathematician Enrico Betti proved the following theorem, also known as Betti's law, which serves as the basis for the theory of elastic system displacements.

Betti's Theorem of Reciprocal Works

In an elastic structure, the virtual work done by external forces of state 1 through displacements in state 2 equals the virtual work done by external forces of state 2 through displacements in state 1.

We clarify the sense of this assertion as follows. We understand the totality of exerted loads and the SSS produced by them as the state of a system. The work of forces in the given loaded state of a system done through infinitesimal displacements allowed by constraints in the system is referred to as *virtual work*. In the theorem's assertion, virtual work is done by external forces in one state of a system through displacements arisen in another state of the system. The displacements are assumed to be small enough for applicability of superposition principle to these two states. While calculating the work of external forces, it is sufficient to take into account the work done by loads only (support reactions do not perform the work because displacements of support connections do not arise in the directions of their reactions).

The states for which the work is calculated are pointed out by subscripts at the denotation of the work. Usually, the first index marks the state, of which the forces do the work, and the second the state for which the displacements are taken in calculation of the work. The formulated theorem in these denotations is represented by the relation:

$$W_{12} = W_{21}. \tag{11.8}$$

Figure 11.6 shows an example of displacements for an overhang beam in two states. In the case presented, equality (11.8) can be written in the form:

$$P_k\Delta_{k(2)} + P_m\Delta_{m(2)} = P_n\Delta_{n(1)},$$

where the first index of the displacement notation indicates the point where displacement is observed, and the second index (in parentheses) indicates the number of the state in which displacement arises.

Proof of Betti's Theorem

We denote by W the work wanted to put a system into superposition of states 1 and 2 (i.e., when the loads of states 1 and 2 are exerted together), and by W_1 and W_2 the work expended for implementation of states 1 and 2 in the system, respectively. Due to transition from an unloaded state to state 1, work W_1 will be done; on the next superimposing of the loads of state 2, the additional displacements will occur in accordance with this state. Through these displacements, the loads of state 2 will perform work W_2, and the loads of state 1 will perform work W_{12}. Thus we obtain:

$$W = W_1 + W_2 + W_{12}. \tag{11.9}$$

Now, let us change the order of the loads' exertion; initially, we shall put the system into state 2 and next additionally apply the loads of state 1. The total expended work will be now as follows:

$$W = W_2 + W_1 + W_{21}.$$

In both cases, the work done for superimposing states 1 and 2 is the same, because, for an elastic system, the internal potential energy can be defined. Out of this, equality (11.8), which was to be proved, holds.

The proved theorem will be required in the following simplified formulation.

Betti's Theorem for the Case When Each State of a System is Created by Single Load

If two external forces are applied at any points of the elastic structure, the work done by first force through displacement caused by second force is equal to the work done by second force through displacement caused by first force.

Let us represent the equality for reciprocal works in this case by involving the displacements.

We denote displacement of the solid body's point m in a specified direction, caused by the load exerted at point n, as Δ_{mn}. We specify the direction of displacement for point m by the load exerted at this point. If there are forces P_m and P_n exerted in the given directions at points m and n of a system, the equality (11.8) can be written in the form (Figure 11.7):

$$P_m \Delta_{mn} = P_n \Delta_{nm}. \tag{11.10}$$

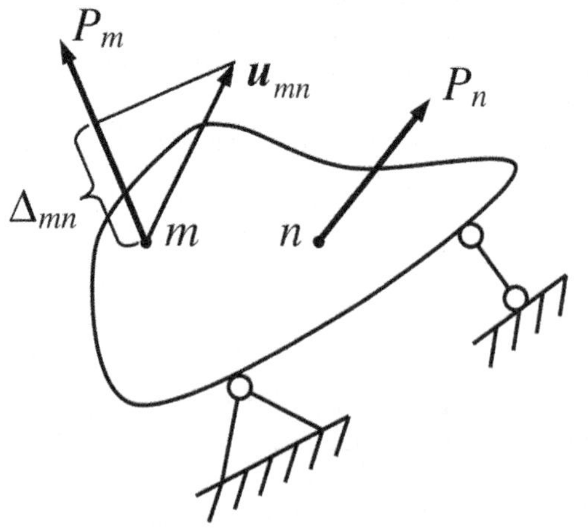

FIGURE 11.7 Displacement caused by force P_n: u_{mn} is the displacement vector; Δ_{mn} is displacement in direction of force P_m.

Betti's theorem remains correct if generalized forces are considered and virtual work is calculated through generalized displacements. For example, in relation (11.10) the generalized force P_m might be the concentrated moment exerted in elementary volume of a body and generalized displacement Δ_{mn} in this case would be the rotation angle for elementary volume in the plane of the moment's action.

In 1864, English physicist James C. Maxwell, best known for his creation of electromagnetic field theory, proved the theorem about reciprocal displacements which still has practical meaning in elastic system analysis. This theorem is studied as the corollary to Betti's theorem, though historically it was proved before the latter.

Maxwell's Theorem of Reciprocal Displacements

In elastic structures, the displacement of point *m* due to a unit load acting at point *n* is equal to the displacement of point *n* due to a unit load acting at point *m*.

Proof

We keep the denotations used in equality (11.10) and denote by δ_{mn} the displacement of point *m* in the direction of force P_m due to the force $P_n = 1$. For any force P_n we have:

$$\delta_{mn} = \frac{\Delta_{mn}}{P_n}. \tag{11.11}$$

We term the ratios of form (11.11) *unit displacements*. Maxwell's theorem states that:

$$\delta_{mn} = \delta_{nm}. \tag{11.12}$$

This equality is readily apparent from (11.10), which was required.

The theorem of reciprocal displacements remains correct for generalized forces and corresponding displacements. The measurement unit of generalized unit displacements (11.11) depends on the sense of the numerator, which may be linear or angular displacement, and of the denominator, which may be a force or concentrated moment of forces. Therefore, the SI measurement unit for quantity δ_{mn} may be $\frac{\mathrm{m}}{\mathrm{N}}, \frac{1}{\mathrm{m}\cdot\mathrm{N}}, \frac{1}{\mathrm{N}}$.

Figure 11.8 clarifies this theorem with an example of beam bending. In the given case, the unit displacements are deflections and the next equality holds: $\delta_{12} = \delta_{21}$.

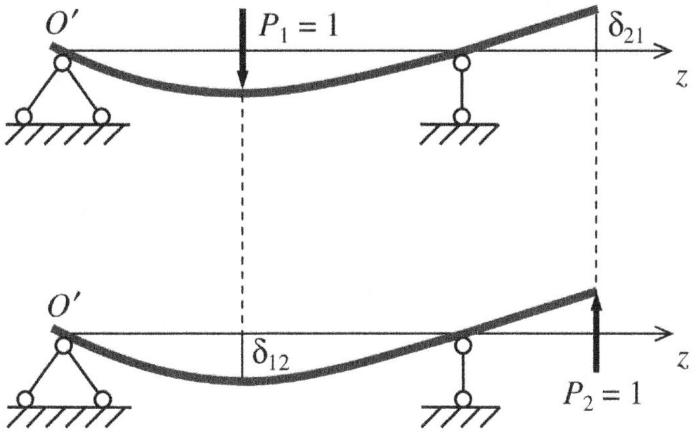

FIGURE 11.8 Two states of a beam caused by unit loads.

11.3 RECIPROCAL WORK CALCULATION AND MOHR INTEGRAL

Further, the dependencies of internal forces in bars versus coordinate z of the cross-section are briefly termed the diagrams of forces. For any bar system, we can deduce the relationship which determines the virtual work through the diagrams of internal forces, and, as a result, we can obtain the expression for displacement at any point of a system through these diagrams.

In formula (11.9) works W, W_1, and W_2 are equal to the system's potential energy in the states attained by doing these works. The potential energy of a plane bar system is determined according to formula (11.6) where, for the sake of brevity, we shall take into account only the first sum. Hence, we obtain the formula for virtual work:

$$W_{12} = W - W_1 - W_2 = \sum \int_0^l \frac{(M_1 + M_2)^2 - M_1^2 - M_2^2}{2EI} dz = \sum \int_0^l \frac{M_1 M_2}{EI} dz, \tag{11.13}$$

where the subscript at denotation of the bending moment points out the number of state for which this moment is defined.

Let it be required to determine the displacement in the given direction for some point m of a bar system. We call the system's state caused by force $P_m = 1$ exerted at this point in the required direction "state 1." Let us denote the bending moment and axial and shear forces at the cross-sections of members in this state as $\overline{M}_m$, $\overline{N}_m$, $\overline{V}_m$, respectively (the dash at the top marks that internal force is caused by the unit load; the subscript marks the point where the load is exerted). We call the system's state due to real loads, in which the sought displacement Δ_{mP} is observed, "state 2." Let us denote internal forces at the cross-sections of members in this state as M_P, N_P, V_P, respectively. The virtual work in the case under consideration happens to be equal to displacement Δ_{mP}, because $W_{12} = P_m \Delta_{mP} = \Delta_{mP}$. Formula (11.13) determines this displacement in the form:

$$\Delta_{mP} = \sum \int_0^l \frac{\overline{M}_m M_P}{EI} dz. \tag{11.14}$$

The obtained expression is known as *Mohr's formula*. If we take into account the addends including axial and shear forces in formula of virtual work (11.13), then Mohr's formula takes the form:

$$\Delta_{mP} = \sum \int_0^l \frac{\overline{M}_m M_P dz}{EI} + \sum \int_0^l \frac{\overline{N}_m N_P dz}{EA} + \sum \int_0^l \frac{\overline{V}_m V_P dz}{GA} \eta. \tag{11.15}$$

The right-hand side of Mohr's formula is referred to as the Mohr integral. The formulas (11.14)–(11.15) are named after German civil engineer and professor of mechanics Otto Mohr, who developed the theory and analysis of statically indeterminate structures by means of these formulas in the 1870s.

REMARK

"Unit" internal forces $\overline{M}_m$, $\overline{N}_m$, $\overline{V}_m$ are correctly determined through corresponding forces M_m, N_m, V_m caused by load P_m of arbitrary value, in accordance with the formulas:

$$\overline{M}_m = \frac{M_m}{P_m}; \; \overline{N}_m = \frac{N_m}{P_m}; \; \overline{V}_m = \frac{V_m}{P_m}.$$

Therefore, unit moment $\overline{M}_m$ has the dimension of length, and unit forces $\overline{N}_m$, $\overline{V}_m$ are dimensionless quantities.

Mohr's formula enables us to calculate generalized displacement coupled to generalized force P_m. The physical dimension of unit internal forces depends on the measurement unit of generalized force.

The displacements in frames are determined mainly by deformations of bending, and they can usually be evaluated by simplified formula (11.14). The displacements in trusses are determined by deformations of tension-compression only, and on the right side of (11.15), we would take into account only the second addend.

In the case of a space bar system, Mohr's formula takes the form:

$$\Delta_{mP} = \sum \int_0^l \frac{\bar{M}_{zm} M_{zP} dz}{GJ} + \sum \int_0^l \frac{\bar{M}_{xm} M_{xP} dz}{EI_x} + \sum \int_0^l \frac{\bar{M}_{ym} M_{yP} dz}{EI_y} + \\ + \sum \int_0^l \frac{\bar{N}_m N_P dz}{EA} + \sum \int_0^l \frac{\bar{V}_{xm} V_{xP} dz}{GA} \eta_x + \sum \int_0^l \frac{\bar{V}_{ym} V_{yP} dz}{GA} \eta_y. \tag{11.16}$$

Six addends at the right side correspond to six internal forces acting upon a cross-section of a bar. In practice, it is sufficient to take into account only three first addends in analysis of frames and only fourth addend in analysis of trusses.

SUPPLEMENTS TO CHAPTER

Strain Energy of Elastic Systems

The potential energy of a system of particles is the function of the particles' position such that variation of this function due to displacements of particles is equal to the work done by external forces through these displacements. (External forces are the forces of a system's interaction with other material objects.)

We consider an elastic system, i.e., a system of elastic bodies jointed by connections of principal types. These connections are perfect ones. *Perfect connection is such a connection that the work done by external forces exerted upon it is equal to the work done by reactions of this connection through its displacement.* (The work done by reactions of a simple rod connecting two bodies of a system – whether in spatial or planar cases – is equal to zero, for the action of loads upon this connection is not allowed, see Chapter 1.)

We assume that in the absence of external forces, a system keeps its undeformed state. We prove that potential energy of an elastic system can be specified as the sum of strain energies of bodies belonging to this system. For this purpose, it is sufficient to prove that **the work done by external forces in the quasistatic process of transition from the unloaded state of a system to the state under given external forces is equal to the sum of the strain energies of system's bodies**.

We consider the quasistatic process of loading the system, in which external forces change from zero values to given values. On the termination of the process, there are some displacements and, possibly, deformations in the system. The work done by external forces in this process is the sum of the works done both by external forces exerted directly upon the bodies and external forces acting upon connections of the system. Further, when we speak of external forces, we imply the forces external with respect to a given system of bodies. In the case of an attached system, external forces are divided into loads (the forces specified in the problem) and reactions of support connections.

In the process under consideration, the work done by forces acting upon an arbitrary body of a system is equal to the strain energy of the body. These forces consist of external forces exerted upon the body being considered and the constraint reactions upon this body. Therefore, the strain energy of the i-th system's body is written in the form of equality:

$$U_{si} = W_{Fb\,i} + W_{Ri},$$

where $W_{Fb\,i}$ is the total work done by external forces exerted upon the i-th body in the quasistatic process, and W_{Ri} is the total work done during this process by constraint reactions acting upon the i-th body.

The sum of the works done by all constraint reactions and all external forces, exerted upon the bodies directly, is the sum of strain energies of bodies belonging to the system; this is written in the form:

$$\sum U_{si} = \sum W_{Fb\,i} + \sum W_{Ri},$$

where summing is done over all bodies of the system.

On the other hand, the sum of the works done by all constraint reactions is the sum of the works done by all external forces exerted upon connections, i.e., the equality holds:

$$\sum W_{Ri} = \sum W_{Fc\,j},$$

where $W_{Fc\,j}$ is the total work done by external forces exerted upon the j-th connection; the summing at the right side is done over all connections of the system.

Consequently, together with the works done by external forces exerted on the bodies, the total work of constraint reactions gives the sum of works performed by all external forces:

$$U_{s\Sigma} \equiv \sum U_{si} = \sum W_{Fb\,i} + \sum W_{Fc\,j}.$$

So, the work done by external forces in the quasistatic process is equal to the sum of strain energies of the system's bodies (which is denoted in the latter formula as $U_{s\Sigma}$). Thus characteristic $U_{s\Sigma}$ determines the potential energy of an elastic system. Because quantity $U_{s\Sigma}$ equals zero in an unstrained system, it complies with the definition of the system's strain energy.

Castigliano's Second Theorem

In the theory of elastic systems, there are three known theorems referred to as Castigliano's theorems. Italian mathematician and railroad engineer Alberto Castigliano published these theorems during 1873–1879 as the basis of his method of elastic system analysis. The most renowned of them is Castigliano's second theorem which is formulated and proved below.

Castigliano's Second Theorem

For an elastic structure, the partial derivative of the strain energy with respect to a load value is equal to the displacement at the point of the load's exertion along its line of action.

Proof

Let a structure be subjected to loads $P_1, P_2, \ldots, P_n$, exerted at arbitrary points and acting in different, generally speaking, directions. A structure is strained because of these loads; the displacements of points of loads' application in the directions of corresponding loads are denoted by $\Delta_1, \Delta_2, \ldots, \Delta_n$, respectively. We shall consider the dependence of strain energy versus these loads at their given directions:

$$U = U(P_1, P_2, \ldots, P_n).$$

Castigliano's theorem states that this dependence determines any k-th displacement through its partial derivative:

$$\Delta_k = \frac{\partial U}{\partial P_k}. \tag{11S.1}$$

Denote the strain energy under fixed load magnitudes by U. Give infinitesimal increment to k-th load dP_k and determine the new value of strain energy $U + dU$. We calculate this energy as the work of external forces due to the quasistatic increase of loading from zero loads up to magnitudes of the ones $P_1,\ldots, P_{k-1}, P_k+dP_k, P_{k+1},\ldots, P_n$. Owing to system elasticity, the sequence of load exertion is of no importance. Firstly, we apply load dP_k only, and, in the next stage, apply loads $P_1, P_2, \ldots, P_n$. Denote the change of displacement Δ_k produced by load's increment dP_k as $d\Delta_k$. At the first stage of loading, the k-th external force does the work $\frac{1}{2}dP_k d\Delta_k$. At the second stage, the loads $P_1, P_2, \ldots, P_n$ do the work through the displacements they produce, equal to the strain energy U. But, as distinguished from the case when only given loads P_i are acting, the load of the first stage dP_k is acting at the second stage as well, and the latter load does the work $dP_k\Delta_k$ through displacement of the second stage. The new value of the strain energy, which is determined by the work of external forces, appears to be equal

$$U + dU = U + \frac{1}{2}dP_k d\Delta_k + dP_k\Delta_k.$$

By neglecting the addend of the second infinitesimal order on the right side of this expression and representing displacement Δ_k through increments dU and dP_k, we come to the required relation (11S.1).

The theorem has been proved.

To exemplify Castigliano's theorem, we use it to derive Mohr's formula (11.14). We proceed from the formula for strain energy of a bar system (11.6), where we shall keep only the first sum (the contribution from the bending moment):

$$U = \sum \int_0^l \frac{M^2 dz}{2EI}. \tag{11S.2}$$

Let us apply load P_m at the point of the bar system where the displacement is sought and in the wanted direction. According to formula (11S.1), the sought displacement is obtained by differentiation of the integrand in (11S.2) as follows:

$$\Delta_{mP} = \sum \int_0^l \frac{M}{EI}\frac{dM}{dP_m}dz. \tag{11S.3}$$

In this formula, we take into account that load P_m is fictitious and must equal zero. Thus, bending moment M in the first multiplicand is equal to moment M_P from real loads. The second multiplicand is the ratio of the bending moment caused by infinitesimal load $P_m = dP_m$ to the magnitude of this load. But this ratio does not depend on load P_m and can be considered as bending moment $\bar{M}_m$ due to the unit load. As a result, expression (11S.3) is converted to the form (11.14).

One can easily ascertain that Castigliano's theorem remains correct for generalized loads and corresponding displacements.

12 Reactions of Constraints in Elastic Systems

12.1 REPRESENTATION OF INTERNAL FORCES BY REACTIONS OF CONSTRAINTS AND DISPLACEMENT IN THE DIRECTION OF THE REMOVED CONSTRAINT

In this chapter, we investigate the constraint reactions in an elastic system affected by displacements in the places of other removed constraints and by exertion of loads upon the system. The chapter includes theoretical provisions, namely, the generalized superposition principle and Rayleigh theorems, being set out for a plane system of elastic bodies bound by connections of the first type (links). From these provisions, the design formulas for constraint reactions in a bar system have been derived.

The universal character of connections of the first type is revealed, at first, by the fact that reactions of connections of the second and third types can be represented through reactions of simple rods (because connections of these types can be implemented to conjoin the bodies by the links). Secondly, internal forces at the cross-section of a member can be represented as reactions of simple rods. For such representation, it is sufficient, in a given system being under loading, to select the element of the member in the neighborhood of the cross-section under investigation and substitute it with a rigid connecting device consisting of several links (Figure 12.1). When such replacement is made, the SSSs of given and replacing systems coincide. The brief substantiation of this affirmation is based on the known result of the elasticity theory, according to which the equilibrium of an elastic body is attained by fulfillment of some compulsory requirements imposed both upon external forces and the field of stresses acting in a body (Timoshenko and Goodier 1951, p. 229). This substantiation is as follows.

We shall consider sections *a* and *b* as rigid joints, i.e., the member's portions partitioned by these sections are taken for separate bars. Let, in the replacing system, the displacements be created the same as in a given system; the loadings on both systems are the same; and at the initial moment of observation time, all points of the replacing system are immovable. It can be seen that in this case, the immovability of the replacing system is kept in time, i.e., the replacing system remains in equilibrium. Really, at the initial moment a connecting device keeps balance, because the totality of internal forces acting on sections *a* and *b* is balanced and the device is rigid. On the other hand, for each member of the replacing system, the external forces (constraint reactions and loads upon the member) and stresses in the member secure its balance, for they coincide with external forces and stresses for the same member in the given system being in equilibrium. So, the replacing system keeps balance being in the SSS of the given system. Assuming the state of equilibrium as unique, we infer that the SSS of the given and replacing systems is the same.

It should be noted that the insertion of rigid arrangement *a–b* instead of elastic bar element *ab* means that after unloading the system, the sections *a* and *b* might not return in the former position, and the replacing system with no loading will happen to be stressed. Nevertheless, we neglect initial stresses in the replacing system, because both the extent of elastic element *ab* and the size of cross-section I–I are small enough compared with the length of the member itself.

We refer to a rigid connection of two bodies in a planar system, formed by several links so that removal of any link converts the connection into a mechanism, as a rigid compound node. By inserting a rigid compound node, we can investigate an internal force at a given cross-section of a member as a reaction of the link. Figure 12.2 (on the left) shows the most simple designs of

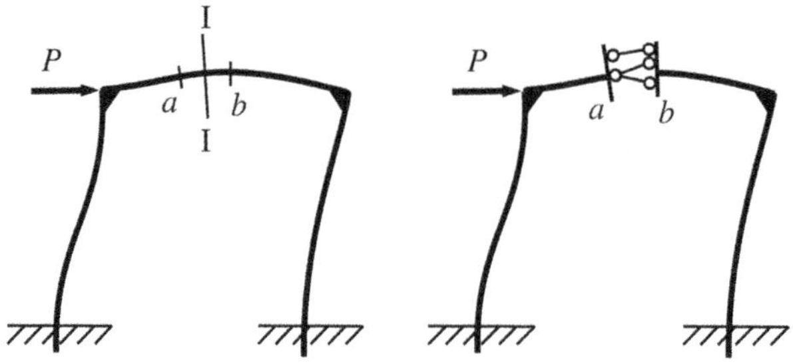

FIGURE 12.1 Addition of new connections forming a rigid joint, with no change of the bar system's SSS: I–I is the cross-section under investigation; *a* and *b* are boundary sections of the removed element.

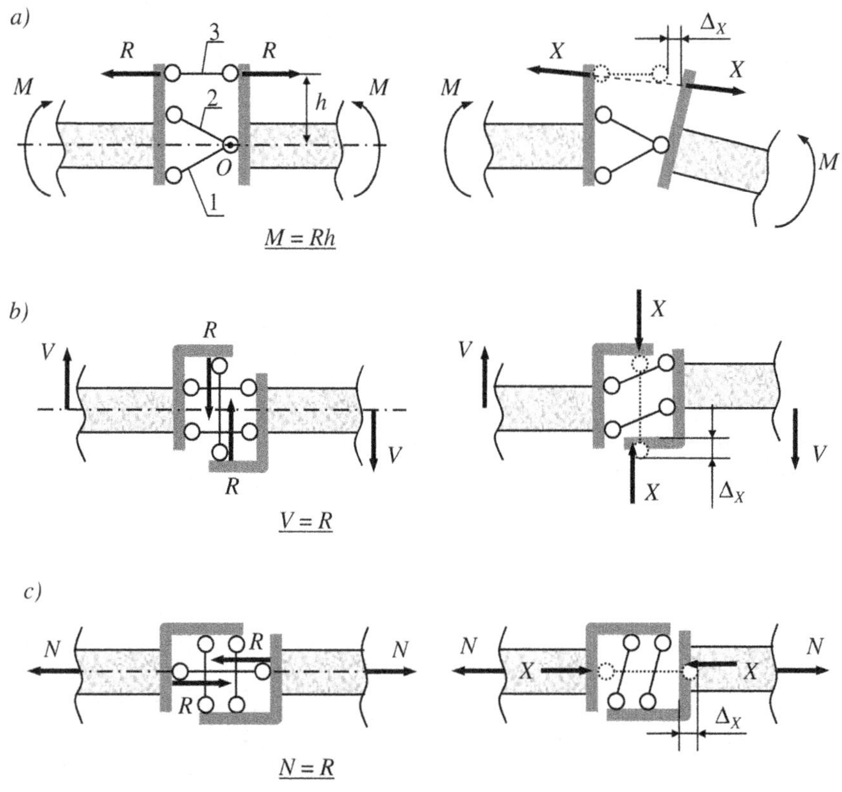

FIGURE 12.2 Representation of internal forces in a member by constraint reactions (on the left) and displacements in the direction of the removed constraint (on the right).

compound nodes, which depend on the wanted internal force. In all presented cases, the dissected parts of the member are connected rigidly, and three reaction forces at either side of the node are statically equivalent to the totality of forces acting at the cross-section of the solid member. In the device in diagram a, point *O* where links 1 and 2 concur, belongs to the axial line of the member, and, as a result, the moment of reaction force developed by link 3 about pole *O* is equal to the bending moment. The same diagram shows reaction *R* made up by two opposite forces and ensuring a balance of moments at either side of the connection. In diagram b, the transverse link of the connecting device develops reaction *R* equal to the shear force, and in diagram c, the longitudinal link of the connecting device develops reaction *R* equal to the axial internal force. *A connection of a rigid compound node developing the reaction which determines internal force at the cross-section*

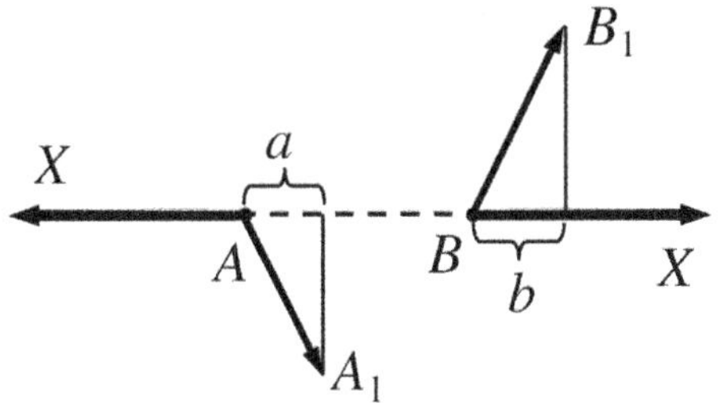

FIGURE 12.3 Displacements of points A and B which were connected by the link.

adjacent to the node is referred to as a coherent connection. Figure 12.2 shows examples of nodes with one coherent connection (in particular, in diagram a, link 3 becomes coherent at $h = 1$).

In the previous chapter, we accepted that the stress state of a system is caused by application of loads. In this chapter, we shall also consider the states caused by displacement at any connection of the first type. It is implied that such a state is created in a system with a removed connection due to external forces substituting for the constraint reaction and causing the definite displacement. The reaction of the link is two equal and opposite forces exerted at the points which are connected by the link. The reaction forces act along the axial line of the link and are defined by algebraic value R and the positive direction taken subjectively. In diagram a in Figure 12.2, the positive direction of the reaction is taken oppositely to the connecting rod; in diagrams b and c, it is taken towards the rod. In a system with a removed connection, substitutive forces X are exerted at the same points and have the same positive directions as the reaction forces; they are equal in magnitude, and may produce displacements of the exertion points. *The sum of displacements determined at the points which were connected by the removed link, in directions of substitutive forces applied at these points is referred to as displacement in the direction of the removed connection.* The usual notation for this displacement is Δ_X where X is the denotation for substitutive forces. The diagrams in Figure 12.2 (on the right) show the action of forces substituting for the reaction of the removed connection and positive displacements corresponding to the directions of these forces. Figure 12.3 presents a more complicated example of defining quantity Δ_X. In this instance, A and B are the points of a removed constraint before displacement; A_1 and B_1 are the same points after displacement; $-a$ and b are the projections of displacement vectors upon directions of substitutive forces. For displacement in the direction of the removed connection, it holds:

$$\Delta_X = b - a.$$

If removal of a constraint results in the permissibility of pivoting motion for one connected member relative to another, then substitutive force can be measured by a moment of this force about the rotation center, and displacement in the direction of the removed connection can be measured by an angle of members' mutual rotation. An example of such a connection is shown in Figure 12.2a. Here we may introduce moment X replacing reaction moment and mutual rotation angle Δ_X for members at the place of removed connection (Figure 12.4).

The magnitude of substitutive forces X together with displacement Δ_X determines the work of substitutive forces by the product $X \cdot \Delta_X$. This means that quantities X and Δ_X are the generalized force and generalized displacement, respectively.

The rigid compound nodes with one coherent connection are used, in particular, in the construction of influence lines for internal forces by the kinematic method. To design statically indeterminate frames, we also shall need the compound node with three coherent connections, which enables us to represent forces N, M, V at the cross-section of a member by reactions of corresponding connections. Such representation is enabled by means of the device in Figure 12.5a. In this device, a new type of connection is employed, which is called the *moment connection.* Moment connection consists

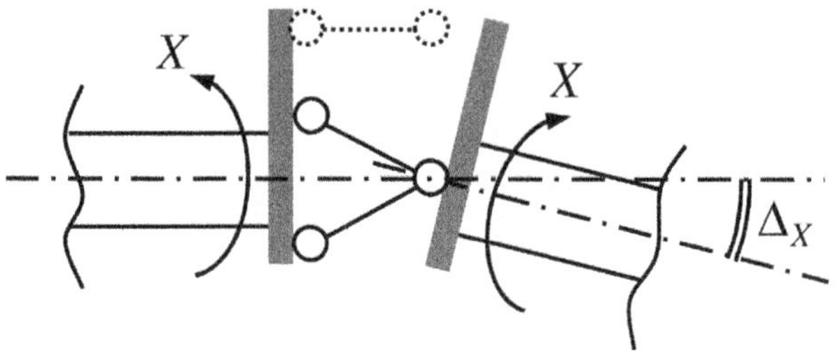

FIGURE 12.4 External moments X substituting for the removed constraint and corresponding displacement Δ_X in the direction of this constraint.

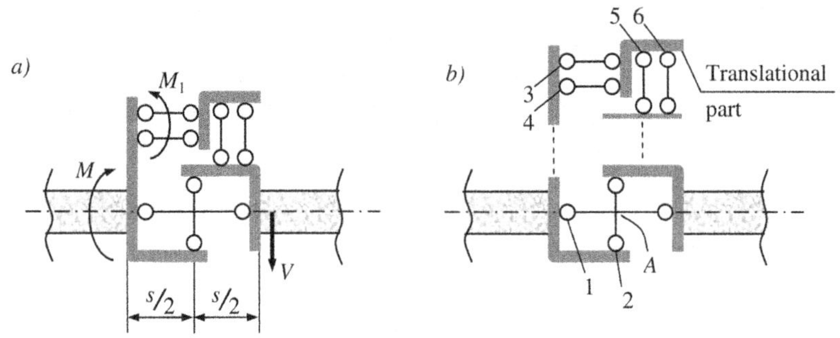

FIGURE 12.5 Rigid compound node with three coherent connections: (a) Scheme of the node (M is the moment at the butt-end of the member to the left of the node; V is the shear force at the butt-end of the member to the right); (b) Scheme of the node with separated moment connection and enumeration of links.

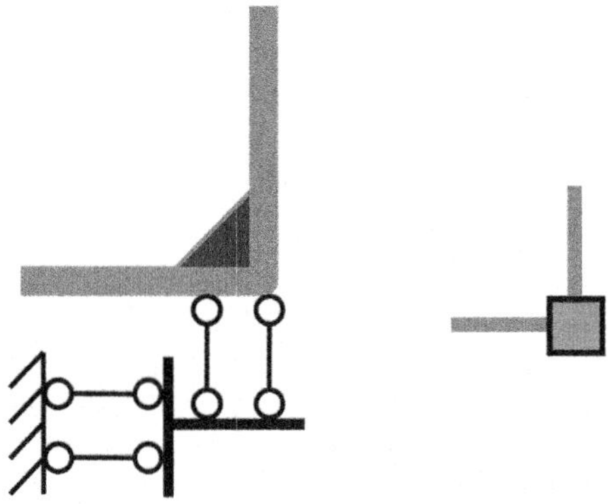

FIGURE 12.6 Floating support (on the right is symbolic notation).

of two couples of parallel simple rods joined by an intermediate translational part (L-shaped, for example) comprising two mutually perpendicular flanges for fastening the hinges. For clearness, in Figure 12.5b the moment connection is depicted separately from the remaining part of the compound node. A support moment connection is called a floating support. Figure 12.6 shows a support moment connection installed at the rigid joint between two members.

A moment connection doesn't allow rotational displacements of one body relative to another but doesn't oppose any translational displacements of bodies connected through it. A moment connection develops a reaction made up by the force couple exerted upon a body from simple rods and measured by a moment of the couple.

In the case of the compound node in Figure 12.5a, link 1 is positioned at the axis of the dissected member and develops the reaction equal to the axial force. Link 2 develops a reaction equal to the shear force, and the moment connection of links 3–6 develops the reaction of a force couple equal to the bending moment. It has to be remembered that the bending moment varies along the axial line of a member subject to simultaneous action of shear force, and, just the same, bending moments at the butt sections of member's parts conjoined by compound node are slightly different. For the reaction of the moment connection in Figure 12.5a it is easy to obtain:

$$M_1 = M + V\frac{s}{2}.$$

If, for example, the connecting device in Figure 12.5a, is inserted instead of element *ab* in Figure 12.1, then the reaction of the moment connection will define the bending moment at section I–I in the middle of the element.

Removal of the moment connection from the node in Figure 12.5 will result in the appearance of a rotation center for one connected bar relative to another at point *A* of the apparent intersection of links 1 and 2. The moment which substitutes for the reaction of moment connection and the angle of bars' mutual rotation at the place of removed connection are the generalized force and generalized displacement, respectively. They are introduced in a similar way as shown in Figure 12.4. Note that in order to eliminate the moment connection, it is sufficient to remove only one of the four simple rods in its composition.

Further on, when discussing connections in elastic systems, we imply simple rods and moment connections.

12.2 RAYLEIGH FIRST THEOREM AND CALCULATION OF REACTIONS DUE TO DISPLACEMENTS IN CONSTRAINTS

Now, we return to the superposition principle stated in the previous chapter. The matter of the considered principle is that superimposing two or more external impacts upon a system – i.e., when two or more totalities of loads are exerted simultaneously – causes superimposing the corresponding responses of a system – i.e., any resulting characteristic of the SSS (stress, strain, displacement) is the sum of values of this characteristic obtained for each totality of loads acting separately. In this case, we may talk about superposition of a system's states, where the term "state" is understood as external impact and responses of a system, i.e., the totality of exerted loads and the SSS produced by them.

The superposition principle is known in different formulations which differ in degree of generality and, maybe, in the objects of applying the principle. For instance, in Chapter 7, this principle was introduced for constraint reactions in statically determinate systems, whereas in Chapter 11, we used *the superposition principle for elastic systems with the state defined by the loads*. In structural mechanics, a few generalizations of the latter principle are also known. In this chapter, we shall employ the generalized principle in which an impact upon a system is defined, besides loads, by given displacements in the directions of removed constraints. In the wording of the principle, these displacements are assumed to be independent, i.e., each displacement can take an arbitrary value in the neighborhood of zero independently from other displacements in the places of removed connections. It is also accepted that an impact defines the SSS of a system unambiguously, whereas the system itself might be geometrically unstable after removal of stipulated constraints. *The superposition principle for elastic systems under the impact of loads and displacements in connections* states: **stresses, strains, and displacements in an elastic system, caused by a number of loads and displacements in the places of removed connections exerted simultaneously, can be determined as the sum of stresses, strains, and displacements caused by separate loads and separate displacements.** Just like in the superposition principle studied before, the displacements are assumed to be sufficiently small. It is allowed that different external impacts may prove to be the displacements exerted in the place of the same connection, and therefore, these displacements are totalized for superimposition of impacts (in the same way the loads are totalized if exerted at the same point).

In Chapter 17, we will study the additional topic of the theory of displacements in elastic systems wherein the generalized superposition principle will be proved for the case when a system with removed constraints remains externally stable. For a structure consisting of a single body, this principle is proved within the elasticity theory.

Note that a displacement in the direction of a removed connection of the first type can be interpreted as the change in length of the rod making up the connection. A displacement in the direction of a removed moment connection can be understood, for example, as the change in angle between flanges of a translational part, which results in the pivoting motion of one connected body relative to another. In all these cases, we speak of displacement in connection itself. If connection is supporting, then such displacements in fact are the support shifts. It is clear, however, that interpretation of an impact of connection upon a system may affect obviousness of reasoning, but doesn't affect the inferences.

On the basis of the generalized superposition principle, we can generalize Betti's theorem proved in the preceding chapter. Let us restate this theorem: "In an elastic structure, the virtual work done by external forces of state 1 through displacements in state 2 equals the virtual work done by external forces of state 2 through displacements in state 1." Formerly, it was sufficient to consider only the loads as external forces, and they alone defined the state of a system. The generalization of the theorem is to interpret its terms in a broader sense: now the state is defined by displacements in the directions of removed connections along with loads, and not only loads, but forces exerted in the places of these connections are taken as external forces. The proof of generalized Betti's theorem is done in the same way as the proof of the "classic" theorem (with no regard to the fact that a structure with removed connections might be an unstable system).

English physicist John W. Strutt, 3rd Baron Rayleigh, in 1874 had published two theorems about reciprocity of reactions and displacements in elastic systems (Lord Rayleigh 1874). These theorems are proved below on the basis of generalized Betti's law. Their significance is that they enable the development of the simple calculation technique for constraint reactions in statically indeterminate systems.

Theorem of Reciprocal Unit Reactions (Rayleigh's First Theorem)

In an elastic structure, the reaction of connection m caused by unit displacement in connection n is equal to the reaction of connection n caused by unit displacement in connection m.

Proof

For clearness of proof, we confine ourselves to support connections. Let us consider hinged movable supports m and n in an elastic system. The first loaded state of a system has arisen from settlement of support m and the second state from settlement of support n (Figure 12.7). Denote the reaction developed by support n in the first state by R_{nm} and the reaction developed by support m in the second state by R_{mn}. The settlements of supports m and n in the directions of the corresponding reactions we denote as Δ_m and Δ_n, respectively. Then the theorem of reciprocal works enables us to establish the relation between these reactions through equality of virtual works:

$$W_{21} = R_{mn}\Delta_m; \; W_{12} = R_{nm}\Delta_n \Rightarrow$$

$$R_{mn}\Delta_m = R_{nm}\Delta_n. \tag{12.1}$$

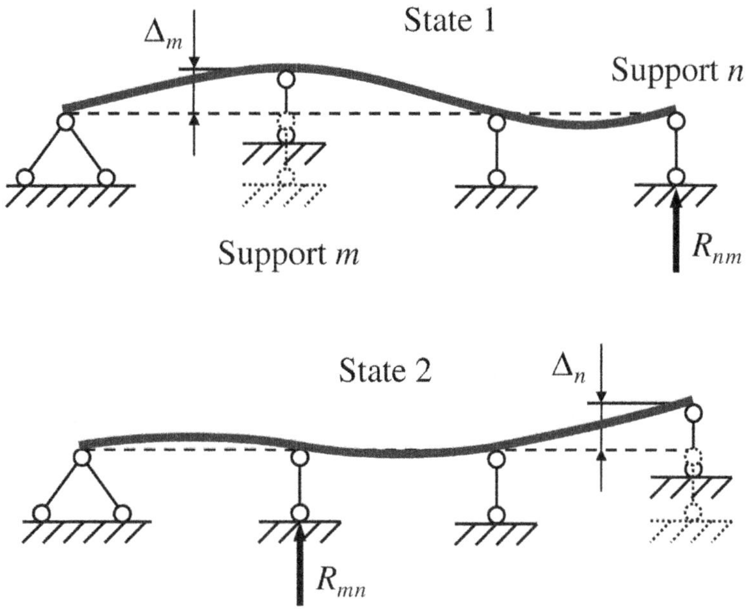

FIGURE 12.7 States of a beam caused by support settlements.

Reactions induced by unit support settlements we denote as follows:

$$r_{mn} = \frac{R_{mn}}{\Delta_n}; \quad r_{nm} = \frac{R_{nm}}{\Delta_m}.$$

From formula (12.1) it holds:

$$r_{mn} = r_{nm}, \tag{12.2}$$

which was to be proved.

The reaction induced by unit displacement in the direction of the removed constraint is referred to as a *unit reaction*. If connections m and n are required for a system, both unit reactions in equality (12.2) are equal to zero.

Constraint reactions induced by displacements in other connections of a bar system are calculated in a similar manner to displacements, through the diagrams of internal forces in members. In order to establish reaction R_{mn} (i.e., the reaction of connection m induced by the known displacement in connection n), we would consider states 1 and 2 of a system caused by displacements in connections m and n, respectively. But, unlike in the proof of Rayleigh's theorem, we have to specify **unit** displacement in connection m for the first state of a system. As a result, the sought reaction will coincide with the virtual work:

$$W_{21} = R_{mn}\Delta_m = R_{mn} \cdot 1.$$

In evaluation of this work, we confine ourselves to frames, for which the SSS is determined, mainly, by bending moments. We introduce the *unit diagram* of bending moments $\bar{M}_m$ for the state of a system evoked by unit displacement in connection m. This diagram is defined by the relation:

$$\bar{M}_m = \frac{M_m}{\Delta_m}, \tag{12.3}$$

where M_m is the diagram of bending moments for the state of the system caused by arbitrary infinitesimal displacement Δ_m in connection m. The diagram of bending moments for the given state (caused by some displacement in connection n) we denote by M_n. For these states, we can approximately calculate virtual work by formula (11.13). Because this work equals the sought reaction, we finally obtain:

$$R_{mn} = \sum \int_0^l \frac{\overline{M}_m M_n dz}{EI}. \tag{12.4}$$

12.3 RAYLEIGH'S SECOND THEOREM AND CALCULATION OF REACTIONS DUE TO LOADING UPON A SYSTEM

Theorem of Reciprocal Unit Reaction and Displacement (Rayleigh's Second Theorem)

In an elastic structure, the reaction of connection m produced by load $P = 1$ is equal to the displacement in the direction of force P, caused by displacement $\Delta_m = 1$ in this connection, and taken with negative sign.

Proof

As before, we confine ourselves to support connections. Let the first state be caused by the action of some load P (not necessarily a unit load) exerted at a known point of a system; the second state of a system arises as a result of some displacement Δ_m of support link m, and this displacement occurs in the positive direction of the reaction's action. The example of these states is shown in Figure 12.8. We denote the reaction of support m produced by load P by R_{mP} and the displacement of the load action point by Δ_{Pm} (in the diagram it is negative). The virtual works of external forces exerted in one state, done through displacements that emerged in another state, are expressed obviously:

$$W_{12} = R_{mP}\Delta_m + P\Delta_{Pm}; \quad W_{21} = 0.$$

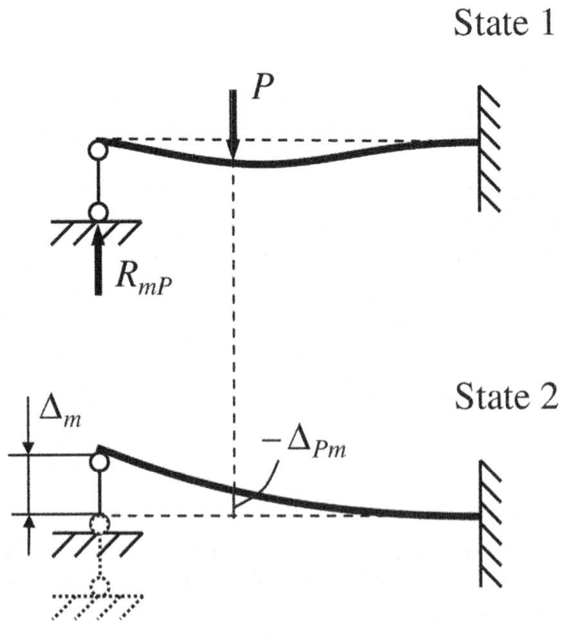

FIGURE 12.8 States of a beam caused by load (state 1) and by support settlement (state 2).

Equating virtual works, we get:

$$R_{mn}\Delta_m = -P\Delta_{Pm}. \tag{12.5}$$

Denote reaction of connection m produced by the unit load P, and displacement of the point of load P's application, induced by the unit displacement in connection m, respectively, as follows:

$$r_{mn} = \frac{R_{mn}}{P}; \quad \delta_{Pm} = \frac{\Delta_{Pm}}{\Delta_m}.$$

From formula (12.5) it follows:

$$r_{mP} = -\delta_{Pm}, \tag{12.6}$$

that was required.

Rayleigh's second theorem enables us to calculate constraint reactions produced by loads in bar systems. To get a handy design formula, displacement δ_{Pm} in (12.6) shall be represented by a Mohr integral. In the next derivation, we restrict ourselves again to analysis of a frame, wherein displacements are defined by bending moments only.

Let us determine displacement Δ_{Pm} of a structure's given point induced by some displacement Δ_m in connection m in the absence of loads. To do that, **we pass to the replacing system which is made by removing connection m from a given system**. We shall assume that the replacing system remains stable. It is allowed to determine displacement Δ_{Pm} for the replacing system wherein there acts an external force at the place of removed connection, which produces displacement Δ_m in the direction of this connection. In the Mohr integral (11.14), the diagrams of bending moments are determined for two states of a structure: a given loaded state and fictitious state when the unit external force is exerted at the point of sought displacement. The given state of the replacing system in our case is defined by displacement in the direction of the removed constraint Δ_m; the corresponding diagram is denoted by M_m.The diagram of bending moments in fictitious state due to the action of force $P = 1$ at a given point will be denoted by $\bar{M}'_P$. Substituting these two diagrams into Mohr's formula, we obtain:

$$\Delta_{Pm} = \sum \int_0^l \frac{\bar{M}'_P M_m dz}{EI}. \tag{12.7}$$

The relation for unit displacement δ_{Pm} is obtained after dividing both sides of this relation by displacement Δ_m in connection m. With the use of denotation $\bar{M}_m$ for the diagram of bending moments in the state of a system caused by unit displacement in connection m, we obtain finally:

$$\delta_{Pm} = \sum \int_0^l \frac{\bar{M}_m \bar{M}'_P dz}{EI}. \tag{12.8}$$

The expression (12.8) is of interest after replacement of displacement by reaction in accordance with (12.6):

$$r_{mP} = -\sum \int_0^l \frac{\bar{M}_m \bar{M}'_P dz}{EI}. \tag{12.9}$$

In the integrand, the diagram $\bar{M}_m$ is a bending moment induced by displacement $\Delta_m = 1$ in connection m of the **given** structure; the diagram $\bar{M}'_P$ is a bending moment due to action of load $P = 1$ at some known point of the **replacing** structure (with removed connection m). The fact that a diagram of internal force is constructed for the replacing structure is marked by a prime symbol next to the denotation of force. We stress that denotation $\bar{M}_m$ has another sense in Mohr's original formula (11.14); this is also a unit diagram of bending moments, but in the state caused by the impact of a unit load upon a structure instead of unit displacement in connection.

So then, we have obtained the expression for dependence of the reaction developed by the m-th connection due to the unit load versus the diagram of bending moments caused by the same load in the replacing structure. If a given structure is subjected to assemblage of several loads, then reaction from them can be established on the basis of superposition principle: this reaction is obtained as the sum of reactions caused by each of the loads exerted separately, in the same manner as the diagram of internal force caused by the totality of loads is obtained by summation of internal force diagrams corresponding to separate loads. After such a summation of diagrams in the integrand of (12.9), we come to *the formula for constraint reaction due to given loads*:

$$R_{mP} = -\sum \int_0^l \frac{\bar{M}_m M'_P dz}{EI}. \tag{12.10}$$

Here M'_P is the diagram of bending moments produced by assemblage of given loads in a system with removed connection m.

In statically indeterminate systems, the construction of bending moments' diagrams may be a laborious problem even after transition to a replacing system with one eliminated constraint. The following assertion makes computing integral (12.10) more expedient:

The replacing structure for calculation of the reaction developed by some constraint m due to loads in an elastic system can be chosen by removal not only of this constraint, but of any other constraints on the condition of geometrical stability of the obtained system. When removing two or more connections from a given structure, diagram $\bar{M}_m$ in (12.10) is calculated for the original structure wherein the unit displacement has been exerted in connection m, whereas diagram M'_P is calculated for a replacing structure.

It is proved as follows. Formula (12.10) is derived from formula (12.8) for displacement in the direction of force P induced by unit displacement in connection m in a given system. The stated assertion will be proved if in the latter formula it is allowed to employ the diagram $\bar{M}'_P$ obtained for a replacing system with several removed connections, subjected to load $P = 1$. We show that it is true by using the example of a continuous beam in Figure 12.9a. This diagram shows the unit load which specifies the point of wanted displacement δ_{Pm} and the positive direction of this displacement, as well as the reaction of support connection m, the direction of which is taken as positive for the support settlement. We choose the replacing system with two eliminated supports m and k – this system is shown in Figure 12.10a under the action of a unit load at the point where the displacement is sought. Figure 12.9b shows the state of the original system with the sought displacement (which is negative in the given case). The same displacement could be produced in the replacing system by external forces exerted in the places of removed connections and equal to support reactions in the given system with unit settlement of support m (Figure 12.10b). In this state, the replacing system is equivalent to the given one in the sense of SSS, and the diagram of bending moments is the same in both systems. The denotation for this diagram was introduced before by formula (12.3).

Thus, it is allowed to determine displacement δ_{Pm} for the replacing system with several removed connections (two in our instance) in the state with the diagram of bending moments $\bar{M}_m$. We find

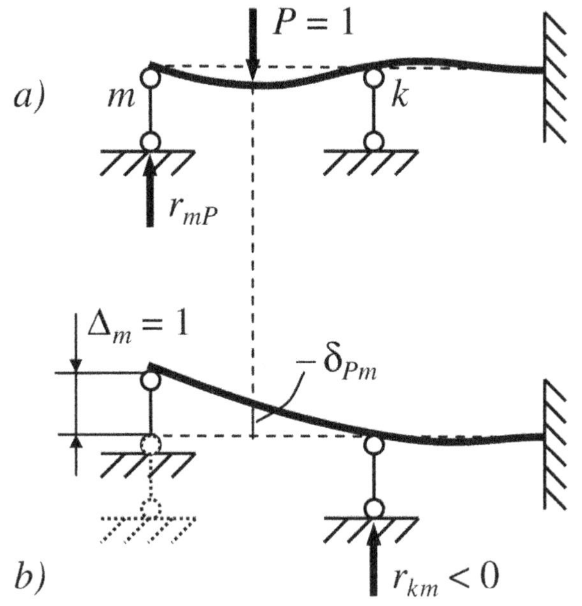

FIGURE 12.9 Reciprocal reaction r_{mP} and displacement δ_{Pm} arisen in two states of a beam.

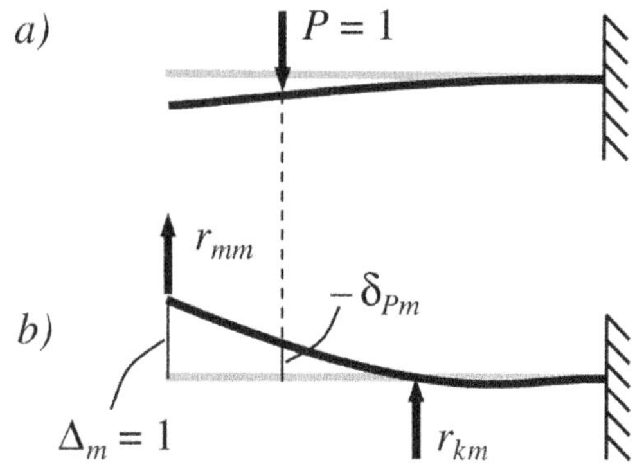

FIGURE 12.10 Replacing beam for calculation of reaction developed by support m of the beam in Figure 12.9.

this displacement by using Mohr's formula, for which we need the diagram of bending moments caused by the unit load, i.e., in the state of replacing system shown in Figure 12.10a. This diagram we denote $\overline{M}'_P$. The substitution of both diagrams in Mohr's formula gives expression (12.8), which was to be proved.

While using formula (12.10), we must remember that it takes into account only the strains of bending. In general case, we have to take into account the strains of tension-compression and shear, by supplementing (12.10) with corresponding integrals.

Section VII

Force Method

This section contains the definitions of primary system and primary unknowns (redundant forces) of the force method and sets out the basic provisions of the force method for calculation of stress state in a statically indeterminate system. The technique of setting up canonical equations of the method is developed, to which end the graph multiplication technique is set out. It is shown in the section that the decomposition of the canonical equation set is possible using the group primary unknowns, and this simplifies the problem of calculation. In order to verify the correctness of solutions obtained by the force method, the special technique has been developed in the section. The section also contains the generalization of the force method for taking into account the temperature changes and support settlements. Further on, the technique of influence line construction is developed for statically indeterminate systems – in particular, the section contains the substantiation of kinematic method for influence line construction by using the force method. At the end of the section, the features of the force method are considered for analysis of space systems.

13 Theoretical Basics of Force Method

13.1 ANALYSIS OF STATIC INDETERMINACY OF BAR SYSTEMS

The subject of the present chapter is the force method, which solves the problem of analysis of elastic statically indeterminate bar systems. This method was introduced by James C. Maxwell in 1864 and refined at the end of the 19th century. It is believed that the greatest contribution to the development of this method was made by Otto Mohr and Heinrich Müller-Breslau (Hibbeler 2012). Nowadays, this method is applied, primarily, as an instrument of qualitative analysis of structures that doesn't require cumbersome calculations subject to a sufficiently simple design scheme.

In this chapter, we consider plane statically indeterminate bar systems. It is appropriate to bring to mind the concept of static indeterminacy from Chapter 1:

If internal forces, acting in a stable mechanical system in equilibrium, for some load assemblage, are not determined from equations of statics, then such a system is called statically indeterminate.

In structural mechanics, it is accepted that the problem of finding internal forces in discs can be reduced to the problem of finding reactions of connections between discs. In the systems under consideration here, a disc is a straight bar in which any elementary portion can be taken as the rigid joint, and thus a bar can be considered as two bars connected by a rigid joint. This means that both the number of discs and number of connections in a bar system are defined subjectively.

The degree of static indeterminacy (or degree of redundancy) is the difference between the number of unknown forces specifying the stress state of a bar system and the number of independent statics equations. **The unknown forces are understood as constraint reactions; the equations of statics are set up for each of the bars interacting through connections.**

In Chapter 1, we obtained the next expression for the degree of static indeterminacy of an attached mechanical system:

$$n = L + 2H + 3R + L_{sup} - 3D. \tag{13.1}$$

The first four addends determine the number of constraint reactions; the fifth term is the total number of statics equations for the number of discs (bars, in our case) D. The obtained difference proves to be the number of the system's redundant connections (being implied as simple rods). The next theorem gives the design formula for the degree of static indeterminacy of a bar system, which is more convenient than (13.1); this theorem also enables us to conclude that the degree of static indeterminacy is an absolute characteristic independent of subjective partitioning of a system into bars.

THEOREM

For a bar system, the degree of static indeterminacy is calculated by the expression:

$$n = 3C - H, \tag{13.2}$$

where C is the number of closed contours, and H is the number of simple hinges in a system.

Proof

The proof consists of verifying coincidence of results obtained by formulas (13.1) and (13.2) for any and all possible systems, which are taken in sequence of increasing complexity. We start from the affirmation that a system attached to the earth by only a clamped bar and shaped as a closed contour without hinges (Figure 13.1a) has degree of redundancy 3. Indeed, in such a system, the number of simple rigid joints is equal to the number of members and there are three support connections, whence by formula (13.1) we have $n = 3$. We obtain the same result by formula (13.2). Next, we discover that attachment of any closed contour without hinges to this system (Figure 13.1b) increases the degree of redundancy by 3, and such a result is obtained by both formulas (13.1) and (13.2). Next, by means of formula (13.1), we establish that addition of a bar jointed rigidly at one end and fixed at another (Figure 13.1c) increases the degree of redundancy by 3. Addition of a bar jointed rigidly at one end only (i.e., cantilever) doesn't change the degree of redundancy. Replacement of an arbitrary simple rigid joint by simple hinge decreases the degree of redundancy by 1 (because the number of constraint reactions decreases). In all these cases, formula (13.2) doesn't contradict formula (13.1). In this way, we can construct an arbitrary bar system in which there are no connections of the first type. One can see, besides, that if in such a system any member with simple hinges at the ends is replaced by connection of the first type, then formulas (13.1) and (13.2) give the same result. Indeed, characteristic n doesn't change due to such replacement: the removal of a member with two hinges decreases the sum (13.1) by one, but after addition of the simple rod, the value of this sum is restored; the sum (13.2) is evidently not changed. Because an arbitrary bar system can be constructed by the listed operations, the assertion of the theorem is proved.

We distinguish external static indeterminacy of a bar system when support reactions are not determined from equations of statics. We distinguish internal static indeterminacy when, under some support reactions, the equations of statics determine reactions of other constraints ambiguously. A system may be only externally indeterminate (Figure 13.2a), only internally indeterminate (Figure 13.2b), and indeterminate both externally and internally (give the example yourself).

Remark

When using formula (13.2), one has to remember that quantity H is the number of simple hinges; if there are multiple hinges in a system, then each of them should be represented by means of the simple ones, and after that, characteristic H should be calculated. For instance, in hinged immovable support of the beam in Figure 13.2a, there is a multiple hinge equivalent to two simple ones. The total number of simple hinges for this beam is equal to 8, and the degree of static indeterminacy is $n = 3{\cdot}3 - 8 = 1$. Make sure that $n = 3$ for the frame in Figure 13.2b.

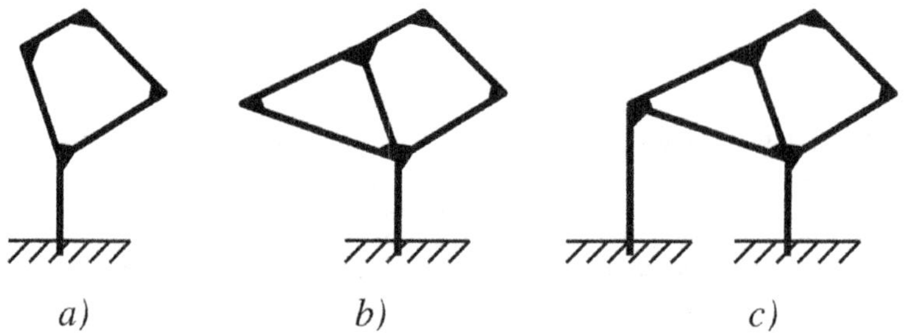

FIGURE 13.1 Gradual complication of redundant structure.

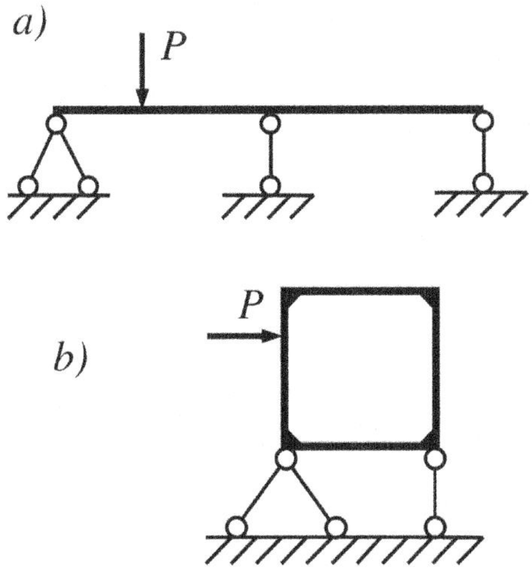

FIGURE 13.2 Externally (a) and internally (b) indeterminate structures.

13.2 FORCE METHOD AND CANONICAL EQUATIONS

Static indeterminacy is caused by redundant (or extra) connections. Below, while speaking of redundant connections, we imply redundant connections of the first type. Connections of the second and third types in a system under investigation we initially represent as the equivalent arrangement of connections of the first type, and after that we proceed with finding redundant connections.

By elimination of redundant connections, one can convert initially indeterminate systems into statically determinate ones. Investigation of a system with removed connections enables us to establish the stress state of a given indeterminate system. Such investigation is made by the *force method*, also known as the *method of consistent deformations*, which is under consideration further on.

In the force method, the bar system obtained from a given indeterminate system by eliminating some redundant constraints and replacing the reactions developed by eliminated constraints with external forces is referred to as a primary system (principal or released structure). In the last chapter, it was explained how to introduce the forces substituting for reactions of removed constraints and the concept of displacement in the direction of the removed constraint was defined. It was pointed out that substitutive force and corresponding displacement have the meaning of generalized force and generalized displacement. *In the force method, the generalized forces exerted in the places of removed constraints are referred to as primary unknowns or redundant forces.*

Before setting forth basic provisions of the force method, we shall introduce some concepts related to the elimination of redundant constraints from the cross-section of a member.

When eliminating redundant constraints, it is acceptable to consider any member as the two separated by cross-section, and one may first insert the rigid compound node into this section and then remove from this node some (or all) connections.

While speaking of elimination of the moment connection in the cross-section of a bar, we imply that the rigid compound node with a coherent connection determining the bending moment is inserted into this cross-section (Figure 12.2a) and then the coherent connection is removed. In fact, this means insertion of the hinge into the given cross-section. Two concentrated moments exerted in the place of the removed connection should be taken in this case as redundant force and an angle of mutual rotation of connected members as displacement in the direction of the removed connection (Figure 12.4).

While speaking of elimination of connections by means of dissecting the bar in given cross-section, we imply that the rigid compound node with three coherent connections is inserted into this cross-section (Figure 12.5a), and after that, all coherent connections are removed. In this case, the redundant forces which substitute for reactions of removed connections are external moments exerted upon cut ends of the bar as well as two external longitudinal forces and two external shear forces. The longitudinal and shear forces produce corresponding displacements in the directions of removed connections, i.e., totalized displacements of cut ends in the directions of longitudinal or shear forces, respectively. As was stated before, the substitutive forces and corresponding displacements at the transverse dissection of the member are generalized forces and displacements. In diagrams, there is no need to display the supporting plates of the compound node from which the links were removed. Assuming the size s of this node is infinitesimal, it would be correct to show substitutive forces and moments as applied to butt-ends of the cut. It is convenient, besides, to consider substitutive shear forces and force couples as exerted at a small distance to the centroid of each butt-end. Owing to this assumption, picturing of the substitutive forces becomes ostensive, and generalized displacements corresponding to these forces are clearly represented in diagrams (Figure 13.3a and b). Such representation doesn't contradict conception of a rigid compound node; it is allowed to situate the links of this node in the neighborhood of the centroid of the cross-section into which it was inserted.

Under denotations of redundant forces accepted in diagrams of Figure 13.3 and in compliance with the sign convention for internal forces in beams, these forces at the sections nigh to butt-ends of the cut are determined by expressions:

$$N = X_1;\ V = X_2;\ M = X_3.$$

The force method is to construct a primary system, determine the primary unknowns which ensure coincidence of the stress state for the primary and given systems, and determine the stress state through established primary unknowns. We say that primary and given systems are equivalent if their stress state is the same. **The primary system of the force method is constructed in such a fashion that displacements in the directions of removed constraints specify primary unknowns unambiguously. Owing to this, the equivalence of primary and given systems is ensured if displacements in the directions of removed constraints in the primary system are absent.**

In the force method, each condition of zero displacement in the direction of the removed connection is represented in the form of an algebraic equation in primary unknowns termed as a *canonical equation*. If the set of canonical equations has a unique solution, then this solution determines the primary unknowns under which the equivalence of primary and given systems is attained. Indeed, the totality of reactions of the removed constraints is the solution of the canonical equation set, because the action of these reactions ensures the absence of displacements in the places of removed constraints. If these reactions in a given bar system are attributed to external forces, then assemblage of external forces exerted upon the primary and given systems will coincide, hence the SSSs of both systems coincide.

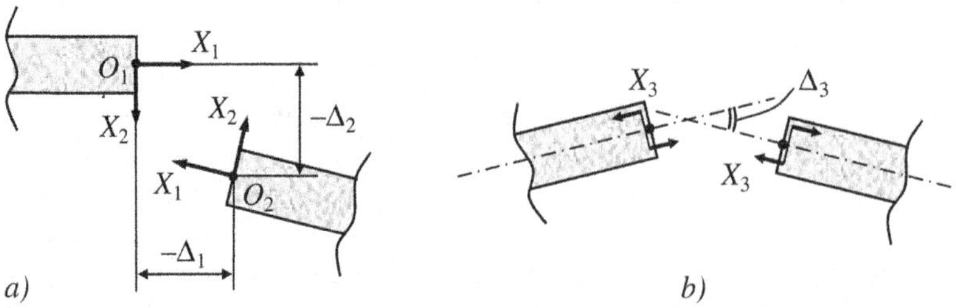

FIGURE 13.3 Substitutive forces at the ends of a dissected member and displacements in the directions of removed constraints: O_1 and O_2 are centroids of sections; X_1 are longitudinal forces; X_2 are shear forces; X_3 are concentrated moments from couples of substitutive forces; Δ_i are generalized displacements (in the shown instance $\Delta_1 < 0$; $\Delta_2 < 0$).

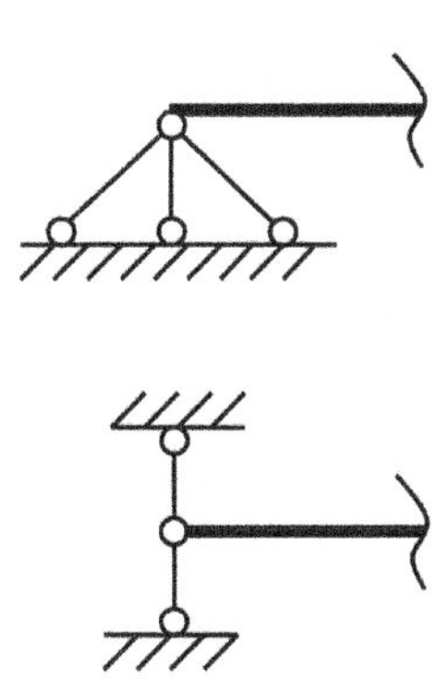

FIGURE 13.4 Incorrect arrangement of redundant connections.

As a rule, a primary system constructed arbitrarily ensures uniqueness of the solution of its canonical equation set. The ambiguity of its solution is evidence of construction flaws or design scheme incompleteness. Two examples of positioning redundant connections when their elimination causes ambiguity of redundant forces are shown in Figure 13.4.*

Usually, one constructs a primary system as statically determinate by eliminating all redundant constraints. The force method will be grounded for this case, which makes substantiation more ostensive.

Firstly, we obtain the set of canonical equations. Let n be the degree of static indeterminacy of a given system. We introduce denotations for the primary system: $X_i,\ i=\overline{1,n}$, are redundant forces in the places of removed connections; Δ_i is the displacement in the direction of the i-th removed connection produced by loads and redundant forces; and Δ_{iP} is the displacement in the direction of the i-th removed connection subject to the action of assemblage of loads only (with no account for redundant forces).

We introduce denotation δ_{ij} for displacement in the direction of the i-th removed connection subject to the action of only the j-th unit redundant force on the primary system. Owing to the superposition principle, displacement is proportional to load. Thus quantities δ_{ij} are determined by equalities:

$$\delta_{ij} \equiv \frac{\Delta_{ij}}{X_j}, \tag{13.3}$$

where Δ_{ij} is the displacement in the direction of the i-th removed connection on the condition that in the primary system only force X_j is exerted, while all other redundant forces and loads are zeroth.

We call displacements Δ_{iP} and δ_{ij} *loaded and unit displacements*, respectively.

The superposition principle enables us to represent displacements from assemblage of both loads and redundant forces in the form:

$$\Delta_i = \sum_{j=1}^{n} \delta_{ij} X_j + \Delta_{iP}. \tag{13.4}$$

We obtain canonical equations expressing requirements of equality to zero for displacements in removed constraints in the form:

$$\sum_{j=1}^{n} \delta_{ij} X_j + \Delta_{iP} = 0, \quad i=\overline{1,n}. \tag{13.5}$$

* Usually it is assumed that all reactions developed by connections in a system are determined uniquely by an impact. Nevertheless, all provisions of elastic systems' theory stated in Chapters 11 and 12 are correct for the systems wherein there may be ambiguity of constraint reactions.

Coefficients δ_{ij} at $i = j$ are termed principal (or main) unit displacements, and at $i \neq j$ secondary unit displacements.

In Chapter 11, we proved the theorem of reciprocal displacements, which is related to quantities δ_{mn} being generalized displacements, the sense of which is specified by subscripts: m points out that the displacement corresponds to some generalized force P_m, and n designates generalized force $P_n = 1$ which evokes the displacement. Coefficients δ_{mn} coincide with unit displacements of the force method by both their denotation and meaning. Pursuant to this theorem, we establish the equality for symmetrically positioned secondary displacements:

$$\delta_{ij} = \delta_{ji},\ i \neq j. \tag{13.6}$$

Figure 13.5 clarifies the sense of canonical equations as conditions of equivalence by example. In diagram a, this figure shows the given system: it is the frame with the degree of static indeterminacy $n = 2$, which is subjected to a distributed load of intensity w and concentrated load P. In diagram b, the corresponding primary system is presented with an indication of external forces, and in diagram c, the same system is shown in a possible deformed state (forces X_i are not depicted). In diagram d, the deformed primary system is shown in the state when it is equivalent to the given system (R_i are constraint reactions).

The generalized displacement in a bar system under the action of a given assemblage of loads can be calculated by means of Mohr's formula (11.15). The loaded displacements are generalized displacements in the primary system and calculated in the form:

$$\Delta_{iP} = \sum \int_0^l \frac{\bar{M}_i M_P dz}{EI} + \sum \int_0^l \frac{\bar{N}_i N_P dz}{EA} + \sum \int_0^l \frac{\bar{V}_i V_P dz}{GA} \eta. \tag{13.7}$$

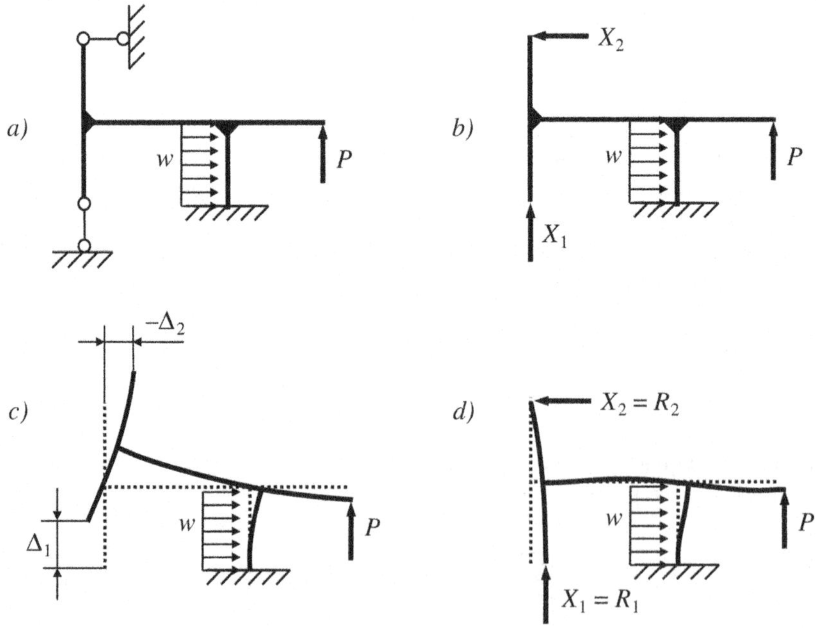

FIGURE 13.5 Given system and primary system of the force method: (a) Given system under loading; (b) Primary system (deformation is not shown); (c) Deformed primary system; (d) Primary system equivalent to the given one.

Here, $\bar{M}_i$, $\bar{N}_i$, $\bar{V}_i$ are the diagrams of internal forces due to action of only the unit generalized force (in given case it is redundant force $X_i = 1$); M_P, N_P, V_P are the diagrams of the same forces due to action of **given** loading. All named diagrams are constructed for the primary system, which is statically determinate. By virtue of this, the problem of computing integral (13.7), as usual, doesn't cause difficulties.

Formula (13.7) is easily transformed for calculation of unit displacements. This formula determines displacements in a primary system under the action of arbitrary loads and redundant forces if the corresponding internal forces are designated as M_P, N_P, V_P. Due to such expanded interpretation of formula (13.7), it holds: $\delta_{ij} = \Delta_{iP}$ – if the single external force $X_j = 1$ acts in the primary system. That is why we have:

$$\delta_{ij} = \sum \int_0^l \frac{\bar{M}_i \bar{M}_j dz}{EI} + \sum \int_0^l \frac{\bar{N}_i \bar{N}_j dz}{EA} + \sum \int_0^l \frac{\bar{V}_i \bar{V}_j dz}{GA} \eta. \tag{13.8}$$

Thus, for evaluation of coefficients and free terms in canonical equations, one has to construct all *unit diagrams*, i.e., the diagrams of internal forces in the primary system produced by each redundant force $X_i = 1$ acting separately; to construct *loaded diagrams*, i.e., the diagrams of internal forces in the primary system produced by loading; and finally, to calculate Mohr integrals (13.7), (13.8) by means of these diagrams.

After solving the canonical equation set, it becomes possible to construct final (or complete) diagrams of internal forces. For example, a final bending moment diagram can be obtained by the formula:

$$M = \sum_{i=1}^{n} \bar{M}_i X_i + M_P. \tag{13.9}$$

For frames, working mainly in bending, in formulas (13.7) and (13.8), it is admissible to restrict ourselves by bending moments, i.e., to take into account the first sum only. For trusses, working mainly in tension-compression, we restrict ourselves in these formulas by axial forces, i.e., take into account the second sum only. The term of shear forces, i.e., the third sum in these formulas, is a small correction and, as a rule, we neglect it in design analysis.

The force method can be developed by means of a statically indeterminate primary system without any changes in equations and design formulas cited above. The only peculiarity is that number n is to be considered as not the degree of static indeterminacy but the number of removed constraints.

13.3 GRAPH MULTIPLICATION TECHNIQUE

Let two physical quantities depend on the coordinate z of the member's cross-section. Let us call the procedure of integrating the product of these quantities along the length of the member *multiplication of diagrams* of these quantities. The determination of displacements by using Mohr integrals assumes construction of diagrams for internal forces produced by some external forces, and multiplication of diagrams. There is a simple technique of diagrams' multiplication in the case when one of them represents linear dependence. This technique was suggested by Russian engineer A. Vereshchagin in 1925 and is stated below.

Let it be required to multiply the diagrams of two functions $f_1(z)$ and $f_2(z)$ on the segment $[0, l]$, i.e., to compute the integral:

$$J = \int_0^l f_1(z) f_2(z) dz. \tag{13.10}$$

We assume that $f_1(z) \geq 0$ and $f_2(z)$ is a linear function:

$$f_2(z) = kz + c. \tag{13.11}$$

The graphs of named functions are depicted in Figure 13.6.

We denote: A_1 is the area of the f_1 diagram, i.e., $A_1 = \int_0^l f_1(z)dz$; $\bar{z}_1$ is the abscissa of the centroid of the f_1 diagram. The wanted integral is computed by the formula:

$$J = A_1 f_2(\bar{z}_1). \tag{13.12}$$

This formula represents the *Vereshchagin rule*:

The result of the multiplication of two diagrams, one of which is the diagram of general form and the other of which is the linear diagram, equals the area of the first of them multiplied by the ordinate of the second one, which is located under the centroid of the diagram of general form.

In order to prove the Vereshchagin rule, we first establish the validity of the formula:

$$\int_0^l z f_1(z)dz = A_1 \bar{z}_1. \tag{13.13}$$

The abscissa of the f_1 diagram's centroid in rectangular CS Ozy is determined by the formula (Hibbeler 2011):

$$S_y = A_1 \bar{z}_1,$$

where S_y is the first moment of the f_1 diagram with respect to axis y. On the other hand, it holds by definition:

$$S_y = \int_{A_1} z dA = \int_0^l z f_1(z)dz,$$

whence follows (13.13).

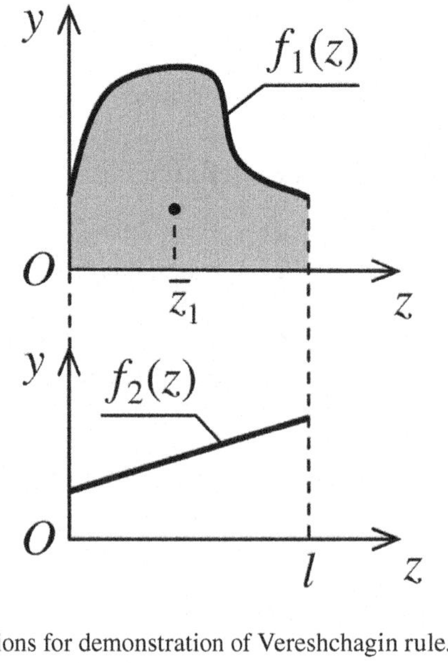

FIGURE 13.6 Graphs of functions for demonstration of Vereshchagin rule.

Now substitute (13.11) into (13.10) and make the replacement (13.13):

$$J = k\int_0^l f_1(z)z\,dz + c\int_0^l f_1(z)\,dz = kA_1\bar{z}_1 + cA_1 = A_1(k\bar{z}_1 + c).$$

Thus, the Vereshchagin rule is proved.

The Vereshchagin rule is effective for computing Mohr integrals (13.7) and (13.8) in the course of finding coefficients and free terms of canonical equations. The matter is that in the case of statically determinate primary system, the unit diagrams of the force method are linear and the calculation of unit and loaded displacements is always reduced to multiplication of the linear diagram and diagram of general form.

Another convenient technique of diagrams' multiplication is based on Simpson's formula. This formula is widely employed for approximate computing a definite integral and has the form:

$$\int_\alpha^\beta f(z)\,dz \approx \frac{\beta-\alpha}{6}\left(f(\alpha) + 4f\left(\frac{\beta+\alpha}{2}\right) + f(\beta)\right). \tag{13.14}$$

It is known that when function $f(z)$ is the polynomial of a degree no greater than 3, the formula (13.14) gives a precise result. In the case of multiplication of diagrams (13.10), Simpson's formula is represented in the form:

$$J = \frac{l}{6}\left(a_1a_2 + 4c_1c_2 + b_1b_2\right), \tag{13.15}$$

where a_i, b_i, c_i are the ordinates of f_i diagrams, respectively, at the left end, at the right end, and in the middle of integration segment, or

$$a_i = f_i(0);\; b_i = f_i(l);\; c_i = f_i\left(\frac{l}{2}\right). \tag{13.16}$$

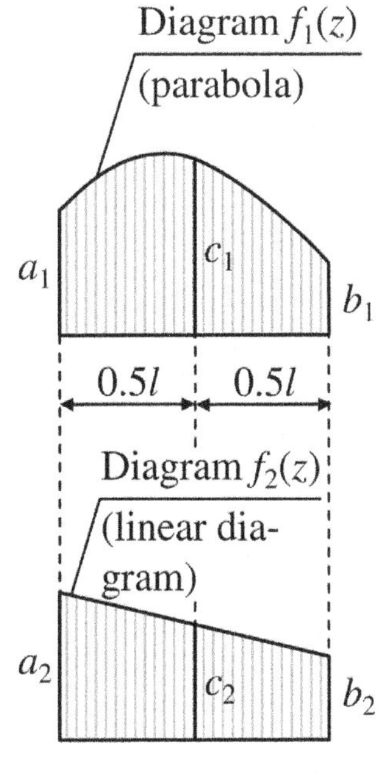

FIGURE 13.7 Example of diagrams to be multiplied by Simpson's formula.

During calculation of Mohr integrals, Simpson's formula is used instead of the Vereshchagin rule if the determination of parameters A_1, $\bar{z}_1$ causes difficulties. But we have to remember that multiplication of diagrams by using Simpson's formula guarantees an exact result only in the case when the function $y = f_1(z)f_2(z)$ is the polynomial of the 3rd degree or less. The example of diagrams permitting multiplication with no error is shown in Figure 13.7.

By means of the Vereshchagin rule and Simpson's formula, one can usually construct a canonical equation set by the graphic-analytical method, i.e., by construction of diagrams and multiplication of the ones with no use of a computer.

14 Simplification Ways of Analysis of Statically Indeterminate Systems

14.1 CONCEPTION OF CANONICAL EQUATIONS' DECOMPOSITION AND TAKING INTO ACCOUNT REQUIREMENT FOR DECOMPOSITION DURING CONSTRUCTION OF PRIMARY SYSTEMS

The usage of the force method in engineering is limited by dimensionality of the problem: even relatively simple design schemes usually have a degree of static indeterminacy of about 10 or greater, which could make manual calculation impossible. At the same time, the employment of the force method in the computer version is not desirable, for this would deprive the method of its main advantage: the ability of quick qualitative estimation of a structure's behavior by means of analytical relations. In the strength analysis on a computer, as a rule, the finite element method is in use, which is studied in one of the following topics.

Consider the example of a steel-monolithic framework of the two-story civic building shown in Figure 14.1. The framework is formed by identical planar two-tiered frames located in rows one after another (the column grid is rectangular in plan view). The transverse connections are fulfilled by the floor slabs and roof purlins, which unite the plane frames into the space system. The simplified design scheme of the plane frame is depicted in Figure 14.2. The simplifications lie in the replacement of roof trusses by hinged chain at the top of the frame and the decrease of the span number to 5. But even such a simple design diagram has the degree of redundancy of 20, and finding a manual solution of the canonical equation set is impossible through ordinary calculus. Meanwhile, this frame has two characteristics: axial symmetry and periodical structure, which could be used to simplify and expedite setting up the canonical equation set and solving it. Such simplification is reached by decomposition of the canonical equation set.

We call transformation and grouping of equations and unknowns *decomposition of the equation set* if every group of transformed equations enables us to determine the corresponding group of unknowns independently from others. Further, we consider the decomposition techniques in the case of a canonical equation set composed for a frame **with axial symmetry**.

The problem of decomposition is sometimes solved by a lucky choice of primary system and corresponding assemblage of redundant forces, but more often, it can be solved only after introduction of a primary system. In the latter case, one passes from the chosen redundant forces to the new set of transformed forces, for which decomposition is fulfilled in an obvious manner.

At the beginning, we consider the technique of canonical equations' decomposition at the stage of construction of a primary system which is used for axial symmetric structures. We take into account only the bending moments in Mohr's formulas while computing unit and loaded displacements. The next integral we shall term the *diagram product for bending moments M_1 and M_2 acting in a bar system*:

$$\left(M_1, M_2\right) \equiv \sum \int_0^l \frac{M_1 M_2 dz}{EI}. \tag{14.1}$$

FIGURE 14.1 Framework of two-story building made up from plane steel-monolithic frames: on the left – general view; on the right – the portion of the plane frame (construction site in Rostov region, RF, 2006).

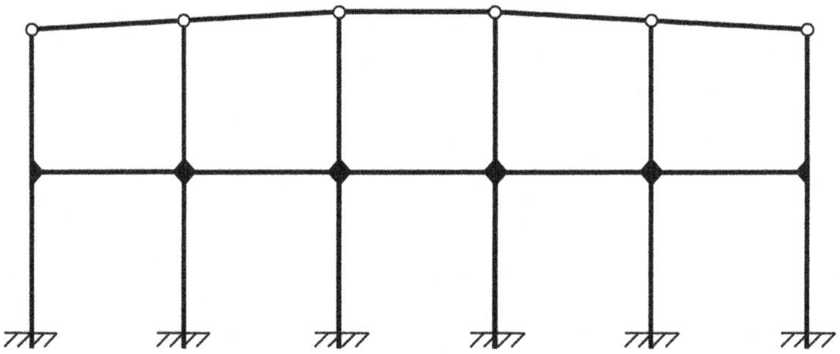

FIGURE 14.2 Simplified design diagram of the plane frame forming space framework of building with the outline of the prototype frame being preserved.

One can see that the diagram product for bending moments is integral (11.13), which determines virtual work due to bending deformation. The diagram in which bending moments at an arbitrary couple of cross-sections located symmetrically are equal in magnitude and fibers tensed by a bending moment are also located symmetrically, whereas members located on the symmetry axis of the structure do not work in bending, is referred to as a *symmetric diagram of bending moments.* The diagram with equal moments at an arbitrary couple of cross-sections located symmetrically when one of two fibers located symmetrically is tensed while another is compressed is referred to as an *anti-symmetric diagram of bending moments.* We call diagrams M_1 and M_2 *orthogonal* if $\left(M_1, M_2\right) = 0$. In particular, symmetric and anti-symmetric diagrams are mutually orthogonal. The simplest trick of decomposition is to construct two groups of primary unknowns with symmetric and anti-symmetric unit diagrams, respectively. The coefficients of canonical equations are determined through the product of unit diagrams:

$$\delta_{ij} = \left(\bar{M}_i, \bar{M}_j\right).$$

That is why the orthogonality of unit diagrams secures equality to zero for corresponding secondary displacements. Due to the orthogonality of symmetric and anti-symmetric diagrams $\bar{M}_i$, the canonical equation set decomposes into two independently solved systems of equations.

Before considering the simple instance of composing a primary system that fulfills the requirement for canonical equations' decomposition, we clarify the rules of the bending moment diagram's construction in frames. In the case of beams, bending moment diagrams are usually constructed "on compressed fiber," meaning that a beam is positioned horizontally and the diagram is drawn on the side opposite to convexity of the elastic curve. For example, if the bending occurs with convexity downward, the diagram is constructed over the axial line of the beam, and the moment is accepted as positive. In the case of frames, we follow other rules of construction: the moment diagram is drawn on the side where the bending moment stretches the outermost longitudinal fiber of the bar; the signs of moments usually are not depicted on their profile. The moment diagram on one side of a bar is taken as positive and on the other side as negative, and usually there is no difference in choice of the positive side. Note that frequent practice is to construct bending moment diagrams for frames on the compressed fiber instead of the tensed one, as is done for beams.

As an example of transition to a rational primary system, we may take the Π-shaped frame with two supporting fixings, wherefrom are removed three connections by dint of dissection in the middle of the girder. Figure 14.3 shows the unit diagrams for the corresponding primary system; two of them are symmetric and one is anti-symmetric. In accordance with the type of diagrams, the system of three canonical equations decomposes into two systems: one enables us to find forces X_1 and X_3 which beget symmetric diagrams and another system consists of one equation in unknown X_2 begetting an anti-symmetric diagram.

14.2 DECOMPOSITION BY MEANS OF LINEAR TRANSFORMATION OF REDUNDANT FORCES

The technique of transformation of given primary unknowns is based on the conceptions of group unknowns and group displacements. *We call an assemblage of primary unknowns taken in definite proportion a group force (group unknown).* In other words, group force Z is the set of redundant forces $X_1, X_2, ..., X_n$ defined by characteristic Z in the form:

$$X_1 = \alpha_1 Z; X_2 = \alpha_2 Z; \; ... \; ; X_n = \alpha_n Z, \tag{14.2}$$

where α_i are the coefficients of group force Z.

Figure 14.4 shows an example of specifying redundant forces X_m, X_n by the group force Z of the form:

$$X_m = Z; X_n = -0.5Z.$$

Diagrams a and b show the given and primary systems, respectively; diagram c represents the action of group force $Z = 1$.

A linear function of displacements in the directions of removed constraints, which is defined by coefficients of a group force, is referred to as the *group displacement* for this force. In other words, the group displacement for the force (14.2) is the term for the quantity:

$$\Delta^Z = \alpha_1\Delta_1 + \alpha_2\Delta_2 + \; ... \; + \alpha_n\Delta_n. \tag{14.3}$$

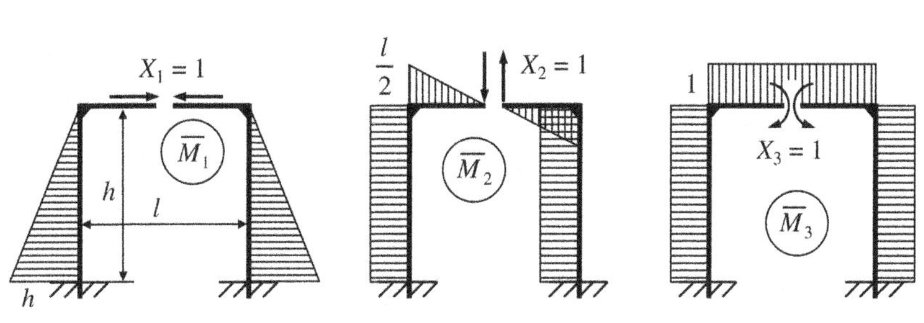

FIGURE 14.3 Unit bending moment diagrams for primary system obtained by dissecting the girder of given portal frame.

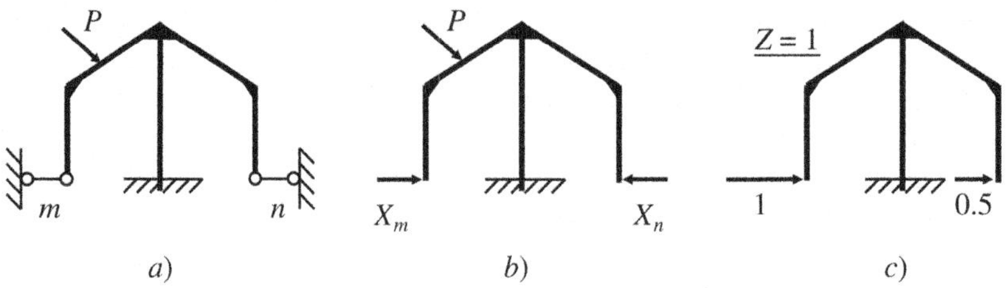

FIGURE 14.4 Redundant forces X_m, X_n defined by the group force Z.

The totality of group forces Z_i, $i=\overline{1,n}$, is considered equivalent to the set of redundant forces X_i, $i=\overline{1,n}$, if these totalities determine each other one-to-one by linear relations, i.e.:

$$\begin{aligned} X_1 &= \alpha_{11}Z_1+\alpha_{12}Z_2+\ldots+\alpha_{1n}Z_n\,; \\ X_2 &= \alpha_{21}Z_1+\alpha_{22}Z_2+\ldots+\alpha_{2n}Z_n\,; \\ &\cdots\cdots\cdots\cdots \\ X_n &= \alpha_{n1}Z_1+\alpha_{n2}Z_2+\ldots+\alpha_{nn}Z_n\,, \end{aligned} \tag{14.4}$$

where (α_{ij}) is a square matrix with a nonzero determinant. This matrix defines the group displacement for group force Z_i as follows:

$$\Delta_i^Z=\alpha_{1i}\Delta_1+\alpha_{2i}\Delta_2+\ldots+\alpha_{ni}\Delta_n. \tag{14.5}$$

After transition to a symmetrical primary system, it might be convenient to replace redundant forces X_i, $i=\overline{1,n}$, by equivalent group forces Z_i, $i=\overline{1,n}$. It is convenient because:

1) For the totality of group forces, one can write down the canonical equation set similar to the set in primary unknowns;
2) The coefficients and free terms of canonical equations in group unknowns are determined in a simple enough way, by means of the technique developed for regular primary unknowns;
3) The group forces often can be selected so that the moment diagrams defined by them are orthogonal for the forces from different groups.

These affirmations are substantiated as follows. Owing to the nonsingularity of matrix (α_{ij}), the set of group displacements Δ_i^Z, $i=\overline{1,n}$, and the set of displacements in the directions of removed constraints Δ_i, $i=\overline{1,n}$, determine each other one-to-one by linear relations. Therefore, the equivalence of the original and primary systems is ensured under the following conditions upon the group displacements:

$$\Delta_i^Z=0,\ i=\overline{1,n}.$$

In turn, the group displacements can be represented in the form:

$$\Delta_i^Z=\sum_{j=1}^{n}\delta'_{ij}X_j+\Delta_{iP}^Z=\sum_{j=1}^{n}\delta_{ij}^Z Z_j+\Delta_{iP}^Z.$$

Here, the first equality is analogous to formula (13.4) and embodies the superposition principle. The second equality is obtained through representation of primary unknowns by the group unknowns.

The coefficient δ_{ij}^{Z} is the unit group displacement – in the given case, the i-th group displacement being caused only by the j-th group force; the term Δ_{iP}^{Z} is the loaded group displacement – in the given case, the i-th group displacement being caused only by given loads. Coefficients δ'_{ij} could be expressed through constituents of matrices (α_{ij}) and (δ_{ij}) (these coefficients will not be required further on, but it is useful to check matrix relation: $\delta' = \alpha^{T}\delta$).

Further, it is no trouble to represent canonical equations by means of group unknowns:

$$\sum_{j=1}^{n} \delta_{ij}^{Z} Z_j + \Delta_{iP}^{Z} = 0, \quad i = \overline{1,n}. \tag{14.6}$$

Let $\overline{M}^{Z}$ be the diagram of bending moments produced in the primary system due to unit group force $Z = 1$ of the form (14.2). It is easy to prove that group displacement (14.3) due to loads is determined as the product of diagrams:

$$\Delta_{P}^{Z} = \left(\overline{M}^{Z}, M_P\right), \tag{14.7}$$

where M_P is the diagram of bending moments produced by loads in the primary system.

Really, from the superposition principle it follows:

$$\overline{M}^{Z} = \alpha_1 \overline{M}_1 + \alpha_2 \overline{M}_2 + \ldots + \alpha_n \overline{M}_n, \tag{14.8}$$

where $\overline{M}_i$ is the diagram of bending moments produced by redundant force $X_i = 1$. The latter relation determines the diagram of bending moments caused by the unit group force in the form of a linear function similar to the function of displacements (14.3), in which regard the diagram $\overline{M}^{Z}$ is called the group one. After substitution of (14.8) into (14.7) and representation of Mohr integrals through displacements, we come to expression (14.3) where displacements on the right-hand side are caused by loads only.

We call expression (14.7) the Mohr integral for group displacement. It determines group displacement in a primary system being under the action of arbitrary forces X_i and loads, but not just of given loads, if the corresponding bending moment is denoted by M_P. The possibility to represent a group displacement by the Mohr integral for arbitrary external forces referred to above means it is possible to calculate the coefficients and free terms of canonical equations in group unknowns as follows:

$$\delta_{ij}^{Z} = \left(\overline{M}_i^{Z}, \overline{M}_j^{Z}\right); \tag{14.9}$$

$$\Delta_{iP}^{Z} = \left(\overline{M}_i^{Z}, M_P\right), \tag{14.10}$$

where $\overline{M}_i^{Z}$ is the group diagram defined by the force $Z_i = 1$, and M_P is the loaded diagram.

On the strength of the superposition principle, the final bending moment diagram can be obtained by the formula:

$$M = \sum_{i=1}^{n} \overline{M}_i^{Z} Z_i + M_P, \tag{14.11}$$

where Z_i are the group forces found by solving canonical equations (14.6).

So, from the point of view of the computing technique, the transition to an equivalent set of group forces doesn't cause difficulties. Figure 14.5a shows the example of such a transition by dividing new variables into two groups with symmetric and anti-symmetric group diagrams, respectively. In the primary system depicted in Figure 14.5b, there are six unknown forces, which can be represented by the group unknowns shown in Figure 14.5c. The group forces specify redundant forces in removed constraints by the relations:

$$X_1 = Z_1 + Z_2; \quad X_4 = Z_1 - Z_2;$$

$$X_2 = Z_3 + Z_4; \quad X_5 = Z_3 - Z_4;$$

$$X_3 = Z_5 + Z_6; \quad X_6 = Z_5 - Z_6.$$

The moment diagrams for unit group forces are shown in Figure 14.6. One can see that these diagrams have symmetry properties.

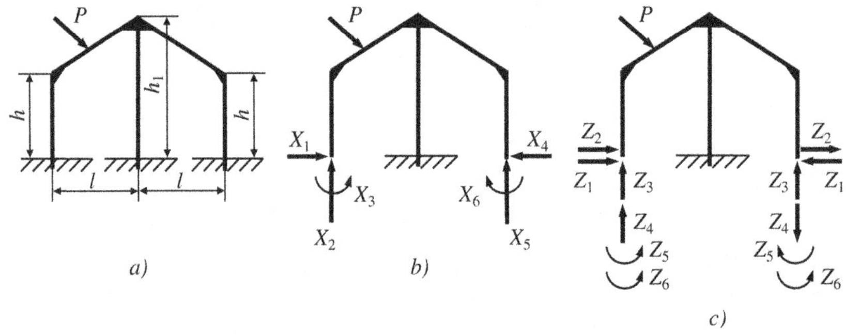

FIGURE 14.5 Transition from the given system to rational primary system: (a) given frame; (b) primary system under action of redundant forces; (c) primary system and group forces.

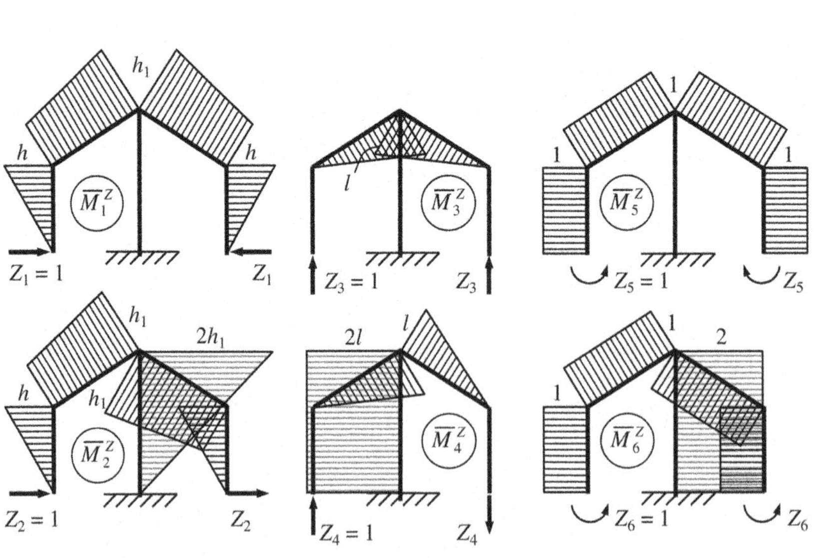

FIGURE 14.6 Group diagrams for the primary system in Figure 14.5c (symmetric diagrams on the top, anti-symmetric on the bottom).

14.3 VERIFICATION OF CALCULATIONS IN THE FORCE METHOD

As before, we take into account only bending deformations in a system. At the completing stage of analysis by the force method, the constructed final (complete) diagrams have to be verified. We attribute the deformation verification and the verification on closed contour to verifications of the completing stage. The *deformation verification* is to calculate the products of final diagram M and each unit diagram $\bar{M}_i$ of a bending moment. These products must be equal to zero. Really, let us write down one of such products:

$$\Delta_i = \left(\bar{M}_i, M\right). \tag{14.12}$$

Quantity Δ_i is the displacement in the direction of the i-th removed constraint, which arises in the primary system due to the action of redundant forces equal to constraint reactions in the given system, together with loads. But equality of redundant forces to constraint reactions ensures the absence of displacements in removed constraints, therefore the products (14.12) must be zero.

If the final diagram is constructed by summing up the diagrams according to formula (13.9), then the deformation check is effective only under the condition that the unit and loaded diagrams were constructed correctly. To raise the reliability of deformation verification, one is recommended to make it by using a new primary system where unit diagrams have another outline than in the system used for analysis.

The *verification on closed contour* assumes calculating the following integral on the frame's closed contour with no hinges:

$$\Delta = \sum \int_0^l \frac{M dz}{EI}. \tag{14.13}$$

This integral we call the *area of diagram M on closed contour.* While computing it, one has to take the sign of the bending moment uniformly. For example, you have to accept $M > 0$ if the moment stretches the member's fiber outside the contour (Figure 14.7). For a properly constructed final diagram, the integral (14.13) must be equal to zero.

Indeed, let us mentally cut out the closed contour from the frame and exert external forces at the places of removed constraints equal to constraint reactions. External forces applied to the contour before cutting off (including reactions of connections with the remaining part of the frame) coincide with forces exerted upon the cutout contour. Thus, in both cases, the bending moment diagrams in the contour are the same. Now attach the contour to the earth by means of three statically determinate connections. Because the assemblage of external forces ensures the contour's equilibrium, the support reactions prove to be zeroth. But due to attachment to the earth, there is an unambiguous relation of the contour's displacements to the exerted external forces. Let us make the cut of contour by elimination of the member's elementary portion (taken for a rigid connection) and apply the

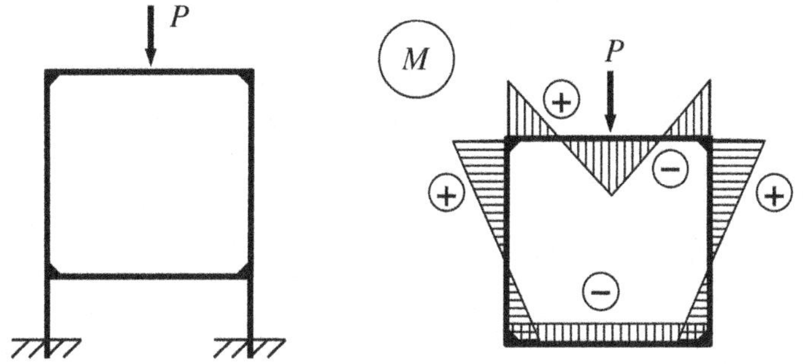

FIGURE 14.7 Diagram M on closed contour in the case EI = const (area of diagram is equal to zero).

reactions of removed constraints upon the ends of the cut. The bending moment diagram does not change as before. To determine the angle of mutual rotation for the ends of the dissected contour, we multiply the final bending moment diagram at the contour and the diagram of bending moments produced by unit force couples exerted on the ends of the cut in opposite directions. The latter diagram equals the unit identically; therefore, the mutual rotation angle is represented in the form (14.13). But elimination of the rigid joint will not change displacements in the contour, because the forces in the places of removed constraints remain the same. That is why the angle of mutual rotation of the ends proves to be zero and, respectively, the integral (14.13) equals zero.

We also classify a contour which is closed through the earth by members clamped in the basement as a closed contour.

In graphic-analytical analyses by the force method, errors often arise in the course of the diagrams' multiplication. In the frames (for which the deformation is determined by the bending moment only), it is possible to discover such errors by making two special controls: by the checking on coefficients of canonical equations and free terms of these equations, respectively. These controls are as follows.

We call the diagram of bending moments produced by simultaneous action of all unit redundant forces in a primary system a *total unit diagram*. According to this definition, the total unit diagram is the sum of the diagrams:

$$\bar{M}_s \equiv \sum_{i=1}^{n} \bar{M}_i. \tag{14.14}$$

We may call quantity (14.14) a total unit bending moment. This moment should be considered a dimensionless characteristic, for the sum (14.14) may include quantities measured in different units.

Verification of unit displacements (coefficients of canonical equations), also called as *universal verification*, is accomplished by multiplying the total unit diagram by itself and checking the equality:

$$\delta_{ss} \equiv \left(\bar{M}_s, \bar{M}_s\right) = \sum_{i,j=1}^{n} \delta_{ij}. \tag{14.15}$$

So, *total unit displacement* defined by the first equality (14.15) must be equal to the sum of all principal and secondary unit displacements in the canonical equation set. The reader can make sure of that by direct calculation.

Verification of the loaded displacements (free terms of canonical equations) is done by multiplying the total unit diagram by loaded diagram and checking the equality:

$$\Delta_{sP} \equiv \left(\bar{M}_s, M_P\right) = \sum_{i=1}^{n} \Delta_{iP}. \tag{14.16}$$

It is expedient to make verifications of unit and loaded displacements right after setting up a canonical equation set. Note that verifications by means of the total unit diagram are specific for the frames only (in contrast to deformation verification, wherein the influence of forces N and V upon the system's deformation could be taken into account).

14.4 EXAMPLE OF CONSTRUCTION OF A BENDING MOMENT DIAGRAM FOR A FRAME BY THE FORCE METHOD

In this item, we employ the force method for the construction of a diagram of bending moments acting in the frame shown in Figure 14.8a. The flexural rigidities of all the members are the same. The given quantities are a and P.

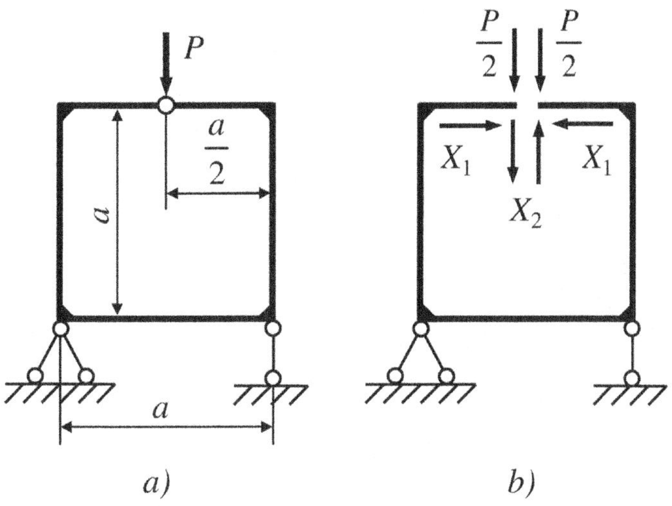

FIGURE 14.8 Given frame (a) and corresponding primary system (b).

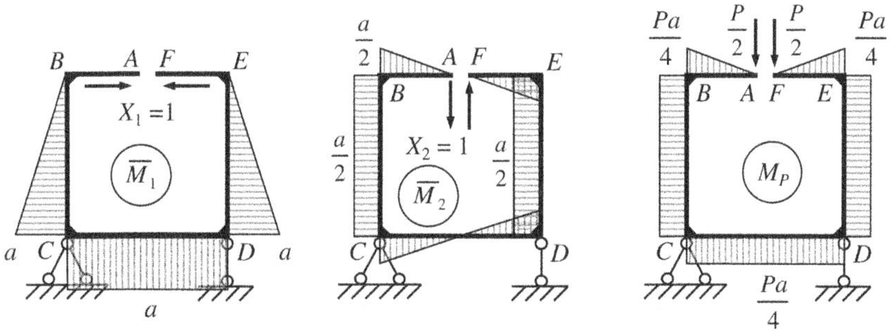

FIGURE 14.9 Unit bending moment diagrams.

By removal of the upper hinge, we come to a statically determinate system. The hinge makes up two connections; therefore, the degree of static indeterminacy for the given system equals 2. The primary system is shown in Figure 14.8b. This system is constructed by replacing load P with two loads $\frac{P}{2}$ exerted near the hinge on the left and right. Then the hinge is removed and instead of reactions developed by the hinged connection, the external forces are applied. The canonical equation set has the form:

$$\begin{cases} \delta_{11}X_1 + \delta_{12}X_2 + \Delta_{1P} = 0; \\ \delta_{21}X_1 + \delta_{22}X_2 + \Delta_{2P} = 0. \end{cases}$$

In order to determine the parameters of these equations, we construct the set of unit and loaded diagrams of bending moments (Figure 14.9). We compute the wanted parameters by multiplication of the diagrams:

$$\delta_{11} = \sum \int_0^l \frac{\overline{M}_1^2}{EI} dz = \int_{BC} + \int_{CD} + \int_{DE} = \frac{1}{EI}\left(\frac{a^2}{2}\frac{2}{3}a \cdot 2 + a^2 a\right) = \frac{5}{3}\frac{a^3}{EI};$$

$$\delta_{22} = \sum \int_0^l \frac{\overline{M}_2^2}{EI} dz = \int_{AB} + \int_{EF} + \int_{BC} + \int_{DE} + \int_{CD} =$$

$$= \frac{1}{EI}\left(\frac{1}{2}\frac{a^2}{4}\frac{2}{3}\frac{a}{2}\cdot 2 + \frac{a^2}{2}\frac{a}{2}\cdot 2 + \frac{a^3}{24}\cdot 2\right) = \frac{2}{3}\frac{a^3}{EI};$$

$$\delta_{12} = \delta_{21} = \sum \int_0^l \frac{\overline{M}_1 \overline{M}_2}{EI} dz = 0 \text{ (as the product of symmetric and anti-symmetric diagrams);}$$

$$\Delta_{1P} = \sum \int_0^l \frac{\overline{M}_1 M_P}{EI} dz = \int_{BC} + \int_{CD} + \int_{DE} = \frac{1}{EI}\frac{Pa^3}{2};$$

$$\Delta_{2P} = \sum \int_0^l \frac{\overline{M}_2 M_P}{EI} dz = 0 \text{ (as the product of symmetric and anti-symmetric diagrams).}$$

The solution of the canonical equation set is obtained in the form:

$$X_1 = -0.3P; \quad X_2 = 0.$$

The final diagram is established by summation:

$$M = \overline{M}_1 X_1 + M_P = -\overline{M}_1 \cdot 0.3P + M_P. \tag{14.17}$$

The values of the total bending moment are not evident only at joints C and D:

$$M_C = M_D = -0.3Pa + 0.25Pa = -0.05Pa \text{ (the internal fiber is stretched).}$$

By applying formula (14.17), we get the diagram shown in Figure 14.10.

Let us make the deformation verification of the final diagram M. To do that, we should make sure that

$$\sum \int_0^l \frac{\overline{M}_i M}{EI} dz = 0.$$

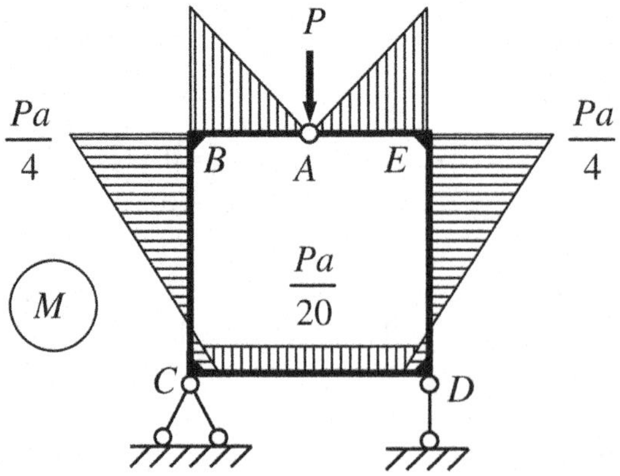

FIGURE 14.10 Final bending moment diagram for the frame in Figure 14.8a.

The product of diagrams ($\bar{M}_2$, M) is obviously zero; for multiplication of diagrams $\bar{M}_1$ and M, we should compute their product over portion BC. We shall use the Vereshchagin rule. The calculation of the ordinate of diagram M located under the centroid of diagram $\bar{M}_1$ is clarified by Figure 14.11. The similarity of triangles gives:

$$KK'' = \frac{1}{3}\left(\frac{Pa}{4} + \frac{Pa}{20}\right) = \frac{Pa}{10} \Rightarrow K'K'' = \frac{Pa}{10} - \frac{Pa}{20} = \frac{Pa}{20}.$$

Hence we obtain:

$$\sum \int_0^l \frac{\bar{M}_1 M}{EI} dz = 2\int_{BC} + \int_{CD} = \frac{1}{EI}\left(2 \cdot \frac{a^2}{2} \cdot K'K'' - \frac{Pa^3}{20}\right) = 0.$$

Thus, the correctness of the analysis has been confirmed.

14.5 TECHNIQUE AND EXAMPLE OF CONSTRUCTION OF SHEAR AND AXIAL FORCES' DIAGRAMS FOR FRAMES

When reactions of removed connections have been determined and the final diagram of bending moments acting in a frame has been constructed, it is required to construct the diagrams of axial and shear internal forces. This problem can be solved by two methods. The first method is to construct diagrams V and N for the primary system wherein the found redundant forces X_i are exerted. Such a system is statically determinate and internal forces in it usually are easily established by the method of sections. We can, however, construct the shear force diagram by using the complete bending moment diagram, to which end uncomplicated expedients are employed. In turn, for the frames of a sufficiently simple outline, we can construct the axial force diagram by means of the shear force diagram.

While constructing the shear force diagram using the bending moment diagram, we represent the frame as an assemblage of members connected only at the ends. Each member of the frame is mentally cut out and, firstly, the shear forces are determined at its ends, for which purpose the statics equations are employed, as was the case when finding the reactions for simple beams. Figure 14.12 shows, for example, the column positioned between joints A and B of some frame and taking up the

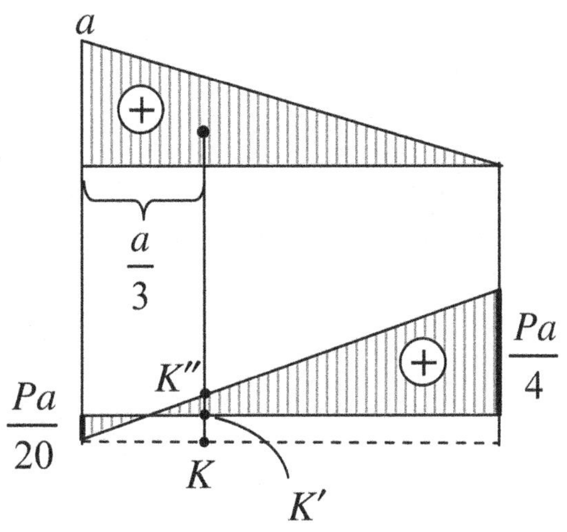

FIGURE 14.11 Multiplication of diagrams $\bar{M}_1$ and M over portion BC of primary system.

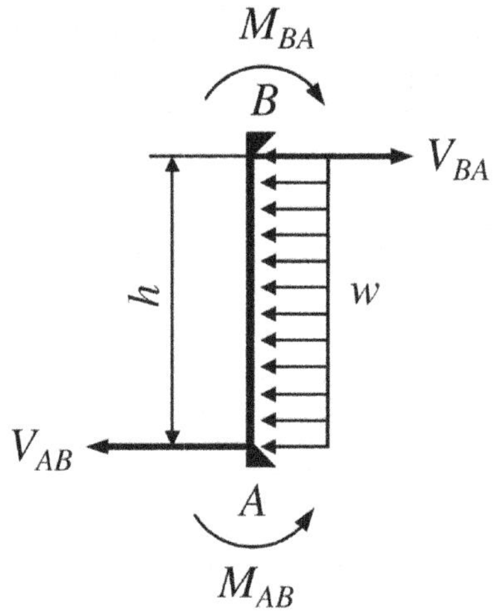

FIGURE 14.12 Free-body diagram of a frame's member.

distributed load of intensity w. During analysis of internal forces, it is convenient to denote each joint of a system by the letter (or mark it by the number), and each characteristic cross-section adjacent to a joint is to be designated by two letters (or numbers) where the first symbol points out the joint located near the section and the second one clarifies at which member the section is taken. In the design diagram of the cutout member, at the ends, we have to designate the positive directions of shear forces (positive force tends to rotate the member clockwise) and the directions of bending moments which for the cutout members are external concentrated moments. The cross-sections are pointed out by subscripts at the denotations of internal forces as in the diagram of Figure 14.12. For instance, M_{BA} is to be read as "bending moment at the section of member BA made in the neighborhood of joint B"; V_{AB} means shear force at the section of member AB near joint A. The represented design diagram suffices to set up the equations of statics. In our example we have:

$$\sum M_A = -V_{BA}h + \frac{wh^2}{2} - M_{BA} + M_{AB} = 0;$$

$$\sum M_B = -V_{AB}h - \frac{wh^2}{2} - M_{BA} + M_{AB} = 0.$$

When the shear forces at the ends of a member have been determined, it is not difficult to construct the profile of shear forces on this member by using the same rules of construction that are employed for beams (the diagram portion sloped to the basic line corresponds to distributed load; the jump corresponds to concentrated load, etc.). In unobvious cases, it is recommended to situate the member on the plot horizontally.

In the particular case when there are no loads upon the member, bending moments at the ends determine the shear force, which is constant along the member. Disposing the member horizontally and accepting that a bending moment is positive if it stretches the upper fiber, we can get shear force by the formula:

$$V = \frac{M^{left} - M^{right}}{l}, \tag{14.18}$$

where M^{left}, M^{right} are the moments at the ends of the member, respectively, left and right; l is the member's length. The sign convention for bending moments in the horizontal member, which declares that positive moments produce the tension of upper fiber, could be named by the sign convention for frames, in contrast to the known convention for beams.

Figure 14.10 shows an example of a frame for which the shear force diagram is easily constructed only by application of formula (14.18). For instance, the bending moment diagram for column BC, constructed on the horizontal basic line in accordance with the sign convention for frames, is shown in Figure 14.11 (at the bottom). Using formula (14.18), we obtain:

$$V_{B-C} = \frac{-\dfrac{Pa}{20} - \dfrac{Pa}{4}}{a} = -0.3P$$

(the subscript "$B-C$" points out to all the portion BC). The complete shear force diagram is shown in Figure 14.13.

The axial force diagram is constructed with the help of the shear force diagram by the method of joints. For each joint, we draft the design diagram of its loading by pointing out the positive directions of the forces. If the frame is not complicated and the sequence of cutting out the joints was chosen properly, then balance conditions for the forces acting upon a joint determine the sought axial forces in an obvious way. Figure 14.14 clarifies the construction of an axial force diagram for the frame in Figure 14.8a. For example, the condition of balance of horizontal forces gives for joint B:

$$N_{BA} = V_{BC} = -0.3P.$$

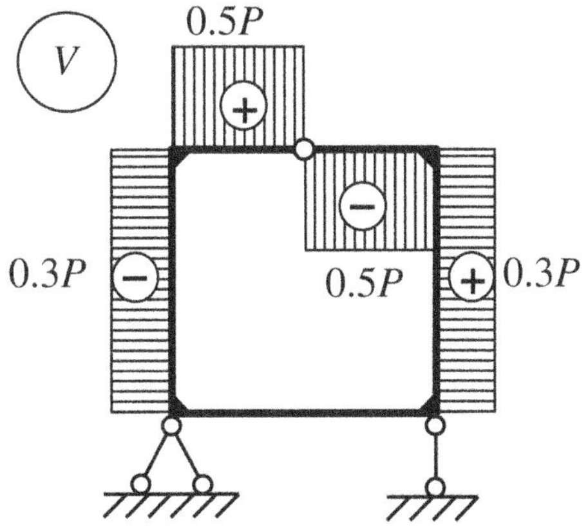

FIGURE 14.13 Shear force diagram for the frame in Figure 14.8a.

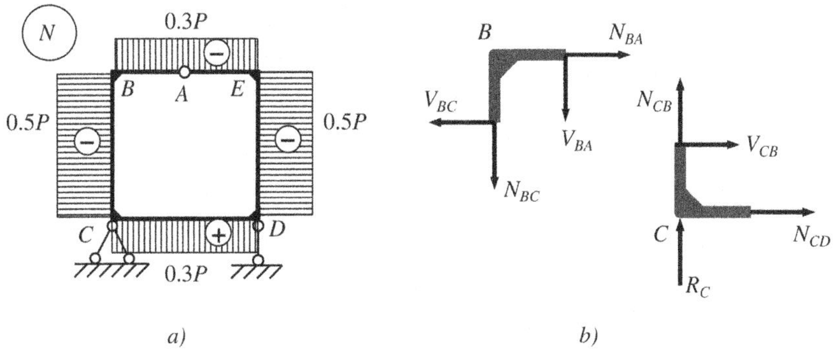

FIGURE 14.14 Axial force diagram (a) and free-body diagrams of joints (b) for the frame in Figure 14.8a.

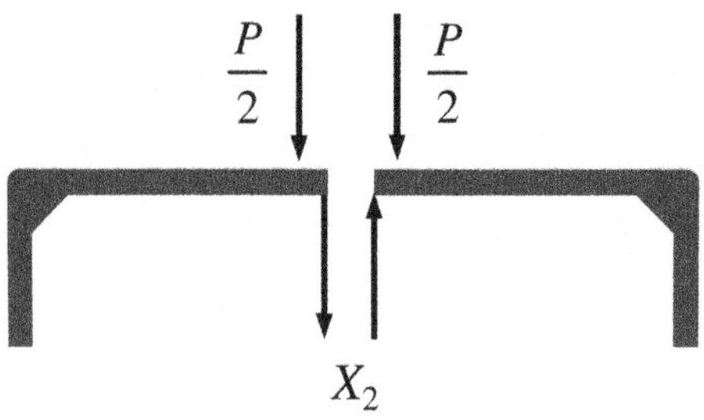

FIGURE 14.15 Location of external forces exerted on the primary system.

The complete diagram of axial forces is shown in Figure 14.14a.

Remark

According to constructed diagram V, a nonzero shear force in the neighborhood of the hinge acts in the top girder of the frame (Figure 14.13). At the same time, force X_2 substituting for the shear force is zero. This is explained by mutual location of external forces exerted on the primary system in the neighborhood of the removed connection (Figure 14.15). In the given case, the shear force is caused not by the redundant force, but by loads exerted at the points biased from the ends of the cut.

15 Force Method for Structures Subjected to Temperature Changes and Support Settlements

15.1 TEMPERATURE DISPLACEMENTS IN BAR SYSTEMS; FORCE METHOD IN CASE OF TEMPERATURE CHANGES

We shall first state the basic points of the analysis of the SSS in elastic systems subjected to temperature impact. The stresses that arise from temperature changes in members of a mechanical system with no loads exerted on it are referred to as *thermal stresses*. A similar definition is adopted for the concept of thermal displacements in an externally stable system. The impact of temperature upon a statically determinate structure does not induce constraint reactions, and the thermal stresses usually are not significant. At the same time, it is thermal stresses which may cause the failure of statically indeterminate structures.

The calculation of an elastic body's SSS in the general case is based on the *generalized Hooke's law with account for temperature changes*. This law is inferred from the affirmation that the total strain at a point of a body is the sum of the strain due to the stresses only and the thermal strain with no stresses. In other words, the *basic relations of thermoelasticity* are obtained by the supplement of the generalized Hooke's law at the uniform temperature with the strain of thermal expansion:

$$\varepsilon_x = \frac{1}{E}[\sigma_x - \nu(\sigma_y + \sigma_z)] + \alpha\Delta t^\circ;$$

$$\varepsilon_y = \frac{1}{E}[\sigma_y - \nu(\sigma_x + \sigma_z)] + \alpha\Delta t^\circ; \qquad (15.1)$$

$$\varepsilon_z = \frac{1}{E}[\sigma_z - \nu(\sigma_x + \sigma_y)] + \alpha\Delta t^\circ,$$

where Δt° is the change in temperature from the initial uniformly distributed one (for which the strains in an unloaded body are taken for zero), and α is the linear coefficient of thermal expansion. Consideration of the temperature effect in the generalized Hooke's law is necessary only in expressions for normal strains (15.1). The relations for shear strains don't change.

Another important provision in analysis of the SSS with account for temperature changes is *the superposition principle for elastic systems being under an impact of loads and temperature*: **stresses, strains, and displacements in an elastic structure, produced by the conjoint impact of loads and temperature, can be determined as the sum of stresses, strains, and displacements caused by loads and temperature acting separately**. The substantiation of this principle is given in the theory of elasticity.

The third provision of the SSS analysis, necessary in further investigation, is the following: **the strains in bodies of externally stable systems, when compatible with sufficiently small displacements, determine these displacements uniquely**. The compatibility of any field of strains with displacements means that for given strains, some field of displacements can be matched. If the given strains are incompatible with displacements, then such strains are not feasible. The assertion about the uniqueness of displacements corresponding to compatible strains is proved in the theory of elasticity for a single externally stable body (Love 1906, p. 50). This assertion could be also substantiated for a structure consisting of several bodies. It is important that substantiation of uniqueness of displacements implies limitation for the magnitude of possible displacements, and the corresponding limit is fixed for each structure and can be minute. There are simple examples of structures in which even comparatively inconsiderable displacements can be ambiguous at given strains. Below, when applying the third provision, we have in mind that displacements in a system are sufficiently small.*

If only temperature acts upon statically determinate system, the constraint reactions remain zero because, in such a system, these reactions are determined by loads from equations of statics, whereas the loads are assumed to be zeroth. At the same time, the stresses in elastic bodies of a statically determinate system could be induced by temperature. The theory of elasticity states that for the absence of thermal stresses in an elastic body, it is sufficient that the temperature is represented by the linear function of the Cartesian coordinates – see, for instance, Timoshenko and Goodier (1951, p. 403).

Now we turn to the problem of finding internal forces in a plane statically determinate bar system that is under the impact of loads and temperature. As before, a system is assumed to have the symmetry plane coincide with the plane of loads. For each member of the system, we introduce the member-bound right-handed CS $O'xyz$ as done in Chapter 8 at the analysis of symmetrical bending of a beam: origin O' is placed at one of the ends of the member, axis z is directed along the axial line of the member, and axis x is positioned perpendicular to the plane of loads (plane of the system) and directed towards the observer. The counting of slope θ of the elastic curve is taken counterclockwise, i.e., the increment of this angle along the length dz of member's portion is positive if the elastic curve is convex in the y-axis direction. For such a direction of convexity of the bent member, the bending moment is also accepted as positive.

The following theorem enables us to determine the displacements in a statically determinate system through the given strains in members. The theorem will be required further for use of the force method in the analysis of statically indeterminate structures with account for temperature changes.

Theorem

Let an externally stable statically determinate elastic bar system undergo planar displacements which result from mutual pivot-motion of the butt-ends of the member's elementary portions by the angle of $d\theta = k_\theta dz$ in conjunction with the axial fiber elongation of $\Delta(dz) = k_z dz$; coefficients k_θ and k_z are given for each member of the system and constant over its length. Then displacement of arbitrary point m of this system in a given direction is determined by the relation:

$$\Delta_m = \sum \int_0^l \bar{M}_m d\theta + \sum \int_0^l \bar{N}_m \Delta(dz), \tag{15.2}$$

where $\bar{M}_m$ and $\bar{N}_m$ are bending moment and axial force arisen from action of unit force exerted at point m in a given direction.

* The brief substantiation of the third provision is as follows: Let two different fields of displacements correspond to given strains. Then the difference of these two fields defines the field of displacements which corresponds to zero strains of the structure. In some structures, the bodies can be displaced in new stable position where they restore their shape and size. But such relocation is impossible if displacements are limited by sufficiently small magnitude. Consequently, the field of limited displacements (if it exists at given strains) is unambiguous.

Proof

Note first that coefficients k_θ and k_z are the curvature and the relative elongation of the axial fiber, respectively. Nevertheless, here we don't use the denotations κ and ε, generally used for these quantities, to emphasize that magnitudes of these quantities are determined by the theorem formulation. In this formulation, the causes of members' deformation are not stated. Let us produce these deformations by means of force couples and axial loads exerted on the members of the system. For all members, we take the Poisson's ratio as $\nu = 0$. Owing to this, the lateral strains in members will be zero and cross-sections of members will keep their shape and size after deformation of the system. Upon every member, we exert the balanced assemblage of forces, which consists of concentrated moments $M_c = EIk_\theta$ acting at both ends in opposite directions in such fashion that convexity of the member would face in the direction of the y-axis, and of axial loads $P = EAk_z$ acting at the ends of the member and causing its tension or compression (Figure 15.1). This loading causes the strains which were specified in the theorem formulation. Really, owing to the balance of loads in the limits of each member, the constraint reactions do not arise in a statically determinate system. Consequently, the strains in the member are determined only by the loads upon it. External moments at the ends produce pure bending of the member with bending moment $M = M_c$. The mutual pivot-motion of two closely located cross-sections due to bending is determined by the formula of elastic line curvature (see Chapter 11), from which it holds:

$$d\theta = \frac{M}{EI}dz = k_\theta dz. \tag{15.3}$$

Axial loads at the ends produce tension-compression of the member with axial internal force $N = P$, which causes elongation of the member's element equal to

$$\Delta(dz) = \frac{N}{EA}dz = k_z dz. \tag{15.4}$$

The longitudinal deformations at the axial fiber are absent in bending; thus elongation of the axial fiber due to the combined action of longitudinal and bending loads is determined by formula (15.4) as before. The expressions (15.3) and (15.4) determine the strains, produced by loads, in the boundaries of any elementary portion of the member and coincide with relations specifying mutual displacements of boundary cross-sections of the element in the theorem assertion.

So, the strains in members stated in the theorem could be created by exertion of known loads upon an elastic system. This means that displacement Δ_m, arisen at given strains, can be considered as the result of loads and computed using Mohr's formula (11.15). In the given case, this formula is written in the form:

$$\Delta_m = \sum \int_0^l \frac{\bar{M}_m M dz}{EI} + \sum \int_0^l \frac{\bar{N}_m N dz}{EA}.$$

FIGURE 15.1 Loads at the ends of member producing given strains: $M_c = EIk_\theta$, $P = EAk_z$.

Making next substitutions in the integrands:

$$\frac{M}{EI} = k_\theta; \quad \frac{N}{EA} = k_z,$$

we come to expression (15.2). The sought displacement can't be another because the strains specify displacements unambiguously. The theorem has been proved.

Note

As all the theorems in the theory of displacements in elastic systems, this theorem remains correct for generalized displacement m.

We shall now consider the thermal displacements in a plane statically determinate bar system. We additionally assume that the cross-sections of each member are symmetrical about the axis perpendicular to the plane of a system. For convenience, we consider the horizontally positioned member of a system. Figure 15.2 shows the element of the member before and after deformation. Let the temperature of an undeformed bar system (initial temperature) be the same at all its points, and after the temperature change, the vertical distribution of temperature in every cross-section becomes linear.* This distribution is specified for every member and the same in all its sections. We denote the increase of the member's temperature t_1° for the upper fiber and t_2° for the lower fiber (Figure 15.2). At given coefficient α of linear thermal expansion, we have the elongation of the upper fiber $\alpha t_1^\circ dz$ and the lower, $\alpha t_2^\circ dz$; thus the angle of mutual rotation of limiting sections equals (Figure 15.3):

$$d\theta_t = \frac{\alpha(t_2^\circ - t_1^\circ)dz}{h}.$$

We obtain the axial elongation by taking into account the symmetry of the cross-section about the horizontal axis:

$$\Delta_t(dz) = \frac{\alpha(t_2^\circ + t_1^\circ)dz}{2}.$$

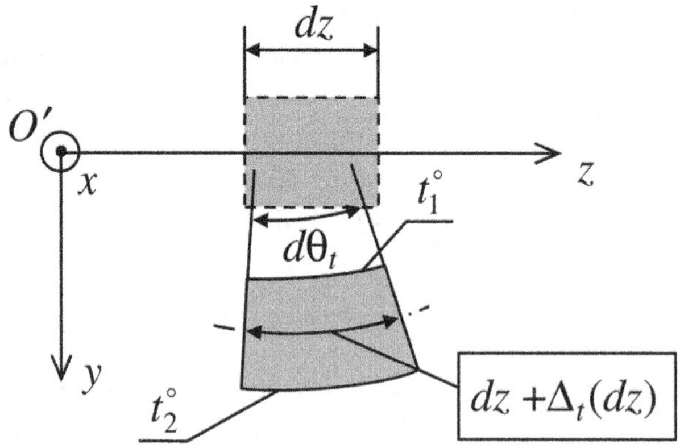

FIGURE 15.2 Element of member deformed by temperature.

* That is, the dependence "temperature versus coordinate of section point" has the form: $t(x, y) = ky + b$.

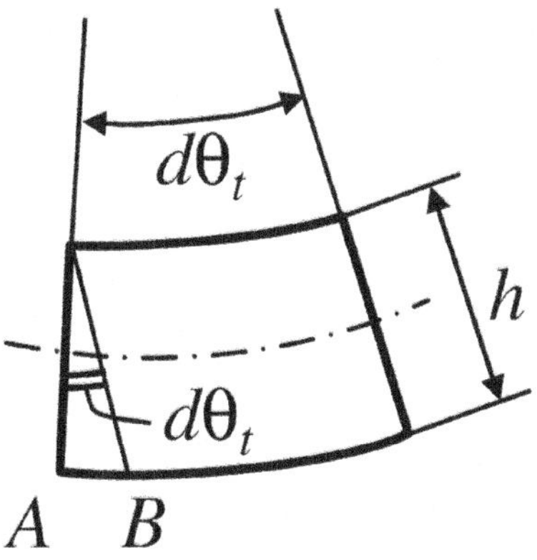

FIGURE 15.3 Concerning the calculation of mutual rotation angle for butt-ends of elementary portion: $AB = \alpha dz(t_2^\circ - t_1^\circ)$.

Note that in derivation of the two latter formulas, the linearity of temperature distribution in a member has been used; due to this linearity, the thermal stresses do not arise, and there are no additional strains from these stresses.

Now we can establish the displacement of point m in a given member induced by temperature changes. For this purpose, we use formula (15.2) where the replacements should be done: $d\theta = d\theta_t$, $\Delta(dz) = \Delta_t(dz)$. Thus we obtain:

$$\Delta_{mt} = \sum \alpha \frac{t_2^\circ - t_1^\circ}{h} \int_0^l \bar{M}_m dz + \sum \alpha \frac{t_2^\circ + t_1^\circ}{2} \int_0^l \bar{N}_m dz. \tag{15.5}$$

Introducing the denotations for areas of unit diagrams:

$$A_{\bar{M}} = \int_0^l \bar{M}_m dz; \quad A_{\bar{N}} = \int_0^l \bar{N}_m dz,$$

we write down the *formula of temperature displacement in a statically determinate system* as follows:

$$\Delta_{mt} = \sum \alpha \frac{t_2^\circ - t_1^\circ}{h} A_{\bar{M}} + \sum \alpha \frac{t_2^\circ + t_1^\circ}{2} A_{\bar{N}}. \tag{15.6}$$

The force method can be completed for analysis of a statically indeterminate system exposed to temperature changes. It is accepted that at some initial temperature, there are no constraint reactions in the given system. To take into account the temperature effect, we use the statically determinate primary system and supplement the canonical equations by the thermal displacements in the places of removed constraints Δ_{it}, $i = \overline{1,n}$. Quantity Δ_{it} is the displacement in the direction of the i-th removed connection induced by the temperature changes only (with no regard for redundant forces and loads). The total displacement, on account of external forces upon the primary system, has the form (compare with formula (13.4)):

$$\Delta_i = \sum_{j=1}^{n} \delta_{ij} X_j + \Delta_{iP} + \Delta_{it}. \tag{15.7}$$

If the primary unknowns equal the constraint reactions in a given indeterminate system, then these displacements turn into zero. Therefore, in the case of temperature changes, we have the following equivalence conditions for original and primary systems (being canonical equations):

$$\sum_{j=1}^{n} \delta_{ij} X_j + \Delta_{iP} + \Delta_{it} = 0, \quad i = \overline{1, n}. \tag{15.8}$$

Temperature term Δ_{it} is calculated by formula (15.6) and its emergence in canonical equations is the single peculiarity of the force method with account for temperature changes. Otherwise, this method doesn't differ from the force method studied formerly, when only loads act upon a system. In particular, the unit and loaded displacements are computed, as before, by Mohr's formulas (13.7) and (13.8).

15.2 STRUCTURE ANALYSIS BY THE FORCE METHOD IN THE CASE OF SUPPORT SETTLEMENTS

The important problem of structural statics is to determine the SSS of statically indeterminate systems with known support settlements. In rigorous terms, the settlement of support means its displacement downward (Davies and Jokiniemi 2008), but the force method has been adapted for analysis in case of any, not only vertical, support displacements. Thus, keeping the terminology accepted in structural analysis, we talk about settlement implying arbitrary displacement.

Below, we term the force method studied in previous chapters, when only loads affect a system, as the basic force method. In composing the basic force method, we used Mohr's formula for computation of displacements caused by loads. Just the same, in analysis of elastic systems affected by support settlements, we need the technique of computing the displacements induced by support settlements. Let us first consider a statically determinate bar system. Let the known displacements have arisen in support connections of such a system, and there is the problem of determining displacement $\Delta_{m\Delta}$ of some point m of a system in a given direction. In order to solve this problem, one can use the principle of virtual work, studied in Chapter 6, as follows:

- Determine the support reactions due to unit load $X_m = 1$ exerted at point m of a system in the required direction;
- Remove support connections which have been shifted, and instead of reactions of removed constraints exert external forces, which ensure equilibrium of system under load $X_m = 1$;
- Set up the equation of virtual work done through displacements arisen from support settlements and solve this equation for wanted displacement $\Delta_{m\Delta}$.

Figure 15.4 exemplifies finding the displacement of point m of the frame in the direction of given axis mn due to displacement of support B. The displacement vector for this support $\boldsymbol{u}_B = (u_B, v_B)$ is given. For support reactions caused by the unit load, we have:

$$R_B^v = \frac{x_m}{l} \sin\alpha; \quad R_B^h = \cos\alpha.$$

Under condition of small enough displacements, we have the equation:

$$\Delta_{m\Delta} \cdot 1 + v_B R_B^v - u_B R_B^h = 0, \tag{15.9}$$

whence the sought displacement is found.

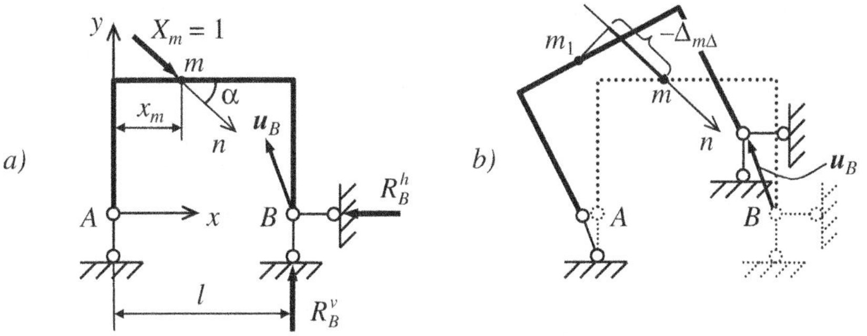

FIGURE 15.4 Example of support displacements: (a) Initial position of frame; (b) Shifted position of frame; m_1 is the location of point m after displacement; u_B is displacement vector for support B; shown the case $\Delta_{m\Delta} < 0$.

By composing the equation of virtual work, one can determine not only translational displacement in a given direction, but also generalized displacements. Besides, this equation may include not only linear support displacements, but angular ones as well (when the clamped end of the member turns by a known angle).

The example provided shows that displacements due to support settlements are easily established for statically determinate systems. In the case of statically indeterminate systems, the calculation of displacements may prove to be more complicated, mainly because of laborious calculation of support reactions caused by the action of unit load X_m in the required direction.

Important comment: The equation of virtual work in unknown displacement $\Delta_{m\Delta}$ for elastic structures (including statically indeterminate ones) could be set up pursuant to Betti's law, in which, for the state 1, we take the case of affecting a structure by unit load X_m, for the state 2 – affecting a structure by support settlements. For work done by external forces in state 2 when the system moves through displacements of state 1, we have the identity: $W_{21} = 0$, from whence we obtain the virtual work equation in the brief form: $W_{12} = 0$. The virtual work equation in developed form (15.9) is easily obtained by application of Betti's law to the instance in Figure 15.4. The example of virtual work analysis, similar in substance, is given in the proof of Rayleigh's second theorem and illustrated in Figure 12.8.

During substantiation of the new version of the force method, we shall firstly confine ourselves to the case when a system is affected only by support settlements, i.e., the loads are absent. In the same manner as in the basic force method, we pass to a primary system by removal of some connections and exertion of external forces (redundant forces) in the places of removed connections; then we find such redundant forces that the primary and given systems would be equivalent. However, unlike in the basic method, the primary system may, instead of loads, be subjected to displacements of support connections that were not removed but have shifted. Another difference is the requirement that in the primary system in the places of removed support connections being displaced in the given system, the displacements in the directions of these connections must be equal to given displacements.

For the i-th removed connection we introduce denotations: $\Delta_{i\Delta}$ – displacement in the direction of this connection induced by support settlements in the primary system and Δ_i^0 – settlement of this connection in the given system. If the connection under consideration is supporting and settled, then $\Delta_i^0 \neq 0$, in other cases $\Delta_i^0 = 0$. If the primary system does not comprise the settled supports, then it holds: $\Delta_{i\Delta} = 0$.

Displacement in the direction of the removed connection produced by the combined action of redundant forces and support settlements in the primary system has the form:

$$\Delta_i = \sum_{j=1}^{n} \delta_{ij} X_j + \Delta_{i\Delta}, \qquad (15.10)$$

where δ_{ij} are the unit displacements of the usual form (13.8). This expression follows from the superposition principle for elastic systems subjected to external forces' action and displacements in connections.

If redundant forces equal sought constraint reactions, then displacement (15.10) corresponds to the state of the given system with shifted supports, i.e., is equal to the given quantity Δ_i^0. Therefore, we get the canonical equation set as follows:

$$\sum_{j=1}^{n} \delta_{ij} X_j + \Delta_{i\Delta} = \Delta_i^0, \quad i = \overline{1,n}. \tag{15.11}$$

The force method is most easily implemented for analysis of the effect of support settlements upon a statically indeterminate system by means of a statically determinate primary system. Displacements $\Delta_{i\Delta}$, induced by support settlements in the primary system, are computed easily enough if the reactions of the support connections are statically determinate. Another reason to use a statically determinate primary system is that support settlements arisen in a statically indeterminate primary system could influence its stress state. In other words, it is not sufficient to find redundant forces for further construction of internal force diagrams, but it is necessary to take into account the component of internal force induced by the support settlements. On the contrary, in statically determinate primary systems the support settlements don't induce stresses, and internal forces depend on the primary unknowns only. For instance, the complete bending moment diagram in analysis of the effect of support settlements by means of a statically determinate primary system could be obtained using the formula:

$$M = \sum_{i=1}^{n} \overline{M}_i X_i. \tag{15.12}$$

If a given system is exposed to the joint impact of loads and support settlements, then in its canonical equation set, new free terms appear: the loaded displacements Δ_{iP}. These addends complement the left-hand sides of equations (15.11). For the rest, the technique of composing the canonical equations remains the same. During construction of the final internal force diagrams, the action of loads is accounted for by loaded diagrams. For example, during construction of a bending moment diagram while taking into account the action of loads, we complement formula (15.12), obtained for statically determinate primary systems, with the loaded diagram, and the final formula takes the form (13.9).

Let us consider the example of a frame with support settlements shown in Figure 15.5a. The vectors of support displacements in the given case have the form:

$$\boldsymbol{u}_A = \begin{pmatrix} a_1 \\ b_1 \end{pmatrix}; \quad \boldsymbol{u}_B = \begin{pmatrix} a_2 \\ b_2 \end{pmatrix}.$$

Besides linear displacements, the supports are subjected to angular displacements θ_1 and θ_2, respectively. If we choose the primary system in Figure 15.5b, then in the given system, we have displacements of constraints which will be removed, as follows:

$$\Delta_1^0 = a_2; \quad \Delta_2^0 = b_2; \quad \Delta_3^0 = \theta_2. \tag{15.13}$$

In the primary system, there arise the following displacements in the directions of removed constraints, induced by support settlements in this system:

$$\Delta_{1\Delta} = a_1; \quad \Delta_{2\Delta} = b_1 + l\theta_1; \quad \Delta_{3\Delta} = \theta_1. \tag{15.14}$$

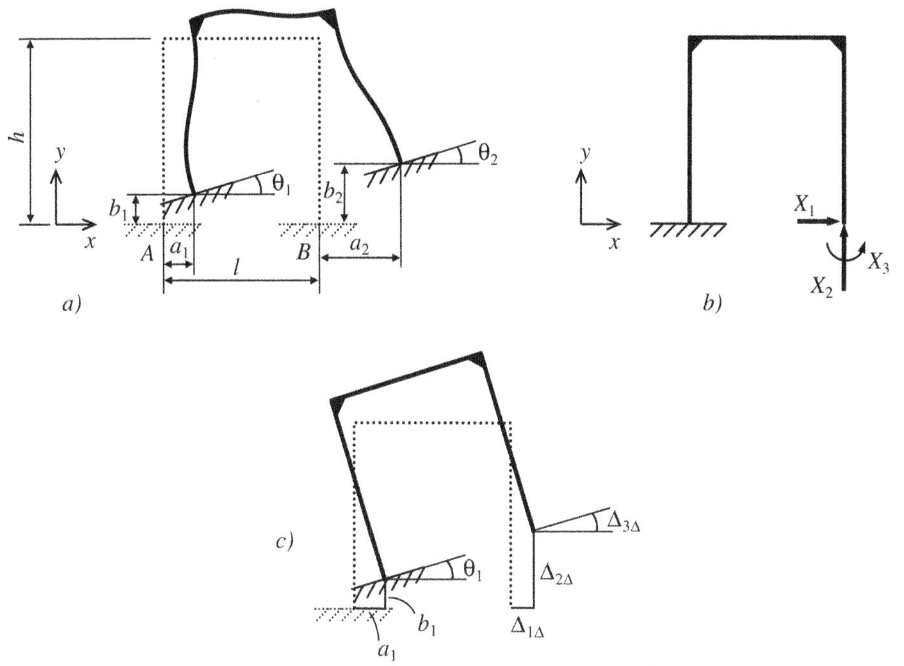

FIGURE 15.5 Frame with shifted supports: (a) Diagrams of the frame before and after the shift of supports; (b) Primary system of force method; (c) Displacements of support in the primary system and corresponding displacements in removed constraints.

In the given case, these displacements are established from geometrical construction in Figure 15.5c, and simplicity of calculation is conditioned by static determinacy of the primary system. From formulas (15.13) and (15.14), we get the free terms of canonical equations (15.11). The reader is recommended to obtain the relations for coefficients δ_{ij} on his own by multiplying the unit bending moment diagrams (assume flexural rigidity EI is the same for all members and take as given the width l and height h of the frame).

After finding the solution of the canonical equations, the final bending moment diagram can be constructed for the frame under consideration by summing up the diagrams in accordance with formula (15.12).

16 Force Method for Construction of Influence Lines and Analysis of Space Systems

16.1 ANALYTICAL METHOD OF CONSTRUCTION OF INFLUENCE LINES

The basics of the theory of influence lines exposed in Chapters 2 and 3 are developed in this chapter with reference to statically indeterminate (or redundant) elastic systems. The construction of an influence line for the response of such a system to an exerted load is most easily done by the force method. By a response, we mean a constraint reaction, an internal force acting on the cross-section of a member, or displacement in a system.

The analytical method of construction of influence lines in statically indeterminate systems by the force method is to identify the dependence of primary unknowns versus the position of the unit load in the primary system and to represent the sought response through components due to redundant forces and the unit load acting separately. It is important for applicability of the method that the response determined in the primary system complies with the superposition principle. The influence function for the response of a redundant system in the general case proves to be nonlinear; that is why for the construction of an influence line, the influence function has usually to be tabulated.

The relationship between primary unknowns and location of unit load is determined from canonical equations. The set of canonical equations in the general case has the form:

$$\sum_{j=1}^{n} \delta_{ij} X_j + \Delta_{iP} = 0, \quad i = \overline{1, n}. \tag{16.1}$$

When a system is subjected to moving load $P = 1$, we denote loaded displacements by δ_{iP} and call them *unit loaded displacements*. If the load of magnitude P has coordinate x on the load line and produces displacement Δ_{iP}, then in accordance with the definition, it holds:

$$\delta_{iP}(x) \equiv \frac{\Delta_{iP}(x)}{P}. \tag{16.2}$$

(Here the functional dependence on x is pointed out explicitly.) In the problem under consideration, we have the canonical equation set:

$$\sum_{j=1}^{n} \delta_{ij} X_j + \delta_{iP} = 0, \quad i = \overline{1, n}. \tag{16.3}$$

Let us represent this set in matrix form:

$$\delta \mathbf{X} + \delta_P = 0, \tag{16.4}$$

where $\boldsymbol{\delta}$ is the coefficient matrix of the canonical equation set; $\mathbf{X}$ is the column vector of primary unknowns; and $\boldsymbol{\delta}_P$ is the column vector of loaded displacements. The solution of equation set (16.4) can be obtained by means of the inverse matrix $\mathbf{r} \equiv \boldsymbol{\delta}^{-1}$:

$$\mathbf{X} = -\mathbf{r}\,\boldsymbol{\delta}_P, \tag{16.5}$$

that may be written in component-wise form:

$$X_i = -\sum_{j=1}^{n} r_{ij}\delta_{jP}, \quad i = \overline{1,n}. \tag{16.6}$$

So, in order to get the influence functions for primary unknowns $X_i = X_i(x)$, one has to invert the coefficient matrix of canonical equations and to establish the influence functions of unit loaded displacements (16.2). To compute the inverse matrix, one can use algebraic adjuncts A_{ij} of corresponding constituents δ_{ij}. Due to the symmetry of matrix $\mathbf{r}$, to compute it, one can use the relations:

$$r_{ij} = r_{ji} = \frac{A_{ij}}{\det \delta}. \tag{16.7}$$

The unit loaded displacements are found by means of the Mohr integral. If only bending moments are taken into account, then this integral enables us to obtain:

$$\delta_{iP} = \sum \int_0^l \frac{\bar{M}_i \bar{M}_P dz}{EI}. \tag{16.8}$$

Here $\bar{M}_i$ is, as usual, the moment diagram due to the force $X_i = 1$, and $\bar{M}_P$ is the moment diagram due to the action of load $P = 1$ (unit loaded diagram). Integral (16.8) allows us to obtain at least the tabular dependence $\delta_{iP} = \delta_{iP}(x)$, and, in simple cases, analytical dependence.

When the influence functions for primary unknowns have been determined, the sought response S can be obtained from the superposition principle in the form:

$$S = \bar{S}_P(x) + \sum_{i=1}^{n} \bar{S}_i X_i(x), \tag{16.9}$$

where $\bar{S}_i$ is the response in the primary system produced by generalized force $X_i = 1$, and $\bar{S}_P(x)$ is the response in the primary system due to the action of load $P = 1$ having coordinate x.

We may see that influence function $S = S(x)$ is obtained as the influence function for the response in the primary system plus the *component of the system's indeterminacy*. The latter is the sum of influence functions for constraint reactions taken with the factors $\bar{S}_i$. If the summation of influence functions is done in the graphical manner, then we talk about summation of corresponding influence lines.

We give two notes in relation to the analytical method stated here:

1. When making analysis of force response S, it is convenient to pass to the primary system which doesn't contain the constraint developing this force. Then formula (16.9) takes the trivial form:

$$S = X_i(x).$$

2. From formulas (16.6) and (16.9) one might get the following expression in order to expedite computing the required influence function:

$$S = \bar{S}_P(x) + \sum_{i=1}^{n} k_i \delta_{iP}(x). \tag{16.10}$$

The reader may establish coefficients k_i by himself.

Example

Construct influence lines for forces R_B and M_C acting in the system represented in Figure 16.1a.

The graphical part of the solution is shown in Figure 16.1; the computational part looks like the following.

Write down the canonical equation:

$$X_1\delta_{11} + \delta_{1P} = 0. \tag{16.11}$$

In the given case, connection 1 removed from the system is support connection B. The positive reading of displacement is taken coherently with the direction of the redundant force, that is, upward. The sought free term of canonical equation (16.11) is the displacement of the beam's free end under the action of load $P = 1$ only (diagram b). In mechanics of materials, such problems were solved. The displacement in the given case is established by integration of an elastic curve equation with zero boundary conditions at the fixing of the beam (Case et al. 1999, p. 302). The solution has the form:

$$\delta_{1P} = -\frac{x^2(3l - x)}{6EI} \cdot 1.$$

This solution at $x = l$ enables us to establish the coefficient of the canonical equation:

$$\delta_{11} = \frac{l^3}{3EI} \cdot 1.$$

Hence:

$$R_B = X_1 = -\frac{\delta_{1P}}{\delta_{11}} = \frac{x^2(3l - x)}{2l^3}. \tag{16.12}$$

The corresponding influence function is represented in Table 16.1.

The influence function of the bending moment under consideration is determined by the formula:

$$M_C = \bar{M}_{CP} + \bar{M}_{C1} X_1,$$

where $\bar{M}_{C1} = \frac{l}{2}$. Figure 16.1 shows the summation of components of the wanted influence line.

16.2 KINEMATIC METHOD OF CONSTRUCTION OF INFLUENCE LINES

In order to substantiate the kinematic method of construction of an influence line for a force (or moment) response of a statically indeterminate system, we remind readers of the basic thesis of the kinematic method developed for statically determinate systems (see Chapter 6). According to this method, a statically determinate system is to be transformed into a mechanism by removal of

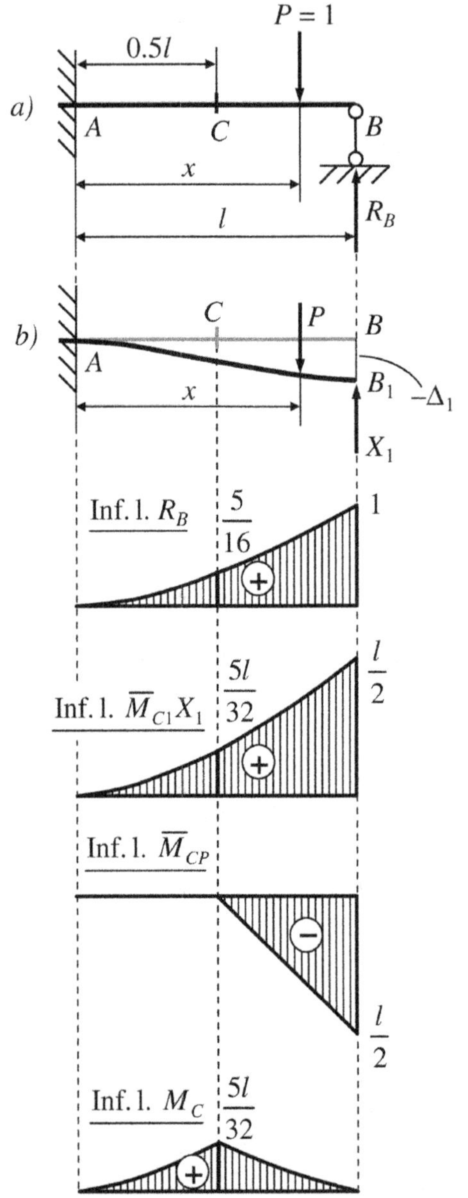

FIGURE 16.1 Example of summation of influence lines: (a) given system; (b) primary system; $\Delta_1 = -BB_1$ – displacement in the direction of removed connection (shown the case when $\Delta_1 < 0$).

the connection which develops a reaction equal to the sought force. Then we have to give virtual displacement to the mechanism and construct the diagram of load line deflections. This diagram should be transformed by dividing its ordinates by displacement in the direction of the removed connection and interchanging the ordinates' signs. The obtained graph coincides with the sought influence line.

In the case of investigation of a required connection in a statically indeterminate system, there are no differences in the kinematic method of influence line construction. As before, the elimination of connection converts a system into a mechanism, where displacements permitted by constraints

TABLE 16.1
Tabular Influence Function for Reaction R_B (Figure 16.1)

x	X_1
0	0
$l/4$	$11/128$
$l/2$	$5/16$
$3l/4$	$81/128$
l	1

resulted from mutual displacements of unchangeable discs. During investigation of the redundant connection, the peculiarities of the kinematic method are as follows:

- Elimination of this connection doesn't cause instability of the system, and virtual displacements in a system are attended by the members' deformations;
- The cause of virtual displacements which determine the sought load line deflection diagram is external generalized force substituting for the reaction of the removed connection.

The other provisions of the method remain valid: the transition from deflection diagram to influence line is done, as before, by dividing the diagram's ordinates by displacement in the direction of the removed connection and interchanging the signs of the ordinates.

So, let there be the problem of the construction of an influence line for some reaction S of a redundant connection in a system. We pass to the primary system where only this connection has been removed, and instead of reaction S, there acts redundant force X_1. First, we shall show that the diagram of load line deflections arisen due to the action of force X_1 in the primary system is proportional to the influence line for force S.

The canonical equation of the force method in the given case has the form (16.11). Let us denote by δ_{P1} the displacement of the point where the load $P = 1$ is exerted, produced by the action of the redundant force $X_1 = 1$ only. Since the unit loaded displacement is the generalized displacement, owing to Maxwell's theorem, we have:

$$\delta_{1P} = \delta_{P1}. \tag{16.13}$$

Hence, we obtain the solution of the canonical equation in the form:

$$X_1(x) = -\frac{\delta_{P1}(x)}{\delta_{11}}. \tag{16.14}$$

Here on the left-hand side, the dependence of force X_1 on the coordinate of the unit load is shown in explicit form; the right-hand side comprises the function of load line deflections produced by unit force X_1. The obtained relation of two functions proves the proportional relation of their graphs.

In the next step, it is easy to establish the rule of transition from deflection diagram to influence line. Rewrite formula (16.14) in the form:

$$S(x) = -\frac{\delta_{P1}(x)X_1}{\delta_{11}X_1} = -\frac{\Delta_{P1}(x)}{\Delta_{11}}, \tag{16.15}$$

where the influence function of the sought constraint reaction is on the left-hand side, and on the right-hand side, we denote the function of deflections produced by any redundant force X_1 as $\Delta_{P1}(x) \equiv \delta_{P1}(x)X_1$ and the displacement produced in the direction of removed connection by this force as $\Delta_{11} \equiv \delta_{11}X_1$. The obtained expression links influence function and deflection function; we see that, really, the sought influence line is obtained by dividing the ordinates of the deflection diagram by displacement in the direction of the removed connection and interchanging the signs of the ordinates.

The main statement of the kinematic method of influence line construction is known as the Müller-Breslau principle. For statically determinate systems, this principle was proved in Chapter 6. We see that it is generalized to statically indeterminate systems by pointing out the cause of virtual displacements. On account of the generalization we have made, the Müller-Breslau principle is formulated as follows:

The influence line for a force (or moment) response of a system is proportional to the diagram of load line deflections in the system obtained by removing the constraint producing this response and by giving virtual displacement at the location and in the direction of the removed constraint. This diagram can be converted into an influence line by dividing ordinates by displacement in the direction of the removed constraint and then interchanging the ordinates' signs. If removal of a constraint makes the system unstable, then there should be considered virtual displacements without members' deformation; if the system remains stable, then there should be considered the displacements produced by generalized force substituting for the reaction of the removed constraint. In the latter case, the system is assumed to be elastic.

Formula (16.15) is correct at infinitesimal displacements in the direction of the removed connection, but at finite displacements, its error may become unacceptable. For instance, according to this formula, **if generalized force X_1 exerted in the place of an eliminated redundant connection produces unit displacement in the direction of this connection, then the influence function coincides with the deflection function taken with the opposite sign.** This is one of the possible formulations of the Müller-Breslau principle in the case of statically indeterminate systems, which should be used with caution. For angular displacements, the involved errors make this assertion doubtful; for translational displacements, the assertion can be taken as correct if the unit of length is sufficiently small.

Note that the positive direction of deflection coincides with the positive direction of the moving load, that is, downward. Therefore, deflection diagram Δ_{P1} becomes a qualitative influence line if positive reading of deflection is taken upward.

The kinematic method is convenient to determine the shape of an influence line. In order to further clarify the influence line, the analytical method is usually employed.

Example

Reveal the approximate profile of the influence line for bending moment M_C acting in the system in Figure 16.1a.

Remove the connection which develops a reaction equal to moment M_C, i.e., insert the hinge in point C. Redundant force X_1 in the given case is two force couples exerted upon the cross-sections adjacent to the hinge, and displacement Δ_{11} is an angle of mutual rotation of these sections. The beam deformed due to exertion of generalized force $X_1 > 0$ is shown in Figure 16.2. The corresponding deflection diagram has the profile of the influence line obtained earlier in Figure 16.1. Taking into consideration the shape of the diagram, one can make some conclusions about the corresponding influence line. For instance, the tangent to the influence line in point A is horizontal, and the influence line is asymmetrical with respect to the vertical axis running through point C.

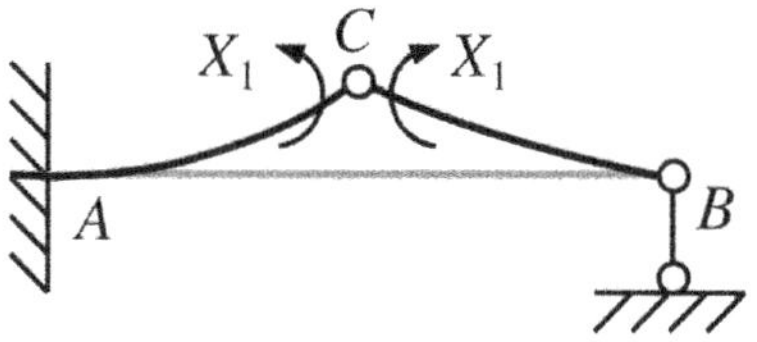

FIGURE 16.2 Beam deformed due to exertion of redundant force X_1.

16.3 CONSTRUCTION OF INFLUENCE LINES FOR STATICALLY INDETERMINATE TRUSSES

During analysis of statically indeterminate trusses by the force method, at the beginning of analysis, as usual, the degree of redundancy should be determined. In the case of plane trusses, the next formula is convenient to obtain this characteristic:

$$n = B + L_{sup} - 2J, \quad (16.16)$$

where J is the number of joints; B is the number of members; and L_{sup} is the number of support connections. In Chapter 7, we obtained the condition of truss stability (7.2) by substitution of a truss's characteristics into the general condition (7.1). Formula (16.16) is obtained by the same substitutions into general formula (1.8) derived for the degree of redundancy of the plane structure.

In the case of the truss in Figure 16.3a, we have one redundant connection: $n = 13 + 4 - 2 \cdot 8 = 1$. Let us construct the influence lines for support reactions R_C and R_A acting upon this truss, at given geometry characteristics d, β, and rigidities of members EA. This example is useful in that it shows how kinematic and analytical methods supplement each other. First, we establish, by means of the kinematic method, that for an arbitrary truss with a redundant support link, the influence line for the reaction of this link is polygonal and continuous, with kinks under the joints of the load chord. To do that, we determine the shape of the load chord deflection diagram, when the truss is subjected to external force in the place of the removed support connection developing the reaction under consideration. **Vertical** displacement at an arbitrary point of the load chord's member is represented by two components: displacement of the member as a whole and displacement due to deformation of tension-compression in this member. The first component depends linearly on the point's coordinate x on the load chord; the second component can be neglected as the second-order infinitesimal.* Thus, the vertical displacement diagram constructed for the separate member is a segment, and the load chord deflection diagram is a continuous polyline, the segments of which are connected at the verticals running through the chord joints. But the diagram displaying the deflections caused by external force in the place of removed connection is proportional to the corresponding influence line. Therefore, the influence line for the reaction of the redundant support link in a truss consists of the segments connected under the joints of the load chord.

Figure 16.3 shows a possible influence line R_C whose outline is established by the kinematic method. Firstly, we notice that the given line is symmetrical relative to the central vertical, in the same manner as the load chord deflection diagram is symmetrical for the truss on the scheme b if only external force X_1 is exerted. Next, the kinematic method enables us to find the ordinates under support joints 1, 5, and 8. There remains to determine the ordinate R_{C3} (or R_{C7}), to which end we use the analytical method.

Let us introduce the primary system of the force method as it is shown in Figure 16.3b. The canonical equation still has the form (16.11), assuming that connection 1 removed from the system

* The load chord of an undeformed truss is assumed to be horizontal.

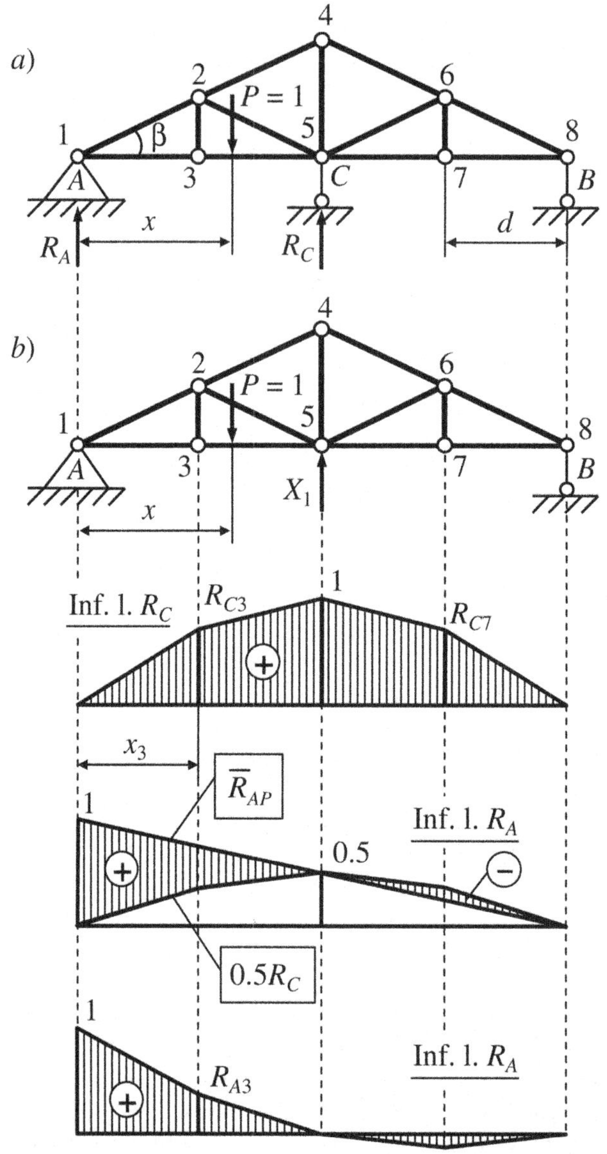

FIGURE 16.3 Statically indeterminate truss and influence lines for support reactions.

is support connection C. The unit displacements in this equation are determined through Mohr integrals:

$$\delta_{11} = \sum \frac{\bar{N}_1^2 l}{EA}, \tag{16.17}$$

where l and A are geometrical characteristics of members (the member number is not displayed), and $\bar{N}_1$ is the axial force in every member due to redundant force $X_1 = 1$;

$$\delta_{1P} = \sum \frac{\bar{N}_1 \bar{N}_P l}{EA}, \tag{16.18}$$

where $\bar{N}_P$ is the axial force in every member due to the unit moving load. In order to determine the ordinate R_{C3}, one has to find coefficient δ_{11} and the deflection at joint 5 due to the action of the unit load upon joint 3, i.e., $\delta_{1P} = \delta_{1P}(x_3)$. The problem is cumbersome enough: it is necessary to compute the forces $\bar{N}_1$ and $\bar{N}_P(x_3)$ in 13 members. Influence line R_A is established by summing up two graphs obtained for the same primary system:

$$R_A = \bar{R}_{AP} + \bar{R}_{A1}X_1,$$

where $\bar{R}_{A1} = -\dfrac{1}{2}$.

In the example being considered, we shall demonstrate how, in the case of a statically indeterminate truss, there could be constructed a primary system of force method with removed internal connections. Insert in post 4–5 of the given truss the rigid compound node with a coherent connection developing the axial force (the node is shown in Figure 12.2c). By removing the coherent connection, pass to the primary system shown in Figure 16.4. The forces substituting for the reaction of the removed connection are now exerted on the plates of the compound node and make up the redundant force X_1. The obtained primary system turns out to be convenient for computing the forces in members and setting up the canonical equation. The primary truss consists of three parts: the left secondary truss member 1–2–5; the right STM 5–6–8; and the three members with a common joint 4. Each STM works as a simply supported truss girder while the members constituting joint 4 should be considered as the source of a load upon the STMs. The load P produces internal forces only in one of the STMs; the generalized force X_1 produces internal forces which are the same in symmetrically located members. Formulas (16.17) and (16.18) still remain in force and enable us to set up the canonical equation, but calculations by these formulas get easier. It is recommended to compute the ordinate R_{C3} of the influence line using the primary system in Figure 16.4, taking as given that truss height is equal to panel length (i.e., $l_{45} = d$) and the rigidities of all members are the same.

Previously it was proved that the influence line for reaction of a redundant support connection is polygonal and continuous, with kinks under the joints of the load chord. The proof provided remains in force in the case of a redundant connection producing axial force in any member of a truss, for example, in the case of the connection which was removed from the central post of the given truss at transition to the principal structure in Figure 16.4. In Chapter 9, it was pointed out that the shape of influence lines for forces in statically determinate trusses has the same peculiarities. Therefore, we may formulate the following property which underlies the construction of an influence line for any force in any truss: **an influence line for the force acting in a truss consists of the segments connected under the joints of the load chord.**

Note

It is possible to construct a graph proportional to the influence line for axial force produced by redundant connection in a member of a truss, with no insertion of a rigid compound node. It is

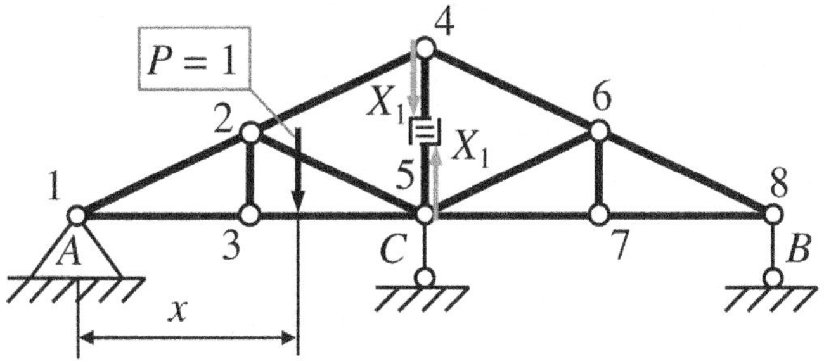

FIGURE 16.4 Primary system for the truss in Figure 16.3a.

allowed to remove the researched member and exert infinitesimal forces replacing the reactions of this member upon corresponding joints; the obtained load chord deflection diagram will be proportional to the sought influence line.

16.4 ANALYSIS OF SPACE SYSTEMS BY FORCE METHOD

Graphic-analytical analysis by the force method can be fulfilled but only for very simple space systems. The analysis of statically indeterminate systems that are sophisticated to some extent requires the application of computer techniques due to the large degree of redundancy in space problems. Such analysis is made, as a rule, by means of the finite element method.

In calculation of a space system by the force method, a statically determinate primary system is used, which is obtained by elimination of redundant constraints from the given system. Canonical equations for a space system redundant n times do not differ in sense and form from equations (16.1) obtained for a plane system. Nevertheless, the force method in relation to space systems has a number of peculiarities:

1. When constructing a statically determinate primary system, we would take into account that in the space, there are 6 independent statics equations for a body – system's member (but not three as in a plane system). Hence, there are consequences:

   1.1. The criterion of static determinacy for an externally stable system in the space has the form:

   $$6B_{sd} = L + L_{sup}, \tag{16.19}$$

   where B_{sd} is the number of statically determinate bodies constituting a space system; L is the number of internal connections (i.e., connections of bodies represented by simple rods); and L_{sup} is the number of the system's connections with the earth. The degree of static indeterminacy of an externally stable system is determined as follows:

   $$n = L + L_{sup} - 6B_{sd}. \tag{16.20}$$

   1.2. The pass to a primary system is done, as usual, by removal of external or internal connections. The lateral dissection of a member eliminates 6 connections; the insertion of a spherical hinge eliminates 3 connections.

   1.3. External static determinacy of a body is ensured by 6 support connections.

2. While computing unit and loaded displacements which specify the canonical equation set, in the general case, we have to employ the six-termed Mohr's formula (11.16) (but not the three-termed version as in planar problems). In analysis of space frames, it is sufficient to take into account, from among internal forces, only bending moments and torque; in analysis of space trusses, it is sufficient to take into account only axial force in members.
3. In analysis of plane frames subjected to space loading, we take into consideration that redundant forces acting in the plane of the frame are determined independently from redundant forces exerted in the direction perpendicular to this plane. In other words, the canonical equation set allows decomposition based on the direction of primary unknowns. We here consider plane frames symmetrical with respect to a plane containing axial lines of the frame's members. By virtue of the frame's symmetry, its deformation under action of a plane assemblage of loads occurs in this same plane.

We clarify statement 1.1 in the example in Figure 16.5a. The given frame is converted into a statically determinate body if one takes it off from supports and dissects the closed contour (Figure 16.5b). This transformation eliminates 4·6 = 24 external connections and 6 internal connections. By using formula (16.20), we get the degree of static indeterminacy $n = 6 + 24 - 6{\cdot}1 = 24$. At

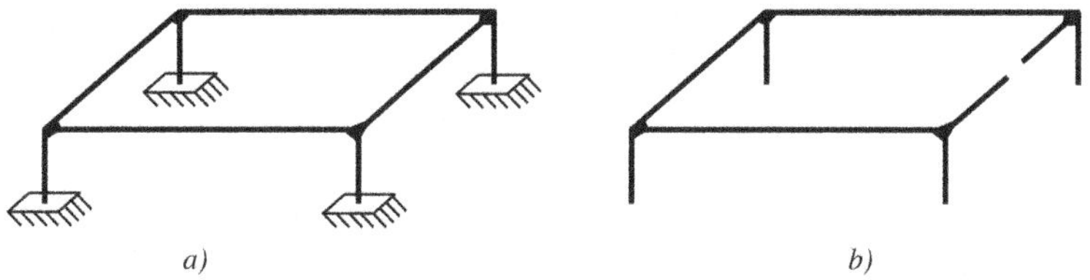

FIGURE 16.5 Statically indeterminate frame (a) and its conversion into statically determinate body (b).

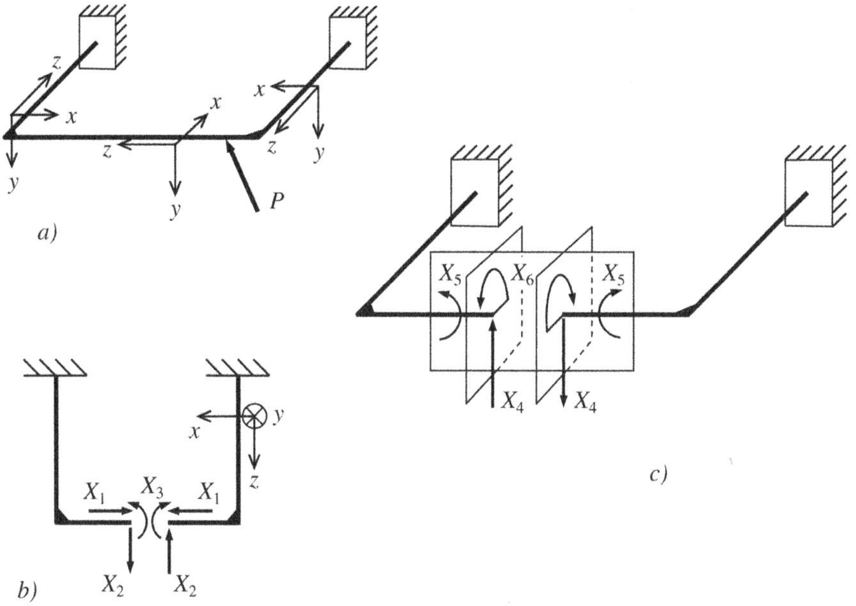

FIGURE 16.6 Plane frame subjected to space loading.

the same time, this frame is a system of 8 members connected by 4 multiple rigid joints. Each rigid joint has the repetition factor of 2, i.e., is equivalent to two simple rigid joints and makes up $6 \cdot 2 = 12$ links. The number of internal connections for the system under consideration proves to be $L = 12 \cdot 4 = 48$, and the degree of redundancy is calculated in the form: $n = 48 + 24 - 6 \cdot 8 = 24$.

We clarify statement 3 in the example of a Π-shaped horizontal frame. In order to specify internal forces, we use the sliding CS with a triplet of axes shown in Figure 16.6a, in three possible locations (the z-axis is directed, as usual, towards the part of member removed from consideration). Let us pass to the primary system by cutting the frame in the middle. The primary unknowns constitute two subsets: unknowns X_1, X_2, X_3 act in the plane of the frame and cause internal forces N, V_x, M_y, respectively, in the neighborhood of dissection (Figure 16.6b); unknowns X_4, X_5, X_6 act in the planes perpendicular to the plane of the frame and cause internal forces V_y, M_x, M_z, respectively, at the adjacent cross-sections (Figure 16.6c). The first group of redundant forces produces displacements only in the plane of the frame. Displacements in the directions of generalized forces X_4, X_5, X_6 do not arise due to the action of forces of the first group, and corresponding secondary unit displacements in canonical equations equal zero. Owing to this, the system of canonical equations decomposes into two systems which enable us to determine each group of unknowns independently from another.

All said about features of the force method in space problems equally applies to frames and trusses, but for analysis of a truss, the suggested approach is not convenient. The matter is that the stress state of a truss is mainly determined by axial forces in members, i.e., by one internal force at a cross-section rather than six, and to take into consideration all totality of internal forces means to complicate the problem without need. We can simplify the design scheme by replacement of rigid joints with spherical hinges, i.e., by transition to a hinged truss. At such transition, the axial forces in members change insignificantly, whereas the other internal forces disappear, and owing to this, the analysis is essentially simplified. Such simplification of the problem makes sense in manual calculations, whereas in computer-based analysis the rigid joints could be simulated without noticeable complication of analysis of a system. Since the manual calculations of a space system are rare in engineering practice, the specificity of analysis by the force method for space hinged trusses is not discussed here.

SUPPLEMENT TO CHAPTER

Peculiarities of Analysis of Space Truss by Force Method

During analysis of a statically indeterminate space truss by the force method, we construct the design diagram in which rigid joints are replaced by spherical hinges. The obtained hinged truss differs from the given truss in that we have removed from the latter the connections which produce the forces in members different from axial forces. Owing to that, the degree of static indeterminacy essentially – sometimes many times over – decreases, and in simple problems, graphic-analytical analysis proves to be possible. But now the members can rotate around their own axes, and, strictly speaking, the space hinged truss is a substatic system. For investigation of such a system, we modify the conceptions of the member's freedom and the system's stability in the following way:

- Rotation of a member around its own axis is not considered as a DOF, i.e., it is accepted that a free bar has 5 DOFs instead of 6;
- Internal (external) stability of space hinged system is understood as an impossibility of mutual displacements of joints (displacements of joints relative to the earth) without deformation of members.

In Chapter 10, we established the condition of stability of a space hinged truss (10.2), which we write down again:

$$B + L_{sup} \geq 3J.$$

Here, the left side is the number of unknown forces specifying the stress state of the truss; the right side is the number of independent equations of statics. The difference between these characteristics is, by definition, the degree of static indeterminacy, which is equal to the number of redundant connections of the system and for a spatial truss has the form:

$$n = B + L_{sup} - 3J. \tag{16S.1}$$

To construct a primary system, the redundant connections are removed. In the example of a plane truss in Figure 16.3a, two primary systems have been constructed; in one case was removed support connection, in another the redundant connection was revealed by insertion of a rigid compound node into the post. In the second case, the rigid node was converted into a translational kinematic pair through removal of the link causing axial force in the member. In order to reveal redundant connections in members of a space truss, one should employ rigid compound nodes of appropriate construction. Just like in planar problems, these nodes should be converted into translational pairs by removal of the coherent connection causing internal axial force. Instead of the reaction forces of the removed coherent connection, the external longitudinal forces are applied. The fact that the

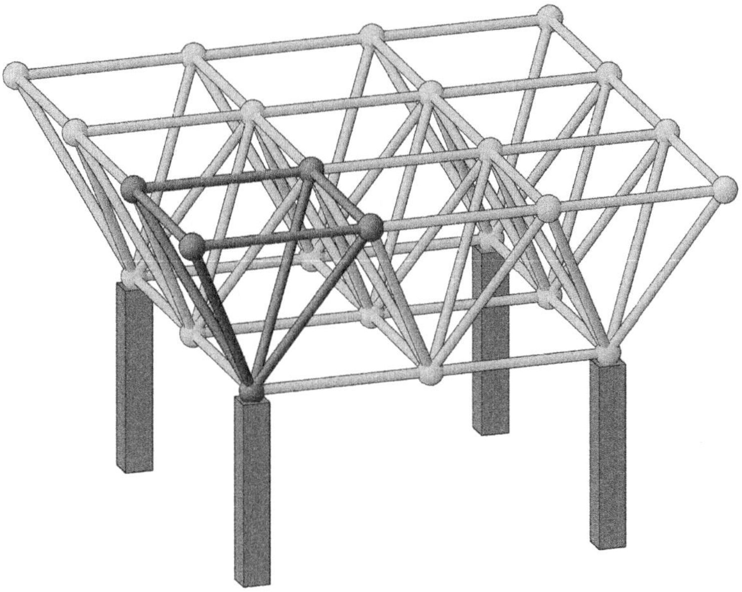

FIGURE 16S.1 Example of space truss of periodical structure (repeated element is emphasized).

member can rotate around its axis has no effect on the procedure of the primary system's construction. All basic provisions of the force method developed for analysis of plane trusses remain in force for analysis of space trusses.

Figure 16S.1 shows a space truss consisting of 72 members connected at 25 joints. The truss is supported by four columns which are considered to be part of the earth. Similar trusses with a greater number of structural elements are used as roof trusses. We shall consider the given truss as a hinged one with immovable support hinges. According to formula (16S.1) we have the degree of static indeterminacy for this truss: $n = 72 + 4{\cdot}3 - 3{\cdot}25 = 9$. One can see that the scheme of a hinged truss enables us to obtain a rather small degree of redundancy, taking into account a large enough number of members in a given system. But further kinematic analysis is difficult; even to verify the stability of this system by hand appears to be a complicated problem, and it is even harder to construct the primary system of force method. So, the analysis of a relatively unsophisticated space truss by the graphic-analytical method causes difficulties even at the stage of kinematic analysis. At the same time, the computer-based analysis by the finite element method in this problem is easily implemented.

Section VIII

Displacement Method and Mixed Method

At the beginning of the section, the theory of elastic systems is developed: the concepts of flexibility matrix and stiffness matrix are defined; Clapeyron's theorem about the strain energy of an elastic system is proved; the conditions of nonsingularity are established for a flexibility matrix. Next, the essence of displacement method is stated, conceptual framework of the method is established, and the stages of analysis through this method are presented. The section set forth the technique of composing canonical equations and explains how – when primary unknowns of the method have been found – the distributions of internal forces should be determined in a system. As a separate issue the verification procedures for calculation results are considered. At the end of the section, Gvozdev's mixed method is exposed, which combines the force method and displacement method.

17 Flexibility Matrix and Stiffness Matrix

Initial Info about Displacement Method

INTRODUCTION FROM LINEAR ALGEBRA

In the introduction to Chapter 7, a linear equation set was under consideration, which we reproduce here in the vector form:

$$\mathbf{Ax} = \mathbf{b}. \tag{17I.1}$$

We consider the vector of unknowns $\mathbf{x}$ and the vector of free terms $\mathbf{b}$ in an n-dimensional real coordinate space; accordingly, $\mathbf{A}$ is a real matrix of order n. The matrix $\mathbf{A}^{-1}$ is referred to as the inverse of a matrix $\mathbf{A}$ if $\mathbf{A}^{-1}\mathbf{A} = \mathbf{AA}^{-1} = \mathbf{I}$, where $\mathbf{I}$ is the identity matrix. The inverse $\mathbf{A}^{-1}$ exists under the compulsory condition that matrix $\mathbf{A}$ is nonsingular. The following stipulations enable us to establish whether this matrix is singular or not:

1. For singularity of matrix $\mathbf{A}$, it is necessary and sufficient that there exists a nonzero solution of equation (17I.1) under condition $\mathbf{b} = 0$.
2. For nonsingularity of matrix $\mathbf{A}$, it is necessary and sufficient that the solution of equation (17I.1) exists at arbitrary vector $\mathbf{b}$.

The function of n-dimensional vector argument, having the form:

$$Q(\mathbf{x}) = \mathbf{x}^{\mathrm{T}}\mathbf{Ax} = \sum_{i,j=1}^{n} A_{ij}x_i x_j, \tag{17I.2}$$

is referred to as the quadratic form of matrix $\mathbf{A}$ with argument $\mathbf{x}$. In equality (17I.2), the quadratic form is represented first in vector notation and next in scalar form. Superscript "T" denotes transposing of the matrix (in the given case of column vector). Matrix $\mathbf{A}$ is referred to as positive semidefinite (positive definite) if its quadratic form (17I.2) is nonnegative (positive) for arbitrary vector $\mathbf{x}$. Positive semidefiniteness (definiteness) of $\mathbf{A}$ is briefly written by inequality: $\mathbf{A} \geq 0$ ($\mathbf{A} > 0$).

The positive definite matrix is nonsingular (which is sufficiently obvious). The symmetric positive semidefinite matrix is positive definite if and only if it is nonsingular (Horn and Johnson 2013, p. 431). If the matrix is not symmetric, then the analogous statement is not correct. For example, a 2D-rotation matrix of the form

$$\mathbf{A} = \begin{pmatrix} 0 & -1 \\ 1 & 0 \end{pmatrix}$$

is positive semidefinite and nonsingular, yet is not positive definite.

17.1 FLEXIBILITY MATRIX AND STIFFNESS MATRIX; EXISTENCE OF STIFFNESS MATRIX

Items 17.1 and 17.2 of this chapter set out the additional topic of the theory of an elastic system's displacements, concerned with representation of the state of the elastic structure through the set of displacements. It will enable us in the third item to start study of a new method of the stress state analysis for a statically indeterminate bar system – the displacement method.

Let an externally stable planar system of elastic bodies be subjected to assemblage of generalized external forces $X_i, i = \overline{1,n}$, which produce corresponding generalized displacements $Z_i, i = \overline{1,n}$. It is assumed that the assemblage of external forces X_i doesn't include reactions of support connections, yet we avoid calling forces X_i as loads, for they might be not given in the problem. Nevertheless, these forces, along with loads, cause displacements in a system and support reactions. By applying the superposition principle for elastic systems, in which the state is defined by external forces, we can represent the relation of displacements and external forces as follows:

$$Z_i = \sum_{j=1}^{n} \delta_{ij} X_j, \quad i = \overline{1,n}, \tag{17.1}$$

where coefficients δ_{ij} are called unit displacements, because quantity δ_{ij} is the displacement in the direction of generalized force X_i produced by only the force $X_j = 1$ exerted upon the system, i.e.:

$$\delta_{ij} \equiv Z_i / X_j. \tag{17.2}$$

In matrix notation, the relation (17.1) has the form:

$$\mathbf{Z} = \delta \mathbf{X}. \tag{17.3}$$

Here $\boldsymbol{\delta}$ is the matrix of unit displacements; $\mathbf{X}$ is the column vector of generalized external forces; and $\mathbf{Z}$ is the column vector of generalized displacements in the places of exertion of external forces.

The matrix, which determines the column vector of generalized displacements $\mathbf{Z}$ *through the column vector of generalized external forces* $\mathbf{X}$, *is referred to as the flexibility matrix for the given totality of generalized external forces.* In the case under consideration $\boldsymbol{\delta}$ is the flexibility matrix. In order for the inverse of the flexibility matrix to exist, the latter must be nonsingular. The inverse matrix is referred to as the *system's stiffness matrix* and denoted by $\mathbf{r} \equiv \boldsymbol{\delta}^{-1}$.

We carry out the investigation of the flexibility matrix properties by means of Clapeyron's theorem, which is proved below. This theorem is formulated here for the general case of an elastic system and was proved by French physicist and railroad engineer Benoît Clapeyron in the year 1833 for the particular case of a single body keeping equilibrium under load (Love 1906, p. 170).

CLAPEYRON'S THEOREM

The strain energy of an elastic system, which is in equilibrium under a given loading, is equal to half the work done by the external forces, acting through the displacements of the system from the unstressed state to the state of equilibrium.

REMARK

We prove the theorem under the assumption that the elastic system is a structure with no support settlements. However, the theorem is correct for arbitrary – maybe unstable – systems of elastic bodies connected by perfect connections if displacements in the systems are small. We do not include support reactions into assemblage of external forces in the proof. This is permissible, since the displacements in the directions of support reactions are not possible in a system being considered.

Proof

We consider the quasistatic process of a system's loading in which the external forces uniformly increase from zero to given values during time segment [0, T]. The strain energy of a system is equal to the total work W done by external forces in such a process (see Supplement to Chapter 11). Denote the vector of generalized external forces exerted upon a system on termination of the process as $\mathbf{X}$; and the vector of corresponding generalized displacements as $\mathbf{Z}$. At the time point t, generalized forces acting upon the system and corresponding displacements are determined by vectors $\mathbf{X}_1(t) = \frac{t}{T}\mathbf{X}$, $\mathbf{Z}_1(t) = \frac{t}{T}\mathbf{Z}$. During a small interval of time dt, the work done by external forces is equal to:

$$dW = \mathbf{X}_1^{\mathrm{T}} d\mathbf{Z}_1 = \frac{t}{T}\mathbf{X}^{\mathrm{T}}\frac{dt}{T}\mathbf{Z} = \frac{tdt}{T^2}\mathbf{X}^{\mathrm{T}}\mathbf{Z}.$$

Total work done during period T is obtained through integration:

$$W = \int_0^T dW(t) = \frac{1}{2}\mathbf{X}^{\mathrm{T}}\mathbf{Z}. \tag{17.4}$$

Because this work equals strain energy of a system, we come to the assertion of the theorem.

Corollary

A flexibility matrix is a positive semidefinite matrix.

Proof

Equality (17.4) allows the representation of the system's strain energy in the form:

$$U_s = \frac{1}{2}\mathbf{X}^{\mathrm{T}}\delta\mathbf{X}. \tag{17.5}$$

This energy is nonnegative at an arbitrary vector of external forces $\mathbf{X}$, and, therefore, the quadratic form on the right side is nonnegative.

The corollary is proved.

The flexibility matrix is symmetric according to Maxwell's theorem. From its symmetry and positive semidefiniteness, it follows that **conditions of nonsingularity and positive definiteness of a flexibility matrix are equivalent**. The conditions of flexibility matrix nonsingularity are established by the following two theorems.

Theorem 1 of Flexibility Matrix

For singularity of a flexibility matrix, it is necessary and sufficient that there exists a nonzero assemblage of generalized external forces (i.e., the such that $\mathbf{X} \neq 0$) which doesn't produce the stresses in a system.*

* This result was published by Smirnov et al. (1981, pp. 271–274).

Proof

If a flexibility matrix is singular, then at some nonzero assemblage of forces $X_i, i=\overline{1,n}$, the set of displacements $Z_i, i=\overline{1,n}$, is zeroth. In this case, the strain energy of the system is zero according to Clapeyron's theorem, and stresses in the system are absent. Conversely, the absence of stresses means the absence of displacements in a stable system, hence $\mathbf{Z} = 0$. If a nonzero totality of forces $X_i, i=\overline{1,n}$, is possible in this state of the system, then we have equality $\boldsymbol{\delta}\mathbf{X} = 0$ at $\mathbf{X} \neq 0$, which means singularity of matrix $\boldsymbol{\delta}$.*

Theorem 2 of Flexibility Matrix

The flexibility matrix is nonsingular if and only if generalized forces can produce only one nonzero displacement Z_k in the set of displacements $Z_i, i=\overline{1,n}$, for every $k = 1,2,\ldots, n$.

Proof

If a flexibility matrix is nonsingular, then a solution of equation

$$\boldsymbol{\delta}\mathbf{X} = \mathbf{Z} \tag{17.6}$$

exists at arbitrary n-dimensional vector $\mathbf{Z}$, including every basis vector

$$\mathbf{Z} = \mathbf{e}_k \equiv (\underbrace{0,\ldots,0}_{k-1},1,0,\ldots,0)^{\mathrm{T}}$$

or proportional to the one. Thus, the theorem's forward assertion (assertion "only if") has been proved. Vice versa: assume that the displacements forming nonzero n-dimensional vector $\mathbf{Z} = \lambda_k\mathbf{e}_k$ can be realized in a system for every $k = 1, 2, \ldots, n$, owing to the action of generalized forces making up vector $\mathbf{X}$. Then the displacements constituting arbitrary basis vector $\mathbf{e}_k$ are implemented under appropriate vector of generalized forces $\mathbf{X}_k$, i.e., every basis vector can be represented in the form:

$$\mathbf{e}_k = \boldsymbol{\delta}\mathbf{X}_k.$$

By means of this equality, it is easy to make sure that the solution of equation (17.6) with an arbitrary vector of absolute terms $\mathbf{Z} = \sum_i Z_i\mathbf{e}_i$ is obtained in the form: $\mathbf{X} = \sum_i Z_i\mathbf{X}_i$. So, an arbitrary vector of absolute terms $\mathbf{Z}$ allows for solution of equation (17.6), which is possible only if coefficient matrix $\boldsymbol{\delta}$ is nonsingular, which was to be proved.

We say that generalized displacements of totality $Z_i, i=\overline{1,n}$, are independent if under appropriate generalized forces $X_i, i=\overline{1,n}$, they can take arbitrary values in some neighborhood of zero independently from each other. In other words, the displacements constituting vector $\mathbf{Z}$ are independent if this vector can be arbitrary in the domain:

$$-\varepsilon < Z_i < \varepsilon, \quad i=\overline{1,n}.$$

The independence of displacement vector components is a necessary and sufficient condition of flexibility matrix nonsingularity. In the proof of Theorem 2, it was pointed out that this condition

* By employing Theorem 1, the reader can easily prove that a nonsingular flexibility matrix is positive definite. To do that, it is sufficient to use expression (17.5) and not employ matrix theory.

is required. The sufficiency of the condition is the corollary of this theorem. The named condition, along with Theorems 1 and 2, is used further on as a criterion of nonsingularity of flexibility matrix.

Figures 17.1 and 17.2 show the examples of generalized forces' action upon an elastic Π-shaped frame. In these examples, every generalized force is specified either by the vector and the joint of exertion or by the concentrated moment from the force couple exerted on a joint or by two equal and collinear forces exerted on different joints of the frame. Figure 17.1 presents two cases where the flexibility matrix is nonsingular. It is easy to see that arbitrary nonzero assemblage of generalized forces (i.e., when $\mathbf{X} \neq 0$) in these instances causes the deformation of the frame. To make sure of that, one has to presume that members of the frame do not oppose the movement of joints upon which the generalized forces are exerted. Then a nonzero assemblage of forces will produce the movement of at least one joint subjected to these forces, i.e., the totality of forces can't be balanced without deformation of the frame. Nonsingularity of the flexibility matrix in these instances follows from Theorem 1. Figure 17.2 shows the instances of generalized forces' application in which the flexibility matrix is singular. In instances *a* and *c*, one can specify the magnitudes of generalized forces, being nonzero in totality, such that they are balanced in all joints; hence matrix $\boldsymbol{\delta}$ is singular according to Theorem 1. In instance *b*, displacements Z_2 and Z_3 arise only simultaneously and singularity of matrix $\boldsymbol{\delta}$ follows from Theorem 2.

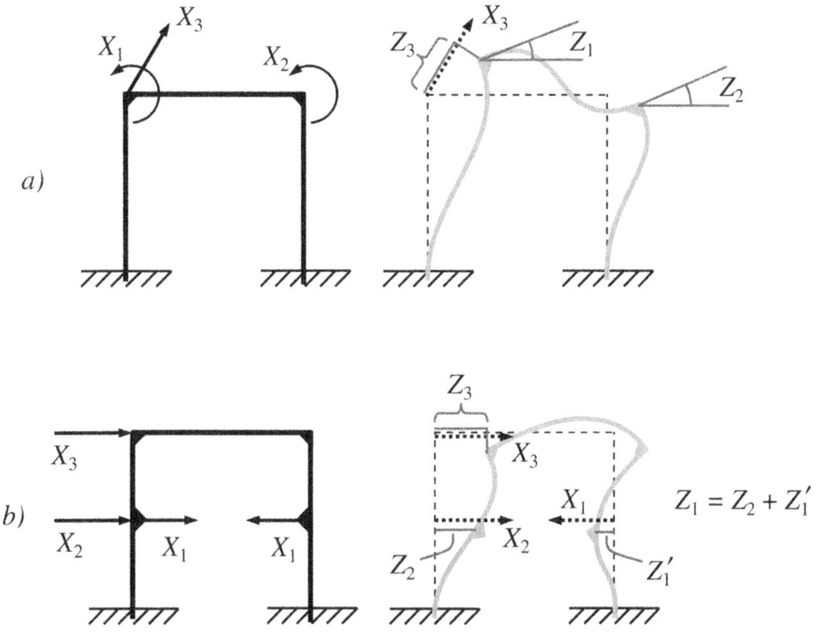

FIGURE 17.1 Instances where stiffness matrix exists: diagrams on the left show exertion of generalized forces; diagrams on the right – generalized displacements in a deformed frame.

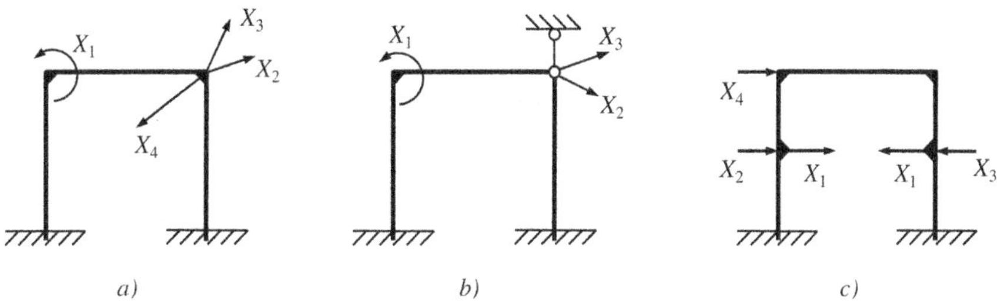

FIGURE 17.2 The cases of singular flexibility matrix.

Note

In the examples considered above, Theorem 2 could be employed instead of Theorem 1: In the direction of every generalized displacement, install a connection which inhibits this displacement. Next, remove each of these connections by turns and verify the possibility of displacement in the direction of the removed connection (return each verified connection into the system before removal of the next connection). If, in the direction of every removed connection, there is the possibility of displacement due to the action of generalized force X_i exerted in the place of this connection, then the flexibility matrix is nonsingular. Otherwise, if at least one displacement in the direction of any removed connection is impossible, the flexibility matrix is singular.

17.2 IMPACT OF DISPLACEMENTS UPON ELASTIC SYSTEM; ACTIVE CONNECTIONS

The established conditions of existence of a stiffness matrix enable us to obtain the following results, which supplement the previously studied topics on the theory of displacements in elastic systems.

Let an elastic structure be affected by loads $P_i, i = \overline{1,m}$, and external generalized forces $X_i, i = \overline{1,n}$, which **independent** generalized displacements $Z_i, i = \overline{1,n}$, correspond to. We shall specify the impact upon a system by the loads and generalized displacements. Let us show that response S of a system, which may be the stress, strain, or displacement, is represented as a linear form of the impact, i.e., in the form:

$$S = \sum_{i=1}^{m} \overline{S}_{Pi} P_i + \sum_{i=1}^{n} \overline{S}_i Z_i, \tag{17.7}$$

where coefficients $\overline{S}_{Pi}$ and $\overline{S}_i$ are the constants for the given system.

In Chapter 11, we have stated the superposition principle, in which an impact is specified by the loads only. This principle we shall briefly call the basic principle. On the strength of this principle, a response can be expressed through external forces using the linear form:

$$S = \sum_{i=1}^{m} \overline{S}_{Pi}^{X} P_i + \sum_{i=1}^{n} \overline{S}_i^{X} X_i. \tag{17.8}$$

Relationship (17.7) we shall obtain by replacement of variables in the latter form on the ground of one-to-one correspondence:

$$\left(X_1,\ldots,X_n,\, P_1,\ldots,P_m\right) \leftrightarrow \left(Z_1,\ldots,Z_n,\, P_1,\ldots,P_m\right).$$

In the problem being solved, the generalized forces $X_i, i = \overline{1,n}$, are applied to a structure together with the loads, and the result of loads' action can be represented by an additional term on the right-hand side of (17.3). We have:

$$\mathbf{Z} = \delta \mathbf{X} + \Delta_P, \tag{17.9}$$

where $\boldsymbol{\Delta}_P$ is the column vector of displacements caused by loads (loaded displacements). The dependence of displacement vector $\boldsymbol{\Delta}_P$ on the loads is linear and can be represented by using some coefficient matrix $\boldsymbol{\delta}^P$ of order $n \times m$ in the form:

$$\Delta_P = \delta^P \mathbf{P}, \tag{17.10}$$

where $\mathbf{P} = (P_1,\ldots,P_m)^{\mathrm{T}}$ is the column vector of the loads. The flexibility matrix in formula (17.9) is nonsingular due to the independence of generalized displacements Z_i, and thus there exists stiffness matrix $\mathbf{r} \equiv \boldsymbol{\delta}^{-1}$. By multiplying the left and right side of equality (17.9) by $\boldsymbol{\delta}^{-1}$, we obtain:

$$\mathbf{X} = \delta^{-1}\mathbf{Z} - \delta^{-1}\Delta_P, \tag{17.11}$$

or:

$$\mathbf{X} = \mathbf{rZ} + \mathbf{R}_P, \tag{17.12}$$

where the denotation is used:

$$\mathbf{R}_P = -\delta^{-1}\Delta_P = -\delta^{-1}\delta^P\mathbf{P}. \tag{17.13}$$

The constituents R_{iP} of column vector $\mathbf{R}_P$ have a meaning of additional generalized forces which oppose generalized displacements Z_i while the loads are exerted. Really, the generalized forces which nullify displacements are obtained by formula (17.12) in the form $\mathbf{X} = \mathbf{R}_P$.

Rewrite relationships (17.12) and (17.13) in the scalar form:

$$X_i = \sum_{j=1}^{n} r_{ij}Z_j + R_{iP}, \quad i = \overline{1,n}; \tag{17.14}$$

$$R_{iP} \equiv -\sum_{j=1}^{m} r_{ij}^P P_j, \quad i = \overline{1,n}; \quad \mathbf{r}^P \equiv \delta^{-1}\delta^P. \tag{17.15}$$

Substituting (17.14) into (17.8) and then making substitution (17.15), we come to the required form (17.7).

Just as linear form (17.8) is the mathematical representation of the basic superposition principle, **form (17.7) expresses the superposition principle for systems subjected to the impact of loads and displacements.** In the case when the system being considered in this item is a structure with removed constraints and generalized forces X_i substitute for constraint reactions, the inference about the possibility of representing the system's response by the form (17.7) confirms the generalized superposition principle presented in Chapter 12 (see also the supplement to the present chapter). Note, however, that in the elasticity theory, the substantiation of the generalized superposition principle doesn't differ in essence from substantiation of basic principle; therefore, the proof of the generalized superposition principle through employing the basic one is not a significant scientific result. Besides, the formulation of the generalized superposition principle allows instability of a system caused by removal of connections (its stability is required only under fixed displacements in the directions of removed connections), whereas in the derivation of form (17.7), we dealt with a stable system.

The conclusion that the superposition of states is ensured under the combined action of loads and displacements upon a system, as well as math treatment (17.7)–(17.15), will be required below in studying the displacement method.

The Activity of Removed Connections as the Condition of Correctness of a Primary System in the Force Method

In order for a primary system of the force method to ensure uniqueness of solution of its canonical equation set, i.e., to be correct, the coefficient matrix of this equation set must be nonsingular. The named matrix is the flexibility matrix for the totality of redundant forces acting in the primary system. *We call a redundant connection, such that its removal from a system makes displacement in the direction of this eliminated connection possible as the effect of external forces replacing reactions of connection, an active connection of a bar system.* If all the connections which were removed from a given system in transition to primary system of the force method are active, then by virtue of Theorem 2, the flexibility matrix is nonsingular and the primary system is correct.

17.3 THEORETICAL BASICS OF DISPLACEMENT METHOD; THESES OF METHOD IN ANALYSIS OF FRAMES

The displacement method serves the same end as the force method: to determine the stress state of a statically indeterminate bar system. This method is known in two forms: the canonical form, which is also referred to as the slope-deflection method, and the iteration form called the moment-distribution method. In the present course, we study only the canonical form of the method.

We remind readers that the force method involves setting up and solving the system of canonical equations to determine the reactions of redundant constraints. In the displacement method, the canonical equation set is also composed, but unknowns in these equations are not constraint reactions but displacements of joints produced by external forces. The internal force diagrams are constructed by means of established displacements. In contrast to the force method, in the displacement method the number of unknown displacements often proves to be small in the case of a large number of redundant constraints in a system, i.e., these two methods supplement each other.

Originally, the displacement method was suggested by French mathematician and bridge engineer C.L. Navier in 1826. At the beginning of the 20th century, this method was enhanced and presented in its modern form by Danish scientists and engineers: Axel Bendixen in 1914 and Asger Ostenfeld in 1926. In the USA, this method has been recognized by dint of investigations of professor G.A. Maney (University of Minnesota), published in 1915.

Let in a given externally stable bar system there be a known set of joints such that their rotational and translational displacements enable us to determine the SSS of a system under loads by means of a sufficiently simple procedure. We denote the set of these displacements as $Z_i, i = \overline{1,n}$. Apply, in addition to loads, external forces in the directions of these displacements (for rotational displacements, apply the force couples upon the joints) $X_i, i = \overline{1,n}$, so that every pair X_i, Z_i represents the generalized force and corresponding generalized displacement. We shall assume that **displacements Z_i make up independent set**. Introduce the flexibility matrix which, in the absence of loads, links the column vectors of generalized forces and displacements in the form (17.3). Take into account the load action by pass to expression (17.9) where, on the right side, we have added the column vector of loaded displacements (i.e., produced by loads). Owing to independence of displacements, there exists the stiffness matrix which determines the n-vector of generalized forces through the n-vector of displacements in the form (17.12), (17.13). Because there are no additional forces exerted in the given system, the displacements of joints must satisfy the system of linear equations as follows:

$$\mathbf{rZ} + \mathbf{R}_P = 0. \tag{17.16}$$

The displacement method is to select n independent displacements of joints which define the SSS of a bar system under loading, determine these displacements from condition (17.16) of the absence of additional forces upon these joints, and then construct the internal force diagrams. Theoretically, this scheme is applicable for arbitrary bar systems; in practice, it is accomplished only for frames.

The displacement method is implemented in several stages:

1. Identify the set of independent displacements of joints which enables us to determine the SSS of a system. According to the displacement method, every rigid joint (excluding rigid supports) can make a pivoting motion, i.e., make rotational displacement in the plane of the system. Besides, the joints can make linear displacements from which the independent ones should be selected. In the method under consideration, independent displacements of joints are sometimes called degrees of freedom of a system – rotational and translational,

respectively – and the joints themselves are called nodes. Linear displacements are identified on the basis of the assumption that **the distance between any two joints connected by a member doesn't change due to the deformation of a system**. This assumption restricts analysis of possible systems by the frames, because it presumes the absence of tension-compression strain in members, which can be neglected only in analysis of frames. Owing to this assumption, the number of independent linear displacements decreases and becomes acceptable for calculation. The totality of independent linear displacements can be established by replacing all rigid joints with hinges and then carrying out kinematic analysis of the obtained mechanism. Example: if in the frame shown in Figure 17.1a, we replace rigid joints with hinges (inter alia, rigid supports with hinged immovable supports), then the obtained mechanism will have one DOF; therefore, there is one linear displacement among independent joint displacements in the frame. It is convenient to specify the directions of the linear joint displacements which determine the position of the mechanism by means of additional support links turning the mechanism into a stable system. The displacements of these additional supports are the required linear displacements.

So, the total number of independent displacements of joints called the *degree of kinematic indeterminacy of a system* is determined in the form:

$$n = n_{piv} + n_{lin}, \tag{17.17}$$

where n_{piv} is the number of rigid joints in a system (excluding support joints), and n_{lin} is the number of DOFs in the mechanism made up through replacement of rigid joints by hinges. These quantities are the numbers of independent rotational and linear displacements of joints, respectively. Examples: for the frame in Figure 17.1a, we get by formula (17.17): $n = 2 + 1 = 3$; for the frame in Figure 14.2 it is easy to obtain: $n = 6 + 2 = 8$.

2. Construct *the primary system of the displacement method which is composed from a given system by installation of additional supports opposing all independent displacements, instead of which the unknown displacements are allowed for these supports.* Linear displacements of joints are inhibited by support links, and rotational displacements are inhibited by floating supports introduced in Chapter 12 (Figure 12.6). The loading both upon the given and primary systems is assumed to be identical. In the primary system, the rotational and linear displacements of additional supports together with loads cause the constraint reactions and the stress state of members. Unknown displacements of additional supports are designated as $Z_i, i = \overline{1,n}$ and termed *primary unknowns of the displacement method.*
3. Establish the relationships between reactions of additional supports and displacements, i.e., we should obtain the next relationship for reaction of the i-th support under given loads:

$$R_i = R_i\left(Z_1,\ldots,Z_n\right).$$

(These relations are represented by linear functions (17.14).)

4. Determine the displacements providing the absence of reactions in additional supports, i.e., we should solve the equation set:

$$R_i\left(Z_1,\ldots,Z_n\right) = 0, \quad i = \overline{1,n}. \tag{17.18}$$

These equations are referred to as *canonical equations of the displacement method.*

5. The primary system can be taken as the given one which is subjected to external forces from additional supports in addition to loads. In the absence of reactions in additional supports, the primary system turns into the given system, i.e., conditions (17.18) ensure the equivalence of the primary system of displacement method and the given system.

The obtained displacements of joints Z_i prove to be real displacements. At the final stage, we construct the internal force diagrams in members of the given system by means of these displacements.

Additional supports create good layout of the object of analysis. At the beginning of this item, we gave a more abstractive substantiation of the displacement method without the use of these supports. We may talk about displacements of additional supports and corresponding support reactions in the primary system of the displacement method, or we may consider the set of independent displacements of joints in a given system together with additional forces exerted on these joints. The difference is only in terminology and degree of visibility. Further on, the column vector of generalized forces **X** is constructed from reactions of additional supports, developed in the primary system of the displacement method. The canonical equation set has the vector form (17.16), where the left side is the vector of reactions of additional supports.

The admissibility of negligence of tension-compression strains in frames is determined by the little effect these strains have upon displacements of frame's joints. Consequently, an account of a member's tension-compression essentially doesn't affect the evaluation of bending moments, whereas precisely the bending moments specify the stress state of frames.

Figure 17.3 shows the frame on two supports as an example of the given system (a) and the primary system of displacement method (b). In the given case, there is one rigid joint which can make a rotational displacement, and one independent linear displacement of joints is possible. These displacements are shown by positive directions for possible displacements of additional supports (respectively Z_1 and Z_2). In Figure 17.3c the primary system is shown in its deformed state.

The presented exposition of the stages of structural analysis requires the following additions to become a method: it is necessary to develop both the technique of composing the canonical equation set (17.16) and the technique of constructing the internal force diagrams at established displacements of additional supports $Z_i, i = \overline{1,n}$. These issues are considered in the following chapter.

SUPPLEMENT TO CHAPTER

Independence Conditions Imposed on Displacements in Directions of Removed Constraints in Stable Systems

In the course of deriving formula (17.7), we used independence of generalized displacements $Z_i, i = \overline{1,n}$. The concept of independence was introduced under the assumption that these displacements are caused only by generalized forces $X_i, i = \overline{1,n}$. In Chapter 12, the generalized superposition principle was formulated, in which the independence of displacements in the directions of removed connections was assumed, i.e., each displacement can take an arbitrary value in the

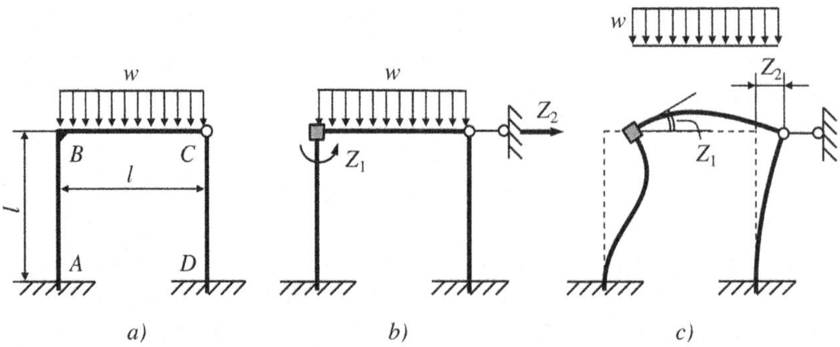

FIGURE 17.3 Given structure (a) and primary system of displacement method (b and c).

neighborhood of zero independently from other displacements in the places of removed connections. In this formulation the action of every possible loading upon a system besides external forces substituting for reactions of removed connections is assumed. In order that formula (17.7) might serve as the proof of the generalized superposition principle under the condition of stability of the system with removed constraints, we have to ascertain that from the independence of displacements in the places of removed connections $Z_i, i=\overline{1,n}$, under the action of arbitrary external forces which ensure system's equilibrium, there follows independence of the same displacements under action only of generalized forces $X_i, i=\overline{1,n}$, substituting for reactions of connections.

From independence of displacements in a broad sense, which is assumed in the generalized superposition principle, we shall deduce independence of the same displacements in a narrow sense, which was defined in the first item of this chapter. This is done by the next plain reasoning. Let in state 1 of a system with removed constraints, the loads cause some displacements in the places of these constraints $Z_i, i=\overline{1,n}$. The same loads exerted upon the original system cause constraint reactions in these very places, which oppose the displacements. If, in the system with removed constraints, subjected to loads of state 1, forces in the directions of removed connections, which are equal to reactions of connections removed from original structure, are applied, then state 2, in which there are no named displacements, will arise. Let us apply the external forces equal to the difference between forces applied in states 1 and 2 upon this system. There will arise state 3, in which the loads upon the system are absent, yet the displacements in the directions of the removed connections are the same as in state 1. These displacements have appeared due to the action of external forces $X_i, i=\overline{1,n}$, in the directions of the removed connections. Therefore, one can match state 3 with the very same displacements caused only by external forces $X_i, i=\overline{1,n}$, to state 1 under arbitrary displacements $Z_i, i=\overline{1,n}$. Thus, it has been proved that the notions of independence of displacements in the directions of the removed connections in a broad and narrow sense are equivalent.

18 Actualization of Displacement Method

18.1 CANONICAL EQUATIONS AND GENERAL FORMULAS FOR CALCULATION OF REACTIONS AND CONSTRUCTION OF DIAGRAMS; DETERMINATION OF REACTIONS BY STATIC METHOD

We remind readers that for analysis by the displacement method, there serves a primary system, which is obtained from a given system by installation of additional support constraints in the places and directions of independent displacements of joints. The reactions of additional supports are represented in linear form (17.14), i.e., as the sum of components determined by displacements of supports and loads. The condition of primary and given systems' equivalence lies in equality to zero of these reactions.

Thus, we come to the canonical equation set of the displacement method:

$$\sum_{j=1}^{n} r_{ij}Z_j + R_{iP} = 0, \quad i = \overline{1,n}. \tag{18.1}$$

Here R_{iP} are the *loaded reactions of additional supports*, which are produced by loads in primary system in the absence of support displacements, and r_{ij} are the *unit reactions of additional supports*, every of which is the reaction of the i-th support due to unit displacement of the j-th support. The latter reactions constitute the stiffness matrix and sometimes are referred to as *stiffness coefficients*. Diagonal entries of the stiffness matrix are termed principal (or main) reactions; off-diagonal ones are called secondary reactions. By virtue of Rayleigh's first theorem, the stiffness matrix is symmetric, i.e.:

$$r_{ij} = r_{ji}. \tag{18.2}$$

In the general case, the computing of unit and loaded reactions can be done by the formulas obtained in Chapter 12. According to formula (12.4), the unit reactions are determined by the products of diagrams:

$$r_{ij} = \left(\bar{M}_i, \bar{M}_j\right) = \sum \int_0^l \frac{\bar{M}_i \bar{M}_j dz}{EI}. \tag{18.3}$$

Here $\bar{M}_i$ is the unit bending moment diagram by the displacement method, i.e., the diagram for the state of the primary system caused by unit displacement of support i. Note that formula (18.3) determines the reactions approximately. The precise formula looks like formula (13.8) of the force method and, besides the product of bending moment diagrams, should include the component due to axial forces and the component due to shear forces. But tension-compression strains are impossible due to the assumptions of the displacement method, and thus a contribution of axial forces into reactions r_{ij} is absent. The contribution of shear forces could be taken into account, but it is negligibly small in comparison with the right-hand side of expression (18.3).

The loaded reactions can be calculated by formula (12.10), which we write down in the form:

$$R_{iP} = -\left(\bar{M}_i, M'_P\right). \tag{18.4}$$

Here M'_P is the diagram of bending moments produced by loads in the stable system made up of the primary system by removal of some connections – among them, mandatorily, support connection i. Naturally, diagram M'_P can be most easily obtained for a statically determinate system.

After solving the canonical equation set, the final bending moment diagram is established, and, by using this diagram, the shear and axial force diagrams are constructed. For construction of the final bending moment diagram, the next formula is usually employed:

$$M = \sum_{i=1}^{n} \bar{M}_i Z_i + M_P, \tag{18.5}$$

where M_P is the loaded diagram of the displacement method, i.e., the diagram of bending moments caused in the primary system by loads only. This relationship in form and sense is similar to relationship (13.9) of the force method. It follows from application of the generalized superposition principle to a primary system with shifted supports. Really, it was noted in Chapter 12 that a displacement of support can be taken as a displacement in the direction of the removed support connection; besides, internal force at the cross-section of a member can be taken as the response of a structure to an impact just as the stress at a point. The generalized superposition principle determines the system response in the form (17.7). In this formula, the response might be the bending moment and the impacts the displacements of supports; after changing denotations, we come to formula (18.5).

Construction of the shear and axial force diagrams is done after construction of the final bending moment diagram by means of the technique substantiated in Chapter 14.

The main limitation in the use of the displacement method is the laboriousness of calculations of unit and loaded reactions by formulas (18.3) and (18.4). In uncomplicated problems, instead of using these formulas, one can set up and solve the statics equations in named reactions for separate parts of a system. This method of finding the parameters of canonical equations is referred to as the static method.

Analysis by the displacement method begins with construction of unit and loaded bending moment diagrams for the primary system. The technique of unit and loaded diagrams' construction lies in consecutive selection of the required diagram for each member from the *set of model diagrams of the displacement method.* Figure 18.1 shows possible schemes of loading upon members in a primary system of the displacement method and the corresponding model diagrams of bending moments. The diagrams were constructed with account of deformations due to bending only, and bending moments are evaluated by using first-order approximation with respect to an impact. In particular, when the impact is the displacement of support Z (diagrams a through d), bending moments are calculated in the form:

$$M(z, Z) \cong \left.\frac{\partial M(z, Z)}{\partial Z}\right|_{Z=0} Z,$$

where z is the coordinate of the cross-section of the member. On the schemes shown in Figure 18.1, on the right, the hinged movable support could be replaced by hinged immovable support, and the bending moment diagrams would not change at that.

The model diagrams, as a rule, suffice to construct the unit and loaded bending moment diagrams through the selection of appropriate diagrams for each member in turn. Let us consider the example of the square frame in Figure 17.3a, where flexural rigidities will be assumed to be

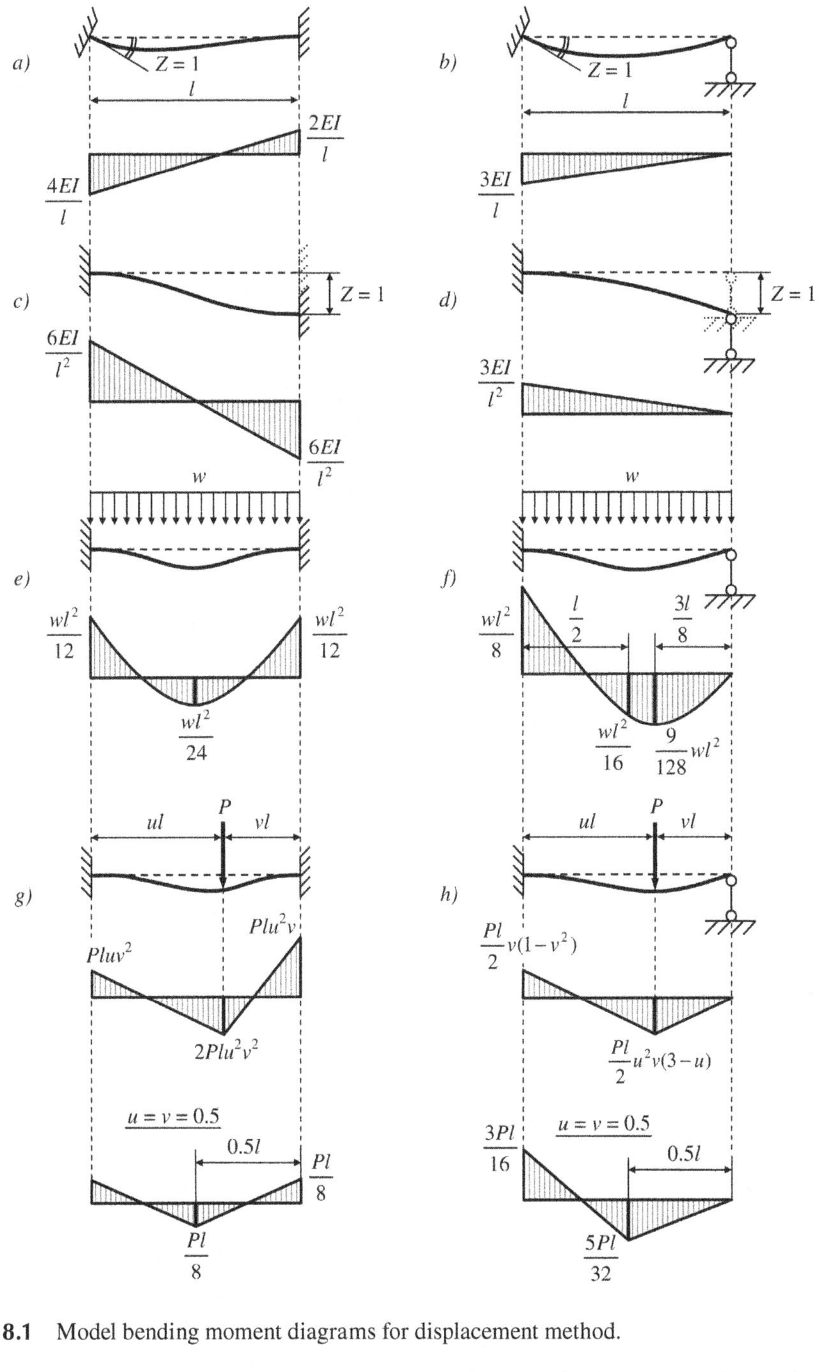

FIGURE 18.1 Model bending moment diagrams for displacement method.

the same for cross-sections of all the members. In this example, by means of selection of model diagrams, we obtain the set of unit and loaded diagrams shown in Figure 18.2. While constructing these diagrams, the assumption of the displacement method, that the distance between the joints of the same member does not change due to deformations, should be employed. For instance, while constructing the loaded diagram in Figure 18.2, this assumption enables us to conclude that translational displacements of the girder's joints do not arise under loading. Besides, rotational displacement is not possible for the left joint. The result is the loading of the girder according to scheme *f* in Figure 18.1, whereas the columns are not subjected to bending.

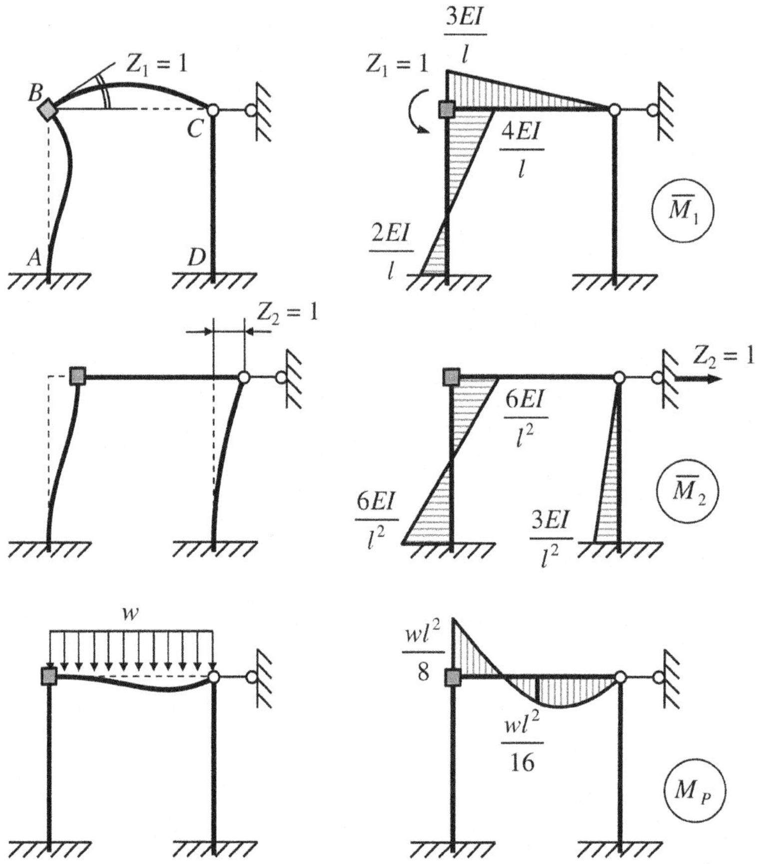

FIGURE 18.2 Possible deformed states of the primary system (on the left) and corresponding diagrams (on the right).

Now, we show how the obtained diagrams can be used for computing the coefficients and free terms of canonical equations by the static method.

Reaction r_{11} is calculated through diagram $\overline{M}_1$ from the equilibrium requirement for joint B, which is subjected to three moments as shown in Figure 18.3a. We get:

$$r_{11} = \frac{7EI}{l}.$$

Reaction r_{21} is established through diagram $\overline{M}_1$ from the equilibrium requirement for girder BC, for which the free-body diagram is shown in Figure 18.3b. For shear force in the given case, we have:

$$V_{BA}^{1} = \frac{M^{left} - M^{right}}{l} = \frac{6EI}{l^2}.$$

Here the superscript at V designates the state of the primary system – for the three diagrams in Figure 18.2 this superscript can take values 1, 2, and P, respectively. The signs of moments in this formula are determined by the sign convention for frames (see Chapter 14). According to the free-body diagram in Figure 18.3b, we get:

$$r_{21} = V_{BA}^{1} = \frac{6EI}{l^2}.$$

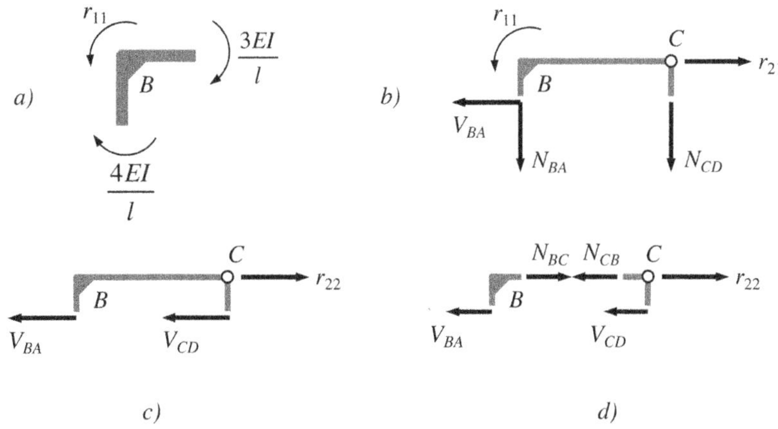

FIGURE 18.3 Free-body diagrams for calculation of unit reactions.

We obtain reaction r_{22} through diagram $\bar{M}_2$ from the equilibrium requirement for girder *BC*, which is subjected to the forces shown in Figure 18.3c (the force couples upon node *B* are not shown). It holds:

$$V_{BA}^2 = \frac{12EI}{l^3}; \quad V_{CD}^2 = \frac{3EI}{l^3}; \quad r_{22} = V_{BA}^2 + V_{CD}^2 = \frac{15EI}{l^3}.$$

Note that in order to determine the latter two reactions, one might consider the equilibrium of joints belonging to girder *BC* instead of the equilibrium of the whole girder; this is clarified by Figure 18.3d.

The loaded reactions are established through diagram M_P in an obvious manner:

$$R_{1P} = \frac{wl^2}{8}; \quad R_{2P} = 0.$$

The canonical equation set takes the form:

$$\begin{cases} \frac{7EI}{l} Z_1 + \frac{6EI}{l^2} Z_2 = -\frac{wl^2}{8}; \\ \frac{6EI}{l^2} Z_1 + \frac{15EI}{l^3} Z_2 = 0. \end{cases} \quad (18.6)$$

18.2 VERIFICATION OF CALCULATIONS IN DISPLACEMENT METHOD

In the completing stage of calculations by the displacement method, the verification of the final bending moment diagram is fulfilled. We regard the static verification and the verification by multiplication of diagrams as verifications of the completing stage. Usually, these verifications are interchangeable, i.e., it is sufficient to make one of them. *Static verification* lies in computing the reactions of additional supports in the primary system under the assumption that these supports are shifted in some fashion, so that the bending moment in the primary system is represented by the diagram being verified. **If the diagram under verification is correct, then all the reactions of additional supports must be zero.** The reactions of floating supports are calculated by solving the equilibrium equations for moments acting upon rigid joints; the reactions of additional support links are calculated by solving the equilibrium equations for forces applied upon the part of the structure containing an additional support link.

In order to substantiate the verification by multiplication of diagrams, we use formula (18.4) for loaded reactions. Diagram M'_P in this formula is constructed for a stable system made from the given one by removal of some connections, including the i-th additional connection. But the given system is obtained from a primary system exactly by removal of all additional constraints, i.e., the diagram M being checked can be taken as the diagram M'_P. Thus, we come to the set of conditions to be met by the final diagram:

$$R_{iP} + \left(\bar{M}_i, M\right) = 0, \quad i = \overline{1,n}. \tag{18.7}$$

The verification by multiplication of diagrams lies in the checking of conditions (18.7). The sense of these conditions is explained by the following theorem.

Theorem about Reactions of Additional Supports

In the state of a primary system of the displacement method caused by loads and displacements of additional supports, the reaction of an additional i-th support is determined by the expression:

$$R_i = R_{iP} + \left(\bar{M}_i, M\right), \tag{18.8}$$

where M is the bending moment diagram in the state under consideration.

Proof

We prove first that, in a primary system, every unit and loaded diagram are orthogonal in pairs (Smirnovh et al. 1981, p. 417):

$$\left(\bar{M}_i, M_P\right) = 0. \tag{18.9}$$

We shall call the state of an unloaded primary system with unit displacement of support connection i state 1; the state of a primary system under loading but with no support displacements we shall call state 2. The virtual work of external forces in the first state done through displacements of the second state is equal to zero, because in the first state, the external forces are exerted by only the supports, and in second state, the displacements of supports are absent. By computing the same virtual work through formula (11.13), we obtain:

$$W_{12} = \left(\bar{M}_i, M_P\right),$$

wherefrom equality (18.9) follows.

Let us calculate the product of diagrams $\left(\bar{M}_i, M\right)$ by means of expression (18.5) for diagram M. We take into account that the diagram multiplication is an operation linear in each co-factor:

$$\left(\bar{M}_i, M\right) = \left(\bar{M}_i, \sum_{j=1}^{n} \bar{M}_j Z_j + M_P\right)$$

$$= \sum_{j=1}^{n} \left(\bar{M}_i, \bar{M}_j\right) Z_j + \left(\bar{M}_i, M_P\right) = \sum_{j=1}^{n} r_{ij} Z_j.$$

The right side of expression (18.8) can now be written in the form:

$$X_i = \sum_{j=1}^{n} r_{ij} Z_j + R_{iP}.$$

So, the formula for the reaction of the additional support, which is to be proved, is represented in the known form (17.14), which was required.

The proved theorem means that conditions (18.7) could be understood as the conditions of the absence of reactions from additional connections in the primary system. Therefore, verifications considered above have the same physical meaning.

Example

For the frame in Figure 17.3a, the system of canonical equations was obtained in the form (18.6). The solution of this system has the form:

$$Z_1 = -\frac{5}{184}\frac{wl^3}{EI}; \quad Z_2 = \frac{wl^4}{92EI}.$$

By summing up the diagrams by formula (18.5), we obtain the final diagram shown in Figure 18.4. The free-body diagrams a and c in Figure 18.3, following minor changes, enable us to make static verification of the final diagram. To make the verification by multiplication of diagrams, we have to use Simpson's formula (13.15).

If the final diagram's verification points out an error, then one has to verify the calculations of coefficients and free terms in canonical equations. For this purpose, in the force method, we used the verifications by means of the total unit diagram. Similar verifications are applicable in the displacement method as well. In the given case, we call the diagram obtained under the condition of unit displacements of all additional supports the *total unit bending moment diagram*. According to this definition, the total unit diagram is the sum of diagrams:

$$\bar{M}_s \equiv \sum_{i=1}^{n} \bar{M}_i. \tag{18.10}$$

Verification of unit reactions (coefficients of canonical equations), also called the *universal verification of displacement method*, is done by multiplying the total unit diagram by itself and checking the equality:

$$r_{ss} \equiv \left(\bar{M}_s, \bar{M}_s\right) = \sum_{i,j=1}^{n} r_{ij}. \tag{18.11}$$

So, the *total unit reaction* defined by the first equality (18.11) must be equal to the sum of all principal and secondary reactions of the canonical equation set.

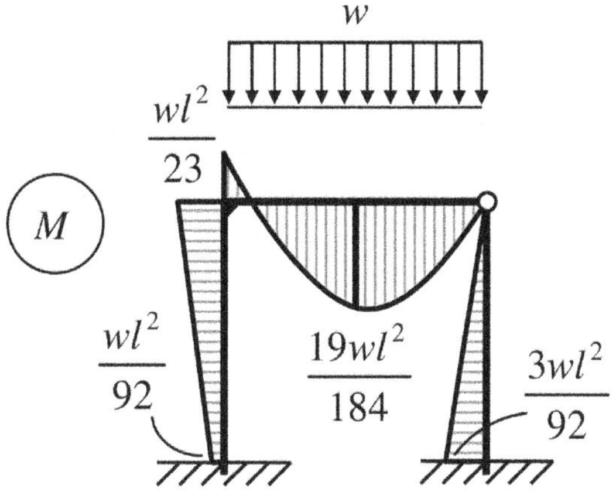

FIGURE 18.4 Final bending moment diagram of displacement method.

For *verification of the loaded diagram* the next criterion is in use:

$$\left(\bar{M}_s, M_P\right) = 0. \tag{18.12}$$

The orthogonality of the total unit and loaded diagrams follows from orthogonality conditions (18.9). If free terms of canonical equations (i.e., loaded reactions) are calculated by means of the static method, then equality (18.12) can be used for verification of these terms. Note, however, that the given criterion confirms the correctness of diagram M_P only. After construction of diagram M_P, in order to calculate the free terms of canonical equations, one has to set up and solve the equations of statics in unknown loaded reactions. The criterion (18.12) doesn't check this part of the calculations.

18.3 MIXED METHOD OF ANALYSIS OF STATICALLY INDETERMINATE FRAMES

The force method is convenient for analysis of systems with a small number of redundant constraints as, for example, it is shown in Figure 18.5a and b. The displacement method is convenient for analysis of systems with a small number of independent displacements of joints as, for example, in Figure 18.5c. Also, there is the *mixed method* developed by Soviet scientist A.A. Gvozdev in 1927. This method is used for analysis of systems where the force method is convenient for analysis of one part and the displacement method for another. Examples of such systems are shown in Figure 18.5d and e.

In the mixed method, a primary system is introduced by removal of constraints in that part of the given system where it is convenient to calculate the system by the force method and by entering additional supports in the part where it is convenient to apply the displacement method. The reactions of removed connections X_i and displacements of additional connections Z_j serve as the primary unknowns of the mixed method. The equivalence conditions of primary and given systems in the mixed method are the requirements of no displacements in the directions of removed connections and the requirements of no reactions in additional support connections. These conditions

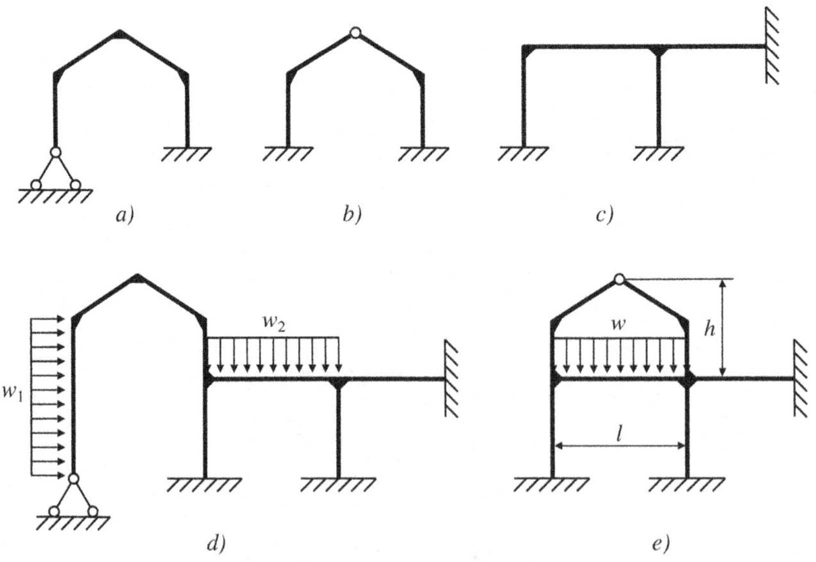

FIGURE 18.5 Example of frames for analysis by force method (a, b), displacement method (c), and mixed method (d, e).

are represented in the form of the canonical equation set in unknowns X_i and Z_j. The established unknowns serve for construction of final internal force diagrams.

Let us consider the example of the given system in Figure 18.5e. The degree of static indeterminacy for this structure is equal to 8; the degree of kinematic indeterminacy is 6. We shall use the mixed method of analysis, for which purpose we introduce the primary system by removal of hinged support on the top of the structure and by entering two floating supports at the second floor (Figure 18.6). In such a system, there are only four primary unknowns: X_1, X_2, Z_3, Z_4. The canonical equation set is written in the form:

$$\begin{cases} \delta_{11}X_1 + \delta_{12}X_2 + \delta_{13}Z_3 + \delta_{14}Z_4 + \Delta_{1P} = 0; \\ \delta_{21}X_1 + \delta_{22}X_2 + \delta_{23}Z_3 + \delta_{24}Z_4 + \Delta_{2P} = 0; \\ r_{31}X_1 + r_{32}X_2 + r_{33}Z_3 + r_{34}Z_4 + R_{3P} = 0; \\ r_{41}X_1 + r_{42}X_2 + r_{43}Z_3 + r_{44}Z_4 + R_{4P} = 0. \end{cases} \quad (18.13)$$

Here, δ_{ij} are the unit displacements, i.e., the displacements in the directions of removed connections produced by forces $X_j = 1$ or displacements $Z_j = 1$; r_{ij} are the unit reactions, i.e., the reactions of superimposed connections caused by forces $X_j = 1$ or displacements $Z_j = 1$; and Δ_{iP} and R_{iP} are the loaded displacements and loaded reactions, respectively, obtained in the state of the load's application and at zero primary unknowns. The first two equations are the conditions of no displacements in removed connections (conditions $\Delta_i = 0$); the last two equations are the conditions of no reactions in superimposed connections (conditions $R_i = 0$). Note that in setting up equations (18.13), the linearity of displacements and reactions in primary unknowns, which follows from the superposition principle, was used.

After the construction of a primary system of the mixed method and writing down the canonical equations in general form (18.13), the set of unit and loaded bending moment diagrams is to be constructed. The unit diagram $\bar{M}_i$ represents the action of the force $X_i = 1$ or displacement $Z_i = 1$ in the primary system. Loaded diagram M_P represents the action of loads in this system. These diagrams enable us to establish the coefficients and free terms of canonical equations. The technique of calculation of the coefficients is clarified as follows.

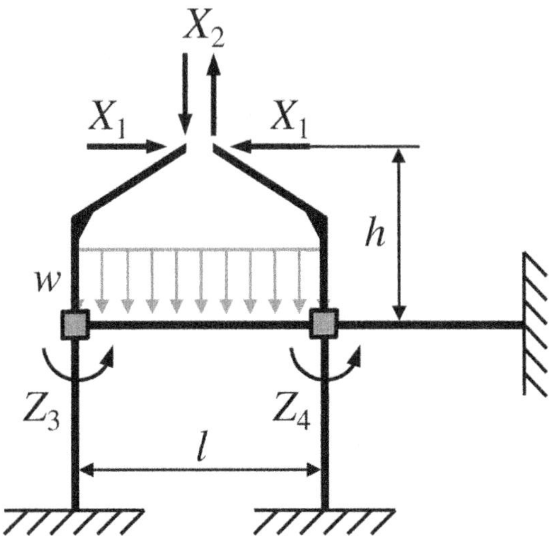

FIGURE 18.6 Primary system of mixed method for the frame in Figure 18.5e.

The coefficient matrix of a canonical equation set of the mixed method (flexibility-stiffness matrix) constitutes of four blocks. For the equation set (18.13), the blocked structure of this matrix has the form:

$$\mathbf{d} = \begin{pmatrix} \boldsymbol{\delta} & \boldsymbol{\delta}' \\ \mathbf{r}' & \mathbf{r} \end{pmatrix} = \left(\begin{array}{cc|cc} \delta_{11} & \delta_{12} & \delta_{13} & \delta_{14} \\ \delta_{21} & \delta_{22} & \delta_{23} & \delta_{24} \\ \hline r_{31} & r_{32} & r_{33} & r_{34} \\ r_{41} & r_{42} & r_{43} & r_{44} \end{array}\right). \tag{18.14}$$

The partition into the blocks is shown by dashed lines. Block $\boldsymbol{\delta}$ contains the displacements from forces; block $\mathbf{r}$ contains the reactions from displacements. The constituents of these blocks are calculated using ordinary techniques (by multiplication of unit diagrams or by static method). Maxwell's and Rayleigh's reciprocal theorems enable us to establish the symmetric link of named constituents:

$$\delta_{ij} = \delta_{ji}; \quad r_{ij} = r_{ji}.$$

Block $\boldsymbol{\delta}'$ contains the displacements from displacements; block $\mathbf{r}'$ contains the reactions from forces. The constituents of these blocks cannot be calculated by multiplication of unit diagrams, but during their calculation, one can use the next anti-symmetric link (wherefrom does it follow?):

$$r_{ij} = -\delta_{ji}. \tag{18.15}$$

It is recommended to establish the reactions from forces by the static method (as was done in the first item of the chapter) and then to calculate the displacements from displacements by using the property of reciprocity (18.15). For instance, the equilibrium conditions for rigid joints of the primary system in Figure 18.6 enable us to obtain:

$$r_{31} = -r_{41} = h; \quad r_{32} = r_{42} = \frac{l}{2}.$$

Sometimes, conversely, we can determine the displacements from displacements by means of simple geometric consideration and then establish reactions.

Loaded displacements and reactions in the mixed method are calculated in the same manner as in "pure" methods.

After determination of the primary unknowns, the final bending moment diagram is constructed as a linear combination of unit and loaded diagrams. In the case under consideration, we have for the final diagram:

$$M = \bar{M}_1 X_1 + \bar{M}_2 X_2 + \bar{M}_3 Z_3 + \bar{M}_4 Z_4 + M_P. \tag{18.16}$$

The verification of the final diagram is made by the same criteria which were grounded in the "pure" methods: the displacements in the directions of removed connections and the reactions of superimposed connections must be equal to zero.

Section IX

Plastic Behavior of Structures

This section is devoted to the statics of bar systems wherein the plastic strains are allowed at longitudinal fibers of members. The section begins with a comparative review of two design methods: the allowable stress design and the load and resistance factor design; the basic provisions of the load and resistance factor design are set out according to construction codes of the USA. The properties of ductile materials are considered, the definition of elastoplastic material is given, and for an elastoplastic member the operability domains are established in the principal cases of loading. Then we pass to elastoplastic bar systems subject to a quasistatic loading process. The notions of step and substep of loading are introduced and the principles of analysis of a one-step loading process are stated. Based on these principles, we construct in the section the models of loading with connections' loss and recovery, which describe the work of elastoplastic systems. The concepts of collapse mechanism and limit equilibrium are introduced, and the method of sequential identification of lost and restored connections is set forth as an instrument for construction of the model of loading. To make this method handy in use, the static principle of elastoplastic system's analysis is formulated and proved. The static and kinematic theorems of plastic analysis are proved, and kinematic method of plastic analysis is substantiated as the alternative to the method of lost-and-restored-connections identification. The adapting of the kinematic method for calculation of continuous beams is given. In the section another field is also considered for employment of the kinematic method; this is the calculation of latticed frames. To make effect in this field, the kinematic method is enhanced up to the method of combining mechanisms. The latter method is substantiated in the section with description of independent collapse mechanisms and recommendations on combining the mechanisms.

19 Load and Resistance Factor Design and Models of Ductile Bar's Collapse

19.1 BASIC PROVISIONS OF LOAD AND RESISTANCE FACTOR DESIGN; LRFD METHOD VERSUS ALLOWABLE STRESS DESIGN METHOD

During the operation of a structure, it is subjected to loads of different types, the magnitudes of which should be specified for structure design. The loads can be often specified by the use of construction codes. Most frequently, the designer must take into account the following types of loads:

- **Dead load** consisting of the weight of structural members and stationary equipment;
- **Live load** which is caused by the objects temporarily placed on a structure in its operation (people, goods, vehicles, and movable equipment);
- **Loads of environmental factors:** wind, snow, earthquake loads.

The codes comprise nominal magnitudes for each type of loads, i.e., expected (close to mean values) magnitudes of loads at dangerous stages of the structure's operation. To take into account a load's random variation from nominal magnitude as well as possible errors of analysis, a nominal load is converted into a factored load by multiplying it with the reliability factor provided in the codes. The factored load is the load of a given type which cannot be exceeded with great probability at a given stage of exploitation. For instance, the factored weight load can be obtained from the corresponding nominal load by the formula:

$$(\text{factored dead load}) = 1.4 \times (\text{dead load}).$$

Nominal load sometimes is termed design load. Note that the pass from nominal loads to factored ones is not always done and is stipulated by the accepted method of analysis.

The codes enable us to specify the loads being exerted in various possible combinations. The load combination is specified by codes as superimposing of factored or nominal loads of several types. For instance, to take into account the combined effect of dead and live loads, the load combination may be used in the form:

$$\text{load combination} = 1.2 \times (\text{dead load}) + 1.6 \times (\text{live load}).$$

Here the sign "+" means superimposing of loads; the values of factors are determined by the codes. The reliability factors in load combinations take into account not only random variations of loads and errors of analysis but also the likelihood that limit loads of different types occur simultaneously.

We shall call the capacity of a structure (member) to support increasing loads "integrity of the structure (or structural member)."* Example: a bar of elastic perfectly plastic material, working in tension, keeps integrity if normal stress in it has not attained yield strength. But if the material yields so that a bar starts to get longer under constant load, then it loses its integrity. Integrity is

* The term "operability" may be used as a synonym to "integrity," yet it is also used in a broader sense.

the condition of serviceability – the latter is the state of an object when its operation meets design requirements.

Nowadays, for analysis of structures, there are two basic methodologies in use:

- The allowable stress design (ASD) is the design method in which elastic stresses produced in a structural member are limited by allowable values specified by the codes. At that, loads exerted upon a structure are assumed to be nominal and, due to their combined action, may be taken with reduction factors. Allowable stresses are determined in the codes through strength characteristics of materials: yield strength, ultimate (rupture or compressive) strength, or – if the structural member is subjected to essential vibrations – fatigue strength. Allowable stresses are determined by dividing the *specified strength characteristics* by the factors of safety. The specified strength characteristics of materials are established by testing and confirmed by certification. Their magnitudes are lower than the mean values on the whole population of test samples because they are established with the exceedance probability close to one. These characteristics are also called *characteristic strength properties* of material. The ASD method, being applied for all members of a structure, ensures the integrity of every member and, consequently, the integrity of the structure as a whole.
- The load and resistance factor design (LRFD) is the design method based on analysis of the *structure's limit states*, i.e., the states in which either sudden destruction (collapse) of the structure can be induced by infinitesimal variation of loads or normal operation of the structure is impossible. This method aims to exclude the occurrence of a structure's limit states. The loads exerted upon a structure are assumed to be factored (most unfavorable). The combined action of loads is presented by their combination with known reliability factors. The LRFD does not exclude the loss of integrity of separate structural members under the condition of the structure's integrity as a whole.

In the codes on design of reinforced-concrete structures, the ASD is called working stress design, and the LRFD is called strength design.

The difference between ASD and LRFD is as follows. While in ASD only elastic strains are allowed, in LRFD the plastic strains are possible. The criteria of a structure's serviceability in ASD are derived from restrictions upon elastic stresses in members, which involve the known strength properties of materials. The same criteria in LRFD are derived from a number of requirements ensuring normal exploitation of a structure, and these requirements are not limited to the strength of members. Both methods allow for the loads to exceed the magnitudes determined by their specified combinations, but in ASD, the probability of such exceedance is greater than in LRFD (the possibility of using understated loads in ASD is due to the fact that analysis of a member within the elasticity stage provides a high enough safety margin). The integrity of a structure in ASD is identified with the integrity of all its members. In LRFD, the integrity of a structure is understood as its ability to function in the way that it is designed to. The failure (the loss of integrity) of individual members in this method doesn't mean the failure of the whole structure.* Both methods allow for occurrence of cracks in structural concrete, but in LRFD, the crack width is determined and limited, whereas in ASD the crack width is not limited with certainty.†

The limit states of a structure in LRFD are divided into two groups as follows:

- *Ultimate limit states* in which the exploitation of the structure becomes impossible due to hazard of its collapse. The recovery of a structure damaged in this state requires the replacement of some structural members or may be impossible. The notion of the ultimate

* The LRFD also allows for the simplistic interpretation of integrity of a structure, which assumes the integrity of all its members.

† The ASD method, as a rule, ensures small enough crack width, but situations when the cracks prove to be significant and require special estimation beyond this method are possible. See comment R10.6.1 of the code *Building Code Requirements for Structural Concrete…* (2011).

limit state as the state preceding destruction is applicable equally to structures and individual members;

- *Serviceability limit states.* If any parameter of structure trespasses the value specifying such limit state, then exploitation of the structure becomes hampered and its serviceable life is shortened, but immediate collapse does not occur. The serviceability requirements, for example, lie in the limitations upon deflections and vertical displacements of a structure, as well as the crack width.

Further on, we consider only the limit states of the first group. The methodology of their analysis applies **to structural members** as follows.

According to the code *Minimum Design Loads for Buildings and Other Structures* (2016), internal force or displacement in a structural member, produced by loads, is referred to as *load effect*. We take the internal force in member, whose magnitude – maybe in conjunction with some other internal forces – enables us to establish the occurrence of ultimate limit state, for the load effect. We consider the set of load effects as the vector in real coordinate space. In this space, some domain corresponds to the totality of states of the member's integrity, whereas on the boundary of this domain, the limit states may arise. The domain of integrity states is determined by the strength properties of the material – the yield strength and ultimate strength. The member is designed under the requirement that the limit state is impossible.

If the condition of a member is specified by single effect, then the absolute value of this effect in the limit state at the boundary of the integrity range is called *resistance* (or *resistance of section*) of the member. The resistance obtained under specified strength characteristics of materials is referred to as *nominal resistance* (or *nominal strength*) and denoted as R_n. The resistance established by the calculation with taking into account the pattern and consequences of the member's collapse as well as errors of analysis is referred to as *design resistance* (or *required strength*). Design resistance is determined through scaling of nominal resistance by the formula $R_d = \phi R_n$, where the factor $\phi \leq 1$ is termed the *resistance factor* and specified by the codes. If the limit state arises at the right-hand boundary of the integrity range, then the criterion of member's integrity has the form:

$$Q < R_d, \tag{19.1}$$

where Q is the load effect.

Factor ϕ is established from the requirement that design resistance would have a high enough exceedance probability while taking account of the difference between actual strength characteristics of materials and their specified values, and the procedure error in calculation of section resistance. Factor ϕ depends on the type of load effect involved for the derivation of integrity criteria.

If the dimension of the space of load effects is 2 or greater, i.e., **the member's condition is determined by two or more effects**, then the domain of integrity states is constructed by means of nominal resistances and resistance factors established for each of the effects. At the beginning, one is to construct this domain in the form of inequalities by using the given characteristics of section geometry and specified characteristics of the materials' strength. Then, into this set of inequalities, the nominal resistances of sections are entered instead of the geometry and strength characteristics. The final criterion of integrity is obtained by formal replacement of nominal resistances with design ones. Resistance factors may be established in the codes specifically for an occasion of combined action of several effects. Besides, the final criterion might contain additional correction coefficients due to conditions of the structure's operation.

Note that when all the resistance factors involved in analysis are the same and equal to ϕ, the criterion of integrity constructed on the methodology carried out is equivalent to the criterion of integrity derived by the use of the *design strength characteristics* of materials defined as $\sigma_d = \phi\sigma_n$, where σ_n are corresponding specified characteristics. This follows from linear dependence of section resistance on strength characteristics. In the stated equivalence of criteria, we see the commonality of the LRFD methods in USA codes and Eurocodes, for the Eurocodes prescribe the use of design strength characteristics.

Now we clarify the rules of specifying the loads in LRFD. Loads of various types can be specified as distributed over the volume or upon the surface of a member, as well as concentrated at separate points of it. The vector of total factored load exerted on the member's volume of small extent in design situations envisaged by the codes is represented in the form:

$$\Delta \boldsymbol{P} = \sum_i \eta_i \gamma_i \Delta \boldsymbol{P}_i. \tag{19.2}$$

Here, $\Delta \boldsymbol{P}_i$ is the vector of a nominal load of the i-th type; the product $\eta_i\gamma_i$ is the reliability factor of the load; γ_i is the load factor determined by general building codes applicable for buildings and structures of different types; and η_i is the load modifier determined by the design codes for individual types of structures. Among the codes specifying factors γ_i, the most important are documents *Minimum Design Loads*... (2016) and *International Building Code* (2015). Coefficients η_i are specified by federal codes and may be corrected by regional codes of states.

If the condition of a structural member is determined by a single effect which linearly depends on loads (for example, the structural member is statically determinate), then integrity criterion (19.1) for this member can be represented in the form:

$$\sum_i \eta_i \gamma_i Q_i < \phi R_n, \tag{19.3}$$

where Q_i is the effect produced by a load of the i-th type.

The errors of this analysis procedure are taken into account by reliability factors $\eta_i\gamma_i$ in calculating the left-hand side of criterion (19.3) and by resistance factors ϕ in calculating the right-hand side of the criterion.

The analysis of structural members with construction of an integrity domain in the space of load effects is implementable only for simple design schemes. The results of these calculations are presented in the codes on LRFD. The obtained results have enabled to develop methods of analysis applicable for multi-member statically indeterminate structures, and in these methods, there is no need to introduce the space of load effects acting in a structure. By employing these methods, an engineer can ascertain the possible schemes of structure collapse in hazardous operating conditions, and this, at every design situation, permits to obtain a criterion of the structure's integrity through estimation of the integrity of individual members.

Examples of Constructing a Criterion for Ultimate Limit State

Example 1

The bar member of plastic material undergoes the action of axial tensile forces. Take the beginning of plastic yielding for the limit state.

In the given case, the axial force at a cross-section is equal to the total axial load; thus, it is allowed to substitute the combination of loads into the left-hand side of criterion (19.3). The nominal section resistance is calculated as $R_n = \sigma_Y A$, where σ_Y is characteristic yield strength and A is the cross-sectional area. The integrity criterion takes the form:

$$\sum_i \eta_i \gamma_i P_i < \phi \sigma_Y A. \tag{19.4}$$

For the rolled section, for instance, the possible value of the resistance factor is $\phi = 0.95$.

Example 2

The short beam-column of a doubly-symmetric* section, made from brittle material, is subjected to compression with pure symmetric bending according to the scheme in Figure 19.1. The specified ultimate tensile-compressive strength is σ_u.

Owing to the small length of the column, we neglect the influence of axial force upon the bending moment and the chance of buckling. The ultimate state arises when the edge normal stress reaches rupture strength. The edge stresses are the stresses at the points of a cross-section most remote from the centroidal axis of the section normal to the load plane. To calculate them, formula (5.6) is employed. The condition of integrity obtained with no account of the random nature of parameter σ_u is represented by two inequalities for edge stresses: $|\sigma_{1,2}| < \sigma_u$. We write down these inequalities using formula (5.6) in the form:

$$\left|\frac{N}{A} \mp \frac{M}{S}\right| < \sigma_u. \tag{19.5}$$

Here, S is the section modulus (in the given case, this modulus is the same for both edge stresses). Let us introduce the nominal resistances of a section separately against compression and against pure bending:

$$N_n = \sigma_u A; \quad M_n = \sigma_u S.$$

It is easy to convert inequality (19.5) into the form:

$$\frac{|N|}{N_n} + \frac{|M|}{M_n} < 1. \tag{19.6}$$

The final integrity criterion is obtained by replacement of the nominal resistances in (19.6) with following design resistances: $N_d = \phi_N N_n$ and $M_d = \phi_M M_n$, where the resistance factors Φ_N and Φ_M are taken from the codes.

In coordinate plane M–N, the domain of integrity states is represented by an isosceles triangle (Figure 19.1). The lateral sides of the triangle are the lines of limit state occurrence. Because $N = -P$ and $M = M_c$, the same triangle (but capsized) defines the domain of integrity states in the plane M_c–P of factored loads.

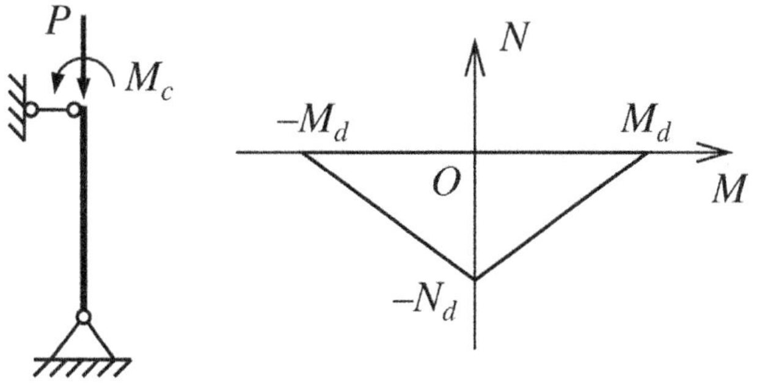

FIGURE 19.1 The domain of integrity states for brittle beam-column.

* A doubly-symmetric section has two mutually transverse axes of symmetry.

In conclusion, we point out the codes most commonly used for structural design. The following *general building codes* specify minimum loads and establish mandatory standards of design:

- ***Minimum Design Loads for Buildings and Other Structures.* 2016. ASCE/SEI 7-16. American Society of Civil Engineers.**
- ***International Building Code.* 2015. International Code Council, Inc.**

Design codes define the detailed rules of design, depending on materials used and types of structures. For reinforced-concrete and steel constructions, federal institutes of the USA developed the codes:

- ***Building Code Requirements for Structural Concrete (ACI318-14) and Commentary.* 2014. American Concrete Institute.**
- ***Steel Construction Manual.* 2011. American Institute of Steel Construction, Inc.**

In bridge design, the next federal codes are employed:

- ***AASHTO LRFD Bridge Design Specifications.* 2016. LRFDUS-7-M. American Association of State Highway and Transportation Officials.**
- ***Manual for Railway Engineering. Vol. 2. Structures.* 2017. American Railway Engineering and Maintenance-of-Way Association.**

The listed codes are regularly renewed by right holders. Besides, they may be corrected and complemented by regional codes of the states.

19.2 ULTIMATE LIMIT STATES OF STATICALLY DETERMINATE ELASTOPLASTIC MEMBER

Mechanical properties of a material are determined by the stress-strain diagram obtained through axial tension and compression of specimens on testing machines. If in such testing a material exhibits a large tensile strain beyond the elastic limit, then it is called ductile; on the chance of a large compressive strain beyond the elastic limit with no splitting or cracking, a material is called malleable. If collapse occurs at the strain about the elastic limit, then the material is called brittle. For example, lead and tin at definite temperatures are malleable materials and at the same time are brittle in tension. As a rule, ductile material is simultaneously malleable. There also are materials – rubber, for instance – the elastic limit of which cannot be established. It is important that structures built from ductile materials can usually bear such loads which Hooke's law is not complied with and elastic analysis is not adequate. It is for such constructions that LRFD is appropriate.

Further on, we confine ourselves to considering highly ductile materials, whose diagrams of tension-compression have yield plateaus (horizontal portions). For example, low-carbon steel belongs to such materials. The typical stress-strain diagram of such material is shown in Figure 19.2, on the left. The yield plateaus B_tC_t (for tension) and B_cC_c (for compression) precede the portions of strain hardening – portion C_tD and portion at the left of point C_c, respectively. Points B_t and B_c correspond to yield stresses. The diagram in Figure 19.2 is depicted schematically: to keep plausible scaling, elastic portions of tension and compression are to be depicted close to the ordinate axis. For example, it is known for a tension diagram:

$$\varepsilon(C_t) \sim 30\varepsilon(B_t),$$

where on the left is the initial strain of hardening; on the right is the yield-point strain; and symbol "~" denotes the same order of magnitude. A similar relation is correct for diagrams of compression.

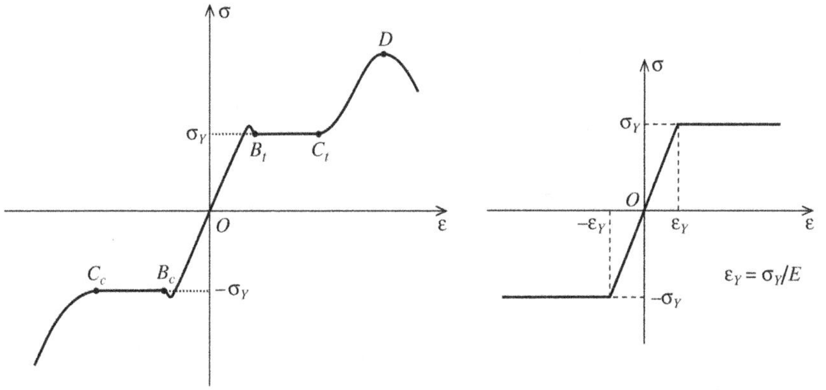

FIGURE 19.2 Diagram of tension-compression for highly ductile material (on the left) and idealized diagram (on the right).

For structural calculations, the piecewise-linear approximation of stress-strain diagrams is used. Unacceptable displacements arise in a structure before longitudinal strains in every member reach the portion of strain hardening; therefore, the possibility of the material's hardening is not taken into account. Usually, in graphic-analytical calculations, one employs a diagram with linear portions of elasticity going into endless horizontal portions of plasticity. This diagram is symmetric with respect to the coordinate origin and determined by yield stress and modulus of elasticity (Figure 19.2, on the right). The diagram is called an *idealized diagram* or *elastic perfectly plastic stress-strain response*. A material characterized by an idealized diagram is referred to as elastic perfectly plastic or elastoplastic.

It is important to mark that diagrams in Figure 19.2 represent the material's behavior in the tests with an increase of the load. If in such a test strains beyond yield strength arose, then the curve of unloading will differ from the shown diagrams. The condition of an idealized diagram's applicability in analysis of bar systems is that normal stresses and corresponding strains rise in longitudinal fibers of the structure member due to the loading process. Further in the chapter, it is assumed that this requirement is fulfilled and materials are elastoplastic.

Let the longitudinal load be exerted upon the butt-end of an elastic perfectly plastic member and causes the same axial internal force (as, for instance, in Figure 19.1 at $M_c = 0$). If the load is compressive, then we assume that there is no buckling of the member. The collapse of such a member, which has a character of unlimited change of length, occurs when the normal stress in a cross-section reaches the yield strength. If the member belongs to a statically determinate hinged truss, then plastic yielding of single member causes the collapse of the whole truss, i.e., in this case, the limit state of the structure is defined by the limit state of each of its members.

The ability of an elastoplastic beam to support the loads is not restricted by the initiation of plastic strain. In mechanics of materials, the plastic bending of a beam was studied and the concept of plastic hinges was introduced (Beer et al. 2012). *The region of plastic deformations in a bar, containing a cross-section entirely, is referred to as a plastic hinge.* The view of a plastic hinge in the longitudinal section of a beam is shown in Figure 19.3. The location of a plastic hinge is specified by a cross-section fully immersed in the plastic region. *A bending moment at a cross-section of a plastic hinge is referred to as plastic moment.* Occurrence of a plastic hinge in a statically determinate beam means an exhaustion of its bearing capacity; when continued, a small increase of load causes collapse.

Let us consider the bending of a statically determinate beam due to transverse loading and obtain the integrity criterion under the assumption that the advent of a plastic hinge is the beginning of collapse. The design diagram is shown in Figure 19.4; the cross-section is assumed to be rectangular; we choose the hazardous section, where the bending moment and shear force reach maximum

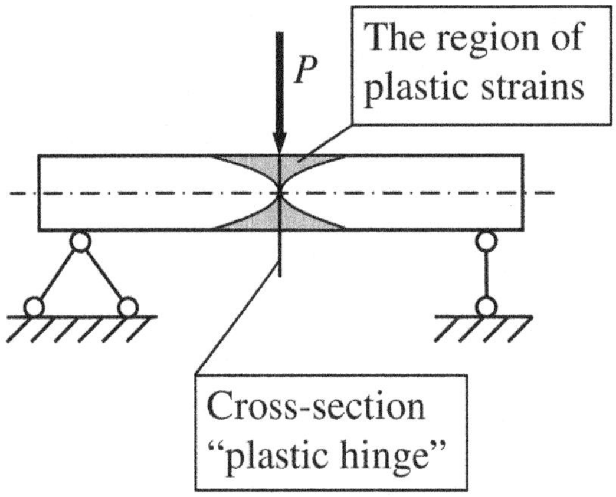

FIGURE 19.3 Plastic hinge in a beam.

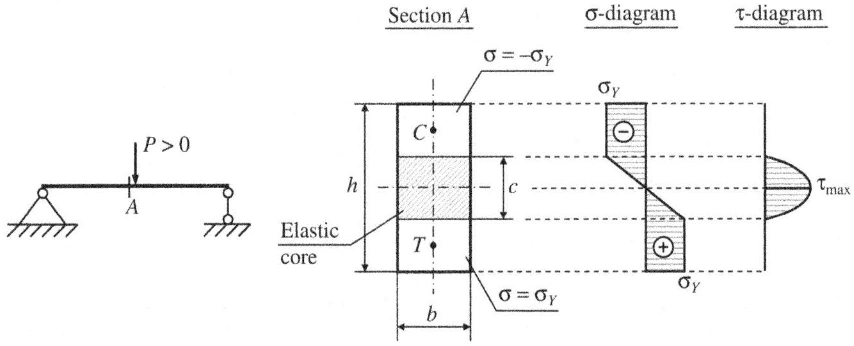

FIGURE 19.4 Stress distribution in cross-section of beam.

values, about the point of load application at either the left or right side of this point. In mechanics of materials, the plastic deformation of a beam has been studied with no account of internal shear force; now, we shall obtain the integrity condition for a beam subjected to a bending moment and shear force simultaneously. When plastic longitudinal strains occur in a beam's cross-section, three regions in it can be distinguished: the zone of plastic compression at the edge of the section (upper or lower); the zone of plastic tension at the opposite part of the section; and, between them, the *elastic core, i.e., the region limited by the lines where the yield stresses* $\pm\sigma_Y$ *are reached, whereas inside it, Hooke's law is fulfilled.* These regions, as well as the diagrams of normal and shear stresses in the cross-section, are shown in Figure 19.4. Shear stress acts only within the boundaries of the elastic core; its diagram is shaped like a parabola and the maximum value τ_{max} of this stress is reached at the neutral axis. For every material, we can specify the ultimate shear stress τ_Y such that a rectangular volume element subjected to pure shear will fail if the shear stress on its facets reaches ultimate magnitude (a brittle element will split while a ductile element will get plastic yielding). **We take for given that the limit state of a beam in bending occurs if shear stress in the cross-section attains ultimate value (when $\tau_{max} = \tau_Y$) or plastic longitudinal strains cover the whole section (when height c of elastic core becomes zero).**

Remark

The attainment of the equality $\tau_{max} = \tau_Y$ at small normal stresses in the cross-section (when $|\sigma| << \sigma_Y$) doesn't mean a collapse of the entire beam but only local destruction about the neutral axis. The usage of shear stress on the neutral axis as the shear-break criterion understates the ultimate shear force at the cross-section. However, if the equality $\tau_{max} = \tau_Y$ is fulfilled and the normal stresses in the section reach the stresses of yielding $\pm\sigma_Y$, then normal stresses at the sites of small shear stresses are close to ultimate values, and, therefore, the state of the material at the whole cross-section is close to ultimate.

The integrity conditions for the beam have the form:

$$\begin{cases} \tau_{max} < \tau_Y; \\ c > 0. \end{cases} \tag{19.7}$$

Let us get the expressions of internal forces *M* and *V* through characteristics of section and stress diagrams and then represent criterion inequalities (19.7) by means of internal forces and corresponding actual resistances of the section.

We presume that an elastic core has formed in the section. The resultants of internal normal forces acting in the regions of plastic compression and tension, respectively, are of the same magnitude and have opposite directions, i.e., make up the force couple. The resultants are exerted upon the centroids of plastic regions *C* and *T*, respectively (Figure 19.4). The arm of the couple equals $CT = \frac{h+c}{2}$, every force of the couple equals $F_p = \sigma_Y b\frac{h-c}{2}$, and the moment produced by the couple of "plastic" forces is expressed as follows:

$$M_{pc} = F_p \cdot CT = \sigma_Y b\frac{h-c}{2}\frac{h+c}{2} = \sigma_Y b\frac{h^2-c^2}{4}. \tag{19.8}$$

The elastic section modulus of a rectangular section equal to the elastic core is defined by the formula (Beer et al. 2012, p. 230):

$$S_{ec} = \frac{bc^2}{6}.$$

This characteristic enables us to determine the moment of internal normal forces acting within the elastic core:

$$M_{ec} = \sigma_Y S_{ec} = \sigma_Y\frac{bc^2}{6}. \tag{19.9}$$

The bending moment at cross-section *A* of the beam is the sum of the plastic component (19.8) and elastic component (19.9):

$$M = M_{pc} + M_{ec} = \sigma_Y b\frac{h^2-c^2}{4} + \sigma_Y\frac{bc^2}{6} = \sigma_Y b\left(\frac{h^2}{4} - \frac{c^2}{12}\right). \tag{19.10}$$

The ultimate bending moment in the absence of shear force (actual flexural resistance of rectangular cross-section) is obtained at $c = 0$:

$$M_p = \sigma_Y b\frac{h^2}{4}. \tag{19.11}$$

The obtained formula defines the *plastic moment* for a beam of rectangular cross-section working in pure bending.

We obtain the maximum moment acting before initiation of plastic longitudinal strains from (19.9) at $c = h$:

$$M_Y = \sigma_Y \frac{bh^2}{6} = \frac{2}{3} M_p. \tag{19.12}$$

We remind readers that in the case being considered, the elastic core has arisen, i.e., $M \geq M_Y$.

The maximum shear stress can be found by using the shear formula for cross-section equal to the elastic core. We have the relation between resultant shear force and maximum shear stress (Beer et al. 2012, p. 388):

$$V = \frac{2}{3} \tau_{\max} bc. \tag{19.13}$$

The greatest internal shear force which could be exerted on the cross-section at the small bending moment (when $M \leq M_Y$) we obtain from formula (19.13) at $\tau_{\max} = \tau_Y$ and $c = h$:

$$V_Y = \frac{2}{3} \tau_Y bh. \tag{19.14}$$

Thus, the actual resistance of the section to shear force is established. It is evident that

$$\frac{V}{V_Y} = \frac{\tau_{\max} c}{\tau_Y h},$$

and, therefore, the condition of the beam's integrity (19.7) is represented in the form:

$$\frac{V}{V_Y} < \frac{c}{h}. \tag{19.15}$$

We obtain the ratio c/h from formulas (19.10) and (19.11). It holds:

$$\frac{M}{M_p} = 1 - \frac{c^2}{3h^2} \quad \text{at} \quad M \geq M_Y,$$

and substituting (19.12) into the latter condition enables us to represent the domain of integrity states in the form:

$$\left[\begin{array}{l} \dfrac{M}{M_p} + \dfrac{V^2}{3V_Y^2} < 1 \quad \text{at} \quad M \geq \dfrac{2}{3} M_p; \\ \dfrac{V}{V_Y} < 1 \quad \text{at} \quad M < \dfrac{2}{3} M_p. \end{array}\right. \tag{19.16}$$

Figure 19.5 shows the domain of integrity states in accordance with this criterion. To take into account the possibility of negative internal forces, they should be taken in absolute value in this criterion. The first of conditions (19.16) determines the ultimate bending moment (with account for shear force acting at the section) as follows:

$$M_V = M_p \left(1 - \frac{V^2}{3V_Y^2}\right) \tag{19.17}$$

According to this formula, the occurrence of shear force which would noticeably decrease the plastic moment at the rectangular section is practically impossible. Assume, for example, $V = 0.5V_Y$. Even at this very noticeable shear force, when hypotheses of bending lead to poor accuracy of

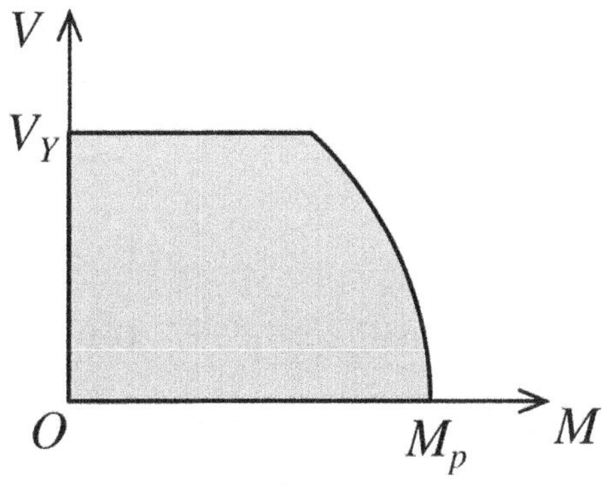

FIGURE 19.5 The domain of integrity states for the beam in Figure 19.4.

analysis, the decrease of the plastic moment is only 8%. Further on, the influence of shear forces upon the ultimate bending moment is not taken into account unless specifically stipulated.

Now, we proceed to the problem of finding the conditions of collapse for a statically determinate member working in pure bending with tension-compression. (This case is shown, for example, in Figure 19.1, on the left, if it is allowed that $P < 0$.) The integrity condition for such a member in the case of brittle material has the form (19.6). Now, we derive a similar condition in the case of elastoplastic material and rectangular cross-section of the member (Figure 19.6). For certainty, we shall consider the tension with convexity downward, i.e., we assume that $P \geq 0$ and $M_c \geq 0$. Internal forces are the same along the whole member, thus, plastic yielding occurs in the whole bar simultaneously, and the limit state of the member could be investigated at cross-section A taken arbitrarily (Figure 19.6). Actual axial resistance (i.e., resistance of member to axial force) is equal to:

$$N_p = \sigma_Y bh. \tag{19.18}$$

Let the section be subjected to given tensile axial force $N < N_p$. Upon increasing of the bending moment, the zone of plastic tension appears and spreads from the lower edge, whereas the upper edge stress decreases. Due to some bending moment $M_Y(N)$, the upper edge stress attains the value $-\sigma_Y$, and in the section there appears an elastic core (Figure 19.6a). Let us consider the state of the member when an elastic core has arisen, i.e., when $M \geq M_Y(N)$ (Figure 19.6b). Select two regions in the section: the region of balanced internal forces comprising the elastic core and the upper edge of the section, and the region of plastic yielding, extended from the opposite edge of the section. On the diagrams, these regions are divided by segment B_1B_2. The first of them is symmetric relative to neutral axis n–n, and the moment of internal forces in the first region can be calculated as a bending moment due to given stresses in the cross-section equal to this region. Elementary internal forces in the second region determine the axial force at the section. **The partition of the section into the region of balanced forces and the region of axial force manifestation makes it easy to represent internal forces M and N through characteristics of section and σ-diagrams.**

The component of the bending moment produced by balanced forces is obtained by formula (19.10), which we apply to the balanced part of the section:

$$M_1 = \sigma_Y b\left(\frac{h_b^2}{4} - \frac{c^2}{12}\right). \tag{19.19}$$

The total internal force acting on the unbalanced part is exerted upon centroid T' of this region and determines the axial force:

$$N = \sigma_Y b\left(h - h_b\right).$$

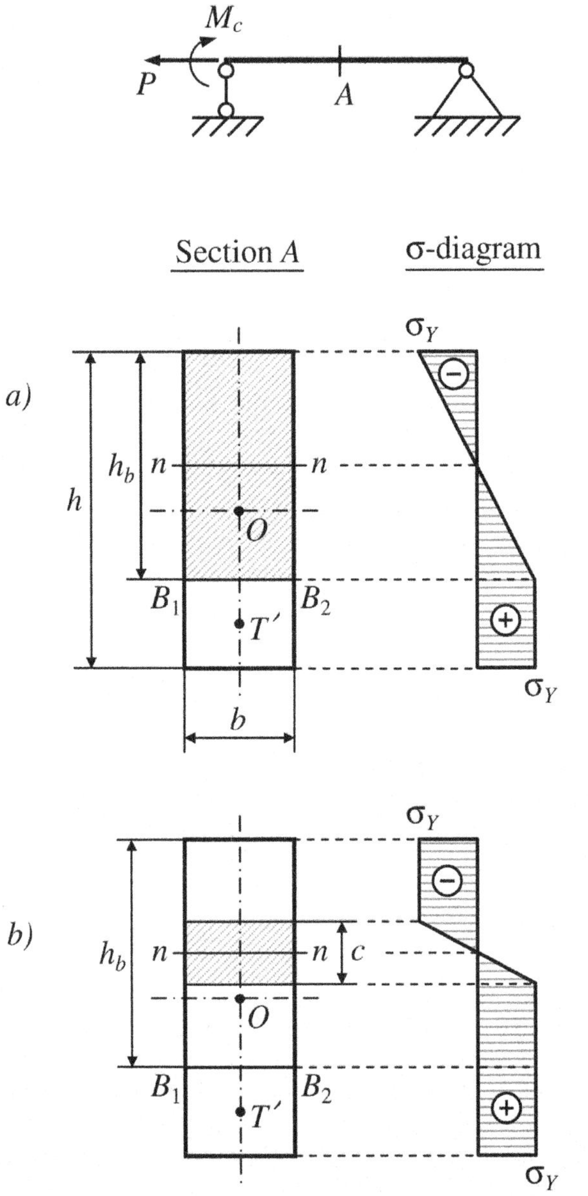

FIGURE 19.6 Stress distribution at the cross-section of tensed-bent member (elastic core is hatched).

The arm of this force about centroid O of whole section equals

$$T'O = \frac{h_b}{2},$$

and the component of the bending moment produced by this force equals

$$M_2 = \sigma_Y b \frac{h_b}{2}\left(h - h_b\right).$$

Thus, we obtain the bending moment at the section in the form:

$$M = M_1 + M_2 = \sigma_Y b\left(\frac{h_b h}{2} - \frac{h_b^2}{4} - \frac{c^2}{12}\right) = \frac{\sigma_Y b}{4}\left(h^2 - (h-h_b)^2 - \frac{c^2}{3}\right). \tag{19.20}$$

Arising of limit state is understood as appearance of plastic longitudinal strains at the entire section; accordingly, the condition of the beam's integrity has the form: $c > 0$. Further, we express this condition by means of internal forces and actual resistances of the section. We have:

$$\frac{M}{M_p} = 1 - \left(1 - \frac{h_b}{h}\right)^2 - \frac{c^2}{3h^2}; \quad \frac{N}{N_p} = 1 - \frac{h_b}{h};$$

wherefrom:

$$\frac{c^2}{3h^2} = 1 - \left(\frac{N}{N_p}\right)^2 - \frac{M}{M_p}.$$

So, the integrity criterion (being the criterion of elastic core preservation) is obtained in the form:

$$\frac{M}{M_p} + \frac{N^2}{N_p^2} < 1. \tag{19.21}$$

The domain of operable states, according to this criterion, is limited by half of the parabola (Figure 19.7). To take into account the possibility of negative loads, internal forces in this criterion should be taken in absolute value. The derivation has been done for the rectangular section, but in the case of the I-section, it is known that the curve of ultimate states is close to the parabola as well (White 2012).

Condition (19.21) determines the ultimate moment with account of axial force at the section as follows:

$$M_N = M_p\left(1 - \frac{N^2}{N_p^2}\right). \tag{19.22}$$

The given formula defines the plastic moment for the rectangular section due to bending with tension-compression.

To take into account the factors of uncertainty in criteria (19.16) and (19.21), we have to substitute design resistances for actual resistances and use factored values of internal forces.

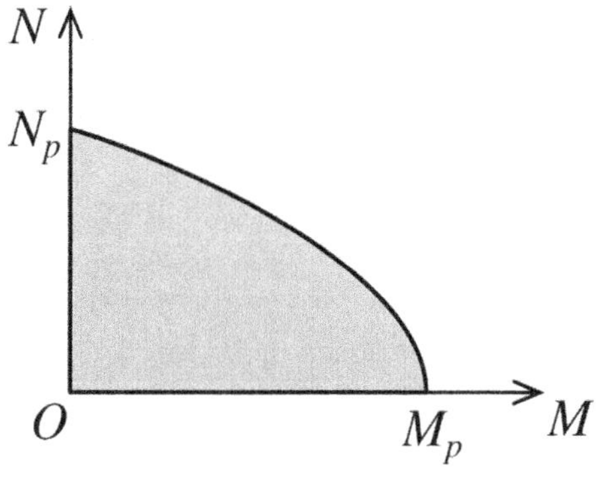

FIGURE 19.7 The domain of integrity states for the beam in Figure 19.6.

20 Principles of Plastic Design of Statically Indeterminate Structures

20.1 PRINCIPLES OF MODELING LOADING PROCESS IN ANALYSIS OF ELASTIC PERFECTLY PLASTIC STATICALLY INDETERMINATE STRUCTURE; EXAMPLES OF CONNECTIONS' LOSS AND RECOVERY

In the cycle of three chapters, starting from this one, we study graphic-analytical methods used for analysis of ductile redundant bar systems under quasistatic loading. These methods have been intensively developed during the 30s, 40s, and 50s of the 20th century. In the 50s, they were supplemented and then gradually superseded by numerical methods of analysis implemented on computers – in the first place, by the finite element method (FEM).

While the analysis methods developed for elastic statically indeterminate structures enable us to obtain exact solutions of applied problems by graphic-analytical means, the exact methods of analysis of statically indeterminate ductile bar structures are exercised on computers only. The models of ductile structures appropriate for graphic-analytical analysis enable us to obtain only a qualitative picture of a structure's operation under loading. Nevertheless, the methods outlined below are used at the early phase of designing, when general construction solutions are taken and the appearance of a structure is determined. Later, the designer must be ready to construct a computer model and implement a precise checking calculation, for which purpose the FEM is usually used.

Further on, we consider planar structures in which the plane of loading is the symmetry plane, and members' cross-sections are doubly-symmetric. We proceed from the given design of a structure, i.e., from the mutual position of members and connections between them being established and engineering materials being assigned. We assume that an out-of-joint (intermediate) load (if the one is exerted upon a member) does not have a longitudinal component. Owing to this assumption, the axial force in every member is uniform over its length, and this significantly simplifies the grounding of the theory.

In structural design, we distinguish *design calculation*, which aims to determine the members' cross-sections in order to secure the operable condition of a structure, and *checking calculation*, which aims to establish either the SSS of a specified structure under loading or the fact of collapse of the structure under loading. In the latter case, it is also important to determine the magnitudes of loads exerted upon a structure at the moment of collapse. Destructive assemblage of loads is referred to as the *bearing capacity of a structure*. Sometimes, it should be determined at the increase of loads beyond the given magnitudes. The checking calculation is done more easily and is sometimes employed as the basis for design calculation (if selecting the sections by the trial-and-error method). Firstly, we shall consider the methods of checking calculation of an elastoplastic bar system with redundant connections.

In Chapter 11, we introduced the concept of the quasistatic process in a mechanical system. Quasistatic loading of a structure is a slow change of loads from zero magnitudes to given values. As a rule, we shall consider the *quasistatic one-step process*. This means that a given system of loads is achieved through uniform and proportional increase of loads during the extended period of time T. Thus, at any time t in the period of load increase $[0, T]$, each exerted load is represented in the form:

$$P_i(t) = k(t)P_i(T); \quad k(t) \equiv \frac{t}{T}. \tag{20.1}$$

The first equality expresses proportionality of loads; the second equality ensures their uniform increase. The denotation of a load with no time argument is used below for the given value of the load, i.e., $P_i = P_i(T)$.

When the bearing capacity of a structure is required, one considers the one-step process in which the time of load increase is not limited to the moment $t = T$ of attaining the giving magnitudes by the loads. The increase of loads in this problem goes on according to law (20.1) up to the collapse of the system. The problem of determining bearing capacity at the combined action of several loads is considered the problem of finding the *factor of bearing capacity* (FBC) k_{col}, which is the ratio of the time of uniform load increase up to collapse to the time of load increase up to given magnitudes of loads. Therefore, it holds: $k_{col} = k(t_{col})$, where t_{col} is the moment of collapse, and it is possible that $k(t_{col}) > 1$.

In more sophisticated problems, the quasistatic process of loading is considered a multistep process. This means that the process is represented as the sequence of several processes called steps. Each step is specified by an assemblage of loads exerted upon a structure to the moment of the step's end. At the beginning of a step, the structure is under the loads of the previous step, which decrease to zero during the current step. At the beginning of the first step, the structure is unloaded. It is taken as given that increase of loads of the given step and decrease of loads of the previous step goes uniformly and proportionally. Further on, we assume the process of loading as one-step if the number of steps is not specified. The dependence of loads versus time in the one-step process is referred to as the *simple law of loading.*

In order to construct a model of the one-step loading process for a statically indeterminate structure, we use an idealized stress-strain diagram supplemented by lines of unloading and reloading. Corresponding portions of σ-ε diagram are shown in Figure 20.1: $A_1A_2A_3$ is the portion of unloading and subsequent compression for the tensed specimen; $B_1B_2B_3$ is the portion of unloading and subsequent tension for compressed specimen. The restoration of material to zero stresses and subsequent deformation operate elastically with the same modulus of elasticity E and in the same boundaries of

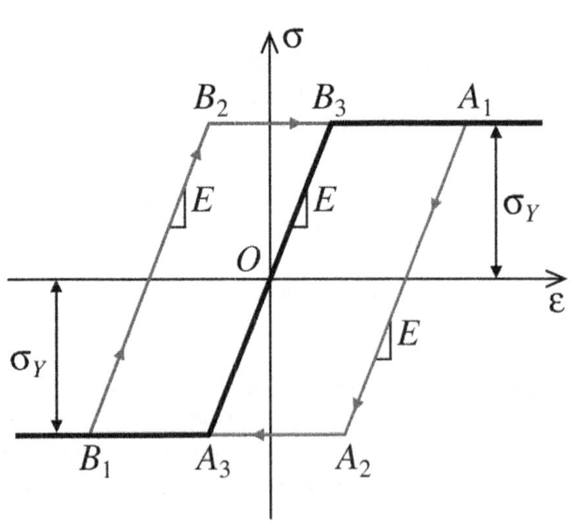

FIGURE 20.1 Idealized stress-strain diagrams with reloading.

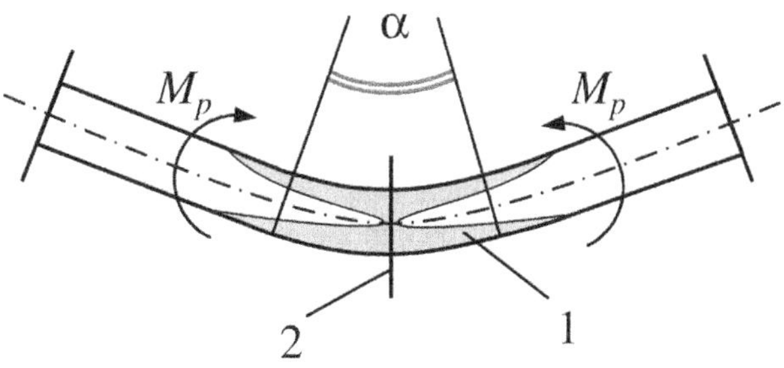

FIGURE 20.2 Bending of a member in the neighborhood of plastic hinge: 1 – region of plastic deformations; 2 – cross-section "plastic hinge"; M_p is a plastic moment.

stresses (from $-\sigma_Y$ to σ_Y) as those in the primary loading. The totality of diagrams in Figure 20.1 is referred to as the *idealized stress-strain diagrams with reloading.*

In calculation of a bending moment at the cross-section with a plastic hinge, the short portion of the member containing the section is considered to be the totality of longitudinal fibers, each of which is deformed according to idealized diagrams with reloading (Hibbeler 2011, p. 338). Let us consider the process of deforming an initially rectilinear member, during which axial force equals zero and within some member portion, bending moments increase until there emerges a plastic hinge. Pursuant to the named diagrams, after emergence of the plastic hinge, the bending moment at the section of this hinge is determined by the dependence on time of the elastic curve's curvature at the point of the section.* This curvature might be measured by angle α of mutual rotation of the sections limiting the short member's portion with a plastic hinge (Figure 20.2). The plastic moment continues to act at the section where the plastic hinge has arisen only if angle α continues to increase or remains constant after the emergence of the hinge. The rise of angle α after attainment of the plastic moment is called the opening of the plastic hinge. If this angle begins to decrease, then the hinge disappears, or, as they say, closes. The plastic moment in these processes can be accepted as a constant. While decreasing an angle between sections limiting the plastic hinge, the linear dependence of the bending moment on this angle is restored, but the restoration of the member's operability has the following features: in a completely unloaded member, there act residual stresses and the unloaded member remains somewhat curved. In practice, the above peculiarities of deforming a member with a plastic hinge also manifest themselves when an axial force acts in a member (the dependence of a plastic moment on this force may need to be taken into account).

The model of the one-step loading process is composed on the basis of the following principles:

- The properties of materials are specified by idealized stress-strain diagrams with reloading;
- With the rise of the load, there occurs an instantaneous reallocation of connections, when at some moment of time there necessarily occurs the loss of some connections and, possibly, the recovery of connections lost before;
- The interval of time between two events of connections' loss or reallocation we shall call the *substep of loading*. As a result of the first substep, the number of connections decreases; at each next substep it may remain unchanged or even increase. After a finite number of substeps, a system turns into a mechanism, which cannot keep equilibrium at further rise of the load. This event means the collapse of the system;
- With the rise of the load, the members instantaneously take on new properties resulting in the loss or reallocation of connections.

* A bending moment is determined by prehistory of the curvature's variation. In math terms, one can say that a bending moment is a functional of the function "curvature versus time."

The properties of members which arise out of load increase we call collapse, damage, or restoration of members' operability (integrity). They are as follows:

- A member wherein there acts only an axial force is considered to be collapsed if the yield stress has been attained in it. This member is taken for the source of external forces exerted upon the joints and equal to the axial force in the member;
- A member wherein there act bending moments is regarded as damaged if a plastic hinge has arisen in it. This member is considered as two members connected by a hinge. The hinge that emerged due to the increase of load is still called a plastic hinge. This one is the source of external moments equal to the plastic moment and exerted on the ends of the divided member;
- The restoration of operability is the closing of the plastic hinge in the bent member or reversal of the sign of the strain in the tensed-compressed member.

The model based on these principles is referred to as the *model of loading with connections' loss and recovery.*

A plastic moment might change both along the length of the member and in time because of its dependence on shear force, but, as a rule, it is allowed to neglect this dependence and determine the plastic moment at zero shear force. If this dependence is nevertheless significant, then we fail to construct a sufficiently simple and reliable model of loading with connections' loss and recovery and must turn to computer modeling on the basis of FEM. A plastic moment might also change in time due to its dependence on axial force. When axial force is significant – for instance, in some columns – the time dependence is to be taken into account.

Sometimes, in cases when the time sequence of load application is important, the principles of analysis developed for the one-step loading process are employed for the multistep process of loading. Below, some examples illustrate the implementation of these principles.

Figure 20.3 shows the example of an elastoplastic hinged system. In diagram a, the design scheme is depicted at the given loading; diagram b shows the collapse of one of the members arisen at the point in time t_1 of load increase; and diagram c shows the state of the system's collapse which occurs at the moment t_2 due to the collapse of the second member during continued increase of load. Since the idealized stress-strain diagram doesn't take into account the portion of strain hardening, load $P(t_2)$ obtained in analysis is less than the actual failure load.

Figure 20.4 shows the example of an elastoplastic beam with a rectangular cross-section. In diagram a, the design scheme is depicted at a given loading; diagrams b and c show the deformed beam at the points in time t_1 and t_2 when the next plastic hinge has appeared (horizontal displacements are small and not shown in diagrams). Plastic hinges emerge where and when bending moment M reaches the magnitude of plastic moment $M_V = M_V(V)$. We shall consider the plastic moment as independent of shear force and take its value according to formula (19.11). Then the first plastic hinge emerges when the following condition is fulfilled:

$$\max_z |M(z)| = M_p, \tag{20.2}$$

where z is the coordinate of the section, and maximum is taken over all sections of the beam. The first hinge appears close to rigid support, as shown in diagram b, because at the beginning of the process, when deformation of the beam is elastic, the bending moment attains maximum value near the rigid support (see the profile M in Figure 18.1f). During further rise of load, the beam is subjected to constant force couple of M_p produced by the plastic hinge. The emergence condition for a second hinge continues to be of the form (20.2) with the feature that the point of maximum shouldn't coincide with the point found before, where the first hinge appeared. The point in time t_2 of emergence of the second plastic hinge corresponds to the ultimate state of the system, shown in diagram c.

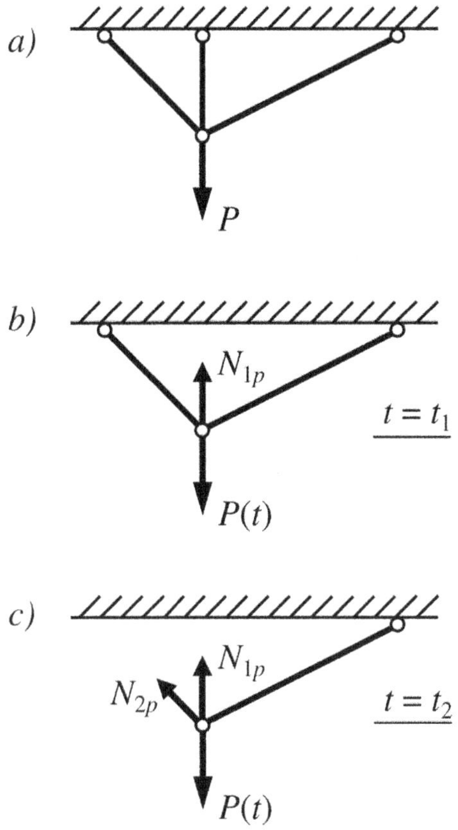

FIGURE 20.3 Loss of connections in a hinged system. N_{1p} and N_{2p} are the axial forces due to plastic tension.

Now, let us consider the development of plastic deformations in the example in Figure 20.4, taking into account the action of shear force. Before plastic deformations emerge, this force is greatest at the rigid fixing; thus, according to condition (19.16), at the section of fixing, the ultimate state of the material is really attained earlier than at other sections. Further increase of load is attended by rotation of the beam's portion near the emerged hinge through a small angle α (Figure 20.4c), and the bending moment at the fixing, at least, does not decrease at that. Besides, at the section with the plastic hinge the shear force increases, which must cause entering the collapse zone, according to the graph in Figure 19.5. In plain words, equilibrium in the neighborhood of rigid support becomes impossible. This contradiction is resolved: the displacements in a beam of elastoplastic material, similar to the kink in the zone of a plastic hinge, occur due to a bending moment somewhat lesser than the plastic moment. In the example under consideration, during time interval (t_1, t_2), the elastic core persists at the fixing of the beam and has an extension sufficient to prevent shearing. The beam's transformation into the mechanism shown in Figure 20.4 and the magnitude of the failure load obtained by the criterion of the connection's loss (20.2) are confirmed by analysis on the FEM basis. But it is right for a rectangular cross-section of a beam, whereas in the case of thin-walled sections, the possibility of shear remains; and for analysis of this possibility at the section of a plastic hinge, modeling by FEM is necessary.

Now let us supplement the design scheme in Figure 20.4 by tensile load as is shown in Figure 20.5. In this case, the plastic moment is determined by formula (19.22) and depends on time. It is easier to take into account axial load in the two-step process of loading, when only the rising axial load $P(t)$ acts at the first step, whereas at the second step, the constant axial load P is complemented by the rising distributed transverse load $w(t)$. In the two-step process, the calculation of the second step

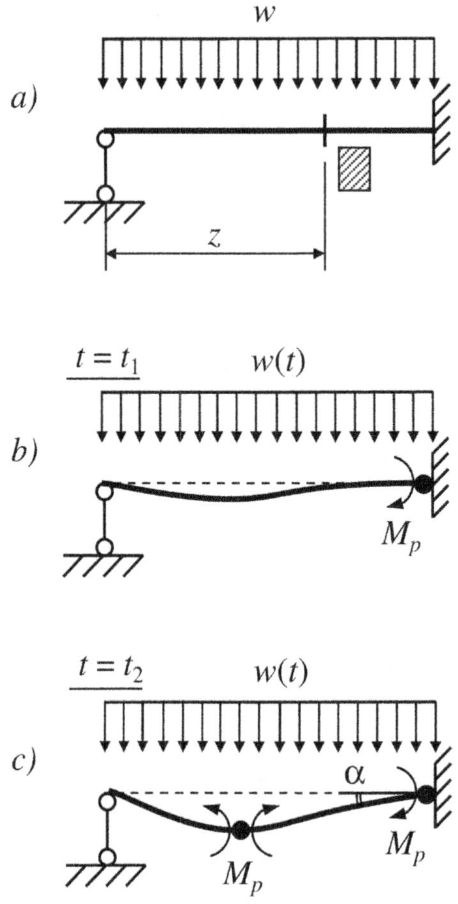

FIGURE 20.4 Loss of connections in a statically indeterminate beam.

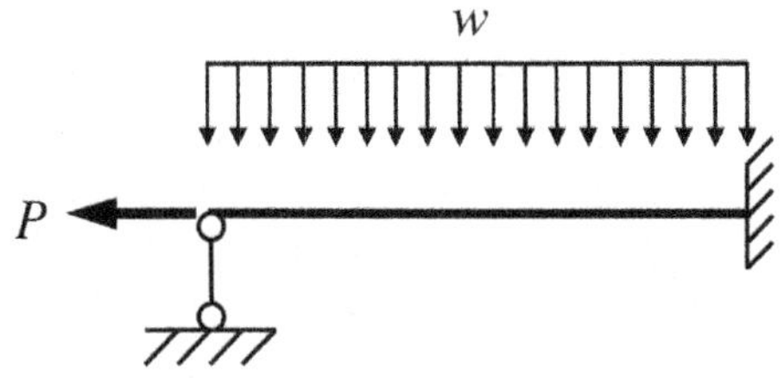

FIGURE 20.5 Beam in bending with tension.

differs from the calculation of the previous example only by the magnitude of the plastic moment. It is understood that on the right-hand side of condition (20.2), instead of actual resistance of section M_p, there should be substituted plastic moment (19.22), obtained with account of axial force being constant in time. If, for instance, $P = 0.7N_p$, then destructive transverse load $w(t_2)$, obtained by using the model of loading with step-by-step loss of connections, constitutes 51% of the failure load in the previous problem, when $P = 0$. In checking calculations by FEM for a beam of square cross-section with the ratio:

$$\frac{\text{span}}{\text{section's hight}} = 20,$$

the same index constitutes 47–49%.

In the case of one-step loading with the design diagram in Figure 20.5, we must take into account that plastic moment M_N decreases at the rise of loads. In particular, the concentrated moment exerted upon the beam by the first plastic hinge decreases in time until the second hinge appears. Nevertheless, the order of occurrence and location of plastic hinges in this problem are the same as in the problem illustrated in Figure 20.4. Introduce denotation w_{0c} for the failure distributed load in the particular case when $P = 0$. If the terminal loads of the step equal $P = 0.7N_p$ and $w = w_{0c}$, then the load $w(t_2)$ obtained by graphic-analytical analysis constitutes 74% of failure load w_{0c}. The calculations of this index by means of FEM in a problem with the pointed above geometry characteristics ($b = h$; $l/h = 20$) give values in the range 72–74%.

So far, we have considered examples with step-by-step loss of connections. Figure 20.6 shows an example of the loading process when a hinged member at first fails and then restores its operability. We presume that both verticals of design scheme a are elastic perfectly plastic and the girder is elastic. At the definite characteristics of the structure, during the rise of load P, the upper vertical is first compressed elastically and then goes into the plastic state and becomes a source of constant load N_{1p} (diagram b). With further increase of load, the same occurs to the lower vertical, which develops load N_{2p} due to plastic compression. It seems that the structure has lost two connections and became a mechanism, but at the point in time t_2 of losing the second connection, the upper vertical starts

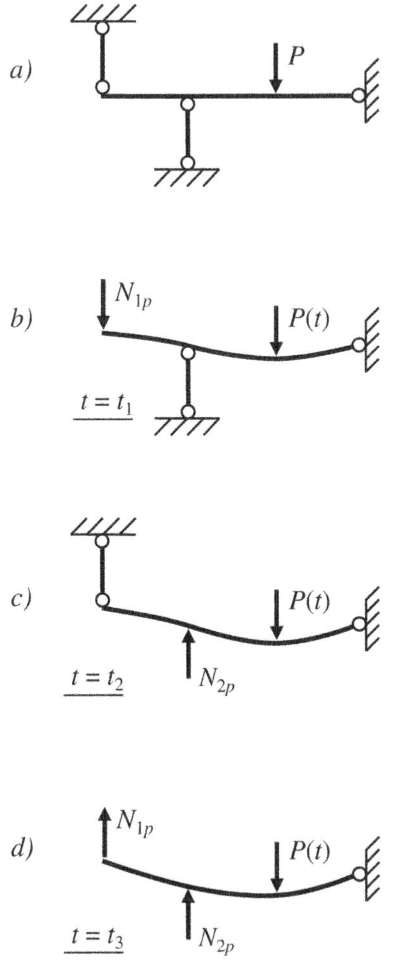

FIGURE 20.6 One-step process: hinged member first fails and then restores its operability.

to work elastically again (diagram c), steadily going from the state of compression into the state of tension. The collapse of the system occurs when the upper hanger comes into a plastically tensed state (diagram d).

In Figure 20.7, we can see the example of a beam's loading through the one-step process, when a plastic hinge at first arises and then closes. The beam consists of elastic portion AB and elastoplastic portion BC (diagram a). The flexural rigidities EI of both portions are the same. The beam has two equal spans; the intermediate support is a connection of the first type, and the left support is an elastoplastic hinged column. It is possible to choose the characteristics of the structure such that, due to the increase of load w, a plastic hinge emerges first in the right-hand span (diagram b), then the column passes into a plastic state, and simultaneously, the hinge closes (diagram c), and then, because of shortening of the column, there opens a new hinge at the junction B of

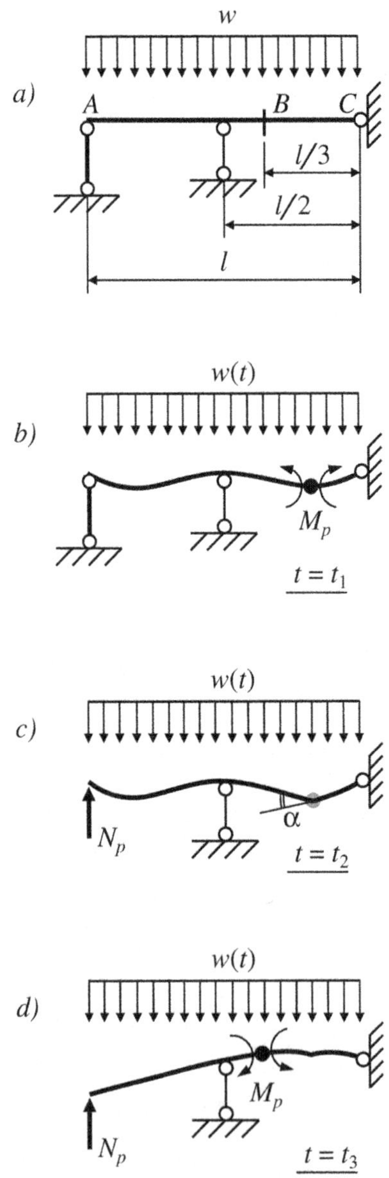

FIGURE 20.7 One-step process: plastic hinge at first arises and then closes.

beam's portions (diagram d). In diagram c, the neighborhood of the plastic hinge is depicted by a grey spot; at time t_2, the plastic deformation of the column has arisen, and from this moment, angle α of mutual rotation of the member's parts jointed by a hinge has started to decrease and the elastic properties of the shown beam's neighborhood have been restored. Diagram d shows the limit equilibrium of the system, when it has turned into a mechanism and will be destructed due to slight increase of load.

20.2 METHOD OF SEQUENTIAL IDENTIFICATION OF LOST AND RESTORED CONNECTIONS; STATIC PRINCIPLE OF PLASTIC ANALYSIS

The model of loading with connections' loss and recovery enables us to determine the bearing capacity of a structure through the *method of sequential identification of lost and restored connections* which is set forth below.

If a system turns into a mechanism and cannot keep equilibrium thereafter due to the quasistatic, maybe multistep, loading process with the loss and reallocation of connections, then such a mechanism is called a *collapse mechanism*. The equilibrium of a system subjected to some loading in the quasistatic process, such that continuation of the process is impossible due to loss of equilibrium, is referred to as *limit equilibrium*.

In accordance with the method under consideration, the analysis is fulfilled in several steps; each step represents the process of load increase until the event of loss or reallocation of connections. The analysis is completed at the step of the structure's transformation into a collapse mechanism. During the whole process, the loads change in time according to formula (20.1). The result of every step is the identification of lost and restored connections and (if required) the point in time of loss-recovery. **As a result of analysis, we establish the moment in time t_{col} of collapse and the corresponding factor of bearing capacity $k_{col} = \frac{t_{col}}{T}$. Also, the results of analysis are the diagram of the collapse mechanism into which a structure is converted and the profiles of internal forces in the state of limit equilibrium.** Further, we consider a general case when a structure has both bent and tensed-compressed members and the effect of axial forces upon the plastic moment is taken into account.

In each step of analysis, a structure is considered as elastic. In each step, bending moments and axial forces produced in members are determined as the functions of factor $k(t)$ multiplying the loads (see formula (20.1)). In the current step, determine the axial resistances for all members and flexural resistances for bent members (which maybe dependent on factor k as well). Acting and ultimate axial forces as well as ultimate bending moments are represented in a tabulated form, whereas acting bending moments are represented in the form of diagrams. For every bent member, discover the extremum points of the bending moment diagram and set up the equations in an unknown time when the moment connection is lost – they are really the equalities between the extreme bending moment and the ultimate moment. For every tensed-compressed member, compose an equation in an unknown time of the connection's loss in the form of the equality between acting axial force and ultimate force. A connection is lost as soon as at least one of the composed equations turns into an identity. The minimal value of time obtained from these equations is the wanted point in time of the connection's loss, and the section of the member or the whole member, for which the solution was obtained, is the place of the connection's loss.

After identification of lost connections, the connections restored to the initiation of the next substep of loading should be discovered. For this purpose, there should be determined the displacements in the places of lost connections due to small rise of load. Depending on pivot direction in the plastic hinge or the sign of longitudinal strain in the tensed-compressed member, we make a conclusion about the connection's recovery.

After turning a structure into a mechanism, it may be required to verify that it is really a collapse mechanism, i.e., that a small increase of load causes the loss of the structure's equilibrium.

More commonly, however, the loss of the system's equilibrium is obvious enough to not do this verification.

The axial forces calculated in this method enable to refine the ultimate bending moments in bent members, as well as can be the cause of collapse for tensed-compressed members.

Important notes:

- In the given method, internal forces in members are determined through the increments in a substep of loading; namely, the sought internal force is computed as the sum of two components, of which the first one is internal force acting at the beginning of the substep, and the second one is the increment of sought force due to increase of load at the interval from initiation of the substep until a given moment in time. Owing to this, we take into account the possibility of occurrence of residual internal forces during the loading process, i.e., the forces which continue to act in the members after the unloading of a system.
- If in a substep under investigation, a system is statically determinate, then internal forces can be computed for the full load at a given point in time (because in a statically determinate system, they are established uniquely from the equations of members' equilibrium). Also, it is allowed to compute internal forces not in increments but for the full load if, in all previous substeps, the connections did not recover. This possibility follows from the fact that in the case of a process without recovery of connections, the residual internal forces are absent in a system at the initial moment of every substep. It is important here that instead of a real elastoplastic system we consider an elastic system constructed according to the model of the loading process with step-by-step loss-recovery of connections. In a real statically indeterminate system, the residual internal forces usually appear with occurrence of plastic deformations.
- The cause of residual internal forces is the displacements in restored connections. As a result of these displacements, a member with a restored moment connection has a kink in the place of the closed plastic hinge, and after recovery, the tensed-compressed member has another length when unloaded.

Let us consider two simple instances of applying the method of identification, when the position of plastic hinges is defined unambiguously, and the recovery of connections does not occur.

Example 1

Determine failure load for the beam in Figure 20.8a at given flexural resistance of section M_p.

Step 1. Construct the bending moment diagram for the initial period of the beam's work, before plastic strains. Every load is specified in the form $P(t) = k(t)P$, and, firstly, the diagrams are constructed for each load separately, which requires the use of the diagram in Figure 18.1h. The complete diagram is constructed by summation of the obtained diagrams. Constituent diagrams and the total diagram of bending moments in the period of elastic work of the beam are shown in Figure 20.8b.

The maximum bending moment acts at the rigidly fixed end (at $z = l$); its value during the initial period of work is $M(l) = \frac{kPl}{3}$. From the requirement $M(l) = M_p$ we may find the factor $k_1 = \frac{t_1}{T}$ for the moment t_1 of the emerging of a plastic hinge, though in the given case it is not necessary. For further analysis, it is sufficient to identify the plastic hinge at the fixing.

Step 2. At the point in time $t = t_1$, the beam becomes statically determinate, and from the side of the plastic hinge, the constant force couple M_p is exerted (Figure 20.8c). At $t > t_1$, the

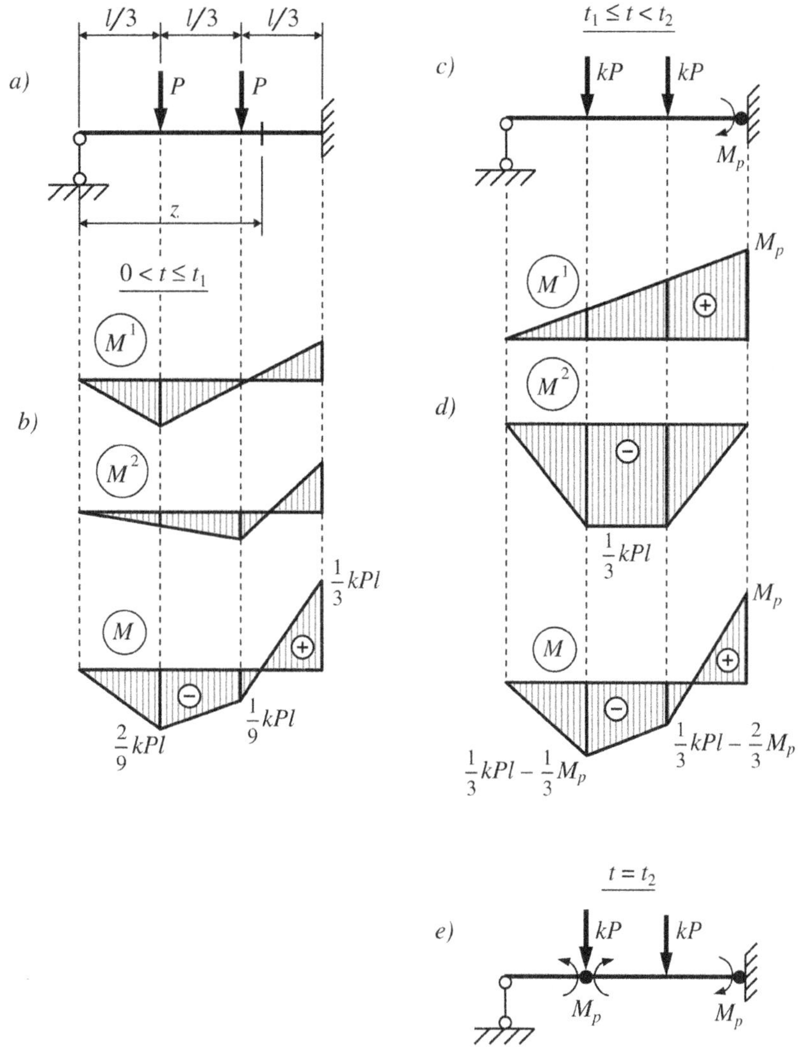

FIGURE 20.8 Beam in the process of loading: (a) Design diagram; (b) Diagrams during the period of elastic work; (c) Beam with one lost connection; (d) Diagrams after the loss of connection; (e) Beam in limit state.

complete bending moment diagram is obtained by summation of the two diagrams: one due to force couple M_p and another due to loads kP. Constituent diagrams and the total diagram are shown in Figure 20.8d. One can see that the second maximum of the bending moment's absolute value will be reached at $z = l/3$ (the bottom profile in plot *d*). This maximum emerges under following condition:

$$\frac{k_2 Pl}{3} - \frac{M_p}{3} = M_p; \quad k_2 \equiv \frac{t_2}{T}.$$

The ultimate state of the structure occurs at time t_2 under load

$$P(t_2) = k_2 P = \frac{4M_p}{l}. \tag{20.3}$$

The scheme of a mechanism in limit equilibrium is shown in Figure 20.8e.

Remark

In the example discussed above, there is no need to introduce FBC, for the loading is determined by only parameter *P*, which can be considered as a load being increasing over time instead of factor *k*. This factor has been introduced into calculation to illustrate the method of identification, and its usage becomes necessary in more complicated problems.

Example 2

Determine the bearing capacity of a tensed-bent member of the rectangular cross-section under loading according to the diagram in Figure 20.9. Loads P_0 are equal to the ultimate magnitude in the absence of tension: $P_0 = \dfrac{4M_p}{l}$; given the coefficient λ which determines tensile load in the form: $P_1 = \lambda N_p$, where N_p is the axial resistance of the section.

Step 1 repeats the first step of the previous example with the distinction that moment in time t_1 when a plastic hinge emerges is determined by the condition: $M(l) = M_N$, where M_N is the ultimate bending moment with account for tensile force. According to formula (19.22), this moment is represented in the form:

$$M_N = M_p\left(1 - \frac{\left(k(t)P_1\right)^2}{N_p^2}\right) = M_p\left(1 - k(t)^2\lambda^2\right). \tag{20.4}$$

Step 2. Just as in the previous example, after the appearance of a plastic hinge at the rigid fixing, the second maximum of the bending moment is reached at the section with coordinate $z = l/3$ (Figure 20.8d). The second plastic hinge arises under the condition which is obtained by substitution of the ultimate moment (20.4) for flexural resistance M_p into equality (20.3) at the load $P = P_0$. Thus, we come to the equation in unknown FBC k_2:

$$k_2 P_0 = \frac{4M_p\left(1 - k_2^2\lambda^2\right)}{l}.$$

After substitution of the ultimate magnitude of load P_0, this equation takes the form:

$$k_2^2\lambda^2 + k_2 - 1 = 0. \tag{20.5}$$

Its solution determines the bearing capacity of the structure:

$$k_2 = \frac{\sqrt{4\lambda^2 + 1} - 1}{2\lambda^2}. \tag{20.6}$$

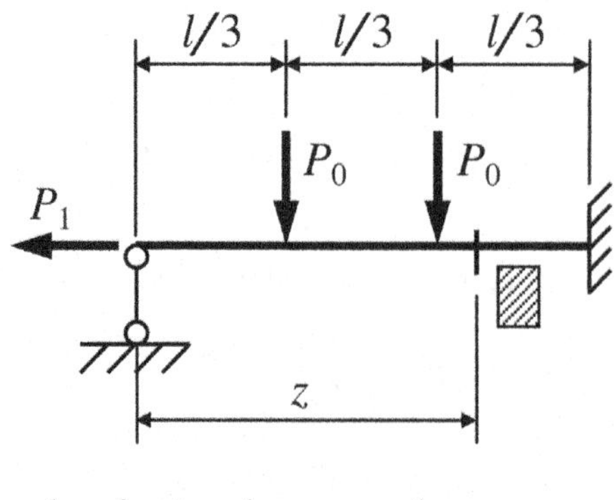

FIGURE 20.9 Tensed-bent member of rectangular cross-section.

The quantity $k_2 \cdot 100\%$ shows what percentage of ultimate load P_0 the member will withstand in the case when longitudinal load $P_1(t)$ is exerted on the beam together with transverse loads and increases parallel to them.

In more complicated problems, the following *static principle of elastoplastic system analysis* is helpful:

If for a given process of loading there exist equilibrium states of a system, dependent on time, in which axial forces and bending moments are limited by ultimate values N_p and M_N, and the most recent of these states is the ultimate one, then this limit state determines FBC. If equilibrium states, in which internal forces are continuous functions of time with the restrictions described above, are determined unambiguously, then they are realized due to quasi-static loading.

This principle assumes, as usual, a one-step process of loading with the initial time-point $t = 0$, when there are no loads. During the process, arbitrary events of loss-recovery of connections are allowed under stipulated requirements upon magnitudes of internal forces N and M. In the formulation, the next denotations are in use: $N_p \equiv \sigma_Y A$ is axial resistance of the section, and M_N is a plastic moment with account for axial force acting in the member.

The static principle of analysis supplements the described method of sequential identification of connections. Sometimes, at a regular substep of loading, there occurs uncertainty in localization of lost connections, or identification of restored connections through analysis of displacements produces difficulties. In this case, one might employ the static principle of analysis by investigating possible variants of connections' loss-recovery and selecting an option which permits the continuation of the process of loading.

Let us consider the design diagram in Figure 20.10a, where there is a peculiarity of loading upon the statically indeterminate beam: in the first substep of the one-step process, at the point in time $t = t_1$, bending moments at the whole portion BC attain the magnitude of plastic moment M_p, while at the rigid fixing plastic moment is not reached. The bending moment diagram for $t = t_1$ is shown on plot b. In the given case, the plastic hinges' location is uncertain, and this uncertainty can be resolved by taking into account the dependence of the ultimate moment versus shear force. The matter is that within the horizontal portion of the M-diagram, the shear force is zero, whereas at the left and right near the portion BC, it equals in absolute value the loads kP_1 and kP_2, respectively. Therefore, the plastic moment is lowered at the sections B and C, and it is there that plastic hinges occur. The corresponding mechanism is presented at the top of diagram b.

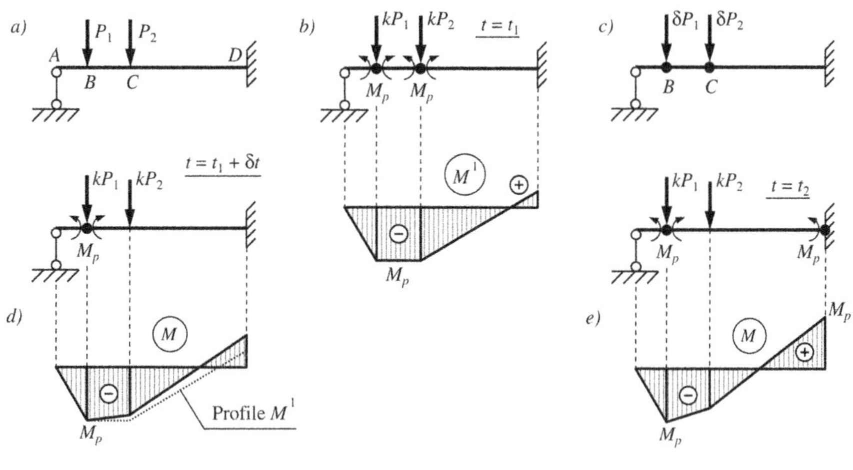

FIGURE 20.10 Process of a beam's loading with uncertainty in the location of the plastic hinge.

The mechanism arising at the moment $t = t_1$ can't keep balance with the subsequent increase of loads. This can be easily seen from consideration of loading at the time $t = t_1 + \delta t$, where δt is a small positive quantity. The increments of loads during interval δt we denote as δP_1 and δP_2, respectively. The assemblage of loads shown in diagram *b* ensures the equilibrium of the mechanism, but the system of two loads δP_1 and δP_2 exerted at points *B* and *C* (see diagram c) doesn't secure its equilibrium. So, the superimposing of increments δP_1, δP_2 upon the assemblage of external forces at $t = t_1$ leads to the loss of equilibrium.

Next, we can analyze the movement of an unbalanced system and find out that the pointed out increments of loads produce the downward motion of hinge *B*, connecting the pieces of mechanism *AB* и *BC*. This means the closing of hinge *C*, i.e., the emergence of a new system, which is in the equilibrium state at the increase of loads.

This cumbersome analysis results in the conclusion of the absence of plastic hinge *C*, because it has appeared and disappeared at the time t_1. However, the uncertainty in the location of plastic hinges at portion *BC* of constant bending moments is easily resolved by means of the static principle. If we obtain the unique allocation of plastic hinges within this portion, which ensures equilibrium with further increase of loads, then this allocation is realized in the process of loading and enables us to solve the problem of analysis.

It is obvious that if the number of hinges at portion *BC* is two or more, the equilibrium of the mechanism under the action of only load increments δP_1 and δP_2 is impossible; thus, the hinge must be single. The system with a single plastic hinge in this portion is statically determinate, and it is easy to construct the bending moment diagram at $t = t_1 + \delta t$ by summing up the diagram in plot *b* with the diagram of bending moments caused by the load increments only. If the hinge does not coincide with point *B*, then it is necessary for equilibrium of the system at $t = t_1 + \delta t$ that absolute value of the bending moment at section *B* should be greater than plastic moment M_p. That is why the equilibrium of such a system is impossible. Thus, the only feasible option remains: the hinge is located at section *B* as it is shown in Figure 20.10d.

The total diagram *M* for this option is shown at the bottom of plot *d*. When $t > t_1$, the minimum of the bending moment along the beam length is reached at the unique point *B* and equal to plastic moment M_p with a minus sign, whereas the maximal bending moment is positive, reached at the fixing, and increases proportionally to the interval of time δt from the magnitude of bending moment at $t = t_1$. Because bending moments are restricted by plastic moment, the diagram on the plot *d* rightly represents the stress state of the system after the point in time t_1.

At some point in time $t = t_2$, the bending moment at the fixing attains the magnitude M_p. Then, there appears a second plastic hinge *D* and the limit state of the system occurs (Figure 20.10e).

21 Theoretical Basics and Kinematic Method of Elastoplastic Structures' Analysis

21.1 STATIC THEOREM OF STRUCTURAL ANALYSIS AND PROOF OF STATIC PRINCIPLE

The theory of elastoplastic structures is grounded in the model of connections' loss and recovery studied in the previous chapter. In its turn, the practical applicability of this model has been confirmed by experimental research carried out over several decades in the middle of the 20th century.

The theory of elastoplastic structures includes two fundamental assertions known as the static and kinematic theorems (Neal 1977). These theorems were formulated and substantiated by Soviet professor A.A. Gvozdev in 1936 (Gvozdev 1949, p. 222). The statement of structural plastic analysis in the works of Gvozdev is very general, but his works demonstrate how to prove these theorems. In 1949, American scientists H.J. Greenberg and W. Prager proved these theorems in more simple and rigorous terms.

Further, when the static and kinematic theorems are formulated and proved, the proofs are given for a system comprising bent members only. The theorems can be reformulated and proved for systems comprising both bent and tensed-compressed members. The reader can supplement the proofs by taking into account systems comprising tensed-compressed members.

The theorems are proved under the assumption of independence of the plastic moment from axial and shear internal forces. While taking into account the dependence of the plastic moment on axial force, the theorems' assertions should be considered hypotheses permitted in solving practical problems.

We shall call plastic hinges and members of structure in the state of plastic yielding *plastic elements*. Let us consider an initial period of system collapse. We make a distinction between plastic elements which at the beginning of collapse restore their elastic properties and the elements in which plastic strains rise in the course of collapse. The first group consists of plastic hinges which are closing and yielding members wherein the time derivative of length changes its sign from the moment of collapse. To the second group, there appertain plastic hinges that are opening and yielding members wherein the variation of length in the period of destruction ensures the rise of plastic strain. We subsume plastic elements, in which additional strains don't arise during collapse, under the first group of the system plastic elements. A distinctive feature of an element belonging to the first group is that it doesn't absorb energy during collapse, i.e., the total work done by external forces exerted upon the element is not positive. An element of the second group, on the contrary, absorbs energy, and forces exerted on it perform work.

The fundamental provision of the limit equilibrium theory is that plastic elements, which during system collapse do not absorb energy and, respectively, work elastically, are the components* or the pieces of components of the **collapse mechanism, whereas plastic elements, in**

* A component of a mechanism is an internally stable part of it. While analyzing displacements in a mechanism, its components are considered as perfectly rigid bodies.

which the energy is absorbed, connect the components of the mechanism and are considered as a source of external forces upon these components.

For the same plastic elements, the collapse process may work in different ways depending upon random effects not covered by the design scheme. In one scheme of destruction, some plastic element might belong to a component of a mechanism as a non-deformable element, but in another scheme, it may appear to be the source of external forces upon the components due to the rising of plastic strains in it.

While proving static and kinematic theorems, the virtual work equation is set up for a collapse mechanism. We remind readers that the virtual work equation is the condition that the work done by external forces through virtual displacements of a mechanism be equal to zero (see Chapter 6). In the case of a collapse mechanism, we attribute, together with loads, the reactions of plastic elements, which connect components of the mechanism, to external forces. While proving the theorems, we consider only these plastic elements, external to the components of the collapse mechanism.

The reactions of a plastic hinge consist of axial forces at the sections adjacent to the hinge on both sides, shear forces at these sections, and the force couples producing plastic moments. The work done by shear and axial forces is equal in total to the work done by the load exerted upon the plastic hinge (this work equals zero if the load is nil). That is why, in the virtual work equation, it is sufficient to take into account the work done by plastic moments only, whereas the work done by the remaining reactions is to be calculated as the work performed by the load upon the hinge. In other words, we may employ the model of loading with connections' loss and recovery, in which the plastic hinge is represented as a common hinge supplemented with two couples of external forces.

Note that the virtual work equation contains only external forces, owing to the assumption of the mechanism components' rigidity. While subsuming plastic elements of the first group under the components of the collapse mechanism, we do not admit their deformations due to virtual displacements.

We shall refer to static loading obtained from a given loading in accordance with the law (20.1) for any moment in time $t > 0$ as *proportional loading*. Proportionality factor k we briefly call the load parameter.

Static Theorem

If an elastoplastic system keeps equilibrium under loading proportional to a given one, then the load parameter k can't be greater than factor k_{col} multiplying the loads in the limit state of this system.

Proof

In the frame of conception of stepwise connections' loss and recovery, generalized upon the multistep loading process, the limit state of an elastoplastic system means emergence of a collapse mechanism at some load. We shall consider this mechanism, called an actual mechanism below, in the state of equilibrium. Denote the load exerted at the i-th point of the system according to design scheme (we confine ourselves to concentrated loads only) as P_i. The bending moment produced by the j-th plastic hinge in the collapse mechanism we denote as M'_j – accordingly, the absolute value $|M'_j|$ is a plastic moment at the same section. Let us set up the virtual work equation in unknown load parameter k_{col}. To this end, we give the mechanism a virtual displacement in the same direction as the displacement arisen due to the system's collapse at the beginning of its movement. We denote additional rotation at the j-th plastic hinge at such displacement as $\delta\theta_j$; virtual displacement in the direction of load P_i is denoted as Δ_i. The virtual work equation for an actual collapse mechanism we write in the form:

$$k_{col}\sum_i P_i\Delta_i = \sum_j M'_j\delta\theta_j. \tag{21.1}$$

In the problem of seeking ultimate loads, this equation usually is interpreted as the condition of equality between the work done by loads and the work done by internal force couples acting on plastic elements of a system.

Now, we consider this system in the equilibrium state under loads kP_i. We keep the numbering of sections where plastic hinges have occurred in the **actual** mechanism the same. Denote bending moments at these sections as M_j (at every possible j's). Insert common hinges into these sections and, to secure equilibrium of the obtained replacing system, exert the moments M_j at the ends of members in the places of insertions. By giving the displacement to the new system, which coincides with displacement of the collapse mechanism being considered, we get the equation of works:

$$k\sum_i P_i\Delta_i = \sum_j M_j\delta\theta_j. \tag{21.2}$$

By multiplying both sides of equation (21.1) by k and both sides of equation (21.2) by k_{col}, we get the coincidence of the left sides of both equations, whence follows the equality for the right sides:

$$k_{col}\sum_j M_j\delta\theta_j = k\sum_j M'_j\delta\theta_j. \tag{21.3}$$

Every addend of the right sum $M'_j\delta\theta_j$ is positive, for this quantity equals absorbed energy due to virtual rotation at the j-th plastic hinge (note that moment M'_j reacts against rotation at collapse, i.e., reacts against the pivot through an angle $\delta\theta_j$). But the moment M_j acting at the j-th section of the structure in the equilibrium state can't be greater than the plastic moment at this section, i.e., $M_j \le |M'_j|$, therefore

$$M_j\delta\theta_j \le M'_j\delta\theta_j.$$

So, we come to the inequality:

$$\sum_j M_j\delta\theta_j \le \sum_j M'_j\delta\theta_j,$$

from which we obtain the necessary condition of equality (21.3) in the form:

$$k_{col} \ge k, \tag{21.4}$$

which was to be proven.

Corollary

The limit state of an elastoplastic system on a set of proportional loadings is possible under the unique proportionality factor k_{col}.

Now, we prove the static principle of an elastoplastic system's analysis:

If for a given process of loading there exist equilibrium states of a system, dependent on time, in which axial forces and bending moments are limited by ultimate values N_p and M_p, and the most recent of these states is the ultimate one, then this limit state determines the FBC. If, besides, equilibrium states in which internal forces are continuous functions of time with pointed above restrictions are determined unambiguously, then they are realized due to quasistatic loading.

Remark 1

Continuous dependence of internal forces versus time is implied for sections which are not subjected to concentrated external forces and force couples, and which doesn't intersect the joints of members. (At these sections, the indefiniteness of internal forces can't arise.) The requirement of continuity in time excludes the equilibrium states with residual internal forces impossible during increase of loads from consideration.

Remark 2

The wording of this principle in the previous chapter is broader, because it comprises the restriction of bending moment by quantity M_N, i.e., allows for an influence of axial force upon a plastic moment. The generalized formulation is the hypothesis applicable for practical calculations.

Proof

The first affirmation of the static principle is right, owing to the corollary to the static theorem. The second affirmation follows from provisions of the model of step-by-step connections' loss and recovery, under which internal forces at every substep of loading are continuous in time, and the process of loading is necessarily accomplished by the state of limit equilibrium. The principle is proved.

21.2 ASSUMED COLLAPSE MECHANISM AND STRENGTHENED SYSTEM; KINEMATIC THEOREM OF STRUCTURAL ANALYSIS

As an *assumed collapse mechanism*, we refer to a mechanism which is obtained from a given system subjected to loading by assignment of some its parts to be plastic elements at the discretion of the researcher but in compliance with following requirements:

- Plastic moments (axial resistances) of bent (tensed-compressed) plastic elements of an assumed mechanism are the same those as in given system at the same cross-sections;
- For an assumed mechanism, there should exist nonzero proportional loading such that an equilibrium of the mechanism is ensured with negligibly small displacements of its components;
- The components of the mechanism can perform the movement from an equilibrium position whereby the energy is absorbed at every plastic element.

The movement of a mechanism specified in the last requirement is called the assumed collapse of a system.

In analysis methods based on hypothesizing about the look of the collapse mechanism and testing the hypothesis by one means or another, the concept of assumed collapse mechanism is in use.

A system obtained from a given system by increasing both plastic moments M_p at some portions of bent members and axial resistances N_p of some tensed-compressed members is referred to as a *strengthened system*. This definition allows increased characteristics $M_p = +\infty$ and $N_p = +\infty$, when some elastoplastic parts of the system become elastic. A given system we also subsume under the strengthened systems. To any assumed mechanism, we can match the strengthened system which turns into this mechanism at the load increase (see Supplement to Chapter). An assumed mechanism is used for calculation of the load parameter from the equilibrium conditions for this mechanism under the assumption of its components' rigidity. In turn, this parameter serves for calculation of FBC and the stress state of the actual system.

Kinematic Theorem

The FBC of an elastoplastic system, obtained from conditions of equilibrium of an arbitrary assumed collapse mechanism made up from this system, is greater than or equal to a true FBC.

Proof

We remind readers that, while proving basic theorems, we consider systems containing bent members only. Let us consider an assumed mechanism in the state of equilibrium. As before, we confine ourselves to concentrated loads only, which are denoted as P_i. The bending moment produced by the j-th plastic hinge in assumed mechanism we denote as M'_j – accordingly, the absolute value $|M'_j|$ is a plastic moment at the same section. We give a virtual displacement to the mechanism in the same direction as displacement in the course of the assumed system's collapse. We denote additional rotation at the j-th plastic hinge for such displacement as $\delta\theta_j$; virtual displacement in the direction of load P_i is denoted as Δ_i. The virtual work equation for the assumed mechanism can be written in the form:

$$k\sum_i P_i\Delta_i = \sum_j M'_j\delta\theta_j, \tag{21.5}$$

where k is the assumed FBC of structure.

Next, we consider a given structure in the state of limit equilibrium. In this state, insert common hinges wherever plastic hinges are located in the assumed system, and exert on the ends of dissected members, in the places of insertions, external moments equal to the actual bending moments at these sections. Internal forces in members of the obtained replacing system have remained the same; therefore, the replacing system keeps equilibrium and the principle of virtual work is applicable for calculation of the load parameter. Keeping the numbering of sections, which was imposed for plastic hinges in the assumed system, denote the bending moment at the j-th section of the given system as M_j. Give the same virtual displacement to the replacing system, which enabled us to obtain the virtual work equation (21.5) for the assumed system. For the replacing system, the virtual work equation will take the form:

$$k_{col}\sum_i P_i\Delta_i = \sum_j M_j\delta\theta_j, \tag{21.6}$$

where k_{col} is the true FBC.

By conversion of equations (21.5) and (21.6), we come to the equality:

$$k\sum_j M_j\delta\theta_j = k_{col}\sum_j M'_j\delta\theta_j. \tag{21.7}$$

Every addend of the right sum $M'_j\delta\theta_j$ is positive, for this quantity equals absorbed energy due to virtual rotation at the j-th plastic hinge (note that moment M'_j reacts against rotation at collapse, i.e., reacts against the pivot through an angle $\delta\theta_j$). But the moment M_j acting at the j-th section of the given structure in the state of limit equilibrium can't be greater than the plastic moment at this section, i.e., $M_j \le |M'_j|$, therefore:

$$M_j\delta\theta_j \le M'_j\delta\theta_j.$$

So, we come to the inequality:

$$\sum_j M_j\delta\theta_j \le \sum_j M'_j\delta\theta_j,$$

from which we obtain the necessary condition of equality (21.7) in the form:

$$k \geq k_{col}, \tag{21.8}$$

which was required.

Corollary

The FBC of a strengthened elastoplastic system can't be less than the FBC of the given system.

Proof

Let us confine ourselves to the case of bent members. Denote the FBC of a system with an increased plastic moment as k^+. This FBC satisfies the virtual work equation:

$$k^+ \sum_i P_i \Delta_i = \sum_j M_j^+ \delta\theta_j, \tag{21.9}$$

where M_j^+ is the bending moment at the j-th plastic hinge of the strengthened system; $\delta\theta_j$ is an additional rotation at the j-th plastic hinge at the virtual displacement of the arisen mechanism in the direction of movement during collapse; and the meaning of denotations P_i and Δ_i is the same as in the proof of the kinematic theorem. The quantity $|M_j^+|$ is a plastic moment at the j-th section. Next, we compose the assumed collapse mechanism from the given system by positioning plastic hinges in the places where they were located at the collapse of the system with increased plastic moments. The virtual work equation for the assumed system has the form (21.5). From the obvious relation between plastic moments $|M_j'| \leq |M_j^+|$, it follows that the right-hand side in (21.9) is not less than the right-hand side in (21.5). Hence $k^+ \geq k$; this inequality, with account of inequality (21.8), proves the corollary.

21.3 KINEMATIC METHOD OF CALCULATION OF BEARING CAPACITY

In the method of identification of connections' loss and recovery, the calculations step by step reproduce the process of changing a system due to increase of load. Even at the degree of static indeterminacy $n = 3$, the analysis becomes cumbersome. There is a less laborious kinematic method of plastic analysis based on the kinematic theorem. *The kinematic method is to construct a collection of assumed collapse mechanisms with one degree of freedom and select among them a mechanism that keeps equilibrium under the lowest load. The collection of mechanisms must include a mechanism representing the true limit state of a system.* The disadvantage of the kinematic method is its informal character, because the collection of mechanisms, if the one is not evident, might be composed on the basis of experimental data or methodical recommendations, or even on the engineer's intuition. However, if such a collection has been composed, then the selection from it by the criterion of the least load acting upon a system in limit equilibrium ensures the correct solution of the problem.

The calculation of FBC k for each of the collapse mechanisms can be done by setting up equations of statics for all members of the mechanism. Because, in such a system, there is the lack of one connection for its stability, the number of statics equations is one greater than the number of constraint reactions. That is why the wanted factor can be included in the number of unknowns and found from this equation set. Frequently, however, the easier way to find the load of limit equilibrium is to set up and solve the virtual work equation for a collapse mechanism with one degree of freedom.

Let us employ the kinematic method to determine the destructive load for the beam in Figure 20.8a. It is evident that collapse occurs due to the emergence of two plastic hinges which

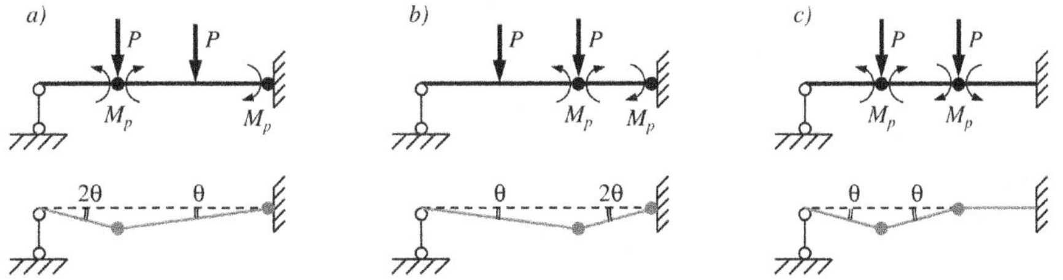

FIGURE 21.1 Assumed collapse mechanisms for the beam in Figure 20.8a.

must emerge in the places of bending moment extremum. The extremum points (besides the nil point at the hinged support) might be positioned in rigid fixing and in the places where load P is exerted. Arranging two plastic hinges in three possible points, we obtain three collapse mechanisms shown in Figure 21.1a, b, and c. Every diagram of loading of the mechanism is supplemented by the scheme of virtual displacements with an indication of DOF – pivot angle θ of one of the members. We determine ultimate loads from the virtual work equation.

In case a, the virtual work equation and its solution have the form:

$$P\frac{2}{3}l\theta + P\frac{1}{3}l\theta = M_p \cdot 2\theta + M_p\theta + M_p\theta. \Rightarrow P = \frac{4M_p}{l}.$$

In cases b and c, it holds, respectively:

$$P\frac{1}{3}l\theta + P\frac{2}{3}l\theta = M_p\theta + M_p \cdot 2\theta + M_p \cdot 2\theta. \Rightarrow P = \frac{5M_p}{l};$$

$$P\frac{1}{3}l\theta = M_p\theta + M_p\theta + M_p\theta. \Rightarrow P = \frac{9M_p}{l}.$$

The least destructive load is obtained for scheme *a*; consequently, this scheme represents actual emergence of the limit state. Earlier, the same load was obtained by step-by-step identification of lost connections; see formula (20.3). Note that in schemes b and c, equilibrium is possible only because of the strengthening of some portions of the beam, because bending moments within the components of corresponding mechanisms are greater than the ultimate values.

The kinematic method enables us to perform not only checking, but also design calculation. In design calculation, the loads are assumed to be constant, the resistances of members' sections are considered proportional to the unknown parameter, and we select the maximal magnitude of this parameter for all possible collapse mechanisms. For example, if for the beam in Figure 20.8a, the flexural resistance is to be found at given loads, then collection of collapse mechanisms and virtual work equations will be the same, but the plastic moment, which must be expressed through load, will be unknown in this equation. The greatest among the obtained solutions is the required one.

To ensure the validity of solutions obtained by the kinematic method, it is supplemented by methodical instructions on construction of collapse mechanisms for structures of various types.

21.4 CALCULATION OF CONTINUOUS BEAMS BY KINEMATIC METHOD

For analysis of continuous beams,* there are methods based on the hypothesis about the failure behavior of such beams, which enables us to construct and investigate the collection of collapse

* A continuous beam is a multi-span solid beam with hinged or rigid supports.

mechanisms. In this chapter item, we consider the kinematic method's implementation for analysis of continuous beams, wherein virtual work equations are set up and solved.

We assume a beam is uniform, i.e., the cross-section and material are the same at all portions. We confine ourselves to the loads exerted upon a beam in the downward direction.

Below, while speaking of the location of the plastic hinge **between** supports of a span, we imply that this hinge coincides with neither support. Under the conditions defined above, the analysis of a continuous beam is based on the following hypothesis:

Under loads exerted upon a beam in the downward direction, only the following collapse mechanisms with one DOF are allowed:

The mechanism with one plastic hinge located on the support of the overhanging portion;

The mechanism with two plastic hinges in a side span with the hinged supports; of them, one hinge is located between supports and another on a support remote from the edge;

The mechanism with three plastic hinges in any other span; of them, two hinges must be on the span supports.

For a span in the limit state, it is known that the top fiber is tensed at the plastic hinge located on the support, and the bottom fiber is tensed at the hinge between supports.

The three cases stated in this formulation are illustrated by examples in Figure 21.2a, b, and c, respectively. The hypothesis may fail if, in the limit equilibrium, there are the spans wherein a plastic moment acts over their length. In this case, the sawtooth collapse mechanism might arise as it does in Figure 21.3. But such a destruction scheme is possible together with one of schemes specified in the hypothesis, and the sawtooth scheme doesn't deliver a new value of the FBC.

The suggested hypothesis about assumed collapse mechanisms of a continuous beam becomes a theorem if we accept following postulate of plastic hinges' localization in the region of a constant bending moment equal to plastic moment M_p: plastic hinges emerge only at the boundaries of the member's portion wherein the plastic moment acts. The reason for this affirmation is the absence of shear force inside the zone of a constant bending moment. Shear force which lessens flexural resistance arises only at the boundaries of this zone. We used this peculiarity of deforming an elasto-plastic member in the example in Figure 20.10. But we ought to remember that the given postulate is the idealization of real processes; in practice, there might be peculiarities of the beam's construction

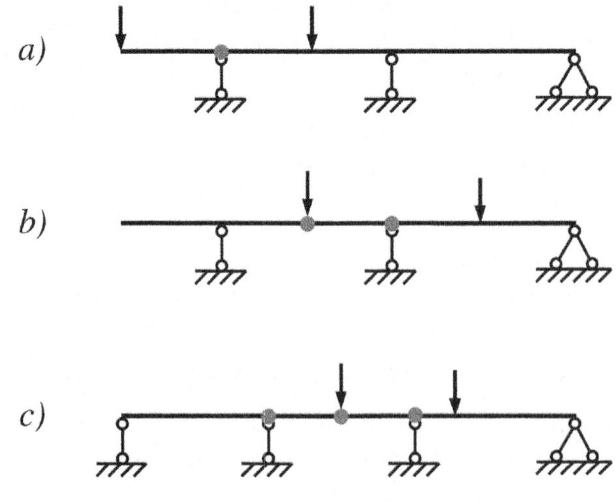

FIGURE 21.2 Examples of collapse mechanisms allowed in continuous beam.

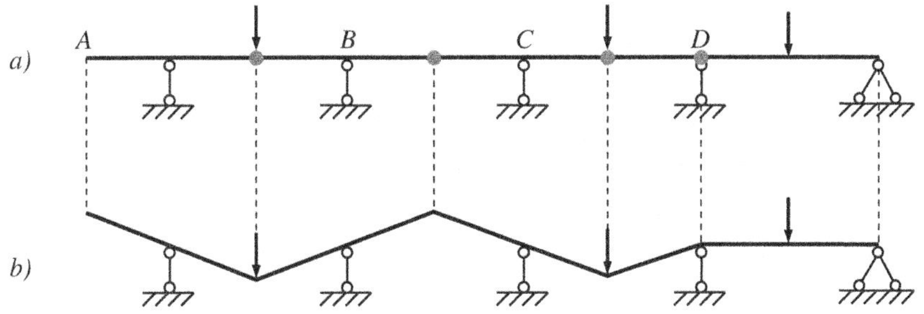

FIGURE 21.3 Sawtooth collapse mechanism: (a) Positioning of plastic hinges being opened in collapse; (b) Displacement in collapse; AD is the destructed portion; BC is the portion where the bending moment is constant, $M = M_p$.

and inhomogeneity of bar's material, owing to which the sawtooth form of destruction can emerge. We avoid the use of this postulate and take for granted that **in the boundaries of plastic moment action (where $M = M_p$), the collapse occurs due to the opening of the plastic hinges at one or several sections randomly located in this zone.** It is allowed to call any other cross-section of this portion an unopened plastic hinge, meaning that, in a repeated test with the same loading, the hinge may open at this section. The theorem about collapse mechanisms of a continuous beam under one-side loading is formulated and proved in the supplement to the chapter.

The calculation of a continuous beam is implemented consecutively, span by span, and at the final stage, the results for all spans are combined to obtain the sought characteristic (whether the FBC or some parameter of the cross-section). If there is an overhanging portion beyond the outermost hinged support, then this portion is analyzed separately, and while calculating the neighboring span, the action of a cantilever upon this span must be taken into account.

Let us use the kinematic method for calculation of bearing capacity of the three-supported beam in Figure 21.4a at a given plastic moment M_p. Locations of plastic hinges and the mechanisms of the span's collapse are shown in Figure 21.4b and c, respectively. Since the position of the plastic hinge inside span AB is not known, diagram c shows the set of collapse mechanisms for span AB, which differ in coordinate z of the hinge, and this coordinate is to be determined. Inside span BC, the position of the plastic hinge is obvious, because at the point where the greatest load P_1 is exerted, the bending moment attains extremal magnitude.

We find the FBC for the case of collapse at the left-hand span, for which purpose we select the mechanism out of the assumed ones, such that it arises under the least load.

Take for the DOF of mechanism ADB deflection v of plastic hinge D. We have the virtual work equation for determining the FBC k as follows:

$$kwz\frac{v}{2} + kw(l-z)\frac{v}{2} = M_p(\theta + \theta_1) + M_p\theta_1.$$

Here, in order to calculate the work done by load kw through virtual displacements, we have replaced the distributed load exerted on each component of the mechanism with the resultant exerted in the middle of a component. We represent rotation angles through deflection:

$$\theta = \frac{v}{z}; \quad \theta_1 = \frac{v}{l-z};$$

and then convert the equation of works to the form:

$$\frac{kwl}{2} = M_p\left(\frac{1}{z} + \frac{2}{l-z}\right), \text{ or: } \frac{kwl}{2} = M_p\frac{l+z}{z(l-z)}. \tag{21.10}$$

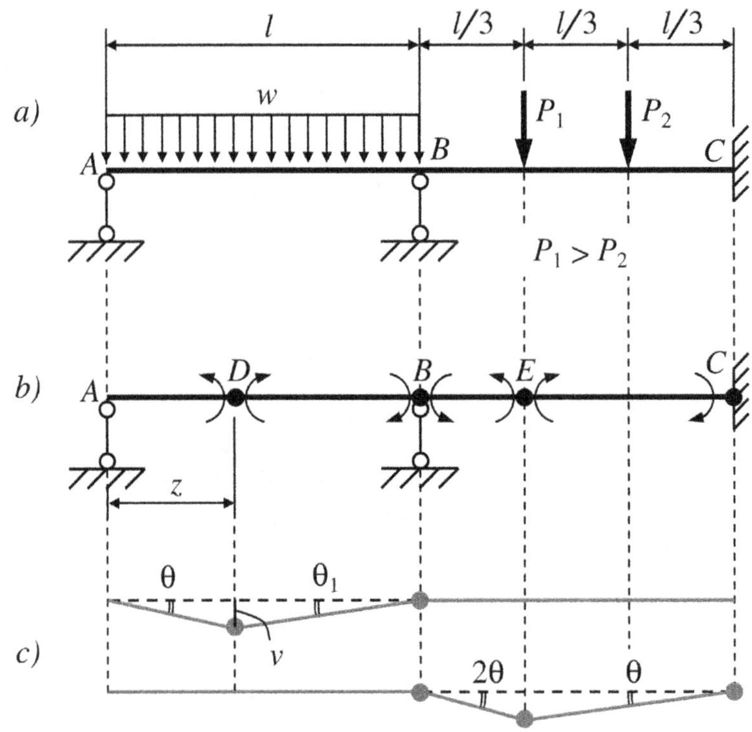

FIGURE 21.4 Example of a continuous beam: (a) Design diagram (A, B, C are the boundaries of spans); (b) Plastic hinges and directions of plastic moments before spans' collapse; (c) Virtual displacements at span's collapse (v is the deflection of hinge D).

The extremum point of function $k(z)$ is determined from condition: $\frac{dk}{dz} = 0$. Derivation gives:

$$\frac{dk}{dz} = \frac{2M_p}{wl}\frac{z(l-z)-(l-2z)(l+z)}{z^2(l-z)^2} = \frac{2M_p}{wl}\frac{z^2+2lz-l^2}{z^2(l-z)^2}.$$

Equate to zero the numerator of the second fraction. The solution of the quadratic equation:

$$z^2 + 2lz - l^2 = 0 \tag{21.11}$$

is obtained in the form: $z_{1,2} = (\pm\sqrt{2}-1)l$. We accept the positive solution which has a physical sense. The parabola which represents the left-hand side of (21.11) has the ordinates negative on the left of point $z_1 = (\sqrt{2}-1)l$ and positive on the right (in some neighborhood of this point). The increase of derivative $\frac{dk}{dz}$ in the neighborhood of zero means that the minimum point of function $k(z)$ has been found. The wanted minimum is established by substitution of $z = z_1$ into formula (21.10):

$$k_1 = k = \frac{2}{(\sqrt{2}-1)^2}\frac{M_p}{wl^2} = 2(\sqrt{2}+1)^2\frac{M_p}{wl^2}. \tag{21.12}$$

The analysis of the left-hand span is complete.

To find out the FBC for assumed collapse mechanism BEC, we have the virtual work equation and its solution:

$$kP_1\theta\frac{2}{3}l + kP_2\theta\frac{1}{3}l = 2\theta \cdot 2M_p + \theta \cdot 2M_p. \Rightarrow k_2 = k = \frac{18M_p}{(2P_1 + P_2)l}. \tag{21.13}$$

The wanted FBC is obtained as the minimum of possible values:

$$k_{col} = \min\left(k_1, k_2\right). \tag{21.14}$$

At some proportion between loads w on one hand and P_1, P_2 on the other, the coefficients k_1 and k_2 become equal. The question emerges: what does the collapse mechanism look like in this case? The answer depends on what property of the real structure the diagram of the obtained mechanism should display. If the picture of destruction is required, then one has to take into account that, in a structure, there are random variations of strength characteristics about specified magnitudes, and collapse occurs in the "weakest" span of the beam. That is why all possible schemes of collapse should be presented at the established FBC as the result of analysis. If the stress state of a system before destruction is a matter of interest, then the diagram of the beam must display all plastic hinges that arise in the limit state of the beam. Thus, the mechanism with several DOFs obtained by conjoining the mechanisms possible in limit state should be shown.

SUPPLEMENTS TO CHAPTER

About the Existence of the Assumed Collapse Mechanism

Theorem

Arbitrary assumed collapse mechanism emerges due to the loading of the corresponding strengthened system.

Proof

Let us confine ourselves to the case of bent members. We introduce a strengthened system by making all parts of the given system elastic except the short portions of members, which contain assumed plastic hinges – in other words, we strengthen these parts by an infinite plastic moment. We assume the elastoplastic portions are small enough not to take into account the uncertainty of the plastic hinges' location within them in calculation of the FBC. We prove that the loading of this strengthened system causes the emergence of the assumed collapse mechanism.

Let us number the plastic hinges in the assumed mechanism, and, in the same way – keeping the same order – number the elastoplastic portions in the strengthened system.

One of the possible equilibrium states of the strengthened system is the equilibrium of the assumed collapse mechanism under loading. Consequently, according to the static theorem, the FBC of the strengthened system can't be less than the parameter of the load upon the assumed collapse mechanism in equilibrium. Further, we proceed to prove a contrario. Suppose that, at the parameter of the load ensuring the equilibrium of the assumed mechanism, the strengthened system doesn't turn into this mechanism, i.e., there is at least one unstrengthened portion wherein a plastic hinge doesn't emerge. Let this portion have a number r. At first, we write down the virtual work equation for the assumed mechanism in the case of virtual displacement when the energy is absorbed by all plastic hinges (this displacement is allowed according to the definition of the assumed mechanism). Keeping the denotations of forces, moments, and displacements in the assumed mechanism, which were introduced in the proof of the kinematic theorem, we come to the equation:

$$k\sum_i P_i\Delta_i = \sum M'_j\delta\theta_j. \tag{21S.1}$$

We choose the positive rotation direction at the j-th plastic hinge such that in this equation additional rotation $\delta\theta_j$ is positive. Then quantity M'_j on the right side of the equation will be plastic moment $M'_j > 0$.

Now, we replace the strengthened system under loads kP_i with the mechanism obtained by inserting hinges into all the unstrengthened portions, and, to keep equilibrium, exert in the places of insertions external moments equal to bending moments M_j at corresponding sections. Give the replacing mechanism the same virtual displacement which enabled us to obtain equation (21S.1). The virtual work equation for the replacing system will have the form:

$$k\sum_i P_i\Delta_i = \sum M_j\delta\theta_j.$$

By subtracting this equation from the equation (21S.1), we get:

$$\sum (M'_j - M_j)\delta\theta_j = 0. \tag{21S.2}$$

Every addend on the left-hand side of (21S.2) is nonnegative, for $M'_j \geq M_j$. Besides, the term of this sum at $j = r$ is positive, because for the r-th section $M'_r > M_r$. So, the left-hand side of (21S.2) is greater than zero, and the obtained contradiction proves the theorem.

Substantiation of Hypothesis about Collapse Mechanisms of a Continuous Beam

A collapse mechanism of a continuous beam, in which every span in a row of several spans has one open plastic hinge between supports, and where this row is limited from either side by the plastic hinge or outermost hinged support, is referred to as a *sawtooth collapse mechanism*.

We define the direction of the pivot of a horizontally positioned cylindrical hinge (Figure 1.4b) as the pivot direction of its right side about the left one. We call the pivot's direction of the right section bounding a plastic hinge in a beam with respect to the left bounding section "the direction of the pivot of a plastic hinge." For instance, the direction of the opening of the plastic hinge in Figure 20.2 is counterclockwise. In Figure 21.4, the hinges D and E open counterclockwise; hinges C and B open clockwise.

Theorem

Subject to the direction of loads upon a continuous beam downward, only the following collapse mechanisms with one DOF are allowed:

1) The mechanism with one plastic hinge located on the support of the overhanging portion;
2) The mechanism with two plastic hinges at the side span with the hinged supports; of them, one hinge is located between supports and another on the support remote from the edge;
3) The mechanism with three plastic hinges in any other span, of them, two hinges must be on the span supports;
4) The sawtooth mechanism.

At a collapsed span of the mechanisms of second and third types, the top fiber is tensed at the plastic hinge arisen at any support, and the bottom fiber is tensed at the hinge between supports. On the supports of collapsed spans belonging to the sawtooth mechanism – except perhaps the outermost hinged support of the beam – a plastic moment is reached and the top fiber is tensed.

The beam's FBC corresponding to a sawtooth mechanism can be obtained from an equilibrium condition for one of the mechanisms of the second and third types.

Proof

The proof is based on the concavity feature of the bending moment diagram,* which is fulfilled for every separate span of a continuous or hinged beam under one-side vertical loads. The latter feature is established from the superposition principle in the following way: The bending moment diagram within a span can be obtained by summation of diagrams representing the action of bending moments at the supports of this span and the action of each separate load inside the span – under the assumption that the span is a simple beam. Every constituent diagram is concave (the rectilinear diagram from the action of moments at the supports is a special case of concavity); consequently, their sum is also concave. All bending moment diagrams shown on the plots of Chapter 20 might serve as examples of concave diagrams.

We call spans which either have one rigid support or are situated not to the side of the beam internal. Spans *BC* and *CD* in Figure 21.3a and span *BC* in Figure 21.4a are examples of such spans.

The case of distraction of an overhanging portion is obvious and not considered below. We shall investigate under which conditions there emerges a collapse mechanism wherein only one span is destructed (or, more precisely, only one span contains plastic hinges).

We consider any internal span. To implement a mechanism with one DOF through dividing a beam by the hinges in the boundaries of one internal span, it is necessary to place three hinges in this span. Besides, the displacements of such a mechanism imply pivots at adjacent hinges in opposite directions. So, we must find where three plastic hinges might emerge in the span, under the condition of opening the adjacent hinges in opposite directions. This means that the bending moment diagram in the span must have three points where ultimate values $\pm M_p$ are attained with sign alternation. A monotonic bending moment diagram does not secure the required sign alternation for bending moments. A nonmonotonic concave diagram has two local maximums at the span supports and attains global minimum inside the span. Consequently, plastic hinges might emerge only at these points. So, the destruction of one internal span means the occurrence of a collapse mechanism of the third type. Besides, if at the indicated points of the span, hinges have arisen, then on the supports – where the bending moment is obviously positive – the top fiber is tensed, and at the hinge between supports – where the bending moment is negative – the bottom fiber is tensed.

The case of a side span with hinged supports differs from the case considered above in that one hinge, required for displacements of components of the arisen mechanism, is already present at the side support. A monotonic bending moment diagram again does not allow the emergence of a collapse mechanism with plastic hinges within the span under consideration; the analysis of a nonmonotonic concave diagram leads to the only possible second type of collapse mechanism.

To implement a mechanism with one DOF through dividing a beam by the hinges in the boundaries of two or more spans, these hinges must be placed according to the scheme of the sawtooth mechanism. Now, we shall consider any sawtooth collapse mechanism in the state of equilibrium and establish the peculiarities of the bending moment diagram within the row of destructed spans in this mechanism.

At the sections of two plastic hinges not lying on supports and belonging to neighboring spans, bending moments have opposite signs, because at the plastic hinge being moved upward due collapse, the top fiber is tensed, and at the plastic hinge being moved downward, the bottom fiber is tensed. A diagram *M* concave within a span, such that maximum value $+M_p$ is attained **between** supports, can be only a horizontal line; that is why diagram *M* is horizontal within a span with a tensed top fiber at the plastic hinge between supports. It therefore follows that at the supports of any span with the kink directed upward during collapse, the bending moment equals the plastic moment. Owing to the alternation of such spans, plastic moment $+M_p$ is attained at all supports inside the destructed portion of the sawtooth mechanism.

* A concave diagram is the graph of a concave function.

In a sawtooth mechanism, either support at the edge of the destructed portion – if it is not the outermost hinged support of the beam – bears a side plastic hinge of the mechanism. Let us establish the sign of the bending moment at a side plastic hinge. A neighboring plastic hinge opens in the opposite direction, and thus the bending moment at the neighboring hinge is of the opposite sign. Different signs of bending moments at these two hinges are possible only if the beam's kink between supports of the span occurs downward. Accordingly, at the point of kink, the bending moment is negative, whereas at the side plastic hinge, the bending moment $+M_p$ is attained and the top fiber is tensed.

Now then, we consider any collapsed span of a sawtooth mechanism, such that at the plastic hinge between supports, the bottom fiber is tensed. If this span is supported by outermost hinged support of the beam, then, at the limit state of the beam, the positioning of plastic hinges at this span is the same as in a collapse mechanism of the second type. Meanwhile, this span of the beam keeps equilibrium under a load of sawtooth mechanism occurrence. We insert into this span common hinges instead of plastic hinges and – taking the beam as being under action of ultimate load – exert equilibrating moments in the places of insertion. Assuming the members, connected by these hinges, are rigid, we obtain the mechanism of second type in equilibrium. Thus, the virtual work equation for virtual displacement of the latter mechanism will determine the ultimate parameter of the load upon the beam, i.e., will determine the FBC of it. If the span with tensed bottom fiber rests on the outermost rigid support or is an internal span of the beam, then, likewise, this span, being part of the sawtooth mechanism, is the collapsed span of the mechanism of the third type. And again, the virtual work equation for components of the mechanism forming this span determines the FBC of the beam. The theorem is proved.

22 Features in Plastic Analysis of Statically Indeterminate Beams and Frames

22.1 ANALYSIS OF CONTINUOUS BEAMS BY METHOD OF ADJUSTING BENDING MOMENTS TO ULTIMATE VALUES

In the previous chapter, we considered the calculation of continuous beams by the kinematic method, using the virtual work equation to determine the FBC. There is an alternative method known as *the method of adjusting bending moments to ultimate values*, which has been specially developed for analysis of continuous beams and often is simpler in implementation. Both of these methods lie in analysis of the hypothesis about assumed collapse mechanisms of a continuous beam in order to select a true mechanism and determine the parameter of the problem. But in the kinematic method, one investigates all possible mechanisms for every span by means of extremum criterion for the required parameters, whereas in the method of adjusting bending moments, the collapse mechanism for every span is established from the conditions which are imposed on the profile of bending moments acting in the span. Further on, we confine ourselves to the case of a continuous beam without overhangs (a cantilevering part, if present, is calculated in the same way in both methods).

We specify coordinate z of the section in every span by counting it from one of the span's supports in the direction of the opposite support. For the bending moment, we employ the sign convention for frames, i.e., we accept the bending moment as positive if the top fiber is tensed. In the method of adjusting bending moments, we investigate the collapse mechanisms in turn for each span, find out the distribution of bending moments in the span, and determine the location of the plastic hinge between supports. The sought parameter λ may be the FBC or geometry parameter of the cross-section. **According to this method, for each span, the assumed magnitude of the sought parameter is established, and, out of obtained magnitudes, the extreme value is selected. The solution of problem will be the minimum of assumed FBC or maximum of assumed cross-section's parameter. During analysis of each span, there should be firstly established a family of functions $M = M_\lambda(z)$ dependent on wanted parameter λ and represented by the family of bending moment diagrams, by using known bending moments at the boundaries of the span; afterwards, the equation for the unknown λ is set up and solved from the requirement that the minimum of function $M_\lambda(z)$ be equal to the plastic moment with the sign reversed.** This requirement follows from the fact that in the limit state of the beam, there is a plastic hinge inside the collapsed span at the section of which the bottom fiber is tensed. Consequently, the plastic moment $-M_p$ is attained at this section, and it is the lower bound for bending moment possible before the beam's collapse.

The condition for a bending moment in a plastic hinge between supports has the form:

$$\min_{0 \le z \le l} M_\lambda(z) = -M_p, \tag{22.1}$$

where l is the length of span.

The finding of a family of bending moment functions $M = M_\lambda(z)$ under prescribed boundary conditions is referred to as adjusting the bending moments to values at the span's boundaries. These functions and corresponding diagrams are referred to as adjusted to boundaries. The finding of parameter λ by means of established family $M_\lambda(z)$ is referred to as adjusting the bending moments to the value at the intermediate hinge. The adjusted bending moment diagram meets the limit state of a span. For calculation of bending moments, the statics equations are used – inter alia, while determining the shear force at the span's boundary.

The distribution of bending moments limited by the range $\pm M_p$ is referred to as *safe* or *admissible*. **The graphic result of calculation is the safe diagram M with ordinates $\pm M_p$ at the sections of plastic hinges.**

Below, we employ the method of adjusting bending moments to solve the problem which was solved in the previous chapter by means of the virtual work equation: namely, we shall calculate the FBC for the beam with the design diagram in Figure 22.1a, which reproduces the diagram in Figure 21.4. Let the limit state be evoked by destruction of the left span. The scheme of loading the span is shown in Figure 22.1d. The bending moment in this span can be determined through shear force at the left boundary of the span, which is equal to reaction R_A of the left support and depends itself on the assumed FBC. The total moment of forces exerted upon the beam's portion [0, z] equals zero; the corresponding equilibrium equation for the moment about the point with coordinate z is transformed to relation:

$$M_k(z) = \frac{kwz^2}{2} - R_A(k)z. \tag{22.2}$$

Reaction R_A is found from the equilibrium condition for moments of forces acting upon span AB, taken with respect to right support of this span. We get:

$$R_A = \frac{kwl}{2} - \frac{M_p}{l}. \tag{22.3}$$

Formulas (22.2) and (22.3) define the family of functions $M_k(z)$ adjusted to the boundaries of the span. They are obtained from equations of statics and from the boundary conditions for bending moment: $M_A = 0$; $M_B = M_p$.

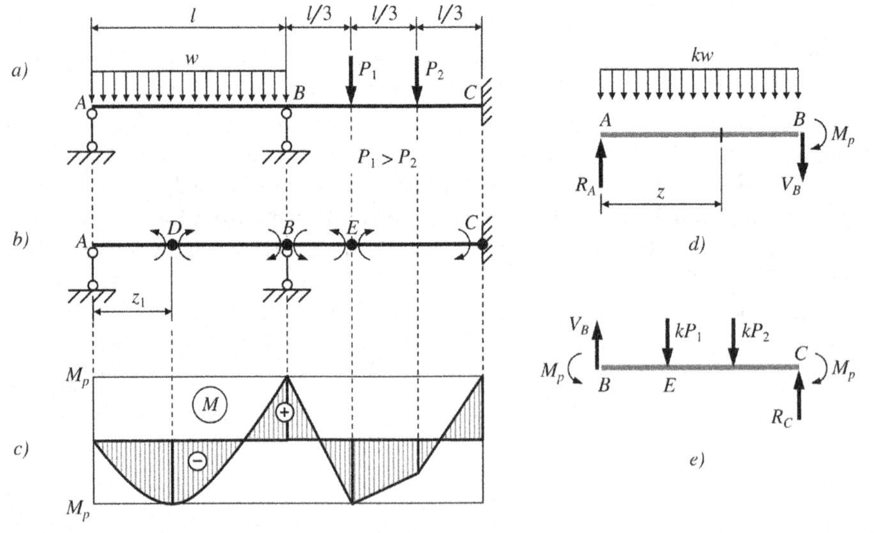

FIGURE 22.1 Example of continuous beam: (a) Design diagram (A, B, C are the boundaries of spans); (b) Plastic hinges and directions of plastic moments before spans' collapse; (c) Adjusted bending moment diagram for spans equal in strength; (d), (e) Schemes of loading the spans in limit state (z is the coordinate of an arbitrary section).

At the point of the plastic hinge between supports, the bending moment is minimal. Coordinate z_1 of this point can be obtained from the condition that shear force equals zero:

$$V_k(z) = R_A - kwz, \Rightarrow R_A - kwz_1 = 0 \Rightarrow z_1 = \frac{R_A}{kw}.$$

Extreme value of the moment is obtained by substitution of z_1 into (22.2) and replacement of R_A according (22.3):

$$M_k(z_1) = -R_A^2 \frac{1}{2kw} = -\left(\frac{kwl}{2} - \frac{M_p}{l}\right)^2 \frac{1}{2kw}. \quad (22.4)$$

The equation for FBC has the form: $M_k(z_1) = -M_p$, or:

$$(kw)^2 - \frac{12M_p}{l^2}kw + \frac{4M_p^2}{l^4} = 0.$$

The solution of this equation $k = k_1$ is obtained in the form (21.12). Coordinate z_1 was also found in the previous chapter in another way. Function $M_{k_1}(z)$ determines the diagram adjusted to the value of the plastic moment inside the span.

Now, we investigate the mechanism of the right span's collapse. In this span, the diagram M adjusted to the moments at the boundaries of the span has the form of a polygonal line with the extremum at the point where load P_1 is exerted. In this case, there is no need to obtain the family of functions $M_k(z)$ in explicit form, because extremum point E is known beforehand; thus, it is sufficient to obtain the dependence of the bending moment versus the unknown parameter only at this point – we denote appropriate function as $M = M_E(k)$. The diagram of loading the span in a limit state is shown in Figure 22.1e. The bending moment at the point of the intermediate hinge is established from equations of statics:

$$V_B = \frac{2}{3}kP_1 + \frac{1}{3}kP_2; \; M_E = M_p - V_B\frac{l}{3}.$$

The equality for FBC has the form: $M_E = -M_p$, or:

$$M_p - \frac{1}{9}kl(2P_1 + P_2) = -M_p.$$

The solution of this equation has the form (21.13). The final FBC is the minimum of the values obtained for each span. If the spans are equal in strength, i.e., both spans are collapsed at the same time, then the bending moment diagram in the limit state has the form shown in Figure 22.1c.

22.2 BASIC PROVISIONS OF THE METHOD OF COMBINING MECHANISMS IN ANALYSIS OF FRAMES; EXAMPLE OF PORTAL FRAME

We call a plane vertical bar grid with rectangular cells, wherein each column is supported by another column or bears on the earth, a *latticed frame*. To latticed frames, we also subsume a frame obtained from a latticed one by changing the shape of some upper cells in such fashion that the top girders become gabled or sloped. The example of a latticed frame with horizontal girders is shown on the design diagram in Figure 22.7. The example of a latticed frame with gabled and sloped girders is shown on the design diagram in Figure 22.9. Further on in the chapter, the methods of plastic analysis are studied with regard to latticed frames.

The framework of multi-story building is often implemented as a spatial steel frame consisting of columns and beams connected in joints at the level of every floor. In a model building framework, the beams are connected into horizontal grids forming the floors, whereas the columns serve as supports for these grids and transmit the load upon foundation. In order to get a crude estimate for the FBC of such a framework, we could consider it as a system of plane vertical frames linked by transverse beams and analyze the plane frames independently of each other. For instance, the Willis building in London is performed in a shape resembling a prawn carapace (29-story block of stepped design consisting of three bodies curved tiered, shown in Figure 22.2, on the left); but even under such an intricate architecture, in its steel skeleton we can identify the almost planar vertical frames of lattice type, one of which is shown in Figure 22.2, on the right.

A particular case of a latticed frame is a portal frame, i.e., a frame with one cell. Portal frames with gabled girders, being linked by purlins and girts into a spatial framework, are used in industrial buildings and hangars (Figure 22.3). In development of elastic perfectly plastic structures' theory, it is a portal frame with a gabled girder that was the main object of analysis.

The basic method of calculation of latticed elastoplastic frames is known as the method of combining mechanisms. This method is grounded on experimental data, which in the 40s and 50s of the 20th century enabled one to establish and classify the types of collapses of latticed frames. Below, we study the common realization of this method, in which the loading process is assumed as one-step and the effect of the plastic moment's decrease in columns due to compression is not taken into account.

A mechanism from some collection of assumed mechanisms of the frame's collapse, which differs from other mechanisms of the given collection in at least one plastic hinge, is referred to as an *independent collapse mechanism.* The *method of combining mechanisms* for analysis of latticed frames is to employ the collection of independent collapse mechanisms, which are converted in

FIGURE 22.2 Willis building in London: On the left – overall view, 2011; on the right – building under construction, 2006. Photo on the left is own work by Colin with contribution of Hugin (License CC BY SA 3.0. Source: https://commons.wikimedia.org/w/index.php?curid=15151843). Photo on the right by Andrew Dunn, http://www.andrewdunnphoto.com (License CC BY SA 2.0. Source: https://commons.wikimedia.org/w/index.php?curid=923589).

FIGURE 22.3 Hangar 39A of SpaceX Company (Photo was donated to the public domain by SpaceX, http://www.spacex.com. Source: https://commons.wikimedia.org/wiki/File:Landed_rockets_in_hangar_39A_(27042449393).jpg).

order to construct a larger collection of mechanisms containing the true mechanism in the limit state of a frame. Each of the independent mechanisms possible in a given frame defines the set of plastic hinges in the limit state of the frame. The full set of plastic hinges, obtained by uniting the hinges belonging to every independent mechanism, serves for construction of the collection of assumed collapse mechanisms. Each collapse mechanism is composed by means of selecting plastic hinges (taking into account a direction of opening) from the full set of hinges, owing to which the frame is transformed into a mechanism with one DOF. A non-independent mechanism obtained through this selection is referred to as a *combined mechanism*.

The collection of combined mechanisms is restricted by the selection of regular mechanisms. In the method under consideration, every independent mechanism is characterized by the degree of freedom, i.e., by the parameter that determines displacements. We call a mechanism with one DOF composed by a selection of hinges from several independent mechanisms in such manner that angular and linear displacements in it could be obtained as the sum of displacements in original mechanisms with the same values of DOFs a *regular combined mechanism*. Building the collection of assumed collapse mechanisms is grounded on the postulate that a **true collapse mechanism is contained in the collection of independent mechanisms and their regular combinations**.

For every mechanism of the formed collection, one sets up the virtual work equation with a view of finding an unknown characteristic of a structure (the FBC or parameter of member cross-sections). For a true collapse mechanism, the assumed characteristic attains minimum or maximum according to the problem being solved.

The completing stage of the method is to verify the correctness of the found collapse mechanism. The verification assumes the construction of a bending moment diagram for a frame in a limit state. **If in the M-diagram constructed for the found collapse mechanism, a bending moment doesn't anywhere exceed the magnitudes of plastic moments of members and attains these magnitudes at the plastic hinge sites of the found mechanism, then this collapse mechanism is correctly established.** We emphasize that one of the verification criteria is the safety of bending moment distribution, i.e., the condition:

$$|M| \le M_{pi},\ i = \overline{1, B},$$

where M_{pi} is the plastic moment in the i-th member, and B is the number of members in a frame.

Firstly, we shall consider the П-shaped frame under the effect of vertical and horizontal loads applied on the girder. For this frame, there are two independent collapse mechanisms: the beam mechanism and the sway mechanism. An example of the frame is shown in Figure 22.4a; independent mechanisms are shown in Figure 22.4b and c, respectively. Plastic hinges about each joint of the girder may arise at cross-sections adjacent to the joint on each side (diagram d). If the plastic moments of the column and girder are different, then the hinge arises in the member with the smallest plastic moment. If the plastic moments are the same, then the hinge in the neighborhood of the joint may arise at any of sections adjacent to the joint. We solve the problem under the assumption that plastic moments of columns and girders are the same, and accept that in the joint's neighborhood plastic hinge arises on the column. We also accept that flexural rigidities of members are the same.

The cross-sections wherein plastic hinges may emerge are designated on diagram a by numbers. To turn the frame into a mechanism with one DOF, 4 hinges are necessary – one greater than the number of redundant connections. Selecting 4 hinges from 5 the possible, we get, in total, 5 collapse mechanisms, of them, 3 are combined, including mechanisms e and f. Among combined mechanisms, it suffices to investigate mechanism e only, since only this mechanism is regular. The mechanisms with one hinge at support don't give new results compared to the beam mechanism, because this hinge is "dead": the rotations in it are the second-order infinitesimal of the DOF.

Thus, we have the collection of three assumed collapse mechanisms and can determine the FBC for every one of them from the virtual work equation. We have for the beam mechanism (diagram b):

$$kP\frac{l}{2}\theta = 4M_p\theta. \Rightarrow k = \frac{8M_p}{Pl}; \tag{22.5}$$

for the sway mechanism (diagram c):

$$kQh\theta = 4M_p\theta. \Rightarrow k = \frac{4M_p}{Qh}; \tag{22.6}$$

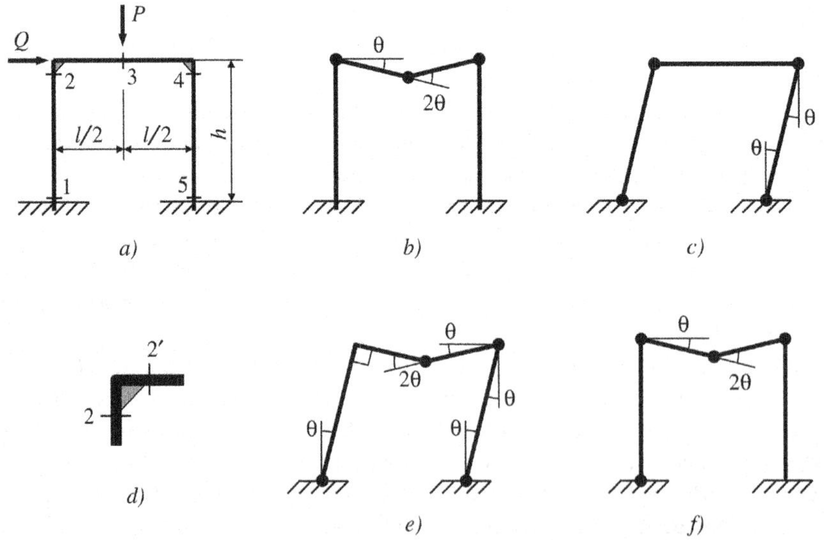

FIGURE 22.4 Portal frame (a, d) and mechanisms of its collapse: (b) Beam mechanism; (c) Sway mechanism; (e) Combined mechanism; (f) Beam mechanism with a dead hinge; 1–5, 2′ are designations of cross-sections where plastic hinges are possible; θ is the degree of freedom of mechanism.

and for the combined mechanism (diagram e):

$$kP\frac{l}{2}\theta + kQh\theta = 6M_p\theta. \Rightarrow k = \frac{12M_p}{Pl + 2Qh}. \tag{22.7}$$

Among assumed collapse mechanisms, the mechanism with minimal FBC is implemented. Depending on the input data, every one of the considered collapse schemes can emerge. Note that for proper calculation, not only the location of plastic hinges must be specified in every mechanism, but also the directions of plastic moments in the hinges, to which end the diagram may present the shape of the mechanism in the process of collapse.

In the above example, one can see the peculiarity of the virtual work equation for a regular mechanism: the sum at the left-hand side consists of works done by loads in each of the mechanisms being combined. Meanwhile, the right side of the equation is not equal to the sum of energies absorbed in plastic hinges of independent mechanisms, because one of these hinges was deleted during combining mechanisms. The right side is less than the sum of energies absorbed in independent mechanisms, owing to which the regular combined mechanism may give an extreme value of k.

It is obvious enough how the virtual work equations are employed in design calculation, when a plastic moment is sought at a given loading. The result is obtained by equating quantity k to unity in (22.5)–(22.7) and representing plastic moment through loads. The greatest moment is the required one.

For complete plastic calculation, it is required to construct the bending moment diagram at the established limit state of the frame. Further, we present the construction of diagram M in the case when collapse goes according to scheme b. We assume that the limit state occurs under loading specified by design diagram (FBC = 1).

In diagram b, the number of connections is equal to the number of equations of statics, but the system is instantaneously unstable and thus is not statically determinate. Instantaneous instability causes the singularity of the coefficient matrix of the statics equation set, i.e., not all these equations are independent. In order to obtain the missing equation, we use the force method. Let us determine axial force N in the girder by analysis of dependence between bending of each column and this force, which enables us to obtain a canonical equation as the condition on displacements due to the presence of the girder. The deformed frame and design diagrams of loading upon the columns are shown in Figure 22.5a and b. Vertical forces are not shown because they do not affect the elastic curve of columns. Horizontal displacement of the upper end of each column is defined by two factors: the total horizontal force exerted on the end of column and the bending moment at this end. Displacement of section 2 has the form:

$$\Delta_2 = \frac{h^2 M_p}{2EI} + \frac{h^3(Q+N)}{3EI}.$$

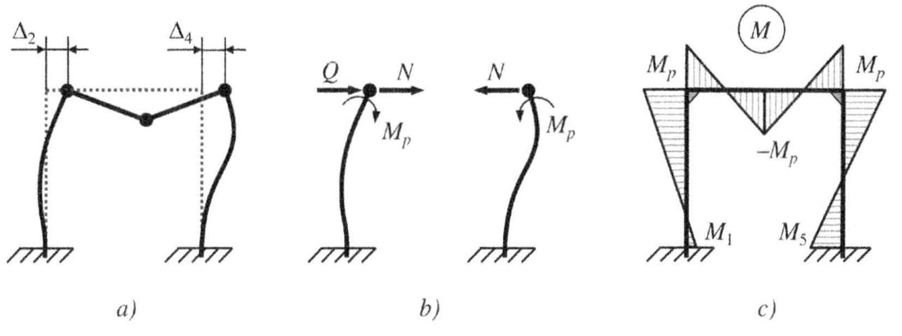

FIGURE 22.5 Construction of a bending moment diagram for beam scheme of collapse (ordinates of the diagram are indicated with account of signs).

The first addend is the contribution from the external moment at the end of the column. It is easily found from Mohr's formula (11.14). The second addend – the contribution from external transverse force – is determined by the known formula for displacement of the cantilever end, which we formerly used in Chapter 16. For displacement of section 4, we get likewise:

$$\Delta_4 = -\frac{h^2 M_p}{2EI} - \frac{h^3 N}{3EI}.$$

The connection developing axial force in the girder secures the equality of these displacements. We have the canonical equation: $\Delta_2 = \Delta_4$; the solution of it has the form:

$$N = -\frac{1}{2}\left(\frac{3M_p}{h} + Q\right).$$

We accept the bending moment as positive if the fiber is tensed outside the outline of the frame. Knowing the axial force in the girder, we get bending moments at support sections 1 and 5 by using diagram b in Figure 22.5:

$$M_1 = M_p + (Q + N)h = -\frac{1}{2}\left(M_p - Qh\right);$$

$$M_5 = M_p + Nh = -\frac{1}{2}\left(M_p + Qh\right).$$

This is enough to construct a bending moment diagram, the possible view of which is shown in Figure 22.5c.

It is useful to compare the problems of construction of bending moment diagrams for the beam mechanism and combined collapse mechanism (Figure 22.4e). The latter problem is the opposite of the problem solved above, for here the number of statics equations is one greater than the number of unknown forces (the number of unknown reactions produced by hinges is 2·4 = 8; the number of statics equations for three components of the mechanism is 3·3 = 9). The design diagram for solving this new problem is depicted in Figure 22.6a. The diagram M for the combined mechanism requires the calculation of a bending moment only at section 2. In order to determine this moment, it is sufficient to find the horizontal reaction of left support R_{1h}. We shall consider the part of the frame to the right side of section 3 (part 3–4–5) as the single dog-leg member. By excluding hinge 4 from consideration, we shall dispose of a redundant equation of statics: the system under examination

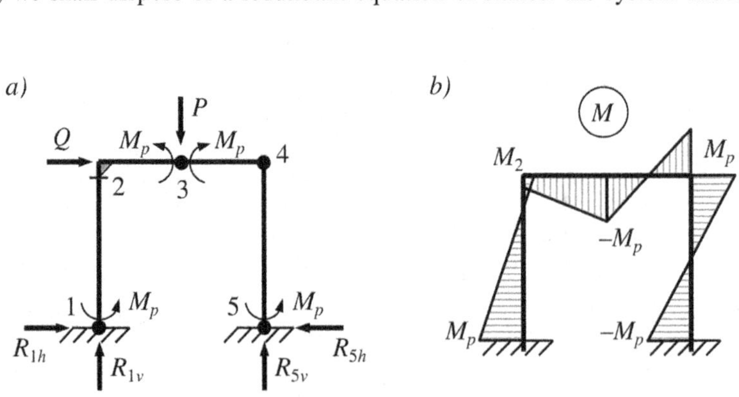

FIGURE 22.6 Construction of a bending moment diagram for combined collapse scheme (ordinates are indicated with account of signs).

will have three hinges and prove to be statically determinate. The support reactions of such a system could be calculated as for the three-hinged arch. The stages of the calculations are as follows:

1) From the equilibrium condition for moments of external forces exerted upon the frame, taken about the right support, we determine the vertical reaction of the left support;
2) From the equilibrium condition for moments of forces acting upon the left component (part 1–2–3), with respect to central hinge 3, we determine the horizontal reaction of the left support;
3) Calculate the wanted moment:

$$M_2 = R_{1h}h + M_p = \frac{Pl}{4} - \frac{Qh}{2}.$$

A typical bending moment diagram is shown in Figure 22.6b. The reader is recommended to fulfill the calculation and verify the obtained value of M_2.

22.3 IMPLEMENTATION OF THE METHOD OF COMBINING MECHANISMS FOR MULTI-STORY AND MULTI-BAY LATTICED FRAMES; FRAME WITH GABLED ROOF

The cells of a latticed frame, located on the same level, constitute a *story of frame*; the cells of a latticed frame, confined with two verticals running through neighboring columns, constitute a *bay of frame*. Now we consider the method of combining mechanisms as it is applied to multi-story and multi-bay frames. We confine ourselves to the case of loads' action upon the girders only; horizontal loads are assumed to be one-sided and vertical loads are directed downwards. For a frame containing rectangular cells only, all possible mechanisms can be constructed from independent mechanisms of three types as follows:

- Beam mechanisms, wherein one of the girders acquires three plastic hinges as in the case of destruction of an internal span in a continuous beam;
- Sway mechanisms, wherein the emerged hinges allow horizontal displacement of each story as an integral whole, except one story on which the columns pivot;
- Joint mechanisms, wherein a joint formed by three or more members can pivot owing to plastic hinges adjacent to the joint.

The types of mechanisms specified in this way are denoted, respectively, as *B*, *S*, *J*. The mechanisms of types *B* and *S* may arise in the limit state of a structure; the mechanisms of type *J* serve only for construction of other mechanisms, i.e., a mechanism of this type doesn't arise in its pure form as a collapse mechanism.

Let us consider the example of the frame presented by a design diagram in Figure 22.7. Vertical loads are assumed to be exerted in the middle of every girder. Figure 22.7 shows all independent mechanisms of this frame: three beam mechanisms, two sway mechanisms, and two joint mechanisms. Note that a plastic hinge at joints *A*, *B*, and *E* arises in the member with smaller plastic moment (if the moments are equal, then the hinge's location is unimportant). The hinges adjacent to these joints we arbitrarily point out at the columns and girders beside these joints, because the problem is solved in general form and numerical values of the plastic moments are not given. It is simple enough to prove that in a latticed frame, for every set of independent mechanisms, one can match no more than one regular combination formed from **all** mechanisms of the set and not containing dead hinges. The latter property permits us to narrow our search for a true collapse mechanism. As the denotation of a regular mechanism, we use the denotations of independent mechanisms connected by the sign "+." Figure 22.8 shows some combined mechanisms. Regular mechanisms are presented on diagrams a–d; among them the mechanisms a and b are obtained by

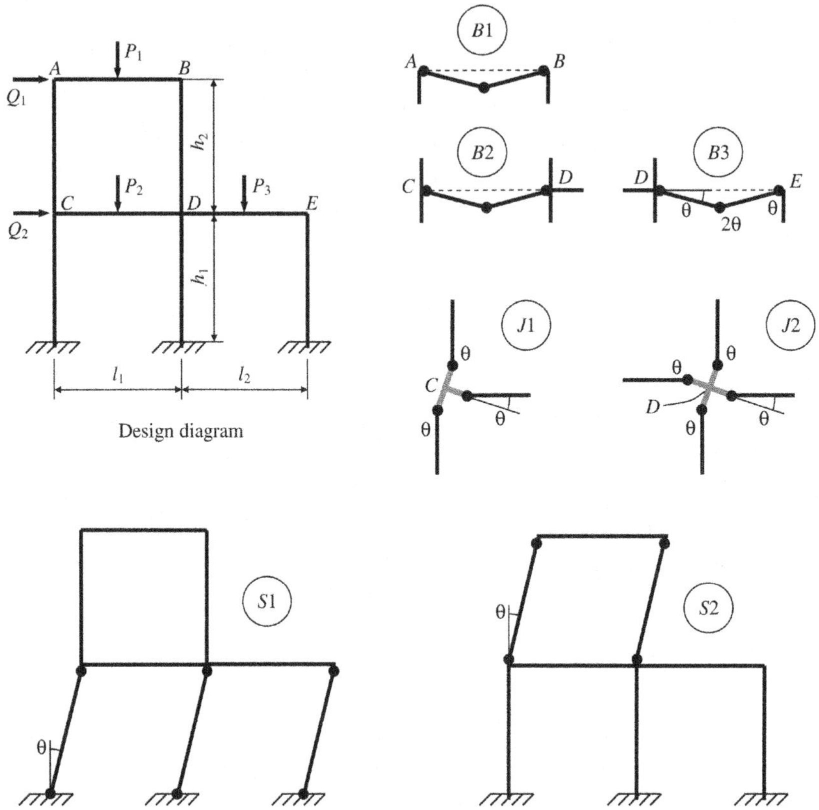

FIGURE 22.7 Design diagram of frame and independent collapse mechanisms: $B1$, $B2$, $B3$ are beam mechanisms; $J1$, $J2$ are joint mechanisms; $S1$, $S2$ are sway mechanisms; θ is the degree of freedom of the mechanism.

only the rotation of joints belonging to independent sway mechanisms, and displacements in these mechanisms are infinitesimally close to displacements in the source sway mechanisms (assuming the joints are infinitesimally small). Examples of incorrect composing of combined mechanisms are shown on diagrams e and f. The combinations crossed by the diagonal are not regular: those mechanisms that are shown in diagrams – yet they are composed by selecting hinges from source mechanisms – have two degrees of freedom. The calculation of assumed FBC at the equality of these DOFs will be an extra operation, because the obtained FBC will be no less than the minimal FBC for constituent mechanisms. For instance, the FBC of mechanism $S1 + B3$ satisfies inequality:

$$k_{S1+B3} \geq \min\left(k_{S1}, k_{B3}\right).$$

Constructing and checking the mechanisms with two DOFs doesn't produce an incorrect final result of analysis but increases its laboriousness.

The upper estimate for total number of independent and regular combined mechanisms can be calculated by the formula:

$$m_{\Sigma} = \sum_{i=1}^{m} \frac{m!}{i!(m-i)!} = 2^m - 1. \tag{22.8}$$

The formula is obtained by summing the total number of combinations of i selected mechanisms in a total of m independent mechanisms.

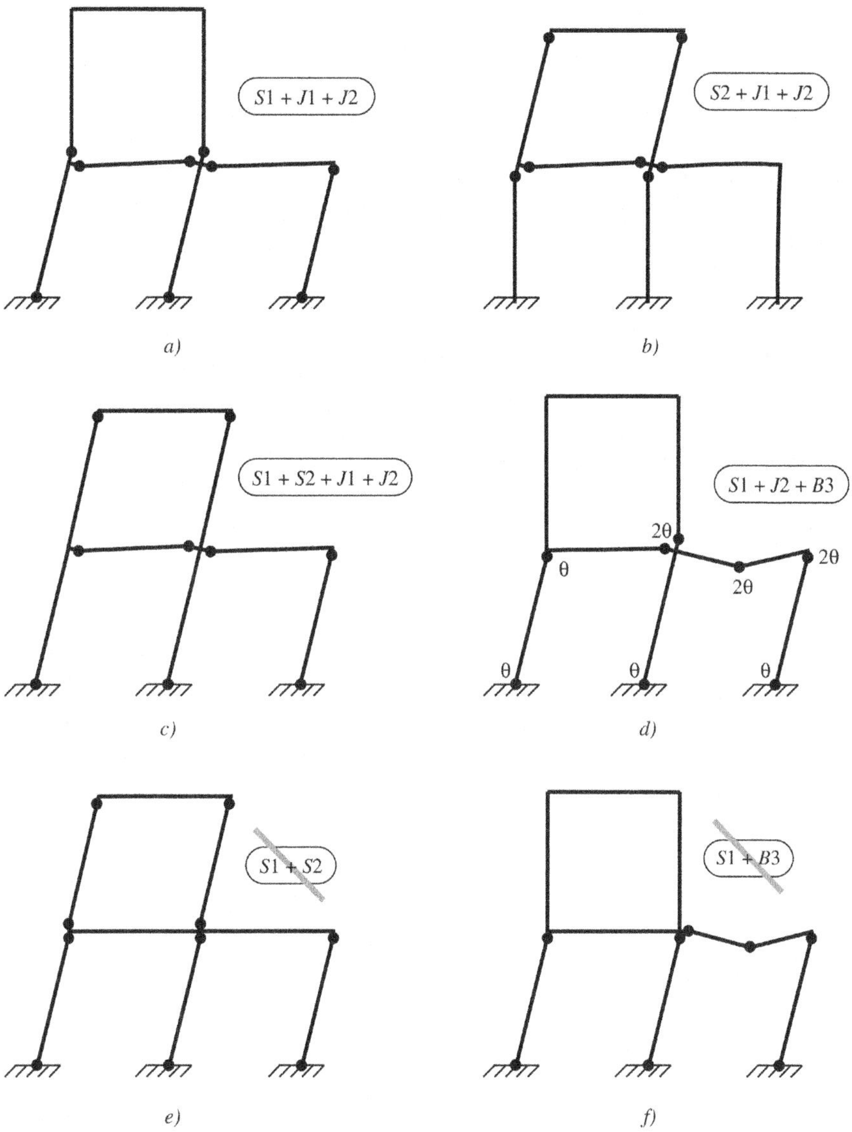

FIGURE 22.8 Combined mechanisms: (a–d) Regular mechanisms; (e), (f) Mechanisms with two DOFs (are not used in analysis).

After composing the complete totality of mechanisms necessary for calculation, the wanted parameter (the FBC or member section characteristic) is evaluated for each of them. To this end, on the diagram of the mechanism, for every plastic hinge, one should point out the total angle of pivot done by the discs connected through the hinge. (It is allowed to indicate the sum of rotation angles at several hinges situated together). The diagram where rotation angles are indicated for given DOF θ is shown in Figure 22.8 for the mechanism $S1 + J2 + B3$. By means of such a diagram, it is easy to obtain the right-hand side of the virtual work equation, i.e., the energy absorbed by plastic hinges. Taking plastic moments of all members as the same, we get the following equation for the FBC in this example:

$$kP_3\frac{l_2}{2}\theta + k(Q_1 + Q_2)h_1\theta = 10M_p\theta.$$

The left side is the sum of works done by loads in each of the mechanisms being combined. The right side is the sum of energies absorbed in plastic hinges of independent mechanisms, minus the energy absorbed by the hinges removed when combining the mechanisms. The reader is recommended to establish which hinges are eliminated in the following sequential combining: $S1 \rightarrow S1 + J2 \rightarrow S1 + J2 + B3$, and how much the elimination of hinges reduces the total absorbed energy.

The described realization of the method of combining mechanisms for the frames with horizontal girders can be also used for latticed frames with sloped top girders. Now we consider the generalization of the method for the frames wherein the top girder has a gabled shape. The example of a frame with gabled and sloped roofs is shown in Figure 22.9. For a gabled girder, there are independent gabled collapse mechanisms (type G) which are used in combinations with beam mechanisms of a girder's collapse. These mechanisms are also shown in Figure 22.9. If in the design diagram the loads are close in value ($Q \approx P$) and the plastic moments of the members are the same, then the collapse is to be expected according to the scheme $B1 + G1$. At least, the experiments with a portal gabled frame have revealed exactly this type of collapse (Lu, Chapman, and Driscoll Jr. 1958). This fact is in accordance with the postulate about selection of true mechanisms from the collection of independent and regular ones. Below, we show that if the DOF is specified for each of the mechanisms in Figure 22.9 by parameter θ which sets angular displacements according to diagrams of the figure, then combined mechanism $B1 + G1$ will be regular.

Figure 22.10 shows combination $B1 + G1$ in the states before displacement and after it (diagram a). Points B, C, D of members take new positions B', C', D' as the result of displacement. During displacement, the gabled component BCD of the mechanism pivots about instantaneous rotation center O. In diagram b, the triangle AOB' is depicted with indication of displacement vectors $\overrightarrow{BB'}$ and $\overrightarrow{CC'}$ and some specific dimensions defined by the length of rafter a. According to this diagram, infinitesimal angular displacement of component BCD is $\angle AOB' = \theta$. Planar displacement of every disc of the mechanism can be specified by either translational displacements of two points belonging to the disc or translational displacement of one point and rotational displacement of the disc. We denote rotational displacement (pivot angle) of a disc as $\varphi(\cdot)$, where denotation of the disc is shown in parentheses. We accept the positive direction for the counting of an angle as

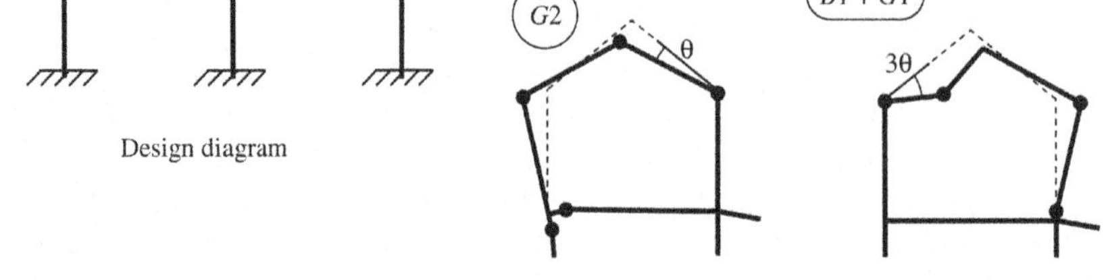

FIGURE 22.9 Frame with gabled roof and mechanisms of roof collapse: $B1$, $G1$, $G2$ are independent mechanisms; $B1 + G1$ is the regular combined mechanism; θ is the degree of freedom.

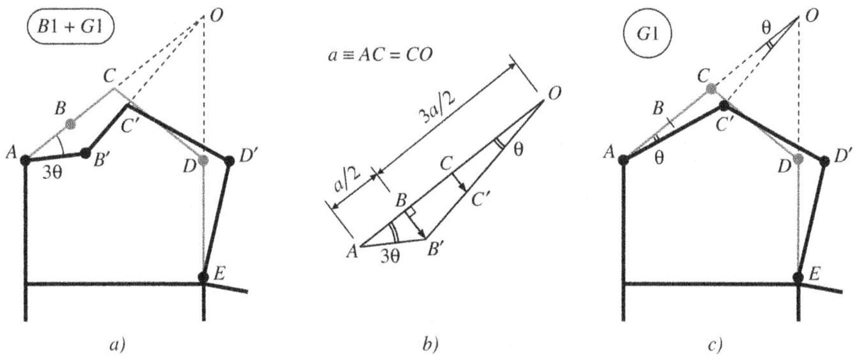

FIGURE 22.10 Displacements in collapse mechanisms.

counterclockwise. Diagrams a and b enable us to obtain infinitesimal displacements of members (discs) in the form:

- Member AB: point A is immovable, $\varphi(AB) = -3\theta$;
- Member BC: $CC' = a\theta$, $CC' \perp AC$ (rotation of $(\cdot)C$ counterclockwise about $(\cdot)O$), $\varphi(BC) = \varphi(BCD) = \theta$;
- Member CD: $CC' = a\theta$, $CC' \perp AC$ (rotation of $(\cdot)C$ counterclockwise about $(\cdot)O$), $\varphi(CD) = \varphi(BCD) = \theta$;
- Member DE: $DD' = OD\cdot\theta$, $DD' \perp DE$ (rotation of $(\cdot)D$ counterclockwise about $(\cdot)O$), point E is immovable.

Here, the direction of displacement is clarified in parentheses; the displacements of points C and D are the results of pivoting these points around point O.

Independent mechanisms $B1$ and $G1$ contain the same four movable discs as the combination $B1 + G1$ (i.e., members AB, BC, CD, DE). It is easy enough to check that the linear and angular displacements established above for discs belonging to mechanism $B1 + G1$ could be obtained by summing up the displacements in mechanisms $B1$ and $G1$ at the DOF's magnitude θ. While doing so, we ought to take into account that right-hand rafter in mechanism $G1$ turns around the same instantaneous rotation center O as in combined mechanism $B1 + G1$ (Figure 22.10c). For example, the sum of angular displacements of member CD in mechanisms $B1$ and $G1$ is

$$\varphi_{B1}(CD) + \varphi_{G1}(CD) = 0 + \theta = \theta,$$

whereas the sum of linear displacements of point C (the roof's gable) is

$$u_{B1}(C) + u_{G1}(C) = 0 + a\theta = a\theta.$$

In these formulas, the subscript shows the mechanism for which the displacement is calculated; $u(\cdot)$ is the denotation for displacement of specified point. The direction of total displacement obtained in last expression coincides with the direction of vector $\overrightarrow{CC'}$ in the combined mechanism.

In conclusion, we make some comments about plastic analysis of trusses. In contrast with problems of calculating the beams and frames, the collapse mechanisms of trusses are not classified, and the main instrument of their analysis is the method of step-by-step identification of lost and restored connections. The peculiarity of calculation is that in every calculation step, for compressed members, one must control the attained compression force so that it does not exceed the buckling force. The factor of buckling is taken into consideration in the same way as the factor of yielding: The point in time when compression force reaches critical value is the moment of connection's loss. From this moment, the member is removed from a system and replaced by a constant reaction equal to the buckling force. To evaluate the latter, it is usually allowed to employ the Euler formula.

Section X

Finite Element Method in Analysis of Elastic Structures

The basics of the finite element analysis of plane systems are set forth. The concepts of node and finite element (FE) are introduced and basic provisions of the finite element method (FEM) are substantiated by means of planar body's model constructed of triangular FEs. The concept of FE's stiffness matrix is introduced as a matrix of quadratic form which determines strain energy of FE, and a general relationship is derived for calculation of this matrix. The concept of elastic system's stiffness matrix is introduced as a matrix of quadratic form which determines strain energy of the system. The properties of stiffness matrix are established, and the relationships for its calculation are derived. It is shown that the stiffness matrix of a structure can be obtained as the block of stiffness matrix constructed for the same system removed from bearings. Next, the equations of the structure's equilibrium state under loading are set up by means of the principle of minimum total potential energy. It is shown that these equations enable determination of the nodal displacements and support reactions in simple matrix form, taking into account the support settlements. Recommendations are given about construction of finite element model to ensure convergence of FEM solutions to exact solutions of the elastic problem. The individual chapter is devoted to the construction of quadrilateral FE of plane body taking into consideration the requirement of FEM solutions' convergence to exact solutions. Further on in the section, the stiffness method is studied as an expansion of FEM to the problems of bar system analysis. The concept of a bar system's stiffness matrix is introduced and its physical meaning is established. The basic provisions of stiffness method are stated, and the substantiation of the method is given for the analysis of trusses. The features of this method are explained when being applied to the analysis of frames – in particular, the section presents the technique of reducing intermediate loads to nodes. As the generalization of theoretical material, the section exposes the principal provisions of FEM being applied to analysis of arbitrary elastic systems, including the frames.

23 Finite Element Method Exemplified in Elastic Plane Body

23.1 ESSENCE OF FINITE ELEMENT METHOD AND CALCULATION OF STRAIN ENERGY OF FINITE ELEMENT

Originally, the finite element method was elaborated for calculation of the SSS of elastic bodies by Canadian mathematician Alexander Hrennikoff in 1941 and American mathematician Richard Courant in 1943. In the 1950s, this method was intensively developed and found considerable practical application owing to the advent of mainframe computers. During this period, the development of the method is attributable, to a large extent, to the work of American mathematician and aviation engineer M. Jonathan Turner, who employed the FEM for the design of aircraft structures. The development and extension of the scope of FEM continued during the second half of the 20th century. The FEM received mathematical grounding in the 70s; inter alia, in the publications of this period, the convergence of FEM solutions to precise solutions of the problem of SSS calculation was proved. Nowadays, the FEM is the main tool for strength calculations on computers, and its development is ongoing.

In mechanics of strained bodies, FEM represents the realization of the principle of virtual work for the SSS analysis. For substantiation of FEM with reference to elastic systems, the principle of minimum total potential energy is used. According to this principle, the equilibrium of a conservative system is attained under displacements which zero variation of total potential energy corresponds to, and this equilibrium is stable if and only if the displacements in the system conform to a rigorous minimum of total potential energy.* Variation of potential energy is its increment due to infinitesimal increments of displacements in a system. If the SSS of a system is determined by a finite number of independent displacements, then energy variation is the differential of the energy as of a function of displacements. The stability of equilibrium of a structure means that infinitesimal additional loads and infinitesimal support shifts produce infinitesimal additional displacements in the structure, uniquely defined by these effects.

In FEM, the state of a structure is approximately defined by a finite number of independent parameters called degrees of freedom. According to this method, the displacements of a known aggregate of the structure's points called *nodes* are taken as the DOFs. A structure is considered as a set of *finite elements* (FEs). FE is the representation of some part of a construction by means of the set of nodes comprised by this part and mathematical relations which enable us to determine (approximately or precisely) its SSS through nodal displacements. In a trivial case, a structure can be specified by only FE.

The FEM, as applied to an elastic externally stable system, lies in representation of the total potential energy of the system, released from connections with the earth, through nodal displacements in order to set up a linear equation system in terms of displacements from the condition of zero variation of the total potential energy. These equations are supplemented with conditions of external stability, and by solving the obtained equation system, nodal displacements are determined, SSS is established, and support reactions are calculated. In order to derive FEM equations, the relations are set up for each type of finite element in a structure, which

* The assertion that the point of stable equilibrium is mandatorily the point of rigorous minimum of potential energy is correct under some stipulations which are fulfilled in practice, see Chapter 27.

relate strains, stresses, and (ultimately) the strain energy of FE to the displacements of element nodes.

In analysis of an externally stable elastic body, the polyhedrons obtained by mental partitioning of a body are often taken as FEs, the nodes of each element being the vertices of a polyhedron. The total strain energy of a body must be determined through nodal displacements with infinitesimal error while the FE's size is reduced. This ensures precision of the solution under conditions of small enough size of FE and the observance of some rules of design scheme construction for the FEM analysis.

Further on, we study the FEM with reference to a plane externally stable elastic body of thickness t in a plane stress state. We consider the body displacements in earth-referenced CS $O'xy$. We have to keep in mind that we interpret space bodies as plane figures; thus, a point represents a segment of length t, and a triangle represents a wedge of the uniform thickness t. Displacements are denoted, as usual, by u and v. We set the aggregate of nodes so that the points of supports and loads' exertion coincide with the nodes. We assume that each support constraint eliminates either DOF u or v, or both of these DOFs. In other words, the constraints imposed on displacements of the support node can be as follows:

$$u = 0, \tag{23.1}$$

or

$$v = 0, \tag{23.2}$$

or

$$u = 0 \text{ and } v = 0. \tag{23.3}$$

We introduce the FEs in the form of triangles with the *linear law* of displacement variation within an element. This supposition means that each triangle holds its triangular shape while being strained, and, therefore, continuity of the deformed body is secured. Below, assuming that the location of an unstrained element is specified by the coordinates of its nodes, we represent the strain energy of an element through nodal displacements (Figure 23.1).

We introduce:

$$\mathbf{u}^{\mathrm{T}} = (u_1, v_1, u_2, v_2, u_3, v_3) \tag{23.4}$$

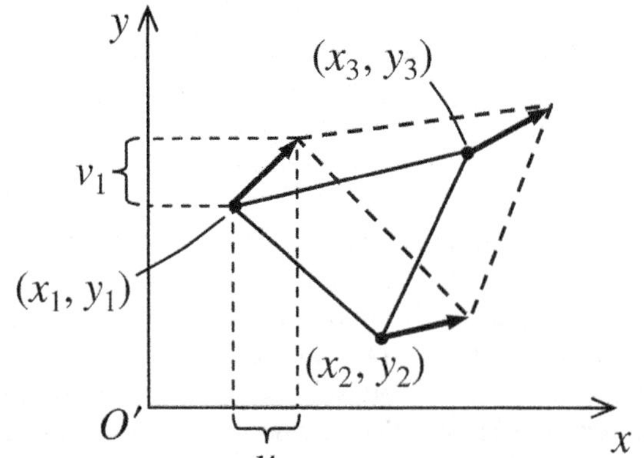

FIGURE 23.1 Finite element before deformation (continuous contour) and after deformation (dashed contour).

– Vector of the element vertices' displacements (which can be termed the element DOFs), the index points to the vertex number;

$$\varepsilon^{\mathrm{T}} = (\varepsilon_x, \varepsilon_y, \gamma_{xy})$$

– Vector of the strains in the element's boundaries;

$$\sigma^{\mathrm{T}} = (\sigma_x, \sigma_y, \tau_{xy})$$

– Vector of the stresses in the element's boundaries.

Because of the linear dependence of displacements versus coordinates, the strain vector is not dependent on coordinates in the boundaries of an element. We denote: **B** – matrix of strains' relation to displacements of element's vertices, i.e., such matrix that $\boldsymbol{\varepsilon} = \mathbf{Bu}$; **D** – matrix of generalized Hooke's law, i.e., such a matrix that $\boldsymbol{\sigma} = \mathbf{D}\boldsymbol{\varepsilon}$. Matrix **D** is given by the relationship:

$$\mathbf{D} = \frac{E}{1-\nu^2}\begin{pmatrix} 1 & \nu & 0 \\ \nu & 1 & 0 \\ 0 & 0 & \dfrac{1-\nu}{2} \end{pmatrix}. \tag{23.5}$$

One can see that this matrix is symmetric.

For the strain energy of an element of volume V it holds:

$$U_1 = \frac{V}{2}\sigma^{\mathrm{T}}\varepsilon = \frac{V}{2}(\mathbf{D}\varepsilon)^{\mathrm{T}}\varepsilon = \frac{V}{2}\varepsilon^{\mathrm{T}}\mathbf{D}^{\mathrm{T}}\varepsilon = \frac{V}{2}(\mathbf{Bu})^{\mathrm{T}}\mathbf{D}(\mathbf{Bu}) = \frac{V}{2}\mathbf{u}^{\mathrm{T}}\mathbf{B}^{\mathrm{T}}\mathbf{DBu}, \tag{23.6}$$

or finally:

$$U_1 = \frac{1}{2}\mathbf{u}^{\mathrm{T}}\mathbf{Ku}, \tag{23.7}$$

where we introduced symmetric *FE stiffness matrix*:

$$\mathbf{K} = V\mathbf{B}^{\mathrm{T}}\mathbf{DB}. \tag{23.8}$$

(Note that the product of the form $\mathbf{B}^{\mathrm{T}}\mathbf{DB}$ is a symmetric matrix on symmetric matrix **D** and arbitrary matrix **B** of allowable dimension.)

Let us find the form of matrix **B**. To do that, we write down linear relationships for displacements in FE's boundaries:

$$\begin{aligned} u &= a_1 x + b_1 y + c_1; \\ v &= a_2 x + b_2 y + c_2. \end{aligned} \tag{23.9}$$

We have:

$$\varepsilon_x = \frac{\partial u}{\partial x} = a_1;\ \ \varepsilon_y = \frac{\partial v}{\partial y} = b_2;\ \ \gamma_{xy} = \frac{\partial u}{\partial y} + \frac{\partial v}{\partial x} = b_1 + a_2. \tag{23.10}$$

We express the coefficients on the right-hand sides through displacements of the triangle vertices by solving two equation sets:

$$\begin{cases} x_1a_1 + y_1b_1 + c_1 = u_1; \\ x_2a_1 + y_2b_1 + c_1 = u_2; \\ x_3a_1 + y_3b_1 + c_1 = u_3; \end{cases} \quad \begin{cases} x_1a_2 + y_1b_2 + c_2 = v_1; \\ x_2a_2 + y_2b_2 + c_2 = v_2; \\ x_3a_2 + y_3b_2 + c_2 = v_3. \end{cases} \tag{23.11}$$

The Cramer rule gives the solutions:

$$\begin{aligned} a_1 &= \Delta^{-1}[u_1(y_2 - y_3) + u_2(y_3 - y_1) + u_3(y_1 - y_2)]; \\ b_1 &= \Delta^{-1}[u_1(x_3 - x_2) + u_2(x_1 - x_3) + u_3(x_2 - x_1)]; \\ a_2 &= \Delta^{-1}[v_1(y_2 - y_3) + v_2(y_3 - y_1) + v_3(y_1 - y_2)]; \\ b_2 &= \Delta^{-1}[v_1(x_3 - x_2) + v_2(x_1 - x_3) + v_3(x_2 - x_1)], \end{aligned} \tag{23.12}$$

where

$$\Delta = \det\begin{pmatrix} x_1 & y_1 & 1 \\ x_2 & y_2 & 1 \\ x_3 & y_3 & 1 \end{pmatrix}.$$

After replacements in (23.10) according to (23.12), we obtain the wanted relationship matrix as follows:

$$\mathbf{B} = \Delta^{-1}\begin{pmatrix} y_2 - y_3 & 0 & y_3 - y_1 & 0 & y_1 - y_2 & 0 \\ 0 & x_3 - x_2 & 0 & x_1 - x_3 & 0 & x_2 - x_1 \\ x_3 - x_2 & y_2 - y_3 & x_1 - x_3 & y_3 - y_1 & x_2 - x_1 & y_1 - y_2 \end{pmatrix}. \tag{23.13}$$

Thus, we finished the investigation of triangular FE: the strain energy is obtained as (23.7) and defined by stiffness matrix (23.8) with substitution (23.13).

Generalizations

All characteristics of the SSS have been considered here in the earth-referenced CS $O'xy$. In FEM such CS, being used for the statement of a problem and representation of final results of analysis, is termed a *global CS*, and, besides this one, *element CS* is introduced for each element, and *nodal CS* for each node. It makes analysis possible in more complicated problems, but in our simple case it can be convenient too. For instance, introduce the element CS with an origin at the first node of the element, axis x directed at the second node, and axis y completing CS to a right one. You will see that in this CS, the relations (23.11)–(23.13) are simplified, and determinant Δ appears to be numerically equal to the double triangle area and positive when the vertices are enumerated counterclockwise.

Formula (23.7) has generality for any type of elastic FE, but formula (23.8) does not; the SSS in FE's boundaries is not always uniform. Usually, the matrix of a strain-displacement relationship depends on coordinates, and it is necessary to integrate the strain energy over the FE's volume. In the general case, instead of formula (23.8), we have:

$$\mathbf{K} = \int_V \mathbf{B}^{\mathrm{T}}\mathbf{D}\mathbf{B}dV. \tag{23.8*}$$

Nodal CS is used to define the displacements of the corresponding node. In general, nodal CS coincides with global CS,* but, for some nodes, the use of nodal CSs pivoted relative to the global one can be convenient. For instance, if a node has an oblique bearing of the "simple rod" type, then, for this node, such a rotation of nodal CS that one of the coordinate axes be directed along supporting link is appropriate, and so one of constraints (23.1)–(23.2) will be applicable. If the pivoted nodal CSs are in use, then basic relations (23.7) and (23.8*) remain in effect, but setting up the relationship matrix is more complicated.

23.2 STRAIN ENERGY OF AN ELASTIC BODY IN APPROXIMATION OF FEM; STIFFNESS MATRICES OF FREE BODY AND STRUCTURE-BODY

In this item, we represent the strain energy of an externally stable elastic body through its DOFs. We solve this problem in two stages: initially, we establish the strain energy of the free body obtained by elimination of supports from the given structure-body; next, in the established expression for the energy as the function of DOFs, we take into account the constraints upon displacements in the structure.

We introduce transposed displacement vector $\mathbf{u}_f^{\mathrm{T}}$ as the row of nodal displacements of a body taken off from bearings:

$$\mathbf{u}_f^{\mathrm{T}} = (u_1, v_1, u_2, v_2, \dots, u_{n_1}, v_{n_1}), \tag{23.14}$$

where n_1 is the total number of nodes. Owing to the absence of supports, these displacements constitute the set of independent parameters determining the SSS, i.e., the set of body's DOFs totaling $2n_1$. We call vector $\mathbf{u}_f$ the *full vector of displacements of the structure's nodes.*

Let the model comprise m FEs. We specify nodal displacements for the i-th element by vector $\mathbf{u}_i$ of the form (23.4). The components of this vector are defined by the DOFs of a body, i.e., for each element, we can establish the relation:

$$\mathbf{u}_i = \mathbf{I}_i \mathbf{u}_f. \tag{23.15}$$

Here $\mathbf{I}_i$ is *the matrix of displacements' selection* for the i-th element. In every row of this matrix, just one component is nonzero; this component equals unity and its serial number defines what component of vector $\mathbf{u}_f$ is the element's DOF u_{ik}, where k is the number of a row. For instance, the model of a rectangular wedge in Figure 23.2 is comprised of two elements. The vectors of element nodal displacements are of the form (with enumeration of vertices in a counterclockwise direction):

$$\mathbf{u}_1^{\mathrm{T}} = (u_1, v_1, u_2, v_2, u_4, v_4);$$

$$\mathbf{u}_2^{\mathrm{T}} = (u_1, v_1, u_4, v_4, u_3, v_3).$$

Mentally remove the supports from the structure. The matrix $\mathbf{I}_2$, defining the pass from DOF vector (23.14) to vector $\mathbf{u}_2$, is obtained as follows:

$$\mathbf{I}_2 = \begin{pmatrix} 1 & 0 & 0 & 0 & 0 & 0 & 0 & 0 \\ 0 & 1 & 0 & 0 & 0 & 0 & 0 & 0 \\ 0 & 0 & 0 & 0 & 0 & 0 & 1 & 0 \\ 0 & 0 & 0 & 0 & 0 & 0 & 0 & 1 \\ 0 & 0 & 0 & 0 & 1 & 0 & 0 & 0 \\ 0 & 0 & 0 & 0 & 0 & 1 & 0 & 0 \end{pmatrix}.$$

* In the FEM software, each nodal CS is depicted with an origin at the corresponding node, but this is done only for visual grasp: the choice of origin is unimportant for computing the displacements.

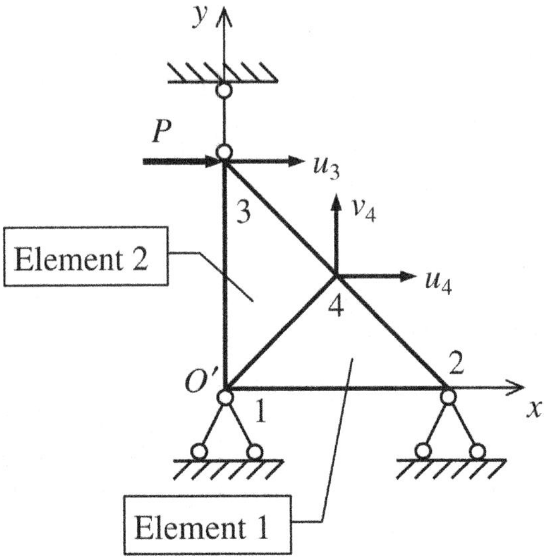

FIGURE 23.2 FE model of the structure-body, comprising two FEs.

For the i-th FE belonging to a model, we shall denote the number of DOFs as n_{ei}. In the present chapter, we consider triangular FEs, for which $n_{ei} = 6$. Matrix $\mathbf{I}_i$ has the dimension $n_{ei} \times 2n_1$. It is convenient to specify the transformation (23.15) by means of vector $\mathbf{S}_i$ comprising the numbers of components belonging to vector $\mathbf{u}_f$ sequentially selected into vector $\mathbf{u}_i$. Under the definite order of these numbers in vector $\mathbf{S}_i$, the components of vector $\mathbf{u}_i$ are obtained in the form:

$$u_{ik} = u_{fS_{ik}}, \quad k = \overline{1, n_{ei}}. \tag{23.16}$$

In order to obtain vector $\mathbf{u}_2$ in the suggested example, instead of matrix $\mathbf{I}_2$, one may use the vector of numbers of selected components $\mathbf{S}_2 = (1, 2, 7, 8, 5, 6)^{\mathrm{T}}$ and employ formula (23.16). Obtain the matrix of displacements' selection and the vector comprising the numbers of selected DOFs for the first element by yourself.

By use of displacement vectors $\mathbf{u}_i$ of the element nodes, we get the expression for the strain energy of a whole elastic body. On a given *FE mesh* (that is a given ensemble of elements defined in space by the mesh of their boundaries), we can calculate stiffness matrix (23.8) for each element. Let us denote the stiffness matrix of the i-th element as $\mathbf{K}_i$, $i = \overline{1, m}$. We obtain the strain energy of the whole body by summation of strain energies of separate elements. The latter are determined by formula (23.7), and thus we have:

$$U_1 = \sum_{i=1}^{m} U_{1i} = \frac{1}{2} \sum_{i=1}^{m} \mathbf{u}_i^{\mathrm{T}} \mathbf{K}_i \mathbf{u}_i.$$

After substitution (23.15), we express the strain energy as a function of the body's DOF vector:

$$U_1 = \frac{1}{2} \sum_{i=1}^{m} \mathbf{u}_f^{\mathrm{T}} \mathbf{I}_i^{\mathrm{T}} \mathbf{K}_i \mathbf{I}_i \mathbf{u}_f = \frac{1}{2} \mathbf{u}_f^{\mathrm{T}} \left(\sum_{i=1}^{m} \mathbf{I}_i^{\mathrm{T}} \mathbf{K}_i \mathbf{I}_i \right) \mathbf{u}_f. \tag{23.17}$$

We refer to a symmetric matrix for which the quadratic form of the DOF vector, taken with the coefficient ½, is equal to the strain energy of the body as the stiffness matrix of the elastic body. By introducing stiffness matrix $\mathbf{K}$ of free body, we can represent formula (23.17) in the form:

$$U_1 = \frac{1}{2}\mathbf{u}_f^{\mathrm{T}}\mathbf{K}\mathbf{u}_f; \tag{23.18}$$

$$\mathbf{K} = \sum_{i=1}^{m} \tilde{\mathrm{K}}_i; \quad \tilde{\mathrm{K}}_i \equiv \mathbf{I}_i^{\mathrm{T}}\mathbf{K}_i\mathbf{I}_i. \tag{23.19}$$

Term $\tilde{\mathrm{K}}_i$ is called the *expanded stiffness matrix* of the *i*-th FE. Stiffness matrix **K** may be called the *stiffness matrix of FEs' ensemble.*

Expanded stiffness matrices have the order $2n_1$, and every *i*-th expanded matrix is obtained from the zeroth matrix of order $2n_1$ by replacement of its constituents according to the rule:

$$\tilde{\mathrm{K}}_{iS_{ik}S_{il}} = K_{ikl}, \; k = \overline{1, n_{ei}}, \; l = \overline{1, n_{ei}}. \tag{23.20}$$

This technique of calculation of an expanded matrix is more economical than multiplying the matrices in the second formula (23.19) by the rule "row by column."

Now, we return to the investigation of a body attached to the earth. In the row of DOFs (23.14) constructed for a body removed from bearings, we make permutation in such manner that at the beginning of the row, there go possible displacements (not eliminated by support connections), and then follow impossible displacements (eliminated by connections). After such permutation one can write down:

$$\mathbf{u}_f = \begin{pmatrix} \mathbf{u} \\ \mathbf{u}_X \end{pmatrix}, \tag{23.21}$$

where **u** is the subvector of the structure's DOFs (i.e., displacements not eliminated by connections), and $\mathbf{u}_X$ is a subvector of zero displacements. The dimension of vector **u** is equal to the number of the structure's DOFs, which we denote $n = \dim \mathbf{u}$; accordingly, $\dim \mathbf{u}_X = 2n_1 - n$. It is obvious that $n < 2n_1$. In the example in Figure 23.2, the DOFs of the structure can be represented by the vector:

$$\mathbf{u}^{\mathrm{T}} = (u_3, u_4, v_4). \tag{23.22}$$

We refer to the vector of nodal displacements in a structure-body, which includes subvector $\mathbf{u}_X$ of displacements eliminated by supports according to formula (23.21), as the *ordered full vector of nodal displacements.*

The permutation of nodal displacements, done above, allows us to single out the stiffness matrix of a structure from stiffness matrix **K** of a free body. Naturally, we calculate matrix **K** with account of this permutation. Decompose the latter matrix into blocks:

$$\mathbf{K} = \begin{pmatrix} \mathbf{K}_{11} & \mathbf{K}_{12} \\ \mathbf{K}_{21} & \mathbf{K}_{22} \end{pmatrix}, \tag{23.23}$$

where block $\mathbf{K}_{11}$ has the order *n*. Substitute the representations of vector (23.21) and matrix (23.23) into formula (23.18). We have:

$$U_1 = \frac{1}{2}\begin{pmatrix} \mathbf{u}^{\mathrm{T}} & \mathbf{u}_X^{\mathrm{T}} \end{pmatrix}\begin{pmatrix} \mathbf{K}_{11} & \mathbf{K}_{12} \\ \mathbf{K}_{21} & \mathbf{K}_{22} \end{pmatrix}\begin{pmatrix} \mathbf{u} \\ \mathbf{u}_X \end{pmatrix} = \frac{1}{2}\begin{pmatrix} \mathbf{u}^{\mathrm{T}} & \mathbf{u}_X^{\mathrm{T}} \end{pmatrix}\begin{pmatrix} \mathbf{K}_{11}\mathbf{u} + \mathbf{K}_{12}\mathbf{u}_X \\ \mathbf{K}_{21}\mathbf{u} + \mathbf{K}_{22}\mathbf{u}_X \end{pmatrix}. \tag{23.24}$$

Take into consideration the condition of attachment to the earth, which has the form:

$$\mathbf{u}_X = 0. \tag{23.25}$$

We obtain the next formula for the strain energy of a structure:

$$U_1 = \frac{1}{2}\mathbf{u}^{\mathrm{T}}\mathbf{K}_{11}\mathbf{u}. \tag{23.26}$$

So, the stiffness matrix of a structure is the block of the stiffness matrix of a body removed from bearings.

Stiffness matrices have *the properties of sign definiteness*: the stiffness matrix of a structure is a positive definite matrix; the stiffness matrix of a free body is a positive semi-definite singular matrix.

The proof of stiffness matrices' definiteness is based on two assertions about displacements of finite elements:

1) The totality of FEs belonging to the model of an externally stable body cannot be displaced as a whole entity;
2) The strains in arbitrary FE equal zero if and only if the element is displaced as a single whole.

Displacement of a point set as a whole entity is understood as a movement of points of a rigid body, i.e., when the points are immovable in some reference frame. The first assertion is an attribute of an adequate FE model. The second assertion is proved by representation of strains through derivatives of displacements with respect to coordinates (first equality in each of the formulas (23.10)). The same assertion is proved in the theory of elasticity for a deformed body (Love 1906, p. 50), and the proof remains valid for finite elements.

From the assertions of FE displacements, it follows that in an FE model of a structure-body, as well as in the body itself, nonzero displacements produce strains. The potential energy of an element's deformation is positive under nonzero strains in accordance with the third equality in (23.6) and positive definiteness of matrix **D**. Therefore, the total potential energy of the deformation calculated by the model at nonzero strains is positive. So, at arbitrary nonzero vector **u** of structure's DOFs, the quadratic form (23.26) is positive. Hence, the **stiffness matrix of a structure is positive definite**.

A body taken off from bearings can be displaced as a single whole, i.e., without strains. The same is correct for the FE model of this body, and therefore, there exists nonzero vector $\mathbf{u}_f$ for which the energy (23.18) is zeroth. Thus, matrix **K** can't be positive definite. But energy (23.18) is obtained by the summation of element energies (23.6), and therefore cannot be negative. Hence, the **stiffness matrix of a free body is a positive semi-definite singular matrix**.

23.3 FEM EQUATIONS AND PHYSICAL MEANING OF STIFFNESS MATRIX

We introduce the transposed vector of external forces exerted on all nodes of a body as follows:

$$\mathbf{F}^{\mathrm{T}} = (F_{x1}, F_{y1}, F_{x2}, F_{y2}, \dots, F_{xn_1}, F_{yn_1}). \tag{23.27}$$

Here, for nodes not subjected to external forces, we assume:

$$F_{xk} = F_{yk} = 0.$$

External forces exerted upon the attached nodes of a structure-body may comprise the loads and support reactions alike. To emphasize the fact that component F_j is the external force exerted on the attached node along the support link, we denote it as X_j. Thus, we can write for this component:

$$X_j \equiv F_j = P_j + R_j, \tag{23.28}$$

where P_j is the load in the direction of the support connection; R_j is the support reaction.

The work of external forces done through displacements of any separate node is of the form:

$$W_k = F_{xk}u_k + F_{yk}v_k.$$

The total work of external forces through the displacements of a free body is obtained by summation of these works and can be written in the form:

$$W = \mathbf{F}^{\mathrm{T}}\mathbf{u}_f,$$

where $\mathbf{u}_f$ is the full displacement vector (23.14). Therefore, the potential energy of a body in the external field has the form:

$$U_2 = -\mathbf{F}^{\mathrm{T}}\mathbf{u}_f. \tag{23.29}$$

For the total potential energy of an elastic body, we have the expression:

$$U = \frac{1}{2}\mathbf{u}_f^{\mathrm{T}}\mathbf{K}\mathbf{u}_f - \mathbf{F}^{\mathrm{T}}\mathbf{u}_f, \tag{23.30}$$

where the displacements making up vector $\mathbf{u}_f$, generally speaking, are not ordered in the form (23.21).

Vector of partial derivatives of a function $U = U(\mathbf{u}_f)$ is denoted as

$$\frac{\partial U}{\partial \mathbf{u}_f} \equiv \left(\frac{\partial U}{\partial u_{f1}}, \ldots, \frac{\partial U}{\partial u_{fn}} \right)^{\mathrm{T}}.$$

One refers to vector $\mathbf{u}_f$ which satisfies equation:

$$\frac{\partial U}{\partial \mathbf{u}_f} = 0, \tag{23.31}$$

as a *stationary point of a function* $U = U(\mathbf{u}_f)$. According to the principle of minimum total potential energy, in a conservative system, the DOF vector which is a stationary point of the function of total potential energy conforms to an equilibrium state of the system. Therefore, equation (23.31), written for total potential energy (23.30), is the equation of equilibrium of a body under consideration.

For the function (23.30), it is easily obtained:

$$\frac{\partial U}{\partial \mathbf{u}_f} = \mathbf{K}\mathbf{u}_f - \mathbf{F}.$$

Here, we used formula (23S.1) of differentiation of quadratic form and have taken into account that matrix $\mathbf{K}$ is symmetric. For the DOF vector, we get the equation:

$$\mathbf{K}\mathbf{u}_f - \mathbf{F} = 0. \tag{23.32}$$

In the composed equation, the vector of absolute terms contains unknown components in the form (23.28). But even if the support reactions are known, the only equation (23.32) does not suffice to establish displacements in a structure, because the stiffness matrix in this equation is singular. In derivation of this equation, we didn't use the information about the attachment of the body to the earth; we developed a vector equation for the displacements of the free body's points in equilibrium under the given external loads, and it is solved ambiguously.

We shall consider equality (23.32) as a system of equations in unknown displacements and support reactions. If we supply this system with conditions of attachment to the earth (23.1)–(23.3), then such an extended system is solved unambiguously. Let make us sure of this.

Further, we assume that full displacement vector $\mathbf{u}_f$ is ordered vector (23.21). Make the permutation of components in the vector of external forces the same as was done in the row of nodal displacements (23.14) during construction of the ordered full displacement vector, i.e., while decomposing the displacement vector into the subvector of the structure's DOFs and the subvector of displacements eliminated by support connections. After such permutation, the vector of external forces is represented in the form:

$$\mathbf{F} = \begin{pmatrix} \mathbf{P} \\ \mathbf{X} \end{pmatrix}, \tag{23.33}$$

where $\mathbf{P}$ is a subvector of loads acting along displacements allowed in a structure; $\mathbf{X}$ is a subvector of unknown external forces (23.28), i.e., the forces including reactions. In the case of the example in Figure 23.2, we have the vector of loads corresponding to the vector of the structure's DOFs (23.22) as follows:

$$\mathbf{P}^{\mathrm{T}} = (P, 0, 0).$$

We refer to the vector of the form (23.33) as the *ordered vector of external forces.*

Equilibrium equation (23.32) continues to be in force. By using block representation of data (23.21), (23.23), and (23.33), we convert this equation to the form:

$$\begin{pmatrix} \mathbf{K}_{11} & \mathbf{K}_{12} \\ \mathbf{K}_{21} & \mathbf{K}_{22} \end{pmatrix} \begin{pmatrix} \mathbf{u} \\ \mathbf{u}_X \end{pmatrix} - \begin{pmatrix} \mathbf{P} \\ \mathbf{X} \end{pmatrix} = 0. \tag{23.34}$$

Next, we substitute the condition of the attachment to the earth (23.25) and get the final form of equilibrium equations:

$$\begin{cases} \mathbf{K}_{11}\mathbf{u} - \mathbf{P} = 0; \\ \mathbf{K}_{21}\mathbf{u} - \mathbf{X} = 0. \end{cases} \tag{23.35}$$

The stiffness matrix of structure $\mathbf{K}_{11}$ is not singular; thus, the first of the equations (23.35) has a unique solution in unknown $\mathbf{u}$. This equation is called the governing, equilibrium, or global equation of FEM. After solving it, we substitute the obtained solution into the second equation in order to calculate vector $\mathbf{X}$ of external forces and establish the support reactions.

So far, the nodal displacements eliminated by connections were assumed to be zero, because we solved the problem of analysis of a structure-body at the given loads and unshifted supports. When a structure undergoes support displacements together with loads, nonzero components in the vector of displacements of support nodes $\mathbf{u}_X$ appear, but derivation of equilibrium equation (23.34) remains correct. This equation, as before, is employed for calculation of displacements and support reactions in a structure. From this equation, we obtain the formula for calculation of the DOF vector through given loads and support shifts:

$$\mathbf{u} = \mathbf{K}_{11}^{-1}(\mathbf{P} - \mathbf{K}_{12}\mathbf{u}_X). \tag{23.36}$$

Further on, we study the FEM for analysis of a structure subjected to loads only; possibility of the support shift will be specifically stated.

Remark

While calculating the load effect upon a structure, it is evident from formula (23.36), that infinitesimal support shifts and increments of loads cause infinitesimal increments of displacements in a system. Hence, the equilibrium state defined by equations (23.35) is stable, and the corresponding stationary point is the point where total potential energy reaches its minimum.

The equilibrium equation for a free body (23.34) can be supplemented not only by the conditions of attachment to the earth, but also other conditions upon displacements of nodes while solving special problems. In the supplement to this chapter, the problem of structure analysis wherein a body has rigid insertions imposing such additional conditions is under consideration.

In order to establish the physical meaning of a structure's stiffness matrix $\mathbf{K}_{11}$, notice that at arbitrary displacements allowed by constraints, this matrix specifies the load vector producing these displacements. The named vector has the form:

$$\mathbf{P} = \mathbf{K}_{11}\mathbf{u} = \sum_{j=1}^{n} \mathbf{K}_{11*j} u_j. \tag{23.37}$$

Here, the denotation $\mathbf{K}_{11*j}$ is used for the j-th column of the stiffness matrix. From the last formula, we can see that the state of externally stable FEs' ensemble with the only displacement $u_j = 1$ is caused due to the vector of loads $\mathbf{K}_{11*j}$. In other words, **every column of a structure's stiffness matrix is the load vector producing unit displacement of one of the nodes.**

The physical meaning of stiffness matrix $\mathbf{K}$ of a free body is established similarly: **every column of a free body's stiffness matrix is the vector of external forces producing unit displacement of one of the nodes**. This affirmation gives us the instruments to verify the calculation of a free body's stiffness matrix: Since external forces on the right side of formula (23.27) make up a balanced system of forces, the sum of components in each column-vector $\mathbf{K}_{*j}$ with even numbers must be equal to the sum of its components with odd numbers and be equal to zero. Similarly, the total moment of forces constituting every column-vector $\mathbf{K}_{*j}$, being calculated with respect to an arbitrary point, should be zero.

23.4 RULES OF FINITE ELEMENT MODEL CONSTRUCTION; EXAMPLE OF ANALYSIS

The practical meaning of FEM is defined by the provision that through the decrease of FE's diameter (maximum size), the result of calculation converges to an exact solution of the elastic problem. In order to ensure this provision is implemented, the next rules of an FE model's construction must be fulfilled.

While constructing an FE model of a solid body, it is important to specify the supports of nodes correctly. We take for granted that a planar body can be supported at some points by support connections of the first or second type as well as rigidly connected to the earth along extended portions of the body's outline. In the latter case, the support line is simulated by hinged immovable supports positioned thickly enough (for instance, on the design diagram in Figure 23.5, at the bottom, three hinged supports are assumed instead of a support along the bottom of the wedge).

A body can be subjected to both concentrated and distributed loads. The distributed loads have limited intensity and can be applied along the outline of a plane body or about its volume. While constructing an FE model, the loads are reduced to nodes. The reduction lies in partitioning a body into neighborhoods of nodes (each includes one node) and calculation of the total load acting in the neighborhood of every node. The obtained load is exerted upon the node inside the neighborhood. The neighborhood of a node is constructed by combining the neighborhoods within the FEs containing the node. In a triangular FE, the neighborhood of every node is cut off by the medians of the triangle. In Figure 23.3a, three neighborhoods are shown in a single FE; Figure 23.3b shows the part of the FE mesh and the neighborhood of node A in this mesh – pentagon obtained by combining

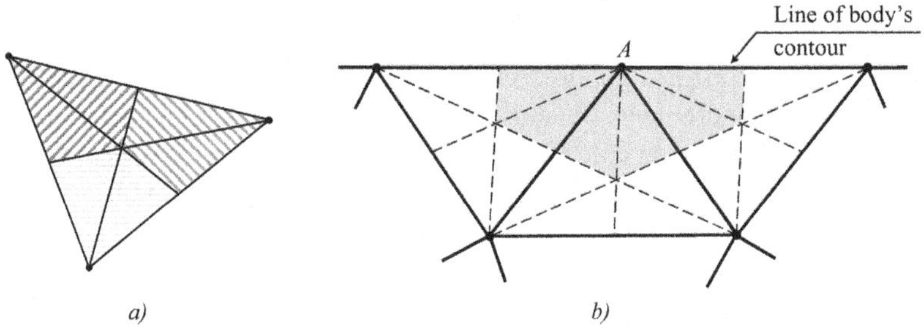

FIGURE 23.3 Neighborhoods of nodes in a model with triangular FEs: (a) Separated FE: the medians cut off three neighborhoods of nodes, each is emphasized by hatching; (b) Fragment of model: the medians of elements are shown by dashed lines; the neighborhood of node *A* is marked in gray.

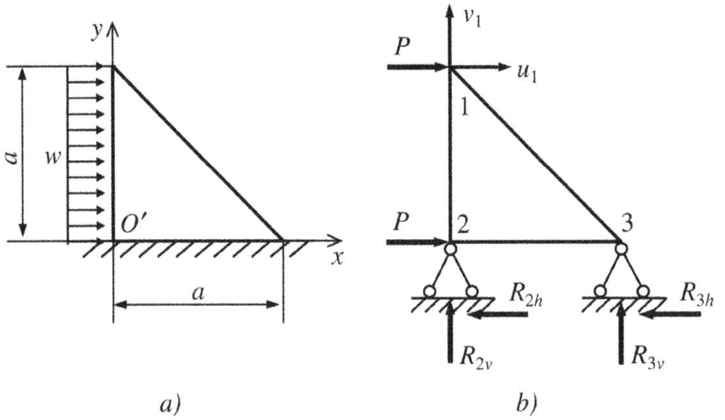

FIGURE 23.4 The given solid body (a) and its one-element model (b). u_1, v_1 are displacements of free vertex.

three portions of elements conjugated with the node. The total load exerted within the boundaries of this region is transferred upon node *A*. In the case of concentrated loads exerted upon a body, it is desirable to construct the FE mesh in such fashion that its nodes were the points of load exertion (because the pass of loads upon nodes increases the error of the simulation).

In order to ensure convergence of the method under the decrease of an FE's diameter, the FE models should meet the requirements for the shape of an element. For a triangular FE, the permitted angles of the triangle must be limited below by fixed value.

While studying the convergence of simulated stresses, strains, and displacements to exact values due to decrease of an FE's diameter, the singular points of a structure, where, according to the design scheme, concentrated external forces are exerted, are excluded from analysis. At these points, the characteristics of SSS are not defined, and in the neighborhood of such point, an SSS characteristic may be not limited. The convergence is studied at the domain of an elastic body wherein there is no ε-neighborhoods of singular points with a fixed radius.

As an example of FEM calculation, we consider the right-angled wedge which has the support along the lower face and is loaded with distributed load (Figure 23.4a). The deformed state of a system, evaluated in FEM, depends on the FEs' number (Figure 23.5). We evaluate the SSS by means of the one-element model in Figure 23.4b. In this model, the support portion of the outline is represented by two hinged immovable supports, and the distributed load is replaced by two identical concentrated loads:

$$P = \frac{wa}{2}.$$

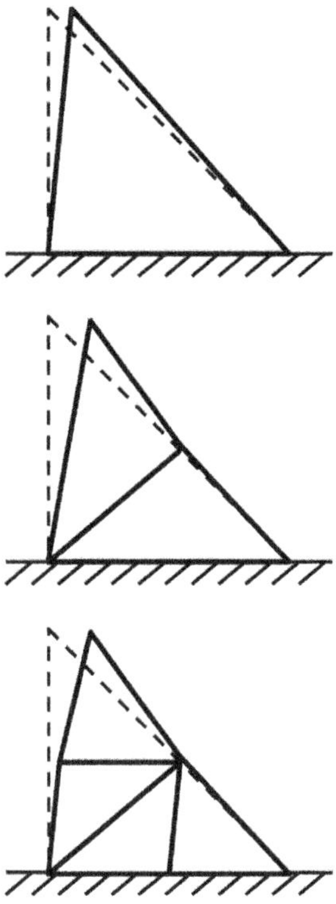

FIGURE 23.5 Deformed state of the body represented by FE model.

The full displacement vector in this problem has the form:

$$\mathbf{u}_f^{\mathrm{T}} = (u_1, v_1, u_2, v_2, u_3, v_3). \tag{23.38}$$

This vector is ordered: the first two components of it are the DOFs of the structure. The subvectors of the corresponding vector of external forces (23.33) have the form:

$$\mathbf{P} = \begin{pmatrix} P \\ 0 \end{pmatrix}; \quad \mathbf{X} = \begin{pmatrix} P - R_{2h} \\ R_{2v} \\ -R_{3h} \\ R_{3v} \end{pmatrix}.$$

Stiffness matrix $\mathbf{K}$ of the free body in the given problem coincides with element stiffness matrix $\mathbf{K}_1$. Since there is no need to determine support reactions, it is sufficient to calculate only block $\mathbf{K}_{11}$ of matrix $\mathbf{K}$, and not calculate all its constituents. In order to establish matrix $\mathbf{K}_1$, in the FEM algorithm, firstly, the relationship matrix $\mathbf{B}_1$ of the form (23.13) is computed, and then formula (23.8) is employed. Rewrite the latter formula indicating the FE's number in the index:

$$\mathbf{K}_1 = V_1 \mathbf{B}_1^{\mathrm{T}} \mathbf{D} \mathbf{B}_1.$$

Further math treatment is fairly obvious:

$$\Delta_1 = a_2; \quad \mathbf{B}_1 = \Delta_1^{-1} \begin{pmatrix} 0 & 0 & -a & 0 & a & 0 \\ 0 & a & 0 & -a & 0 & 0 \\ a & 0 & -a & -a & 0 & a \end{pmatrix} = a^{-1} \begin{pmatrix} 0 & 0 & -1 & 0 & 1 & 0 \\ 0 & 1 & 0 & -1 & 0 & 0 \\ 1 & 0 & -1 & -1 & 0 & 1 \end{pmatrix};$$

$$\mathbf{DB}_1 = \frac{Ea^{-1}}{1-\nu^2}\begin{pmatrix} 1 & \nu & 0 \\ \nu & 1 & 0 \\ 0 & 0 & \dfrac{1-\nu}{2} \end{pmatrix}\begin{pmatrix} 0 & 0 & -1 & 0 & 1 & 0 \\ 0 & 1 & 0 & -1 & 0 & 0 \\ 1 & 0 & -1 & -1 & 0 & 1 \end{pmatrix}$$

$$= \frac{Ea^{-1}}{1-\nu^2}\begin{pmatrix} 0 & \nu & -1 & -\nu & 1 & 0 \\ 0 & 1 & -\nu & -1 & \nu & 0 \\ \dfrac{1-\nu}{2} & 0 & -\dfrac{1-\nu}{2} & -\dfrac{1-\nu}{2} & 0 & \dfrac{1-\nu}{2} \end{pmatrix}.$$

Next, the matrices $\mathbf{B}_1^{\mathrm{T}}$ and $\mathbf{DB}_1$ are multiplied. Omitting the elements of matrices not needed for calculation of block $\mathbf{K}_{11}$, we get:

$$\mathbf{K} = \mathbf{K}_1 = V_1\mathbf{B}_1^{\mathrm{T}}\mathbf{DB}_1 = \frac{a^2t}{2}\cdot\frac{Ea^{-2}}{1-\nu^2}\begin{pmatrix} 0 & 0 & 1 \\ 0 & 1 & 0 \\ & \vdots & \end{pmatrix}\begin{pmatrix} 0 & \nu & \cdots \\ 0 & 1 & \\ \dfrac{1-\nu}{2} & 0 & \end{pmatrix}$$

$$= \frac{Et}{2(1-\nu^2)}\begin{pmatrix} \dfrac{1-\nu}{2} & 0 & \vdots \\ 0 & 1 & \\ \cdots & & \ddots \end{pmatrix}.$$

We have the stiffness matrix of the structure and the inverse of it:

$$\mathbf{K}_{11} = \frac{Et}{2(1-\nu^2)}\begin{pmatrix} \dfrac{1-\nu}{2} & 0 \\ 0 & 1 \end{pmatrix};\quad \mathbf{K}_{11}^{-1} = 2\frac{1-\nu^2}{Et}\begin{pmatrix} \dfrac{2}{1-\nu} & 0 \\ 0 & 1 \end{pmatrix}.$$

Further, we calculate the displacements by formula (23.36):

$$\begin{pmatrix} u_1 \\ v_1 \end{pmatrix} = \mathbf{K}_{11}^{-1}\mathbf{P} = 2\frac{1-\nu^2}{Et}\begin{pmatrix} \dfrac{2}{1-\nu} & 0 \\ 0 & 1 \end{pmatrix}\begin{pmatrix} P \\ 0 \end{pmatrix} = 2\frac{1-\nu^2}{Et}\begin{pmatrix} \dfrac{2}{1-\nu}P \\ 0 \end{pmatrix} = 4\frac{1+\nu}{Et}\begin{pmatrix} P \\ 0 \end{pmatrix}.$$

In the given problem, formula (23.15) appears to be the identity:

$$\mathbf{u}_1 = \mathbf{u}_f = (u_1, v_1, 0, 0, 0, 0)^{\mathrm{T}}.$$

The strain state is obtained through the found displacements:

$$\begin{pmatrix} \varepsilon_x \\ \varepsilon_y \\ \gamma_{xy} \end{pmatrix} = \mathbf{B}_1\mathbf{u}_1 = a^{-1}\begin{pmatrix} 0 & 0 & \\ 0 & 1 & \cdots \\ 1 & 0 & \end{pmatrix}\begin{pmatrix} u_1 \\ v_1 \\ \mathbf{O} \end{pmatrix} = a^{-1}\begin{pmatrix} 0 \\ v_1 \\ u_1 \end{pmatrix} = \frac{4}{at}\frac{1+\nu}{E}\begin{pmatrix} 0 \\ 0 \\ P \end{pmatrix}.$$

Here, **O** is zero vector of necessary dimension.

Thus, by means of the one-element model, we have determined the strains:

$$\varepsilon_x = \varepsilon_y = 0; \quad \gamma_{xy} = \frac{4}{at}\frac{1+\nu}{E}P.$$

The corresponding stress state is defined by generalized Hooke's law:

$$\sigma_x = \sigma_y = 0; \quad \tau_{xy} = G\gamma_{xy} = \frac{P}{0.5at}.$$

Mark that shear stress is obtained as the ratio of the load upon node 1 to the area of the horizontal section in the middle of the wedge.

The reader is recommended to make a complete calculation of the FE model in Figure 23.4b; namely: to fully write out stiffness matrix **K** and determine the support reactions from the second equation (23.35). Note that in this problem, the analysis by means of one element though is illustrative but doesn't give results applicable in practice.

SUPPLEMENTS TO CHAPTER

Differentiation of Quadratic Form

We prove that the vector of partial derivatives of quadratic form:

$$y(\mathbf{x}) = \mathbf{x}^{\mathrm{T}}\mathbf{A}\mathbf{x}$$

with any matrix **A** of order n has the form:

$$\frac{\partial y}{\partial \mathbf{x}} = (\mathbf{A} + \mathbf{A}^{\mathrm{T}})\mathbf{x}. \tag{23S.1}$$

Indeed, for any given $k = \overline{1,n}$, it holds:

$$y = \sum_{\substack{i\neq k\\ j\neq k}} A_{ij}x_ix_j + \sum_{j\neq k} A_{kj}x_kx_j + \sum_{i\neq k} A_{ik}x_ix_k + A_{kk}x_k^2.$$

Therefore, for arbitrary k, we get:

$$\frac{\partial y}{\partial x_k} = \sum_{j\neq k} A_{kj}x_j + \sum_{i\neq k} A_{ik}x_i + 2A_{kk}x_k = (\mathbf{A}\mathbf{x})_k + (\mathbf{A}^{\mathrm{T}}\mathbf{x})_k,$$

that can be written in the form (23S.1).

Perfectly Rigid Bodies as Part of a Structure

In FEM simulation, it is sometimes necessary to take into consideration the presence of perfectly rigid bodies (PRB) working together with elastic elements of a structure. The model of PRB is an aggregate of an elastic body's nodes which is displaced as an entity, with no mutual displacements of nodes. The requirement, imposed upon displacements of an aggregate of nodes being a whole entity, can be represented in the form of a constraint on the ordered displacement vector:

$$\mathbf{A}^{\mathrm{T}}\mathbf{u}_f = 0, \tag{23S.2}$$

where **A** is the constraint matrix having dimension $2n_1 \times k$ (k is a number of scalar conditions for nodal displacements). Example: if the nodes No. 1 and 2 of a plane body's FE model are linked with a rigid bar, then the scalar condition of invariant distance between them is of the form:

$$(x_2 - x_1)(u_2 - u_1) + (y_2 - y_1)(v_2 - v_1) = 0.$$

Hereinafter, we set up the global equation of FEM in a more complicated problem, when, besides conditions of attachment to the earth (23.25), there is vector constraint (23S.2) upon displacements. While considering a system removed from the bearings, the function of total potential energy $U = U(\mathbf{u}_f)$ is represented as before in the form (23.30). In the state of stable equilibrium, the displacements are determined by the point of minimum of the function $U(\mathbf{u}_f)$ under condition (23S.2). We find this point by Lagrange's method of undetermined multipliers. Set up the Lagrangian function in the form:

$$L = \frac{1}{2}\mathbf{u}_f^{\mathrm{T}}\mathbf{K}\mathbf{u}_f - \mathbf{F}^{\mathrm{T}}\mathbf{u}_f - \boldsymbol{\lambda}^{\mathrm{T}}\mathbf{A}^{\mathrm{T}}\mathbf{u}_f, \tag{23S.3}$$

where $\boldsymbol{\lambda}$ is the vector of undetermined multipliers. Differentiating this function with respect to $\mathbf{u}_f$, we obtain the equilibrium conditions:

$$\begin{cases} \mathbf{K}\mathbf{u}_f - \mathbf{F} - \mathbf{A}\boldsymbol{\lambda} = 0; \\ \mathbf{A}^{\mathrm{T}}\mathbf{u}_f = 0. \end{cases} \tag{23S.4}$$

The unknowns here are the vectors $\mathbf{u}_f$, $\boldsymbol{\lambda}$, and also the components of subvector **X** belonging to vector **F** of the form (23.33).

The equation system (23S.4) is to be solved on condition $\mathbf{u}_X = 0$. By using the block structure of the stiffness matrix, we come to the set of equations:

$$\begin{cases} \begin{pmatrix} \mathbf{K}_{11} & \mathbf{K}_{12} \\ \mathbf{K}_{21} & \mathbf{K}_{22} \end{pmatrix}\begin{pmatrix} \mathbf{u} \\ \mathbf{u}_X \end{pmatrix} - \begin{pmatrix} \mathbf{P} \\ \mathbf{X} \end{pmatrix} - \mathbf{A}\boldsymbol{\lambda} = 0; \\ \mathbf{u}_X = 0; \\ \mathbf{A}^{\mathrm{T}}\begin{pmatrix} \mathbf{u} \\ \mathbf{u}_X \end{pmatrix} = 0. \end{cases} \tag{23S.5}$$

The obtained set is similar in form to the system of equations (23.34) and (23.25) and can be solved in a similar way: firstly, determine the unknowns **u** and $\boldsymbol{\lambda}$, then vector **X**, comprising support reactions.

24 Quadrilateral Finite Element of Plane Body

24.1 APPROXIMATION OF DISPLACEMENT FIELD BY SQUARE FE OF PLANE BODY; SHAPE FUNCTIONS OF DISPLACEMENT FIELD

The main shortcoming of triangular FE in modeling a plane body is the slow convergence of simulated SSS characteristics to the exact solution of an elastic problem, i.e., the FEs of very small size should be employed to get a satisfactory analysis. To increase the accuracy of calculations under the restriction of the number of nodes in a model, a quadrilateral FE could be used.

To start with, we give the derivation of relations which define a square FE. We specify the side of the square by the quantity $2a$. In order to represent displacements in this element, we introduce *natural element CS*, which is determined as rectangular CS $\xi\eta$ with an origin in the center of the square and axes parallel to the sides of the element (Figure 24.1). The purpose of FE is to represent the real field of displacements in the boundaries of FE through nodal displacements of an element with the greatest possible precision. In doing so, it is important that the approximation of the field have an error which tends to zero as the size of the FE decreases. We denote the displacement field in a plane body as $\phi = \phi(x, y)$, where the function arguments are global coordinates. Function ϕ can be one of displacements u or v and will be considered either in a global CS of a solid body or an element CS (we use denotation ϕ for this function with no respect to CS wherein it is represented). The approximate field function constructed on square FE mesh of pitch $2a$ is denoted $\phi^{(e)} = \phi^{(e)}(x, y)$. The convergence of approximation error to zero on the planar closed domain Ω, which a body occupies, can be written in the form:

$$\max_{(x,y)\in\Omega}\left(\phi(x,y)-\phi^{(e)}(x,y)\right)\xrightarrow[a\to 0]{}0. \tag{24.1}$$

It can be proved that under the continuity of field ϕ, the sufficient condition of convergence (24.1) is the next inequality, which should hold for every FE of the model:

$$\min_{\xi,\eta}\phi(\xi,\eta)\le\phi^{(e)}(\xi,\eta)\le\max_{\xi,\eta}\phi(\xi,\eta). \tag{24.2}$$

Here, the minimum and maximum are taken in the confines of the FE; arguments ξ and η of function $\phi^{(e)}$ accept arbitrary values in the segment $[-a, a]$. In plain words, the magnitudes of an approximating function within the bounds of every FE must fall in the range of values of the true field function.*

In addition to the precision requirement upon representation of displacements, it is natural to demand that the approximating function for displacements secure the continuity of displacements on the lines of the FE mesh, i.e., on passing through the border separating adjacent elements, the jump of displacement won't arise. Owing to this, the continuity of a material medium is ensured in the FE model. The meaning of this requirement in finite element analysis is that by the use of it, one proves the convergence of the FEM solution to a true solution of the elastic problem as the diameter (maximum size) of the FE decreases.

* The convergence (24.1) follows from condition (24.2) due to uniform continuity of a field function.

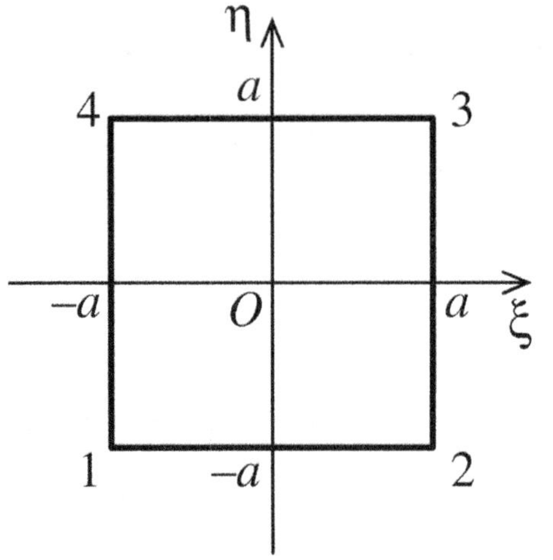

FIGURE 24.1 Natural CS and numbering of nodes in square FE.

In order to construct a sufficiently simple and exact approximation of a field on a square FE mesh with the pitch of $2a$, the requirements to approximating function $\phi^{(e)} = \phi^{(e)}(x, y)$ are usually applied as follows:

- At the nodal points, the magnitudes of given function ϕ and its approximation $\phi^{(e)}$ coincide;
- $\phi^{(e)}$ is a polynomial in natural coordinates ξ, η within the boundaries of each FE;
- The values of function $\phi^{(e)}$ in the confines of an arbitrary FE on a given mesh fall in the range of magnitudes of field function at the nodes of element:

$$\min_i \phi_i \le \phi^{(e)}(\xi, \eta) \le \max_i \phi_i, \tag{24.3}$$

where for every nodal point (ξ_i, η_i), we relate the denotation $\phi_i \equiv \phi(\xi_i, \eta_i)$, $i = \overline{1,4}$;

- The approximation function is continuous at the lines of the FE mesh;
- The order of the approximating polynomial $\phi^{(e)}(\xi, \eta)$ is the least which meets the previous requirements for this function.

The requirement (24.3) ensures the condition of convergence (24.2).

For a square FE, the approximating polynomial must have the second or higher order to satisfy the condition $\phi^{(e)} = \phi$ at the four nodal points, subject to an arbitrary field. This polynomial is entered as a bilinear function of coordinates ξ, η:

$$\phi^{(e)}(\xi, \eta) = \alpha_1 + \alpha_2 a^{-1}\xi + \alpha_3 a^{-1}\eta + \alpha_4 a^{-2}\xi\eta. \tag{24.4}$$

Four unknown coefficients α_i are determined uniquely through magnitudes of field function ϕ_i, $i = \overline{1,4}$, at the nodal points of a square element. The equation system for coefficients of field function is of the form:

$$\phi_i = \alpha_1 + \alpha_2 a^{-1}\xi_i + \alpha_3 a^{-1}\eta_i + \alpha_4 a^{-2}\xi_i\eta_i, \; i = \overline{1,4}.$$

The convenience of a bilinear approximating function is that it inherently meets the requirement of continuity on the mesh's lines, because this function, at the borders of an element, is represented by the graphs shaped as the line segments connecting the points of the field's magnitudes at the nodes. Now, we are to solve this set of equations, establish explicit form of function (24.4), and verify the requirement of convergence (24.3).

We introduce the numbering of an FE's nodes according to the diagram in Figure 24.1. This permits us to express the equation system for unknowns α_i in the matrix form:

$$\begin{pmatrix} \phi_1 \\ \phi_2 \\ \phi_3 \\ \phi_4 \end{pmatrix} = \begin{pmatrix} 1 & -1 & -1 & 1 \\ 1 & 1 & -1 & -1 \\ 1 & 1 & 1 & 1 \\ 1 & -1 & 1 & -1 \end{pmatrix} \begin{pmatrix} \alpha_1 \\ \alpha_2 \\ \alpha_3 \\ \alpha_4 \end{pmatrix}.$$

The solution of this system is easily obtained through the inverse of the coefficient matrix:

$$\begin{pmatrix} \alpha_1 \\ \alpha_2 \\ \alpha_3 \\ \alpha_4 \end{pmatrix} = \frac{1}{4} \begin{pmatrix} 1 & 1 & 1 & 1 \\ -1 & 1 & 1 & -1 \\ -1 & -1 & 1 & 1 \\ 1 & -1 & 1 & -1 \end{pmatrix} \begin{pmatrix} \phi_1 \\ \phi_2 \\ \phi_3 \\ \phi_4 \end{pmatrix}. \tag{24.5}$$

By substituting the obtained coefficients into (24.4), we get the polynomial which can be represented in the next form:

$$\phi^{(e)}(\xi,\eta) = \sum_{i=1}^{4} N_i(\xi,\eta)\phi_i. \tag{24.6}$$

Here, we introduced the functions of natural coordinates:

$$N_i = \frac{1}{4}\left(1 + a^{-2}\xi\xi_i\right)\left(1 + a^{-2}\eta\eta_i\right), \tag{24.7}$$

Functions N_i have the next properties:

$$N_i(\xi,\eta) \geq 0, \quad \sum_i N_i(\xi,\eta) = 1. \tag{24.8}$$

Linear combination of some variables, wherein coefficients are nonnegative and equal in sum to one, is referred to as a weighted sum, and the coefficients of the combination are called weighting coefficients, or weights. If an approximating field function specified in the natural CS of FE is a weighted sum of field magnitudes at an element's nodes, and approximate and true magnitudes of the field coincide at the nodal points, then the weighting coefficients of this sum are referred to as *shape functions of a field in finite element.* From relations (24.6)–(24.8), it follows that functions $N_i(\xi,\eta)$ are the shape functions of a field. The feasibility of representing an approximating function through the shape functions of a field means that the condition of convergence (24.3) is fulfilled.

Remark 1

Approximation (24.6) is built from bilinear function (24.4). The choice of a bilinear function is not random. If we proceed from the polynomial of second order of general form:

$$\phi^{(e)}(\xi,\eta) = \alpha_1 + \alpha_2 a^{-1}\xi + \alpha_3 a^{-1}\eta + \alpha_4 a^{-2}\xi\eta + \alpha_5 a^{-2}\xi^2 + \alpha_6 a^{-2}\eta^2,$$

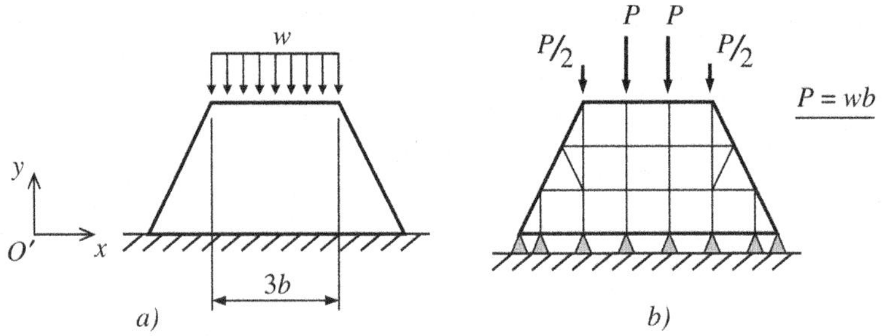

FIGURE 24.2 Trapezoidal body under vertical load (a) and its FE model (b).

then its coefficients are not determined uniquely from the magnitudes of field function ϕ at four nodes. Having obtained linear relations which determine the set of possible vector-functions $\boldsymbol{\alpha}(\phi_1, \ldots, \phi_4)$, $\boldsymbol{\alpha}^T \equiv (\alpha_1, \alpha_2, \ldots, \alpha_6)$, we can narrow this set by imposing the requirement of continuity at the boundaries of FEs. It turns out, that to secure continuity of the approximating function at the boundaries of FEs, coefficients α_5, α_6 must be arbitrary constants. Next, it can be established that at nonzero coefficients α_5 and (or) α_6, requirement (24.3), generally speaking, is not fulfilled; thus, we come to a bilinear function as the only possible form of an approximating function of the second order.

Remark 2

Requirement (24.1) of uniform convergence of an approximating field function to a given function does not, in itself, ensure the convergence of approximate FEM solutions to exact solutions of an elastic problem. The matter is that the proximity of approximate and given functions doesn't imply a small difference between their first derivatives, i.e., the displacement fields might differ a little, whereas the difference between corresponding strains may be noticeable. Nevertheless, the fulfillment of the five requirements stated above ensures required convergence of FEM solutions. The study of problems involved in FEM convergence goes beyond the present course.

Triangular and square FEs can be used in conjunction as, for example, in the case of modeling the trapezoidal body shown in Figure 24.2. The incorporation of square FEs into a model enables us to increase the accuracy of analysis with a reduced number of elements.

24.2 QUADRILATERAL CONVEX FE OF A PLANE BODY: SHAPE FUNCTIONS OF FE AND NATURAL CS

A square FE is used in the construction of an arbitrary quadrilateral convex FE. The approximating field function in a general case of quadrilateral FE is constructed in two stages: initially, the one-to-one correspondence is set between the points of a given quadrilateral and square FE with the size parameter $a = 1$; whereupon, the given field mapped on the square FE is approximately represented by this element.

A FE of known type is used for construction of a new, more complex, type of FE not only in the case of quadrilateral FEs for solving planar problems of the elasticity theory. In the similar cases, an FE of known type, upon which the field of displacements is mapped from the domain of an FE of a new type, we call a *template element*, while we call an element of a new type a *generalized FE*. A type of generalized FE includes a type of template FE.

We specify the coordinates of a point belonging to generalized quadrilateral FE in arbitrary Cartesian CS xy; we consider the square template element in natural CS $\xi\eta$. For definiteness, CS xy is taken as a global one. While approximating the field in a generalized element by means of the template element, the global coordinates are replaced by natural coordinates, to which end

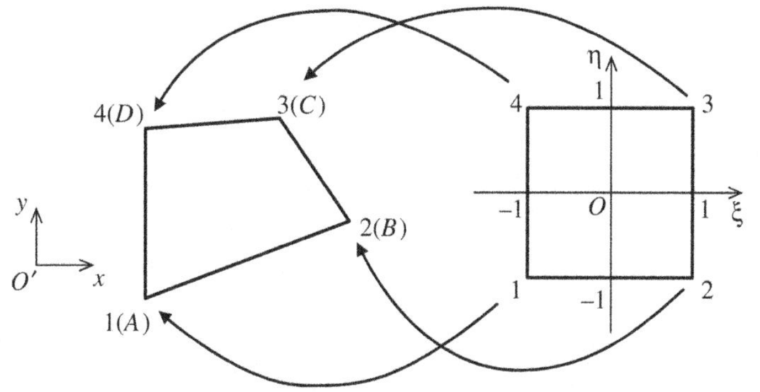

FIGURE 24.3 Mapping of nodes belonging to square template FE upon nodes of generalized quadrilateral FE.

the mapping $(x, y)^T = \boldsymbol{r}(\xi, \eta)$ must be specified. To ensure the continuity of strains in the limits of FE as the functions of coordinates, **we confine ourselves to considering continuously differentiable mapping $\boldsymbol{r}(\xi, \eta)$. Besides, this mapping should be represented in analytical form, and its parameters should be the coordinates of nodes of a generalized FE.**

If in the mapping $\boldsymbol{r}(\xi, \eta)$ of the template's domain upon a generalized FE's domain, each global coordinate is represented by the weighted sum of nodal coordinates, and a node is mapped to a node, then the weighting functions of this sum are referred to as *shape functions of generalized FE*. In generalized quadrilateral FE, the coordinates are converted by means of field shape functions (24.7) at $a = 1$. Further, we number the vertices of a convex quadrilateral (nodes of generalized FE) counterclockwise as shown in Figure 24.3, wherein the alternative letter designation is given in parentheses for every vertex. Global nodal coordinates in generalized FE will be denoted as x_i, y_i, $i = \overline{1,4}$. The transformation of natural coordinates into global ones has the form:

$$x(\xi,\eta) = \sum_{i=1}^{4} N_i(\xi,\eta)x_i; \;\; y(\xi,\eta) = \sum_{i=1}^{4} N_i(\xi,\eta)y_i. \tag{24.9}$$

The image of any node of the square template in this mapping is the node of generalized FE with the same number (Figure 24.3).

Now, we show that transformation (24.9) maps one-to-one the domains of the template and generalized FE. While making an analysis of this transformation, we shall employ the graphic method. We shall call the vertices formed by three sides of a quadrilateral as adjacent vertices. We call two sides of a quadrilateral not having common endpoints opposite sides. The line segments connecting opposite sides of a quadrilateral are referred to as girdle segments. Further derivation is based on the next sufficiently obvious statement (the property of girdle segments):*

Let two opposite sides of a convex quadrilateral be connected by line segments 1 and 2, and on each side, the distance from the endpoint of segment 2 to the adjacent vertex is greater than the distance from the endpoint of segment 1 to the same vertex. Then segments 1 and 2 do not have common points.

This property is clarified by the diagrams a, b, and c in Figure 24.4. On all the diagrams, segment 1 is denoted $A'D'$, and segment 2 is denoted $A''D''$; the segments connect opposite sides AB and CD of quadrilateral $ABCD$. The location of the segments meets the stipulated conditions: $A''A > A'A$; $D''D > D'D$. Diagram a shows the case wherein the extended intersection of opposite sides of a convex quadrilateral is observed beyond the half-plane which is cut off by the sideline AD and

* The brief (but not rigorous) formulation of this property: If the endpoints of a girdle segment are moved away from corresponding adjacent vertices of a convex quadrilateral, then the girdle segment in its new position doesn't have common points with the same segment in its initial position.

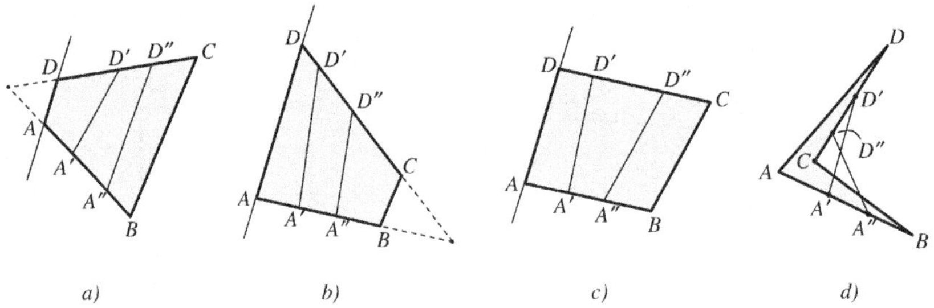

FIGURE 24.4 Girdle segments in quadrilateral: (a), (b), (c) Quadrilateral is convex, girdle segments $A'D'$, $A''D''$ do not intersect; (d) Quadrilateral is nonconvex, segments $A'D'$, $A''D''$ intersect.

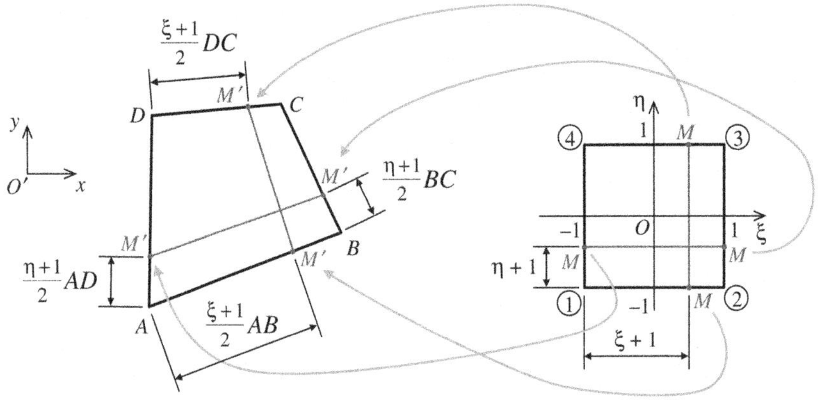

FIGURE 24.5 Mapping the endpoints of coordinate lines in template FE upon generalized FE: M is the denotation of coordinate line endpoints at the borders of template FE; M' is the image of point M in four possible positions.

contains the quadrilateral. Diagram b shows the case wherein the extended opposite sides intersect in the half-plane with the frontier AD, which includes the convex quadrilateral. In the case shown on diagram c, the convex quadrilateral has parallel opposite sides. Lastly, diagram d gives the example of a nonconvex quadrilateral with the analogous positioning of segments 1 and 2. One can see that assertion of the absence of common points of segments 1 and 2 is fulfilled in the case of a convex quadrilateral but may fail for a nonconvex quadrilateral.

So, we return to analysis of conversion (24.9). Since each of functions (24.9) is bilinear, arbitrary coordinate line ξ = const or η = const, limited by template FE, is mapped into some curve having Cartesian coordinates which are linear functions of the natural coordinate (η or ξ, respectively). In other words, the coordinate line of the template FE is mapped upon a generalized FE as a line segment. Since every vertex of the template square is transformed into the vertex of a generalized FE with the same number (Figure 24.3), **the image of every side of the template square is the side of a generalized quadrilateral with the same numbering of the endpoints.**

We establish the image's position for an arbitrary coordinate line in the template with respect to the vertices of a generalized quadrilateral. Firstly, we consider the line ξ = const. We denote the image of arbitrary point $M(\xi, -1)$ at the bottom side of the square as M' (Figure 24.5).The length of segment AM' is linearly dependent on coordinate ξ and thus is proportional to the length of its preimage (i.e., segment "1–M").Therefore, the next proportion holds:

$$\frac{AM'}{AB} = \frac{\xi+1}{2}.$$

The similar proportion can be obtained in the case of positioning point M at the top side of the square; for corresponding mapping $M(\xi, 1) \to M'$, it holds:

$$\frac{DM'}{DC} = \frac{\xi + 1}{2}.$$

These proportions enable us to construct the image of coordinate line ξ = const of the template square in the form of a girdle segment on a generalized quadrilateral as shown in Figure 24.5.

In order to position the image of coordinate line η = const relative to the vertices of the quadrilateral, we will affix the images of point M located both on the left and right sides of the square to these vertices. We have proportions:

$$\frac{AM'}{AD} = \frac{\eta + 1}{2} \quad \text{— for the mapping} : M(-1, \eta) \to M';$$

$$\frac{BM'}{BC} = \frac{\eta + 1}{2} \quad \text{— for the mapping} : M(-1, \eta) \to M'.$$

Just as in the previous case, the corresponding image is represented by the girdle segment (Figure 24.5).

According to the property of girdle segments, two coordinate lines in a generalized quadrilateral which correspond to different values of natural coordinate ξ do not intersect. The same is correct for coordinate lines obtained for different values of η. Therefore, for an arbitrary point of the template square's image, the unique pair of natural coordinates (ξ, η) matches. Consequently, for mapping (24.9), there exists inverse mapping.

For applicability of the given mapping to approximating the field at the domain of a generalized FE, it remains to be proved that the image of the template square coincides with a generalized quadrilateral. This assertion is almost obvious from the fact that, for any point of a generalized quadrilateral, we can construct an arbitrarily exact approximation, which is the node of the nonuniform coordinate mesh obtained by mapping the square reference grid of the template upon a generalized quadrilateral. (To ensure the proximity of the node to the selected point, this reference grid must be sufficiently thick.) The proof of coincidence for named domains is given in the supplement to the chapter.

Unique natural coordinates ξ, η, which determine the global coordinates of the arbitrary point belonging to a generalized quadrilateral in accordance with (24.9), can be considered generalized coordinates and called natural coordinates of this point. To establish natural coordinates of any point in generalized FE, it is convenient to use the coordinate mesh obtained by mapping the square grid of template FE. The example of a nonuniform mesh of natural coordinates on a generalized FE is shown in Figure 24.6. Among reference lines, there are coordinate axes ξ and η mapped from template FE. These images are the *bimedians* of the quadrilateral, i.e., the line segments connecting the midpoints of its opposite sides. An origin of natural coordinates O on the generalized quadrilateral is its *vertex centroid*, i.e., the intersection of two bimedians. Note that point O may differ from the centroid (center of aria) of the quadrilateral.

One of the problems arising in construction of FE models lies in reduction of loads to nodes. The technique of loads' reduction was exposed in the previous chapter using an example of a model with triangular FEs. The reduction procedure is to split the FE ensemble into neighborhoods of nodes and calculate a load upon every node as a sum of loads exerted in the neighborhood of the node. The neighborhoods of nodes in a quadrilateral FE are constructed through the partitioning an element into four parts by means of bimedians of the element as shown in Figure 24.7. (Compare to construction of nodal neighborhoods in the triangular FE in Figure 23.3a.) The nodal neighborhood of an

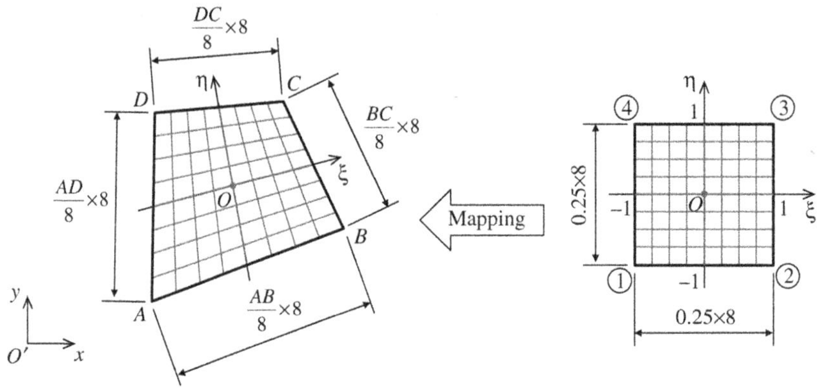

FIGURE 24.6 Mesh of natural coordinates in template and generalized FEs.

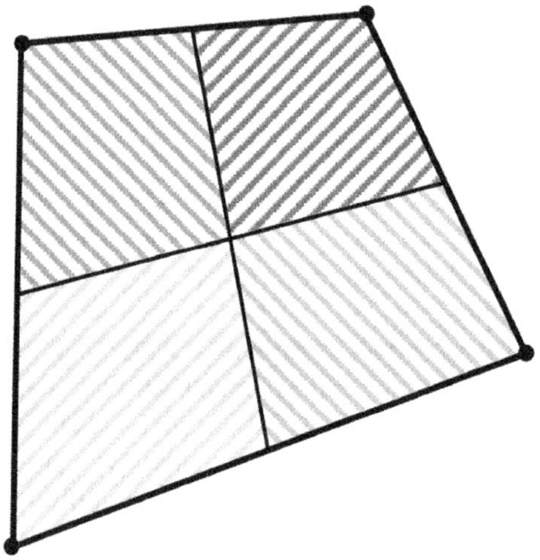

FIGURE 24.7 Neighborhoods of nodes in quadrilateral FE cut off by bimedians (every neighborhood is a quadrilateral emphasized by hatching).

FE model is a union of nodal neighborhoods of elements containing the node. Figure 24.2 shows an example of reducing the load, distributed at the body's outline, to nodes of square FEs.

In conclusion, we name two examples of constructing new types of FEs by using template FE of known type (Kohnke 2001):

- Quadrilateral FE of an elastic shell is constructed by means of a square template. Unlike the quadrilateral element on the plane, considered above, the shell element can be arbitrarily positioned in the space. The insignificant warping of the element's surface is permitted when the nodes of the element do not lie in the same plane, though its edges remain rectilinear;
- A spatial FE in the shape of a convex hexahedron with quadrilateral faces is constructed from a cubic template. Some warping of the element's faces is permitted when the nodes of a face do not lie in the same plane, but its edges remain rectilinear. In this case, the hexahedron may become nonconvex.

In these and similar cases, the position of a point in an FE of new type is specified by means of natural CS, in which the coordinates are established in template FE through the inverse image of the point.

24.3 STIFFNESS MATRIX OF QUADRILATERAL FE OF PLANE BODY

For a quadrilateral FE, we introduce the vector of vertices' displacements as follows:

$$\mathbf{u}^{\mathrm{T}} = (u_1, v_1, u_2, v_2, u_3, v_3, u_4, v_4). \tag{24.10}$$

(Earlier in this chapter, we analyzed the displacement fields along axes $O'x$ and $O'y$ separately and, accordingly, denoted $\phi_i = u_i$ or $\phi_i = v_i$.) The definitions of strain vector $\boldsymbol{\varepsilon}$ and stress vector $\boldsymbol{\sigma}$, given in the previous chapter, remain in force. The components of vectors $\mathbf{u}$, $\boldsymbol{\varepsilon}$, and $\boldsymbol{\sigma}$ are, as before, determined in global CS. In distinction to triangular FE, now vectors $\boldsymbol{\varepsilon}$ and $\boldsymbol{\sigma}$, generally speaking, depend on coordinates in the boundaries of the element.

We introduce for quadrilateral FE the relationship matrix $\mathbf{B}$ such that $\boldsymbol{\varepsilon} = \mathbf{Bu}$ and the symmetric stiffness matrix defined by condition: $2U_1 = \mathbf{u}^{\mathrm{T}}\mathbf{Ku}$. For calculation of the stiffness matrix, formula (23.8*) is employed, which we rewrite in the form:

$$\mathbf{K} = t\int_{\Omega} \mathbf{B}^{\mathrm{T}}\mathbf{DB}dA. \tag{24.11}$$

We replaced integration over the volume in (23.8*) by integration over the area of the middle surface of a body, which is allowed for a body of uniform thickness; Ω is the denotation of the middle surface of an FE.

The calculation of a stiffness matrix is done in three stages:

- The relationship matrix is established as a function of natural coordinates;
- The matrices in integrand (24.11) are multiplied;
- Integral (24.11) is calculated with transformation of the integrand necessary for passing from global coordinates to natural ones.

Below, the calculation procedure is set forth in general form and clarified by the example of a rectangular FE with the sides parallel to the global coordinate axes. For such an FE, it is not a complicated problem to establish the product $\mathbf{B}^{\mathrm{T}}\mathbf{DB}$ in analytical form and calculate the sought integral exactly. In the general case of quadrilateral FE, the product $\mathbf{B}^{\mathrm{T}}\mathbf{DB}$ is computed at fixed points of coordinate plane $\xi\eta$, and the stiffness matrix is computed approximately by means of the Gaussian quadrature rule.

The relationship matrix $\mathbf{B}$ is calculated in the following manner. Approximating fields of displacements and global coordinates are represented as known functions of natural coordinates, which we write down in general form:

$$u = u(\xi, \eta); \quad v = v(\xi, \eta). \tag{24.12}$$

$$x = x(\xi, \eta); \quad y = y(\xi, \eta). \tag{24.13}$$

These functions is parametrical representation of displacement fields $u = u(x, y)$ and $v = v(x, y)$. We introduce a Jacobian matrix for coordinate conversion (24.13):

$$\mathbf{J} = \begin{pmatrix} x'_\xi & x'_\eta \\ y'_\xi & y'_\eta \end{pmatrix}.$$

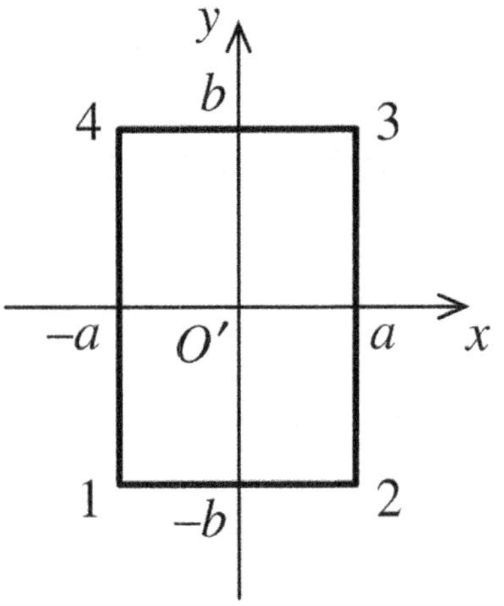

FIGURE 24.8 Rectangular FE.

The determinant of this matrix depends on the global coordinates of the element's nodes x_i, y_i, $i = \overline{1,4}$, and natural coordinates ξ, η, this dependence being linear in variables ξ, η. Taking into account the specific form of coordinate conversion (24.9) and convexity of a generalized FE, it is possible to prove that inside the template square, the Jacobian det **J** is not equal to zero.* The partial derivatives of function $u = u(x, y)$ are determined by derivatives of the functions in parameterization (24.12) and (24.13) as ratios of determinants:

$$u'_x = \frac{\begin{vmatrix} u'_\xi & u'_\eta \\ y'_\xi & y'_\eta \end{vmatrix}}{\det \mathbf{J}}; \quad u'_y = \frac{\begin{vmatrix} x'_\xi & x'_\eta \\ u'_\xi & u'_\eta \end{vmatrix}}{\det \mathbf{J}}. \tag{24.14}$$

Analogous formulas can be written for derivatives v'_x and v'_y. These four partial derivatives determine the strain vector. Each of them is a linear combination of nodal displacements with coefficients dependent on the global coordinates of nodes and natural coordinates ξ, η. These coefficients make up the constituents of matrix **B**. The relationships happen to be cumbersome in general case but relatively simple in the case of a rectangular FE.

Now, we calculate the relationship matrix for the rectangular element with the sides $2a$ and $2b$ (Figure 24.8). In the given case, we have the coordinate conversion: $x = a\xi$, $y = b\eta$, and Jacobian: det **J** = ab. Formulas (24.14) take the form:

$$u'_x = u'_\xi / a; \; u'_y = u'_\eta / b. \tag{24.15}$$

Through differentiation of relation (24.6) wherein shape functions N_i are being taken with parameter $a = 1$, we obtain:

$$\varepsilon_x = u'_x = \frac{1}{4a}\sum_{i=1}^{4}(1+\eta\eta_i)\xi_i u_i; \tag{24.16}$$

---

* By comparing the coordinate mesh in template FE and its mapping upon generalized FE (Figure 24.6), one can see that an elementary square is mapped into an elementary parallelogram. But the ratio of areas of these figures is Jacobian of conversion. Thus, it is clear that det **J** ≠ 0.

$$\varepsilon_y = v'_y = \frac{1}{4b}\sum_{i=1}^{4}(1+\xi\xi_i)\eta_i v_i; \qquad (24.17)$$

$$\gamma_{xy} = u'_y + v'_x = \frac{1}{4b}\sum_{i=1}^{4}(1+\xi\xi_i)\eta_i u_i + \frac{1}{4a}\sum_{i=1}^{4}(1+\eta\eta_i)\xi_i v_i. \qquad (24.18)$$

Mark that these expressions remain correct on the parallel transference of global coordinate axes; therefore, the relationship matrix does not depend on the position of initial point O'. Taking into account the sequence of displacements in the element's DOF vector (24.10), we obtain:

$$\mathbf{B} = \frac{1}{4}\begin{pmatrix} -\frac{1}{a}(1-\eta) & 0 & \frac{1}{a}(1-\eta) & 0 & \frac{1}{a}(1+\eta) & 0 & -\frac{1}{a}(1+\eta) & 0 \\ 0 & -\frac{1}{b}(1-\xi) & 0 & -\frac{1}{b}(1+\xi) & 0 & \frac{1}{b}(1+\xi) & 0 & \frac{1}{b}(1-\xi) \\ -\frac{1}{b}(1-\xi) & -\frac{1}{a}(1-\eta) & -\frac{1}{b}(1+\xi) & \frac{1}{a}(1-\eta) & \frac{1}{b}(1+\xi) & \frac{1}{a}(1+\eta) & \frac{1}{b}(1-\xi) & -\frac{1}{a}(1+\eta) \end{pmatrix}. \qquad (24.19)$$

The constituents of the relationship matrix (as well as Jacobian $\det \mathbf{J}$) depend on both global coordinates of the element's nodes x_i, y_i, $i = \overline{1,4}$, and natural coordinates ξ, η. In the case of a rectangular FE, the dependence on variables ξ, η is linear; in general case, the constituents of the relationship matrix are linear fractional functions of natural coordinates ξ, η, and the denominator of every function is Jacobian $\det \mathbf{J}$.

The next stage is the calculation of matrix product $\mathbf{B}^{\mathrm{T}}\mathbf{DB}$. For arbitrary quadrilateral FE, the elements of matrix $\mathbf{B}^{\mathrm{T}}\mathbf{DB}$ are rational functions of variables ξ, η with a polynomial of second order in the numerator and a squared Jacobian in the denominator. The calculation of these elements in analytical form is laborious. In the case of rectangular FE, this calculation is simplified; we calculate, as an example, two of the first elements in the first row of matrix $\mathbf{B}^{\mathrm{T}}\mathbf{DB}$. The corner element $(\mathbf{B}^{\mathrm{T}}\mathbf{DB})_{11}$ is calculated as follows:

$$(\mathbf{DB})_{*1} = \mathbf{DB}_{*1} = \frac{E}{1-\nu^2}\begin{pmatrix} 1 & \nu & 0 \\ \nu & 1 & 0 \\ 0 & 0 & \frac{1-\nu}{2} \end{pmatrix}\frac{1}{4}\begin{pmatrix} -\frac{1}{a}(1-\eta) \\ 0 \\ -\frac{1}{b}(1-\xi) \end{pmatrix}$$

$$= \frac{E}{4(1-\nu^2)}\begin{pmatrix} -\frac{1}{a}(1-\eta) \\ -\frac{\nu}{a}(1-\eta) \\ -\frac{1-\nu}{2b}(1-\xi) \end{pmatrix};$$

$$(\mathbf{B}^{\mathrm{T}}\mathbf{DB})_{11} = \mathbf{B}^{\mathrm{T}}_{*1}(\mathbf{DB})_{*1} = \frac{E}{16(1-\nu^2)}\left(\frac{1}{a^2}(1-\eta)^2 + \frac{1-\nu}{2b^2}(1-\xi)^2\right). \qquad (24.20)$$

Element $(\mathbf{B}^{\mathrm{T}}\mathbf{DB})_{12}$ is calculated in a similar manner:

$$(\mathbf{DB})_{*2} = \mathbf{DB}_{*2} = \frac{E}{1-\nu^2}\begin{pmatrix} 1 & \nu & 0 \\ \nu & 1 & 0 \\ 0 & 0 & \dfrac{1-\nu}{2} \end{pmatrix}\frac{1}{4}\begin{pmatrix} 0 \\ -\dfrac{1}{b}(1-\xi) \\ -\dfrac{1}{a}(1-\eta) \end{pmatrix}$$

$$= \frac{E}{4(1-\nu^2)}\begin{pmatrix} -\dfrac{\nu}{b}(1-\xi) \\ -\dfrac{1}{b}(1-\xi) \\ -\dfrac{1-\nu}{2a}(1-\eta) \end{pmatrix};$$

$$(\mathbf{B}^{\mathrm{T}}\mathbf{DB})_{12} = \mathbf{B}^{\mathrm{T}}_{*1}(\mathbf{DB})_{*2} = \frac{E}{16(1-\nu^2)}\frac{1+\nu}{2ab}(1-\xi)(1-\eta). \tag{24.21}$$

At the stage of calculating the integral (24.11), it is transformed into natural coordinates and represented in the form of an iterated integral:

$$\mathbf{K} = t\int_{\Omega}\mathbf{B}^{\mathrm{T}}\mathbf{DB}dA = t\iint_{TS}|\mathbf{J}|\mathbf{B}^{\mathrm{T}}\mathbf{DB}d\xi d\eta = t\int_{-1}^{1}\left(\int_{-1}^{1}|\mathbf{J}|\mathbf{B}^{\mathrm{T}}\mathbf{DB}d\xi\right)d\eta. \tag{24.22}$$

Here, *TS* is the denotation of the template square. The transition to natural coordinates requires the multiplication of the integrand upon Jacobian, and this reduces the order of the denominator in every constituent of the matrix under the integral: now denominator is equal to Jacobian, i.e., it is the linear function of variables ξ, η. Further, an arbitrary constituent of the integrand in (24.22) is denoted $R_{ij}(\xi, \eta)$, where indices point out the position of the constituent in the matrix. The problem of finding the elements of the stiffness matrix is solved in the form:

$$K_{ij} = t\int_{-1}^{1}\left(\int_{-1}^{1}R_{ij}(\xi,\eta)d\xi\right)d\eta. \tag{24.23}$$

In the general case, $R_{ij}(\xi, \eta)$ is a rational function with a polynomial of the second order in the numerator and a linear function in the denominator. Integrals (24.23) can be calculated in elementary functions by means of the standard technique of integration. In the case of the rectangular FE in Figure 24.8, there is no need for cumbersome manipulations because all functions $R_{ij}(\xi, \eta)$ are polynomials of the second order. The stiffness matrix of this FE can be established in the form:

$$\mathbf{K} = \frac{Etab}{8(1-\nu^2)}\begin{pmatrix} A_1^K & B_1^K & A_3^K & B_7^K & A_5^K & B_5^K & A_7^K & B_3^K \\ A_2^K & B_2^K & A_8^K & B_4^K & A_6^K & B_6^K & A_4^K & B_8^K \\ A_3^K & B_3^K & A_1^K & B_5^K & A_7^K & B_7^K & A_5^K & B_1^K \\ A_4^K & B_4^K & A_6^K & B_2^K & A_8^K & B_8^K & A_2^K & B_6^K \\ A_5^K & B_5^K & A_7^K & B_3^K & A_1^K & B_1^K & A_3^K & B_7^K \\ A_6^K & B_6^K & A_4^K & B_8^K & A_2^K & B_2^K & A_8^K & B_4^K \\ A_7^K & B_7^K & A_5^K & B_1^K & A_3^K & B_3^K & A_1^K & B_5^K \\ A_8^K & B_8^K & A_2^K & B_6^K & A_4^K & B_4^K & A_6^K & B_2^K \end{pmatrix}, \tag{24.24}$$

where notations are used as follows:

$$\left.\begin{array}{ll}A_1^K=\frac{4}{3}\left(\frac{2}{a^2}+\frac{1-\nu}{b^2}\right); & A_2^K=\frac{1+\nu}{ab};\\ A_3^K=\frac{2}{3}\left(-\frac{4}{a^2}+\frac{1-\nu}{b^2}\right); & A_4^K=\frac{3\nu-1}{ab};\\ A_5^K=-\frac{2}{3}\left(\frac{2}{a^2}+\frac{1-\nu}{b^2}\right); & A_6^K=-\frac{1+\nu}{ab};\\ A_7^K=\frac{4}{3}\left(\frac{1}{a^2}-\frac{1-\nu}{b^2}\right); & A_8^K=\frac{1-3\nu}{ab};\end{array}\right] \tag{24.25a}$$

$$\left.\begin{array}{ll}B_1^K=\frac{1+\nu}{ab}; & B_2^K=\frac{4}{3}\left(\frac{2}{b^2}+\frac{1-\nu}{a^2}\right);\\ B_3^K=\frac{1-3\nu}{ab}; & B_4^K=\frac{4}{3}\left(\frac{1}{b^2}-\frac{1-\nu}{a^2}\right);\\ B_5^K=-\frac{1+\nu}{ab}; & B_6^K=-\frac{2}{3}\left(\frac{2}{b^2}+\frac{1-\nu}{a^2}\right);\\ B_7^K=\frac{3\nu-1}{ab}; & B_8^K=\frac{2}{3}\left(-\frac{4}{b^2}+\frac{1-\nu}{a^2}\right).\end{array}\right] \tag{24.25b}$$

In the case of a non-rectangular FE, the coefficients of polynomials in functions $R_{ij}(\xi, \eta)$ are bulky, and the integration procedure is somewhat complicated. Probably for this reason, in the FEM software, integrals (24.23) are calculated approximately, by using the Gaussian quadrature rule (Kohnke 2001, pp. 14–152, 13–1). Two-point quadrature approximation is employed:

$$\int_{-1}^{1} f(x)dx \approx f\left(-\frac{1}{\sqrt{3}}\right)+f\left(\frac{1}{\sqrt{3}}\right). \tag{24.26}$$

In order to evaluate integral (24.23), this formula is used twice: firstly for integration with respect to variable ξ, and then with respect to variable η. Finally, we obtain the design formula:

$$K_{ij} \approx t\sum_{i=1}^{4} R_{ij}\left(\frac{1}{\sqrt{3}}\xi_i, \frac{1}{\sqrt{3}}\eta_i\right), \tag{24.27}$$

where ξ_i, η_i are the coordinates of the i-th vertex of the template square (Figure 24.3, on the right).

It is known that formula (24.26) gives an exact result when function $f(x)$ is a polynomial of a degree not greater than 3. This condition is secured if integral (24.23) is calculated for the case of the rectangular FE in Figure 24.8. The reader is recommended to calculate constituents K_{11} and K_{12} as iterated integrals (24.23) through substitutions (24.20) and (24.21), and then to verify that formula (24.27) gives identical results.

If the shape of a quadrilateral element differs from rectangular, then quadrature approximation (24.27) contains an error which affects the precision of the SSS evaluation. Besides, for a non-rectangular element, the inherent FEM error may be significant, the latter being not related to approximation (24.27). It is recommended to avoid in FE models the use of quadrilateral FEs wherein the angles between adjacent sides are essentially different from the right angle (it is advisable to keep the range of angles 60°–120°).

SUPPLEMENT TO CHAPTER

Image of Template Square Box upon Generalized Quadrilateral

Theorem

The image of the template square coincides with a generalized quadrilateral.

Proof

We consider an arbitrary point M' in generalized FE and prove that its inverse image exists in the template FE. To do that, we construct the sequence of reference grids in the template square in the following manner: the first grid contains two coordinate lines coinciding with coordinate axes ξ and η, along with the sides of the square; every next grid is produced by addition of intermediate lines to the coordinate lines of the previous grid in such a fashion that the grid pitch is halved. Therefore, the pitch of the n-th grid is equal to 2^{1-n}. Let us map each n-th grid upon a generalized quadrilateral and denote the node of grid's image nearest to point M' as M'_n. Since the diameter of an arbitrary cell of the mapped grid tends to zero as $n \to \infty$, we have: $M'_n \xrightarrow[n\to\infty]{} M'$. It is known that the continuous image of a closed bounded set in Euclidean space is a closed set too. In the given case, the image of the template square is a closed set, and therefore, limit point M' of this image also belongs to it. The theorem is proved.

The correctness of the proved assertion can be established by a more subtle instrument. Namely, this assertion is the corollary of the following theorem, appertaining to the theory of finite-dimensional linear spaces:

The closed subset of E^n, which has a convex boundary and at least one point inside the boundary, is determined by this boundary unambiguously.

A convex boundary is a boundary of some convex set; the point inside the boundary is the internal point of this set. The image of the template square and convex generalized quadrilateral have the same boundary, inside which, for instance, the image of the natural CS's origin is contained. Hence, the named sets coincide.

25 Stiffness Method and Its Implementation for Analysis of Trusses

25.1 DEGREES OF FREEDOM AND STIFFNESS MATRIX OF FREE BAR SYSTEM

In Chapters 17–18, we studied the displacement method, which one may consider a graphic-analytical realization of the finite element method to cover analysis of elastic bar systems. Just like in the calculation of an elastic body by means of FEM, wherein global equations are set up with respect to nodal displacements and the found displacements are used to determine the SSS, in the displacement method, the canonical equation set is composed in unknown displacements of the nodes of a bar structure. By solving it, we can construct the profiles of internal forces in members. In the 1930s, the matrix methods come into use in development of the methods of solving mechanics problems. In 1944, American engineer Gabriel Kron suggested the generalized form of the displacement method, in which tension-compression strains are taken into account as well as deformations of bending, and basic relations are represented in matrix terms. This generalized method was enhanced by German professor of mathematics John Argyris in 1954. In 1959, M. Turner states the generalized displacement method in modern matrix form and calls it as the direct stiffness method. The displacement method in Turner's formulation is known as the *stiffness method* and is the FEM with regard to bar systems.

Whereas the displacement method in its classic forms, known up to the 1940s, enables graphic-analytical calculations, the method stated by Argyris and Turner requires computer-based implementation. Manual calculations by the stiffness method are possible for extremely simple structures and serve the purposes of learning and demonstration.

The stiffness method is based on the notion of a stiffness matrix. According to the definitions given, a stiffness matrix of an FE model of an elastic body is a symmetric matrix for which the quadratic form of the vector of nodal displacements, taken with the coefficient ½, is equal to strain energy of the FE system. In the case of a free body, this matrix determines the vector of external forces exerted upon nodes through the full vector of nodal displacements; in the case of a structure-body, the stiffness matrix determines the vector of loads through the vector of the structure's DOFs (these properties were obtained in Chapter 23 in the form of relation (23.32) and the first equation (23.35), respectively). Now, we introduce the notion of degrees of freedom for a bar system and generalize the definition of a stiffness matrix to such a system.

We refer to a joint of members and also a free or support end of a system's member as a node of a bar system. *The totality of DOFs of a bar system is the totality of independent nodal displacements which determine the SSS of the system subjected to the action of nodal external forces.* We shall consider a plane elastic **free** bar system (for example, a structure removed from bearings) which is subjected to the balanced totality of external forces **exerted on nodes**. The system may be internally unstable, but it is important that connections between members be only of the second and third types (see Chapter 1). We allow the exertion of concentrated external force upon an arbitrary node, specified by the vector; upon a rigid node, besides, we allow the exertion of the force couple specified by the moment's magnitude. Under the action of these forces, the small linear and angular displacements of nodes occur. We specify displacement of a hinged node by the vector of linear displacement, and for a rigid node we also establish an angle of rotation. Therefore, displacement of every hinged node is characterized by two parameters and displacement of every rigid node by three parameters. The displacements of a mixed node, when the hinge is positioned close to a rigid joint, we specify in the same way as for rigid nodes, i.e., by three parameters. These parameters are the DOFs of a system.

We prove the property of independence of a bar system's nodal displacements. This property means that every DOF can take arbitrary value in the neighborhood of zero independently from the others in the equilibrium state of a system. Really, under arbitrary small displacements of a system's nodes, the equilibrium of each of its members is secured owing to the action of some internal forces at the end sections (i.e., sections adjacent to nodes). These internal forces determine the reactions of the corresponding nodes required for equilibrium. The total vector of a node's reaction force determines the external force which should be exerted for equilibrium of the node; and, just the same, the total reaction moment developed by a node determines the external force couple wanted for equilibrium of the node. Owing to external forces determined in this way for every node, the equilibrium of nodes is ensured as well as the equilibrium of members connected by the nodes, i.e., the system keeps equilibrium. It is crucial that there are no connections of the first type in a system under consideration, for perfectly rigid rods do not permit arbitrary displacements of their endpoints.

As an example of finding external forces exerted upon a system's nodes and required for equilibrium under given displacements of nodes, we consider the portal frame in Figure 17.4a (load w is accepted as zero, flexural rigidity EI is one and the same for all members). We establish the external nodal forces wanted for the unit rotation of the frame's node B, as shown in Figure 25.1a. A similar problem was solved in Chapter 18 during the analysis of this frame by the displacement method. Figure 25.1b and c shows the action of external and internal forces upon node B which are required for its equilibrium. At cross-sections of members constituting this node, axial forces are absent because the length of the members is not affected by displacements. Shear forces and bending moments at the given displacement are found by the calculation technique studied in Chapter 18. Vector of external force $\boldsymbol{F}_B$ and force couple M_B are determined from statics equations:

$$M_B = \frac{7EI}{l}; \ \boldsymbol{F}_B = -(\boldsymbol{V}_{BA} + \boldsymbol{V}_{BC}).$$

The reader is recommended to find external forces acting upon nodes A and C in the equilibrium state of the system. Note that in the stiffness method, external nodal forces at given displacements of nodes are determined unambiguously, while in the displacement method, they are determined ambiguously. The latter is related to the negligence of axial deformations of members. For instance, in the problem illustrated by diagrams in Figure 18.2, at the top, we can prevent translational displacements of nodes by entering the additional support rod not at node C but at node B.

The totality of DOFs of a bar system attached to the earth can be obtained from the totality of DOFs of the same system released from support connections. This can be done by removal of all displacements of support nodes in the directions of support connections from the set of nodal displacements of the named free system.

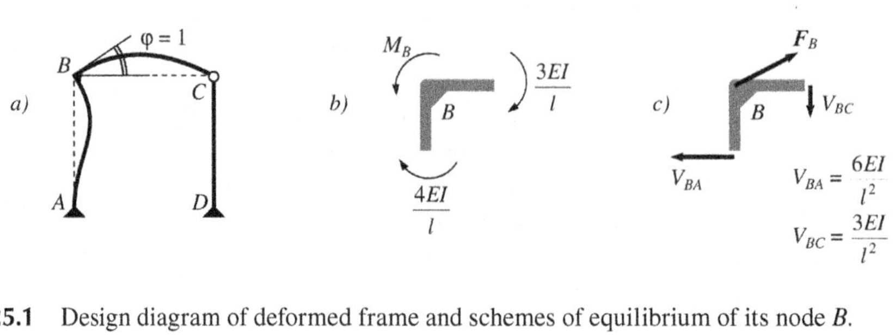

FIGURE 25.1 Design diagram of deformed frame and schemes of equilibrium of its node B.

We refer to a symmetric matrix for which the quadratic form of the DOF vector, taken with the coefficient 1/2, is equal to strain energy of a bar system as a stiffness matrix of a system. The stiffness matrix of a bar structure enables us to establish its SSS. While constructing the stiffness matrix of a structure, the same free system is considered (i.e., the structure removed from bearings), and the stiffness matrix is constructed for it. The stiffness matrix of a free system contains the stiffness matrix of the structure in the form of a submatrix.

Let for every node of a **free** bar system the rectangular Cartesian CS (nodal CS) be assigned, in which both the components of external force exerted upon a node and the components of linear displacement of the node are established. We assume the positive directions of pivot angles and force couples are the same (for instance, counterclockwise). We introduce vector **u** of the system's DOFs, composed of its nodal displacements, and vector **F** constructed from components of external forces and moments exerted upon nodes. Let the sequence of external forces and moments in vector **F** coincides with the sequence of linear and angular displacements in vector **u**. So, each couple of components F_i, u_i is the generalized force and corresponding generalized displacement.

We introduce the vector of a system's DOFs and the vector of external forces exerted on nodes in such a manner that the work done by *constant* forces and moments through displacements of nodes is calculated in the form of a dot product:

$$W = \mathbf{F}^{\mathrm{T}}\mathbf{u}. \tag{25.1}$$

According to Clapeyron's theorem, half this work is the potential energy of deformation of elastic system U_1:

$$U_1 = \frac{1}{2}\mathbf{F}^{\mathrm{T}}\mathbf{u}. \tag{25.2}$$

The factor 0.5 in this formula is obtained at the calculation of the work done by external forces in the quasistatic process of loading, when displacements of nodes rise due to an increase of external forces up to specified magnitudes – in this case, the work of external forces is equal to the accumulated internal energy of system U_1.

For stiffness matrix **K** it holds, according to definition:

$$U_1 = \frac{1}{2}\mathbf{u}^{\mathrm{T}}\mathbf{K}\mathbf{u}. \tag{25.3}$$

The potential energy of a deformation of an elastic system can't be negative at arbitrary displacements. Besides, at the displacements of a system as a whole this energy is zero. So, quadratic form (25.3) is nonnegative and can be equal to zero at $\mathbf{u} \neq 0$. Therefore, **the stiffness matrix of a free bar system is positive semi-definite but is not positive definite and thus singular.**

In Chapter 23, we stated the principle of minimum total potential energy, according to which the equilibrium of a conservative system is attained at the stationary point of total potential energy. The potential energy of a bar system in external field U_2 is equal to work (25.1) with the opposite sign. Finding the stationary point of total potential energy represented as the next function of displacements:

$$U = U_1 + U_2 = \frac{1}{2}\mathbf{u}^{\mathrm{T}}\mathbf{K}\mathbf{u} - \mathbf{F}^{\mathrm{T}}\mathbf{u},$$

we obtain the vector of forces exerted upon nodes of a bar system in equilibrium as follows:

$$\mathbf{F} = \mathbf{K}\mathbf{u}. \tag{25.4}$$

Relation (25.4) enables us to establish a physical meaning of a stiffness matrix: **any column-vector $\mathbf{K}_{*j}$ of stiffness matrix K of a free bar system is the vector of external forces which corresponds to displacement $u_j = 1$ while other nodal displacements are zero.**

From the latter affirmation follows the defining property of a stiffness matrix: **matrix K, which determines the vector of external forces through displacement vector in the form (25.4), is a stiffness matrix**. Really, if equality (25.4) is fulfilled, then constituent K_{ij} is the component F_i of the external forces' vector, under which single nonzero displacement $u_j = 1$ is possible. Out of this, in accordance with Rayleigh's first theorem,* it follows that matrix **K** is symmetric. By substitution of (25.4) into (25.2), we come to expression (25.3) which determines the stiffness matrix.

From the physical meaning of the stiffness matrix also follows the equilibrium property for its constituents: **any column-vector of a stiffness matrix defines the balanced assemblage of external forces exerted upon the nodes of the system**. This means that in plane free bar systems, the components of the arbitrary column-vector of the stiffness matrix satisfy two conditions of force balance and one condition of moment equilibrium.

Note

If connections of the first type are allowed in a system, then nodal displacements in the totality of displacements, as was introduced above, lose independence; consequently, symmetric matrix **K**, which ensures the equality (25.3) for strain energy of system, does not ensure equality (25.4).

25.2 THE STAGES OF CALCULATION IN STIFFNESS METHOD; MEMBER OF TRUSS AS FINITE ELEMENT

In the present chapter, we study the stiffness method with regard to trusses; in the next chapter this method will be generalized for frames. For an arbitrary bar system, the stiffness method is implemented in three stages:

- In a structure given for analysis, the exertion of loads is possible both upon nodes and upon some members out of nodes. If some members support out-of-node loads (intermediate loads), then, at the first step of analysis, they are reduced to nodes. Therefore, a major part of calculation is done for a structure which is subjected to nodal loads only;
- For the system released from bearings, the stiffness matrix is constructed. By using this stiffness matrix, the equations of the structure's equilibrium are set up in the form specific for the method. Through solving these equations, one establishes the displacements of nodes due to nodal loads and support reactions due to loads specified in the problem;
- Through the displacements of nodes, the internal forces in all members are determined with taking into account the action of intermediate loads.

In order to construct the stiffness matrix of a bar system, we have to obtain the stiffness matrices of its subsystems (individual members), and then to transform and combine them into the stiffness matrix of system.

In the statement of a problem, the position of a structure in space is specified in the reference system, which is called global CS. With respect to global CS, the subsidiary CSs are specified, which are nodal CSs and element CSs. An element CS is introduced for every member and enables us to operate with forces and displacements while the member's stiffness matrix is constructed. Nodal CS is used for representation of forces exerted on a node and displacements of the node while the system's stiffness matrix is constructed.

* Rayleigh's theorem establishes the relation between constraint reactions in a structure, whereas in the given case, we consider external forces acting upon a free system. In order to apply Rayleigh's theorem, one should attach this free system to the earth by eliminating all nodal displacements by support links, and then analyze support reactions induced by support shifts.

Further, we expose the FEM with regard to analysis of a plane truss, i.e., a bar system wherein the nodes are cylindrical hinges and connections of the first type are possible only to the earth. We represent every member of the truss as a finite element. This means that we have to obtain the relations determining the SSS of the hinged member, caused by displacements of its nodes – in the given case, the endpoints upon which only the axial forces are exerted (see definition of FE in Chapter 23). These relations are constructed by means of the element's stiffness matrix.

For tensed-compressed member *AB*, we introduce the element rectangular CS Ax_ey_e such that the coordinate plane coincides with the plane of loading of a structure comprising this element, and axis Ax_e coincides with the axial line of member (Figure 25.2). We point out the denotations of nodes by subscripts at the displacements and coordinates of nodes; vectors of forces acting on the member through end sections *A* and *B* we denote, respectively, $\boldsymbol{F}_A$, $\boldsymbol{F}_B$. Let us construct the stiffness matrix for this member in the element CS. If we adhere to the common rule of construction of a bar system's stiffness matrix, the vector of the element's DOFs is introduced in the form:

$$\mathbf{u}^{\mathrm{T}} = (u_{eA}, v_{eA}, u_{eB}, v_{eB}),$$

where subscript "*e*" indicates that displacements are determined in the element CS. Forces acting on the member through the nodes in the directions of the pointed out displacements are represented by vector:

$$\mathbf{F}^{\mathrm{T}} = (F_{Ax_e}, F_{Ay_e}, F_{Bx_e}, F_{By_e}).$$

The stiffness matrix links these two vectors by relation (25.4). But in the given case, the wanted matrix will be sparse, i.e., most of its constituents will be equal to zero. This is because the equilibrium of a tensed-compressed member in the chosen CS is possible only under condition: $F_{Ay_e} = F_{By_e} = 0$. In order to exclude zeroth constituents from consideration, we use reduced vectors of displacements and forces in the form:

$$\mathbf{u}_{red}^{\mathrm{T}} = (u_{eA}, u_{eB}); \tag{25.5}$$

$$\mathbf{F}_{red}^{\mathrm{T}} = (F_{Ax_e}, F_{Bx_e}). \tag{25.6}$$

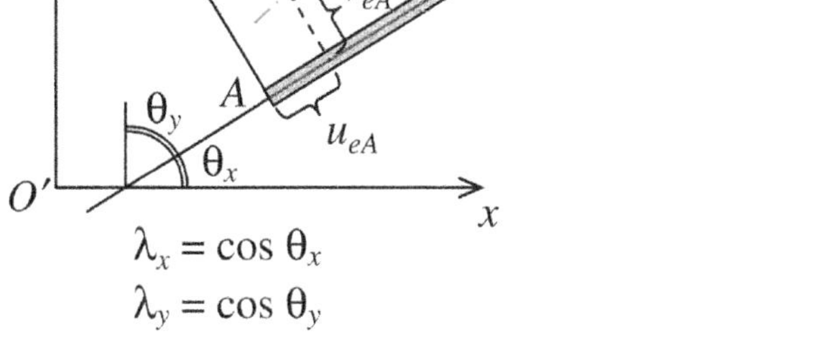

FIGURE 25.2 Hinged member before deformation (continuous contour) and after deformation (dotted contour). $Ax_e\,y_e$ is element CS; $O'xy$ is global CS.

Taking into account the smallness of displacements, internal axial force can be represented through DOFs in the form:

$$N = EA\frac{u_{eB} - u_{eA}}{l}. \tag{25.7}$$

where EA is the axial rigidity of the member, and $l = x_{eB}$ is the length of the member. Vector (25.6) of balanced forces external with respect to the member, which cause this internal force, has the form:

$$\mathbf{F}_{red}^{\mathrm{T}} = (-N, N). \tag{25.8}$$

Two last formulas can be represented in the form:

$$\mathbf{F}_{red} = \mathbf{K}_e \mathbf{u}_{red}, \tag{25.9}$$

where the stiffness matrix of the element has been introduced:

$$\mathbf{K}_e \equiv \frac{EA}{l}\begin{pmatrix} 1 & -1 \\ -1 & 1 \end{pmatrix}. \tag{25.10}$$

Relations (25.9)–(25.10) define the FE "tensed-compressed member." (It can be also termed a "hinged member.")

We consider the truss wherein hinged moveable supports, if present, are parallel to axes of global CS $O'xy$, and loads are exerted on the nodes only. Displacements of all nodes and external forces exerted on a truss we consider in global CS. In order to construct the stiffness matrix of a system, we specify displacements and forces acting upon any element AB of a truss by vectors:

$$\mathbf{u}^{\mathrm{T}} = (u_A, v_A, u_B, v_B); \tag{25.11}$$

$$\mathbf{F}^{\mathrm{T}} = (F_{Ax}, F_{Ay}, F_{Bx}, F_{By}), \tag{25.12}$$

components of which are determined in global CS. We introduce the denotations for direction cosines of coordinate axis Ax_e in global CS as follows (Figure 25.2):

$$\lambda_x \equiv \cos(x\hat{,}x_e); \quad \lambda_y \equiv \cos(y\hat{,}x_e), \tag{25.13}$$

and call the next rectangular matrix the *matrix of element CS's rotation*:

$$\Lambda = \begin{pmatrix} \lambda_x & \lambda_y & 0 & 0 \\ 0 & 0 & \lambda_x & \lambda_y \end{pmatrix}. \tag{25.14}$$

The vector's components of the force acting upon any endpoint of a hinged member are determined in global CS through the components of this force in element CS:

$$F_{x(y)} = F_{x_e}\lambda_{x(y)},$$

whence there follows the conversion:

$$\mathbf{F} = \Lambda^{\mathrm{T}}\mathbf{F}_{red}. \tag{25.15}$$

The component of the vector of endpoint displacement in element CS is determined through the components of this vector in global CS:

$$u_e = u\lambda_x + v\lambda_y,$$

whence we get the conversion:

$$\mathbf{u}_{red} = \Lambda\mathbf{u}. \tag{25.16}$$

Multiply both sides of equality (25.9) by matrix Λ^{T}, writing the multiplier on the left, and make substitutions (25.15) and (25.16). We obtain:

$$\mathbf{F} = \Lambda^{\mathrm{T}}\mathbf{K}_e\Lambda\mathbf{u}.$$

Hence, the stiffness matrix of the FE, which links forces and displacements in accordance with (25.4), in global CS has the form:

$$\mathbf{K} = \Lambda^{\mathrm{T}}\mathbf{K}_e\Lambda. \tag{25.17}$$

Note that conversion (25.17) of some symmetric matrix $\mathbf{K}_e$ by means of arbitrary rectangular matrix Λ keeps the symmetry of the converted matrix.

By using displacement vector (25.11), internal force in a member is calculated as the second component of the reduced vector of external forces exerted upon the member. Formulas (25.8), (25.9), (25.10), and (25.16) enable us to obtain:

$$N = F_{red\,2} = \frac{EA}{l}\begin{pmatrix}-1 & 1\end{pmatrix}\Lambda\mathbf{u}.$$

By calculating the matrix product $\begin{pmatrix}-1 & 1\end{pmatrix}\Lambda$, we come to the design formula:

$$N = \frac{EA}{l}\begin{pmatrix}-\lambda_x & -\lambda_y & \lambda_x & \lambda_y\end{pmatrix}\mathbf{u}, \tag{25.18}$$

which will be required at the final stage of analysis, when displacements of all nodes have been established.

25.3 STIFFNESS METHOD FOR TRUSS ANALYSIS: STRUCTURE STIFFNESS EQUATIONS; CASE OF INCLINED SUPPORTS

By means of stiffness matrices of individual FEs, we can obtain the stiffness matrix of a system released from bearings and set up the equations in displacements and support reactions in a given truss. The solving of these problems mainly repeats the FEM technologies of elastic body analysis presented in Chapter 23.

At first, we consider a structure taken off from the bearings. Let this free – maybe internally unstable – system comprise n_1 nodes. The row of the system's DOFs in global CS has the form:

$$(u_1, v_1, u_2, v_2, \ldots, u_{n_1}, v_{n_1}), \tag{25.19}$$

where the subscript points out the number of the node. In this row, we make permutation in such manner that at the beginning of the row, the displacements which are not eliminated by connections of a structure with the earth, i.e., the DOFs of a **structure**, are positioned. We denote the number of these DOFs n. The displacements eliminated by support connections and equal to zero (of total number $2n_1 - n$) complete the row. The DOF vector of a system with no supports, obtained by such permutation (and subsequent transposition of the row), we denote as $\mathbf{u}_f$ and call an *ordered full vector of a truss's nodal displacements*. For every system's FE, four components can be found in this vector, which are the displacements of the first and second nodes of the FE. We assume that all finite elements (hinged members) of a system are numbered from 1 to m. The vector of DOFs of the i-th FE will be represented in the form (25.11) and denoted $\mathbf{u}_i$. This vector can be obtained by transformation of vector $\mathbf{u}_f$ by using some matrix $\mathbf{I}_i$ as follows:

$$\mathbf{u}_i = \mathbf{I}_i \mathbf{u}_f. \tag{25.20}$$

Matrix $\mathbf{I}_i$ has the dimension $4 \times 2n_1$ and determines the rule of selecting the components of vector $\mathbf{u}_f$ into vector $\mathbf{u}_i$: every row of this matrix has a single nonzero element which equals unity, and all rows are different. We call this matrix a *matrix of DOFs' selection for the i-th FE*. Note that it is convenient to specify transformation (25.20) by using vector $\mathbf{S}_i$ constituted by the numbers of the components of vector $\mathbf{u}_f$ sequentially selected into vector $\mathbf{u}_i$. Under appropriate order of these numbers in vector $\mathbf{S}_i$, the components of vector $\mathbf{u}_i$ are obtained in the form:

$$u_{ik} = u_{fS_{ik}}, \quad k = \overline{1,4}. \tag{25.21}$$

Stiffness matrix $\mathbf{K}$ of a system released from bearings, according to its definition, allows us to represent the strain energy of a system by means of the quadratic form:

$$U_1 = \frac{1}{2}\mathbf{u}_f^{\mathrm{T}}\mathbf{K}\mathbf{u}_f. \tag{25.22}$$

In order to get this matrix, for every system's FE, we introduce an expanded stiffness matrix in the form:

$$\tilde{\mathbf{K}}_i \equiv \mathbf{I}_i^{\mathrm{T}}\mathbf{K}_i\mathbf{I}_i, \tag{25.23}$$

where $\mathbf{K}_i$ is the stiffness matrix of the i-th FE obtained by conversion (25.17). We represent the strain energy of an arbitrary element through the vector of DOFs of a system removed from the bearings:

$$U_{1i} = \frac{1}{2}\mathbf{u}_f^{\mathrm{T}}\tilde{\mathbf{K}}_i\mathbf{u}_f.$$

We obtain the strain energy of a system by summing up the energies of elements:

$$U_1 = \sum_{i=1}^{m} U_{1i} = \frac{1}{2}\mathbf{u}_f^{\mathrm{T}}\left(\sum_{i=1}^{m}\tilde{\mathbf{K}}_i\right)\mathbf{u}_f,$$

whence we get for the sought matrix the design formula:

$$\mathbf{K} = \sum_{i=1}^{m}\tilde{\mathbf{K}}_i. \tag{25.24}$$

Note that expanded matrices have the order of $2n_1$ and every i-th expanded matrix is obtained from the zeroth matrix of order $2n_1$ by replacement of its constituents according to the rule:

$$\tilde{K}_{iS_{ik}S_{il}} = K_{ikl}, \quad k = \overline{1,4}, \quad l = \overline{1,4}. \tag{25.25}$$

This technique of calculation of an expanded matrix is more economical than multiplying the matrices on the right side of (25.23) by the rule "row by column."

In order to set up the equations' set in DOFs of a structure and support reactions, we decompose the obtained stiffness matrix of free system into blocks:

$$\mathbf{K} = \begin{pmatrix} \mathbf{K}_{11} & \mathbf{K}_{12} \\ \mathbf{K}_{21} & \mathbf{K}_{22} \end{pmatrix}, \tag{25.26}$$

where block $\mathbf{K}_{11}$ has the order n. Firstly, we show that submatrix $\mathbf{K}_{11}$ is the stiffness matrix of the structure.

Let us consider the bar system in which the support connections have been restored. The full vector $\mathbf{u}_f$ of nodal displacements in a structure has dimension $2n_1$, whereas the DOFs of the structure are positioned at the beginning of vector $\mathbf{u}_f$ and make up the subvector of dimension n, $n < 2n_1$, which we denote $\mathbf{u}$. Further, there follow zero displacements, which we represent by subvector $\mathbf{u}_X$ of dimension $2n_1 - n$. So, we represent the vector of nodal displacements in a structure in the form:

$$\mathbf{u}_f = \begin{pmatrix} \mathbf{u} \\ \mathbf{u}_X \end{pmatrix}; \tag{25.27}$$

at that, the condition of attachment to the earth has the form:

$$\mathbf{u}_X = 0. \tag{25.28}$$

By means of relations (25.26) and (25.27), we repeat the treatment (23.24). Taking into account condition (25.28), we come to equality:

$$U_1 = \frac{1}{2}\mathbf{u}^{\mathrm{T}}\mathbf{K}_{11}\mathbf{u}, \tag{25.29}$$

from which it follows that $\mathbf{K}_{11}$ is the stiffness matrix of a structure.

The stiffness matrix of a structure is positive definite. Really, in a system stable with respect to the earth, the displacements mandatorily produce strains; in its turn, the strain energy of a system is positive at arbitrary nonzero strains. So, at arbitrary nonzero vector $\mathbf{u}$, the quadratic form (25.29) is positive, which means positive definiteness of the matrix appertaining to this form.

Setting up the equations in unknown displacements and support reactions in a structure lies in the transformation of relation (25.4), which is correct with no regard to a system either free or attached to the earth. In the latter case, the support reactions are included in an assemblage of external forces, and some displacements are known beforehand: namely, the displacements in the directions of support connections are equal to zero.

In order to set up the wanted equation set, we introduce ordered vector $\mathbf{F}$ of external forces exerted upon the system's nodes, in such manner that the sequence of forces' components in this vector coincide with the sequence of displacements in vector $\mathbf{u}_f$. This can be done by making the permutations in the row of external forces

$$(F_{x1}, F_{y1}, F_{x2}, F_{y2}, \dots, F_{xn_1}, F_{yn_1})$$

exactly the same as those required when displacement vector $\mathbf{u}_f$ was constructed from the row (25.19).

We make the separation of the external forces' vector into two subvectors, of which the first has the dimension of the structure's DOF vector $\mathbf{u}$. More precisely, we represent the vector of external forces in the form:

$$\mathbf{F} = \begin{pmatrix} \mathbf{P} \\ \mathbf{X} \end{pmatrix}, \tag{25.30}$$

where **P** is the subvector of loads acting in the directions of displacements possible in the structure, dim $\mathbf{P} = n$ at that; **X** is subvector of external forces comprising reactions.

By using block representation of data (25.26), (25.27), and (25.30), we write down equation (25.4) in the form:

$$\begin{pmatrix} \mathbf{K}_{11} & \mathbf{K}_{12} \\ \mathbf{K}_{21} & \mathbf{K}_{22} \end{pmatrix} \begin{pmatrix} \mathbf{u} \\ \mathbf{u}_X \end{pmatrix} - \begin{pmatrix} \mathbf{P} \\ \mathbf{X} \end{pmatrix} = 0. \tag{25.31}$$

In the last equation, we substitute the condition of attachment to the earth (25.28) and obtain the set of equations:

$$\begin{cases} \mathbf{K}_{11}\mathbf{u} - \mathbf{P} = 0; \\ \mathbf{K}_{21}\mathbf{u} - \mathbf{X} = 0, \end{cases} \tag{25.32}$$

where the unknowns are vectors **u** and **X**. Matrix $\mathbf{K}_{11}$ is not singular as a positive definite matrix, therefore, the first equation (25.32) has a unique solution under given loads; it can be found by known methods, and then established displacements allow us to find support reactions from the second equation. Equations (25.32) are referred to as *structure stiffness equations.* They can be called *equations of bar system's equilibrium in the terms of displacements*, because they define the conditions of equilibrium for such a system according to the principle of minimum total potential energy.

The analysis is completed by determination of internal forces in truss members. For the *i*-th FE, we establish the vector of displacements of the element's nodes in the global CS by using formula (25.20). The axial force is established according to (25.18).

So far, all nodal displacements were considered in global CS; if displacements of every node should be considered in nodal CS, then the presented exposition of the method covers the case when all nodal CSs coincide with global. Now, let the directions of some support rods not coincide with the directions of global coordinate axes, as shown, for example, in Figure 25.3. In this case, for nodes which are the endpoints of inclined support rods, we introduce nodal CSs wherein support rods are directed along one of the coordinate axes; the displacements of other nodes we shall consider in the nodal CSs coinciding with the global CS. All presented formalism remains in force if we

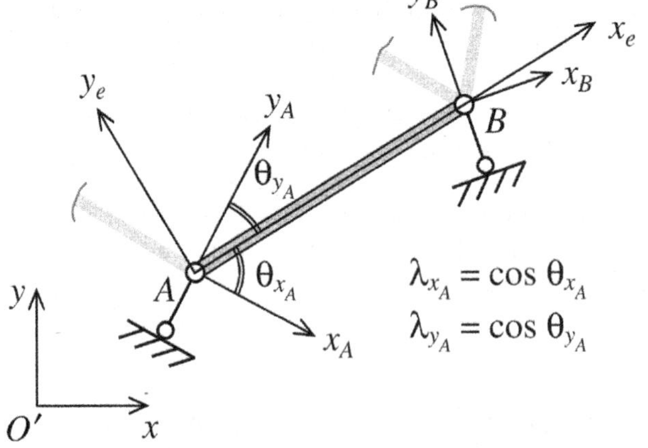

FIGURE 25.3 Member of truss with inclined supports: $O'xy$ is global CS; x_Ay_A, x_By_B are nodal CSs; λ_{x_A}, λ_{y_A} are direction cosines of element axis x_e in nodal CS x_Ay_A.

change design formula (25.14) for a rotation matrix. For every element, we introduce the direction cosines of axis x_e in coordinate systems of the first (A) and second (B) nodes of an element:

$$\lambda_{x_A} \equiv \cos(x_A \hat{,} x_e), \quad \lambda_{y_A} \equiv \cos(y_A \hat{,} x_e);$$
$$\lambda_{x_B} \equiv \cos(x_B \hat{,} x_e), \quad \lambda_{y_B} \equiv \cos(y_B \hat{,} x_e). \tag{25.33}$$

Here $x_{A(B)}$, $y_{A(B)}$ are reference axes of the element's node with a subscript specifying the node. The choice of an origin for every CS is of no importance: this point might be an origin of global CS or, as shown in Figure 25.3, coincide with a reference node. Usually, nodal coordinate axes coincide with global ones, and formulas (25.33) are reduced to (25.13). But in the case when an element's node has inclined support, we employ inclined nodal CS for calculation of direction cosines (25.33) and replace formula (25.14) by the next more general formula for calculation of a rotation matrix:

$$\Lambda = \begin{pmatrix} \lambda_{x_A} & \lambda_{y_A} & 0 & 0 \\ 0 & 0 & \lambda_{x_B} & \lambda_{y_B} \end{pmatrix}. \tag{25.34}$$

All displacements are now determined in nodal CSs. Owing to the orientation of nodal reference axes in the directions of the support rods, each support rod eliminates one DOF of the system, and thus, we can make necessary permutations of zero displacements in the full displacement vector, represent it in the form (25.27), and obtain the equation system of equilibrium in the form (25.32).

The generalized form of rotation matrix (25.34) assumes according generalization of formula (25.18) for calculation of internal forces in members. These forces are now calculated as follows:

$$N = \frac{EA}{l}\begin{pmatrix} -\lambda_{x_A} & -\lambda_{y_A} & \lambda_{x_B} & \lambda_{y_B} \end{pmatrix}\mathbf{u}. \tag{25.35}$$

25.4 EXAMPLE OF TRUSS ANALYSIS BY STIFFNESS METHOD

Further on, in design diagrams, we use the denotations usual in the stiffness method. All the nodes of a structure are numbered and their numbers are circled; all the elements are also numbered, and their numbers are enclosed within rectangles. Nodal reference axes are depicted with their origins at corresponding nodes. If the nodal CS coincides with the global one, then it is not designated on the diagram.

Below, we calculate by the stiffness method the bracket shown in Figure 25.4, made up by an isosceles hinged triangle supported with the connection of first type. All strained members have the same axial rigidity EA; the bracket length l is given.

It is convenient to designate an FE by numbers of nodes, because this way the order of nodes is assigned and the direction of element axis x_e is established. In the given structure, there are three FEs: 1–2; 1–3; and 2–3. We obtain their stiffness matrices. Take into account that the CS of node 3 is rotated 45° with reference to global CS. Substitution of (25.34) into (25.17) gives the matrix:

$$\mathbf{K} = \frac{EA}{l}\begin{pmatrix} \lambda_{x_A}^2 & \lambda_{x_A}\lambda_{y_A} & -\lambda_{x_A}\lambda_{x_B} & -\lambda_{x_A}\lambda_{y_B} \\ & \lambda_{y_A}^2 & -\lambda_{y_A}\lambda_{x_B} & -\lambda_{y_A}\lambda_{y_B} \\ & & \lambda_{x_B}^2 & \lambda_{x_B}\lambda_{y_B} \\ & & & \lambda_{y_B}^2 \end{pmatrix}. \tag{25.36}$$

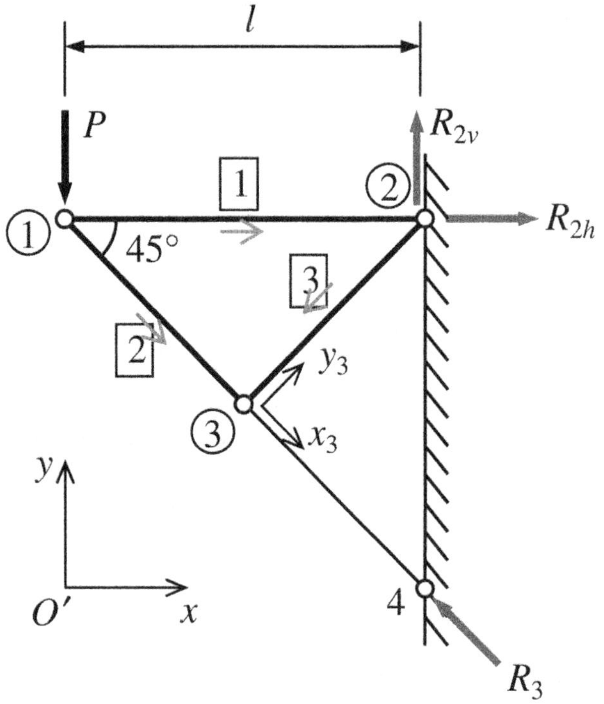

FIGURE 25.4 Design diagram of bracket under loading (arrows near numbers of FEs show the directions of element axis x_e).

For member 1–2 (element 1), we have the direction cosines: $\lambda_{x_A} = \lambda_{x_B} = 1$; $\lambda_{y_A} = \lambda_{y_B} = 0$. Thus, we get the stiffness matrix:

$$\mathbf{K}_1 = \frac{EA}{l}\begin{pmatrix} 1 & 0 & -1 & 0 \\ & 0 & 0 & 0 \\ & & 1 & 0 \\ & & & 0 \end{pmatrix}.$$

For member 1–3 (element 2), the direction cosines: $\lambda_{x_A} = \frac{1}{\sqrt{2}} = 0.707$; $\lambda_{y_A} = -0.707$; $\lambda_{x_B} = 1$; $\lambda_{y_B} = 0$. Stiffness matrix:

$$\mathbf{K}_2 = \frac{EA\sqrt{2}}{l}\begin{pmatrix} 0.5 & -0.5 & -0.707 & 0 \\ & 0.5 & 0.707 & 0 \\ & & 1 & 0 \\ & & & 0 \end{pmatrix} = \frac{EA}{l}\begin{pmatrix} 0.707 & -0.707 & -1 & 0 \\ & 0.707 & 1 & 0 \\ & & 1.414 & 0 \\ & & & 0 \end{pmatrix}.$$

For member 2–3 (element 3) the direction cosines: $\lambda_{x_A} = -0.707$; $\lambda_{y_A} = -0.707$; $\lambda_{x_B} = 0$; $\lambda_{y_B} = -1$. Stiffness matrix:

$$\mathbf{K}_3 = \frac{EA\sqrt{2}}{l}\begin{pmatrix} 0.5 & 0.5 & 0 & -0.707 \\ & 0.5 & 0 & -0.707 \\ & & 0 & 0 \\ & & & 1 \end{pmatrix} = \frac{EA}{l}\begin{pmatrix} 0.707 & 0.707 & 0 & -1 \\ & 0.707 & 0 & -1 \\ & & 0 & 0 \\ & & & 1.414 \end{pmatrix}.$$

We introduce the full displacement vector in the form:

$$\mathbf{u}_f^{\mathrm{T}} = \left(u_1, v_1, v_3, u_2, v_2, u_3\right).$$

The three first components of this vector are the displacements not eliminated by connections. The displacements of node 3 are specified in the inclined nodal CS, and other displacements are specified in nodal CSs coinciding with the global one.

External forces (loads and support reactions) acting in the directions of these displacements make up the vector:

$$\mathbf{F}^{\mathrm{T}} = \left(0, -P, 0, R_{2h}, R_{2v}, -R_3\right).$$

Numbers of components selected into displacement vector $\mathbf{u}_i$ of each element make up the following vectors:

- For member 1–2, $\mathbf{S}_1^{\mathrm{T}} = (1, 2, 4, 5)$;
- For member 1–3, $\mathbf{S}_2^{\mathrm{T}} = (1, 2, 6, 3)$;
- For member 2–3, $\mathbf{S}_3^{\mathrm{T}} = (4, 5, 6, 3)$.

Further, we apply formula (25.25) to get the expanded stiffness matrices. Of the three matrices of the given problem, we consider the constructing of the expanded matrix of the FE No. 3 which has the form:

$$\tilde{\mathbf{K}}_3 = \frac{EA}{l}\begin{pmatrix} 0 & 0 & 0 & 0 & 0 & 0 \\ & 0 & 0 & 0 & 0 & 0 \\ & & 1.414 & -1 & -1 & 0 \\ & & & 0.707 & 0.707 & 0 \\ & & & & 0.707 & 0 \\ & & & & & 0 \end{pmatrix}.$$

The constituent in row 1 of the initial matrix is placed in the row $S_{31} = 4$ of the expanded matrix, and, in a similar way, the constituent in column 1 of the initial matrix is placed in the column $S_{31} = 4$ of the expanded matrix. Therefore, $K_{311} = \tilde{K}_{344} = 0.707$. The fourth diagonal constituent of the initial matrix is placed in the row (column) $S_{34} = 3$ of the expanded matrix. Therefore, $K_{344} = \tilde{K}_{333} = 1.414$. The constituent in row 2 of the initial matrix is placed in the row $S_{32} = 5$ of the expanded matrix, whereas the constituent in column 4 of the initial matrix is placed in the column $S_{34} = 3$ of the expanded matrix. Therefore, $K_{324} = \tilde{K}_{353} = \tilde{K}_{335} = -1$. These three clarifications suffice to verify the correctness of placing the other constituents of the written matrix.

We write down the sum (25.24) pointing out all addends in each cell containing nonzero addends:

$$\mathbf{K} = \frac{EA}{l}\begin{pmatrix} 1+0.707+0 & 0-0.707+0 & 0 & -1+0+0 & 0 & 0-1+0 \\ & 0+0.707+0 & 0 & 0 & 0 & 0+1+0 \\ & & 0+0+1.414 & 0+0-1 & 0+0-1 & 0 \\ & & & 1+0+0.707 & 0+0+0.707 & 0 \\ & & & & 0+0+0.707 & 0 \\ & & & & & 0+1.414+0 \end{pmatrix}.$$

Three addends at any cell are written according to the numbering of expanded matrices. By means of such recording, we can restore the expanded matrix for every FE. It is recommended to restore the expanded matrix of the first or second element and verify that the constituents from the initial matrix $\mathbf{K}_i$ have been correctly arranged in it. Having some practice, one can construct expanded matrices without writing them down separately but instead through filling the wanted stiffness matrix by constituents as shown in the presented example.

In the given problem, we obtained the stiffness matrix of a free hinged triangle as follows:

$$\mathbf{K} = \frac{EA}{l}\begin{pmatrix} 1.707 & -0.707 & 0 & -1 & 0 & -1 \\ & 0.707 & 0 & 0 & 0 & 1 \\ & & 1.414 & -1 & -1 & 0 \\ & & & 1.707 & 0.707 & 0 \\ & & & & 0.707 & 0 \\ & & & & & 1.414 \end{pmatrix}.$$

The system of equilibrium equations (25.32) has the form:

$$\begin{cases} \dfrac{EA}{l}\begin{pmatrix} 1.707 & -0.707 & 0 \\ & 0.707 & 0 \\ & & 1.414 \end{pmatrix}\begin{pmatrix} u_1 \\ v_1 \\ v_3 \end{pmatrix} - \begin{pmatrix} 0 \\ -P \\ 0 \end{pmatrix} = 0; \\ \\ \dfrac{EA}{l}\begin{pmatrix} -1 & 0 & -1 \\ 0 & 0 & -1 \\ -1 & 1 & 0 \end{pmatrix}\begin{pmatrix} u_1 \\ v_1 \\ v_3 \end{pmatrix} - \begin{pmatrix} R_{2h} \\ R_{2v} \\ -R_3 \end{pmatrix} = 0. \end{cases}$$

By solving the system of the first three scalar equations, we get displacements:

$$u_1 = -P\frac{l}{EA}; \quad v_1 = -2.414P\frac{l}{EA}; \quad v_3 = 0.$$

After substitution of these solutions into the equation subsystem in unknown external forces (consisting of three lust equations), we obtain support reactions:

$$R_{2h} = P; \quad R_{2v} = 0; \quad R_3 = 1.414P.$$

Verification of obtained solution is possible from equilibrium conditions: the sum of loads and support reactions in the projection upon each of axes $O'x$, $O'y$ must be equal to zero.

In the given problem, the support reactions determine internal forces in members in an obvious manner. Nevertheless, as an example, we calculate axial force N_{13} by formula (25.35):

$$N_{13} = \frac{EA\sqrt{2}}{l}\begin{pmatrix} -\lambda_{xA} & -\lambda_{yA} & \lambda_{xB} & \lambda_{yB} \end{pmatrix}\begin{pmatrix} u_1 \\ v_1 \\ u_3 \\ v_3 \end{pmatrix}$$

$$= \frac{EA\sqrt{2}}{l}\begin{pmatrix} -0.707 & 0.707 & 1 & 0 \end{pmatrix} P\frac{l}{EA}\begin{pmatrix} -1 \\ -2.414 \\ 0 \\ 0 \end{pmatrix}$$

$$= P\sqrt{2}(0.707 - 2.414 \cdot 0.707) = -P\sqrt{2}.$$

In the example we have taken, the forces in members and support reactions can be found "in mind," with no records. But analytic treatment in stiffness method enables us to solve the problems of analysis of bar systems with no regard to their degree of redundancy or kinematic indeterminacy. Sometimes, this cumbersome technique of analysis happens to be a single instrument of solving. Besides, for computer-based modeling, such a scheme is simple and convenient.

26 Stiffness Method for Analysis of Frames and Generalization of FEM for an Arbitrary Elastic System

26.1 PARTICULARITIES OF FRAME ANALYSIS BY STIFFNESS METHOD: REDUCTION OF INTERMEDIATE LOADS TO NODES; BENDING MEMBER OF FRAME AS FINITE ELEMENT

The frames under consideration in this chapter differ from trusses in that the nodes in them may be not only hinged but rigid and mixed, and, besides, the loads may act upon members directly, beyond the nodes. In the case of exertion of out-of-node loads, the first stage of calculation is to reduce these loads to nodes. The procedure of reduction is as follows.

All nodes of a structure are being immovably fixed by additional constraints: the hinged node should be turned into a hinged immovable support, and the rigid node should be fitted with rigid support (the rigid node may be part of the mixed node). All nodal loads are assumed to be zero, and only intermediate loads are allowed. We come to *a substitute system of the stiffness method*, in which, for every member under load, we find the distribution of internal forces. Internal forces at the end sections of a member determine the reactions of supports fixing this member. The named reactions are to be summed up over all members forming a node in order to establish the total reaction of each support connection. We call the established SSS of a substitute system the state with immovable nodes, or state 1.

From the substitute system, we return to the given system loaded in the special manner. Upon all nodes of the substitute system, we apply loads equal to the reactions of the supports of these nodes. Loads exerted on the support nodes in the directions of support reactions don't affect the SSS of a system; thus, the substitute system will keep state 1. But now, all the support reactions of this system become zero, for every support node will keep balance without interaction with the earth. Thus, we can remove all additional supports from the system, keeping its SSS unchanged. We come to the given system, which is subjected to given intermediate loads and nodal loads equal to support reactions in the substitute system. We refer to this loading of the system as loading with immovable nodes, or loading 1. **Under loading 1, the members of the system keep state 1, and all support reactions are equal to zero.**

Next, we consider a given system under action of nodal loads obtained by the following correction of the loading of the problem being solved: instead of out-of-node loads, we apply the loads upon system's nodes, which are equal to the support reactions in the substitute system considered above, but oppositely oriented. These additional loads can be referred to as loads reduced to nodes. They act together with given nodal loads. We call such loading of a system loading adapted to nodes, or reduced to nodes, or loading 2; the corresponding SSS of the system we call state 2. The procedure of reduction is finished by construction of design scheme of a structure under loading 2.

If, for a system affected by loading 2, the distribution of internal forces in members has been determined, then the true distribution of internal forces can be obtained by means of the superposition principle. The load upon an arbitrary node of a system affected by loading with immovable nodes, added to the load reduced to the same node, produces zero external force upon the node. Thus, the superimposing of loadings 1 and 2 results in action of given loads only. So, the summation of internal forces in the first and second states of a system enables us to obtain the distribution of internal forces in the given system due to given loads.

Because the support reactions in the state of loading 1 are equal to zero, **the support reactions in the state of loading 2 are equal to the support reactions in the given system at the given loading**.

The analysis of a substitute system in state 1 is simple enough and usually lies in the construction of the loaded diagram of bending moments by the displacement method, and also in construction of the shear force diagram by using the statics equations (see Chapters 18 and 14, respectively). The main part of calculations is done for loading 2, where the FEM is employed.

Figure 26.1 shows an example of reducing intermediate loads to nodes for the frame comprising mixed (*A*), rigid (*B*), and hinged (*C*) nodes. Diagram b illustrates both a substitute system and a given system under loading 1. External forces denoted as $R_{\bullet}$ cancel out when loading 1 and 2 are superimposed. The result of superposition is the loading in accordance with the design diagram. Note that loads upon node *D*, acting in state 2, do not influence the SSS of the system, but these loads secure coincidence of support reactions under loading 2 and given loading.

The second stage of analysis is to determine the nodal displacements of a frame in which the loads have been reduced to nodes. We represent a frame member, which has rigid nodes on both sides, as a finite element. We obtain the stiffness matrix of bending member *AB* in the element CS Ax_ey_e wherein axis Ax_e coincides with the axial line of member (Figure 26.2). In distinction from the tensed-compressed members studied in the previous chapter, nodes *A* and *B* are subjected not only to external forces $\boldsymbol{F}_A$, $\boldsymbol{F}_B$, but also to external moments M_A and M_B, respectively. (These forces and moments are external with respect to the member.) Besides translational DOFs, each node of bending member has a rotational DOF which is angle φ of rotation of the end cross-section with

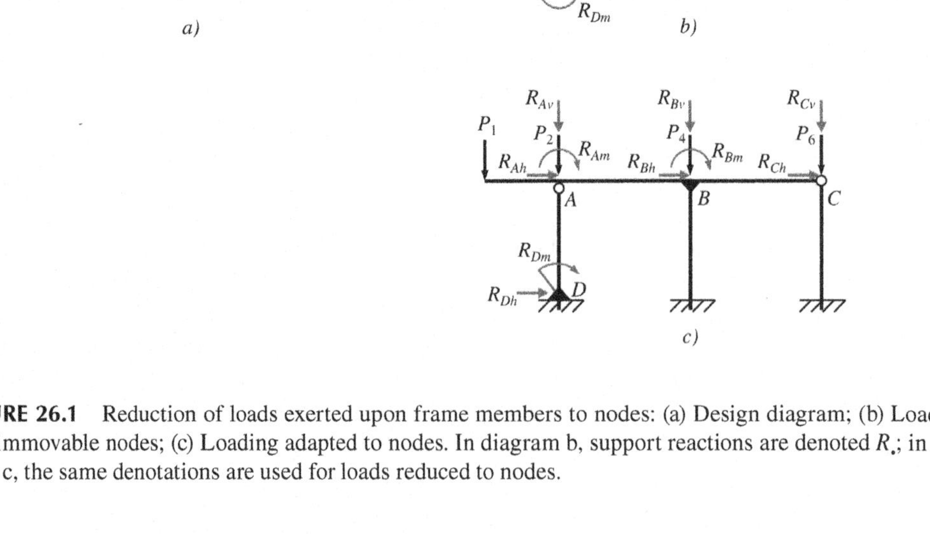

FIGURE 26.1 Reduction of loads exerted upon frame members to nodes: (a) Design diagram; (b) Loading with immovable nodes; (c) Loading adapted to nodes. In diagram b, support reactions are denoted $R_{\bullet}$; in diagram c, the same denotations are used for loads reduced to nodes.

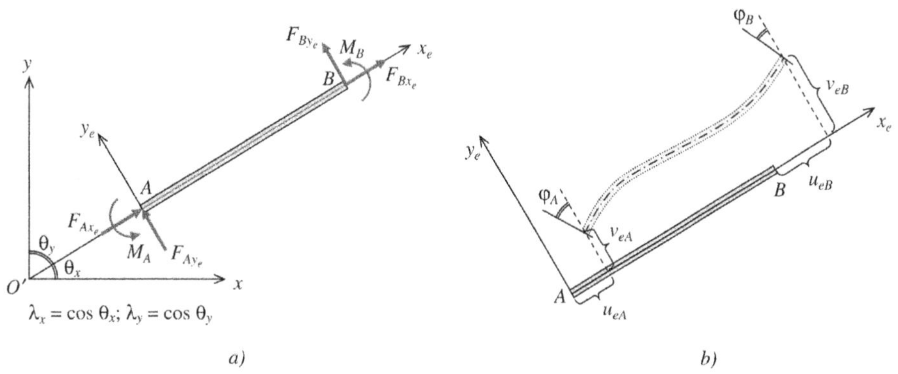

FIGURE 26.2 Bending member under the action of external forces upon nodes: (a) Positive directions of external forces; (b) Linear and angular displacements of nodes in element CS. Ax_ey_e is element CS; $O'xy$ is global CS; φ_A, φ_B are angles of rotation of member's end cross-sections; λ_x, λ_y are direction cosines of element axis x_e in global CS.

positive sense counterclockwise (Figure 26.2). The vector of the element's DOFs is introduced in the form:

$$\mathbf{u}_e^{\mathrm{T}} = (u_{eA}, v_{eA}, \varphi_A, u_{eB}, v_{eB}, \varphi_B), \tag{26.1}$$

where subscript "e" indicates that displacements are determined in the element CS. External forces and force couples acting on the nodes in the directions of the pointed out displacements are represented by the vector:

$$\mathbf{F}_e^{\mathrm{T}} = \left(F_{Ax_e}, F_{Ay_e}, M_A, F_{Bx_e}, F_{By_e}, M_B\right). \tag{26.2}$$

Stiffness matrix $\mathbf{K}_e$ of the bending member in element CS is easily obtained from its physical meaning; the arbitrary i-th column of this matrix $\mathbf{K}_{e*i}$ is the vector of external nodal forces producing unit displacement u_{ei} whereas all other displacements of element nodes are zero. Matrix constituents which correspond to longitudinal displacements $u_{eA} = 1$ and $u_{eB} = 1$ were obtained earlier in the form (25.10) (now, the positioning of constituents in the matrix is different). The constituents which correspond to unit transverse and angular displacements can be obtained by means of diagrams a and c in Figure 18.1 as well as by using formula (14.18). For instance, we shall obtain the constituents of the second column of the stiffness matrix, which are the forces and force couples exerted upon nodes due to transverse displacement $v_{eA} = 1$.

Since the unit displacement v_{eA} doesn't produce axial forces, we have:

$$K_{e12} = F_{Ax_e} = K_{e42} = F_{Bx_e} = 0.$$

The diagram of bending moments caused by the displacement being considered is presented on diagram c in Figure 18.1. Bending moments at the borders of this diagram are the external moments exerted on nodes, thus

$$K_{e32} = M_A = K_{e62} = M_B = \frac{6EI}{l^2},$$

where EI is flexural rigidity, and l is the length of the member. Note that positive external moment M_B causes tension of the lower longitudinal fiber of the member, which corresponds to negative bending moments on the diagram.

The shear force along the member is determined by formula (14.18), whence we get:

$$V = \frac{\frac{6EI}{l^2} - \left(-\frac{6EI}{l^2}\right)}{l} = \frac{12EI}{l^3}.$$

From this we obtain, taking into account positive directions of forces in Figure 26.2:

$$K_{e22} = F_{Ay_e} = V = \frac{12EI}{l^3}; \qquad K_{e52} = F_{By_e} = -V = -\frac{12EI}{l^3}.$$

In a similar way, we can find other columns of the stiffness matrix and establish it in the form:

$$\mathbf{K}_e = \begin{pmatrix} \frac{EA}{l} & 0 & 0 & -\frac{EA}{l} & 0 & 0 \\ 0 & \frac{12EI}{l^3} & \frac{6EI}{l^2} & 0 & -\frac{12EI}{l^3} & \frac{6EI}{l^2} \\ 0 & \frac{6EI}{l^2} & \frac{4EI}{l} & 0 & -\frac{6EI}{l^2} & \frac{2EI}{l} \\ -\frac{EA}{l} & 0 & 0 & \frac{EA}{l} & 0 & 0 \\ 0 & -\frac{12EI}{l^3} & -\frac{6EI}{l^2} & 0 & \frac{12EI}{l^3} & -\frac{6EI}{l^2} \\ 0 & \frac{6EI}{l^2} & \frac{2EI}{l} & 0 & -\frac{6EI}{l^2} & \frac{4EI}{l} \end{pmatrix}. \tag{26.3}$$

The next step is to transform the stiffness matrix to nodal coordinate systems. Let all nodal CSs coincide with the global one. We introduce six-dimensional vectors of displacements and forces acting upon the nodes of the bending member, with representation in nodal CSs. To this end, in formulas (26.1) and (26.2), we replace the projections of $2D$-vectors on the element axes by the projections of the same vectors on the nodal (in our case global) coordinate axes:

$$\mathbf{u}^{\mathrm{T}} = \left(u_A, v_A, \varphi_A, u_B, v_B, \varphi_B\right), \tag{26.4}$$

$$\mathbf{F}^{\mathrm{T}} = \left(F_{Ax}, F_{Ay}, M_A, F_{Bx}, F_{By}, M_B\right). \tag{26.5}$$

We introduce the matrix of element CS's rotation in block-diagonal form:

$$\Lambda = \begin{pmatrix} \Lambda_A & \mathbf{O} \\ \mathbf{O} & \Lambda_B \end{pmatrix}; \quad \Lambda_A = \Lambda_B \equiv \begin{pmatrix} \lambda_x & \lambda_y & 0 \\ -\lambda_y & \lambda_x & 0 \\ 0 & 0 & 1 \end{pmatrix}. \tag{26.6}$$

This matrix enables us to represent the conversion of the displacement vector and force vector due to transition from one CS to another in a convenient form:

$$\mathbf{F} = \Lambda^{\mathrm{T}}\mathbf{F}_e; \tag{26.7}$$

$$\mathbf{u}_e = \Lambda\mathbf{u}. \tag{26.8}$$

Each of submatrices $\mathbf{\Lambda}_A$ and $\mathbf{\Lambda}_B$ implements the transformation of three components of displacements and forces appertaining to one of the element's nodes. In our case, the submatrices coincide, but in the case of different nodal CSs, they will differ. The last two formulas enable us to obtain the conversion of the stiffness matrix to global CS in the form:

$$\mathbf{K} = \mathbf{\Lambda}^{\mathrm{T}}\mathbf{K}_e\mathbf{\Lambda}. \tag{26.9}$$

This formula repeats formula (25.17) and is obtained likewise, from the conditions:

$$\mathbf{F} = \mathbf{K}\mathbf{u}; \quad \mathbf{F}_e = \mathbf{K}_e\mathbf{u}_e, \tag{26.10}$$

which determine stiffness matrices in corresponding CSs. However, now, in contrast to the previous chapter, all the matrices in (26.9) have the order 6.

Besides bending members with rigid nodes at both endpoints, a frame may include bending members with rigid and hinged nodes at the endpoints. We introduce the FE "bending member AB with hinged node B," which differs from the FE in Figure 26.2 in that the moment M_B is identically equal to zero. Respectively, angle φ_B of the section's rotation near node B is not an independent parameter which determines the SSS of the member. In order to determine the SSS of the bending member with a hinged node, five DOFs are sufficient, which are represented by the vector obtained by removal of the last component from vector (26.1). Just the same, the vector of external forces (force couple) for this element is obtained by removal of the last component from vector (26.2). In element CS Ax_ey_e with axis Ax_e on the axial line of member (Figure 26.2a), the stiffness matrix of the new element can be obtained as follows:

$$\mathbf{K}_e = \begin{pmatrix} \frac{EA}{l} & 0 & 0 & -\frac{EA}{l} & 0 \\ 0 & \frac{3EI}{l^3} & \frac{3EI}{l^2} & 0 & -\frac{3EI}{l^3} \\ 0 & \frac{3EI}{l^2} & \frac{3EI}{l} & 0 & -\frac{3EI}{l^2} \\ -\frac{EA}{l} & 0 & 0 & \frac{EA}{l} & 0 \\ 0 & -\frac{3EI}{l^3} & -\frac{3EI}{l^2} & 0 & \frac{3EI}{l^3} \end{pmatrix}. \tag{26.11}$$

The technique of finding the matrix constituents is the same as in the case of matrix (26.3), with the distinction that diagrams b and d in Figure 18.1 are being used. The rotation matrix for representation of this matrix in global CS is obtained by removal of the last row and column from matrix $\mathbf{\Lambda}$ of the form (26.6).

In the general case, a frame may contain the FEs of all three types: members with rigid nodes; members with rigid and hinged nodes; and the hinged members studied in the previous chapter.

At the final stage of analysis, while the internal forces in members are determined through found displacements of nodes, these forces in the state of loading 2 are determined by means of the second of formulas (26.10). The DOF vector on the right side of this formula is calculated by transformation (26.8). Internal forces in a bending member in state 2 are defined uniquely by the vector of forces acting upon end sections of the element: axial and shear forces are constant along the length of the member and are equal, respectively, to the fourth and second components of vector $\mathbf{F}_e$; the bending moment is linearly dependent on coordinate x_e, and in the case of a member with rigid nodes, is defined by the third and sixth components of this vector (in the case of a member with a single hinged node, this moment is defined by only the third component).

26.2 BASIC PROVISIONS OF FEM FOR ANALYSIS OF ARBITRARY ELASTIC STRUCTURE

Reviewed implementations of FEM for analysis of elastic plane bodies and trusses enable us to state basic provisions of FEM for analysis of arbitrary elastic systems – of the frames in that number. These provisions are set forth below and are correct in the most general case of a space system including bars, massive bodies, and shells connected by rigid connections or ball-and-socket joints. The reader may interpret these provisions with regard to plane systems of bodies and bars joined by connections of the second and third types.

FEM is to represent an externally stable system, attached to the earth by connections of principal types, in the form of the totality of interacting elements called finite elements. Every FE has a number of properties which enable us to obtain the SSS of a given system under loading. FE has the same principal properties as a system including it. Indeed, every FE taken separately is the specific case of a structure if being provided by supports to ensure external stability of the element. Further, when giving the definitions of degree of freedom and node of a system, we imply that they relate to finite elements as well.

A fundamental concept of FEM is the degree of freedom. For a system of elastic bodies in equilibrium, the concept of DOF is defined taking into account possible deformation of bodies (in contrast to Chapter 1, where the notion of DOF was defined for a system of discs). We call a necessary parameter in a set of independent parameters which determine the location of a system in the space at known requirements to the totality of external forces acting upon system the DOF of an elastic system (in particular, FE). The location of a strained system in the space specifies the field of displacements for every body of the system, i.e., the SSS of its elastic elements. The necessity of a parameter means that the location of a system in the space depends on it, and consequently, the SSS of the system as well.

Example

If in the totality of external forces (26.2) acting on a bending member, the first five components can take arbitrary values while moment M_B is identically equal to zero; then six components of displacement vector (26.1) are not the set of DOFs of this element, because their independence is not secured. For instance, the sixth component – angular displacement φ_B – is uniquely determined by the previous five components of the displacement vector. In the case when $M_B \equiv 0$, displacement φ_B is removed from the number of DOFs, and we come to the FE "bending member with one hinged end."

In FEM, the DOFs are introduced as translational and rotational displacements of nodes. We refer to some points of a system upon which (and only on them) external forces (force couples) may be exerted, and the displacements of which constitute in totality the DOFs of the system removed from bearings as *nodes*. Displacements and forces acting upon a node should make up pairs of generalized forces and displacements; at that, a force of a known direction is matched to every linear displacement, and a force couple is matched to every angular displacement. The totality of nodes is introduced in such manner that their displacements constitute the totality of DOFs. Displacements of a node are sometimes referred to as DOFs of the node, meaning the DOF of a system or element which includes the node.

Example

Let the external force exerted upon node A of a system be defined by its components F_ξ and F_η in oblique CS $A\xi\eta$ (Figure 26.3). Skew-angle projections Δ_ξ, Δ_η of displacement vector $\mathbf{\Delta}$, shown in Figure 26.3, cannot be associated with forces F_ξ, F_η as the system's DOFs, because the products $F_\xi\Delta_\xi$ and $F_\eta\Delta_\eta$ are not equal to the works done by pointed out forces through displacement $\mathbf{\Delta}$. At the

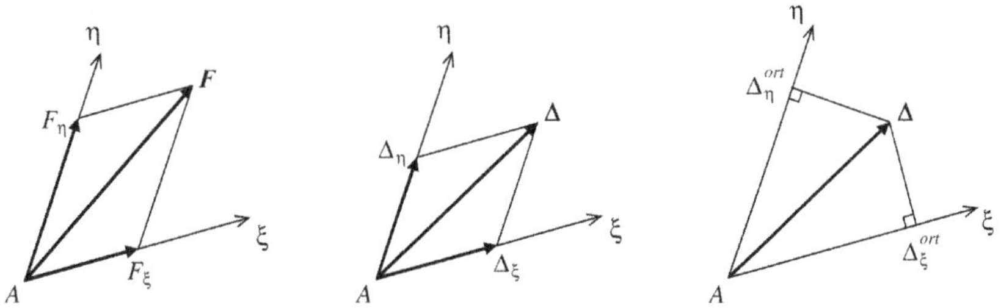

FIGURE 26.3 Vectors of force and displacement in oblique CS: F_ξ, F_η are skew-angle vector projections of force vector $\boldsymbol{F}$; Δ_ξ, Δ_η are skew-angle vector projections of displacement vector $\boldsymbol{\Delta}$; Δ_ξ^{ort}, Δ_η^{ort} are orthogonal scalar projections of displacement vector $\boldsymbol{\Delta}$.

same time, orthogonal projections Δ_ξ^{ort} and Δ_η^{ort} of displacement vector on corresponding coordinate axes allow such matching and can be taken as the DOFs.

When dividing a system into a set of FEs, one must secure that the DOFs of system unambiguously determine the DOFs of each element.

Example

Instead of FE "bending member AB with hinged node B" we can employ the element "bending member AB with two rigid nodes" if the design scheme of a system will be complicated in the following manner. We are to include rotation's angle φ_B of the end section of the member with rigid nodes into the set of the system's DOFs, and into the totality of loads upon the system, accordingly, should be included moment M_B of force couple exerted on this section. This force couple is fictitious, and moment M_B in the vector of loads is equated to zero, but now the FEM formalism enables us to determine unambiguously the DOFs of the system through loads. If hinge B forms a mixed node with a single rigid connection, then we interpret node B as the second rigid node attached by a hinge to a rigid node of the design scheme.

The system of FEM equations is the system of equations of a structure's equilibrium, wherein the unknowns are displacements of nodes and support reactions. Setting up and solving this equation system is the main part of calculations in FEM. In order to set up the system of equilibrium equations, initially, we introduce the vector of a system's DOFs $\mathbf{u}_f$ and corresponding vector $\mathbf{F}$ of external forces exerted on nodes; then the stiffness matrix of a system is found. The DOFs of a system and nodal external forces are introduced in nodal CSs (usually coinciding with global CS).

Vector $\mathbf{u}_f$ is introduced for a system removed from bearings, and its components are positioned in such manner that subvector $\mathbf{u}$ of the structure's DOFs (i.e., the subvector of displacements not eliminated by support constraints) goes first, and next follows subvector $\mathbf{u}_X$ of displacements eliminated by support constraints. The sequence of force components in vector $\mathbf{F}$ must correspond to the sequence of displacements in vector $\mathbf{u}_f$ in such manner that the components of these vectors with the same numbers u_{fi} and F_i make up generalized displacement and generalized force. (Vector $\mathbf{u}_f$ may be called the *ordered full vector of the structure's nodal displacements* and vector $\mathbf{F}$ the *ordered vector of forces.*)

According to the division of vector $\mathbf{u}_f$ into two subvectors $\mathbf{u}$ and $\mathbf{u}_X$, the vector of external forces is separated into subvector of loads $\mathbf{P}$ and subvector $\mathbf{X}$ of external forces comprising support reactions. Therefore, we have representation:

$$\mathbf{u}_f = \begin{pmatrix} \mathbf{u} \\ \mathbf{u}_X \end{pmatrix}; \quad \mathbf{F} = \begin{pmatrix} \mathbf{P} \\ \mathbf{X} \end{pmatrix}. \tag{26.12}$$

Further, the dimension of vectors $\mathbf{u}_f$ and $\mathbf{F}$ is denoted n_0; the dimension of vectors $\mathbf{u}$ and $\mathbf{P}$ is denoted n.

Stiffness matrix $\mathbf{K}$ of a system is a symmetric matrix of the quadratic form, which determines the strain energy of the system through the vector of DOFs:

$$U_1 = \frac{1}{2}\mathbf{u}_f^{\mathrm{T}}\mathbf{K}\mathbf{u}_f. \tag{26.13}$$

This matrix also links the vector of external forces in the state of equilibrium with the vector of DOFs, that can be written in the form:

$$\mathbf{K}\mathbf{u}_f - \mathbf{F} = 0. \tag{26.14}$$

The latter expression, supplemented by condition:

$$\mathbf{u}_X = 0, \tag{26.15}$$

represents the wanted system of equations of the structure's equilibrium. In the previous chapter, equation (26.14) was obtained for a bar system in the form (25.4). Just the same, derivation can be done in the case of an arbitrary elastic system with established totality of nodes.

Further math manipulations are aimed to find the stiffness matrix and convert equilibrium equations to standard form. This treatment repeats the technology of setting up global equations of FEM for analysis of a truss or planar elastic body, as studied in previous chapters.

The stiffness matrix of a system is constructed from stiffness matrices of its finite elements. The vector of DOFs of the i-th FE in nodal CSs can be obtained by selecting components from the DOF vector of a system removed from bearings as follows:

$$\mathbf{u}_i = \mathbf{I}_i\mathbf{u}_f, \tag{26.16}$$

where $\mathbf{I}_i$ is the matrix of selection, the rows of which are linear-independent unit row vectors.* Stiffness matrix $\mathbf{K}_i$ of any FE, by definition, is a symmetric matrix of quadratic form representing strain energy of FE through the vector of the element's nodal displacements as follows:

$$U_{1i} = \frac{1}{2}\mathbf{u}_i^{\mathrm{T}}\mathbf{K}_i\mathbf{u}_i. \tag{26.17}$$

Making substitution (26.16) into the last expression and summing strain energies of all the elements of a system, we come to expression (26.13) wherein the stiffness matrix is represented in the form:

$$\mathbf{K} = \sum_{i=1}^{m}\tilde{\mathbf{K}}_i; \quad \tilde{\mathbf{K}}_i \equiv \mathbf{I}_i^{\mathrm{T}}\mathbf{K}_i\mathbf{I}_i. \tag{26.18}$$

Addend $\tilde{\mathbf{K}}_i$ is referred to as the expanded stiffness matrix of the i-th FE.

Note that transformation (26.16) can be specified more briefly; if we introduce vector $\mathbf{S}_i$ made up by numbers of components of vector $\mathbf{u}_f$ sequentially selected into vector $\mathbf{u}_i$, then the components of the latter are obtained in the form:

$$u_{ik} = u_{fS_{ik}}, \quad k = \overline{1, n_{ei}}, \tag{26.19}$$

---

* Unit row vector has the form: $(\underbrace{0,\ldots,0}_{k-1},1,0,\ldots,0)$, $k \geq 1$.

where n_{ei} is the number of DOFs of the i-th element. Just the same, the transformation of matrices defined by the second formula (26.18) is represented in the form:

$$\tilde{K}_{iS_{ik}S_{il}} = K_{ikl}, \quad k = \overline{1, n_{ei}}, \quad l = \overline{1, n_{ei}}. \tag{26.20}$$

Formulas (26.19) and (26.20) were already used in solving particular problems in Chapter 23 for analysis of structure-bodies and in Chapter 25 for analysis of trusses.

The finding of stiffness matrices $\mathbf{K}_i$ of FEs making up a system is a stand-alone research problem. FEM assumes the formation of a library of FEs for engineering calculations. In this library, for every FE there is some stiffness matrix associated, presented in element CS (maybe as a simple algorithm of searching this matrix).

In order to set up the system of global equations of FEM in a form convenient for calculation, we introduce the block representation of the obtained stiffness matrix of a system:

$$\mathbf{K} = \begin{pmatrix} \mathbf{K}_{11} & \mathbf{K}_{12} \\ \mathbf{K}_{21} & \mathbf{K}_{22} \end{pmatrix}, \tag{26.21}$$

where block $\mathbf{K}_{11}$ has the order n. This block is the stiffness matrix of a structure, which is proved by reproducing the treatment (23.24) and representation of strain energy of an elastic structure in the form (23.26). The results of Chapter 23, being obtained for an elastic body, are applicable also for analysis of determinacy of stiffness matrices in the more general case under consideration. Namely, one can establish – on condition of the FE model's adequacy – that the stiffness matrix of a free system of elastic bars and bodies is positive semi-definite and singular, whereas the stiffness matrix of an elastic structure is positive definite.

By substituting relations (26.12) and (26.21) into equilibrium equation (26.14) and making uncomplicated matrix transformations with account for conditions of attachment to the earth (26.15), we come to the final set of equilibrium equations:

$$\begin{cases} \mathbf{K}_{11}\mathbf{u} - \mathbf{P} = 0; \\ \mathbf{K}_{21}\mathbf{u} - \mathbf{X} = 0, \end{cases} \tag{26.22}$$

where the unknowns are the vectors $\mathbf{u}$ and $\mathbf{X}$.

After solving the obtained equation system, the SSS of elements is determined through established displacements of their nodes. In the case when, in the first step of calculation, the intermediate loads exerted upon bars was reduced to nodes, the SSS of corresponding members is determined on the basis of the superposition principle, as the result of superimposing the SSS obtained at given out-of-node loads and fixed nodes, with the SSS established by means of global equations (26.22) at the loads reduced to nodes.

Whilst the stiffness method gives exact distributions of internal forces in members (in the case of frames, insofar as the hypotheses of bending are fulfilled), the peculiarity of FEM calculations of systems including planar of spatial bodies is their approximate character. The SSS of bodies is represented by displacements of the finite totality of nodes with an error which may be essential. In order to decrease the error of FEM, one should use body FEs of a size as small as possible. The analysis of errors in FEM is beyond the frames of this course.

26.3 EXAMPLE OF FRAME ANALYSIS

As an example, we make calculation of the frame in Figure 26.4a, by the stiffness method. Accept that the flexural rigidity of members is given and equal to EI, and axial and flexural rigidities are linked by the relation:

$$EA = \frac{3000}{l^2} EI. \tag{26.23}$$

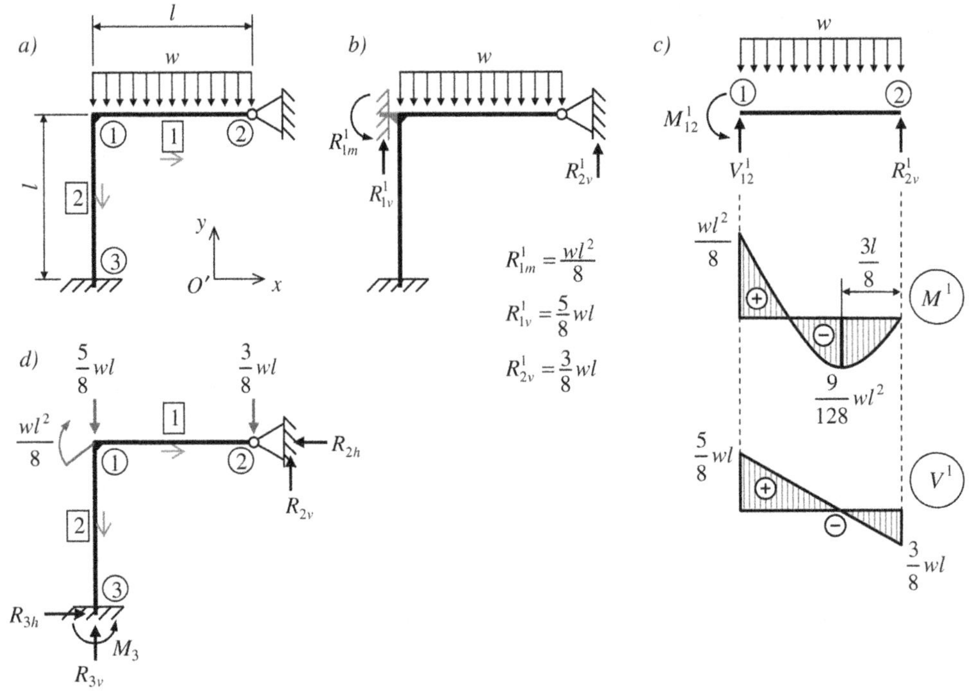

FIGURE 26.4 Design diagram of frame (a), the substitute frame of stiffness method (b, c), and given frame with loads reduced to nodes (d).

Given the characteristics w and l of the design diagram.

Let us reduce load w to nodes; to do that, we determine internal forces in the members and support reactions in the substitute system with immovable nodes (this system is in state 1). We accept bending moments as positive if they stretch a fiber on the outside of the contour of the frame. The loading of a substitute system is shown in Figure 26.4b; the corresponding design diagram of loading member 1 is shown on plot *c*. Superscript "1" of notations on these plots points to the number of the state. Internal forces in member 2 are zeroth. In member 1, axial force is absent; bending moment diagram M^1 is shown in Figure 26.4c, and reproduces the diagram in Figure 18.1f. The shear force diagram is rectilinear and determined by the shear forces at the ends of the member. Moment M^1_{12} is determined by the bending moment diagram on plot *c*: $M^1_{12} = \dfrac{wl^2}{8}$; lateral forces V^1_{12} and R^1_{2v} are established from equations of statics, then diagram V^1 is constructed (plot c, at the bottom). The support reactions in the substitute system are determined by constructed diagrams of internal forces. Obtained reactions taken with the opposite sign specify the loads reduced to nodes.

The loading adapted to nodes (loading 2) is shown on the design diagram in Figure 26.4d. In the substitute frame, the support node 2 is subjected to reaction force R^1_{2v}. This force is exerted with the opposite sign in the state of loading 2 and considered to be the load, thus support reactions caused by loading 2 are true reactions for the given system. At the same time, the internal force diagrams in states 1 and 2 must be summed up in order to obtain final diagrams.

We proceed to the calculation of the structure in state 2. We introduce the full vector of nodal displacements in the form:

$$\mathbf{u}_f^{\mathrm{T}} = \left(u_1, v_1, \varphi_1, u_2, v_2, u_3, v_3, \varphi_3\right). \tag{26.24}$$

The numbers of nodes are pointed out in subscript; all linear displacements are given in global CS; and the positive sense of angular displacements is, as usual, counterclockwise. The displacements not eliminated by constraints constitute three first components of this vector.

External forces (loads and support reactions) acting in the directions of these displacements make up the vector:

$$\mathbf{F}^{\mathrm{T}} = \left(0, -\frac{5}{8}wl, -\frac{wl^2}{8}, -R_{2h}, X_{2v}, R_{3h}, R_{3v}, M_3\right);$$

$$X_{2v} \equiv R_{2v} - \frac{3}{8}wl.$$

Here, X_{2v} is the vertical component of the total external force exerted on node 2.

The directions of coordinate axes of element 1 coincide with the directions of the same name global axes. Accordingly, stiffness matrix $\mathbf{K}_1$ has the form (26.11). In this matrix, we reduce axial rigidities to flexural rigidities by substitution (26.23). We get:

$$\mathbf{K}_1 = \frac{EI}{l^3}\begin{pmatrix} 3000 & 0 & 0 & -3000 & 0 \\ 0 & 3 & 3l & 0 & -3 \\ 0 & 3l & 3l^2 & 0 & -3l \\ -3000 & 0 & 0 & 3000 & 0 \\ 0 & -3 & -3l & 0 & 3 \end{pmatrix}.$$

The sequence of components in the DOF vector of element 1 coincides with the sequence of the first five components of vector (26.24); therefore, expanded stiffness matrix $\tilde{\mathbf{K}}_1$ includes matrix $\mathbf{K}_1$ as a corner block.

In the case of a bar element with two rigid nodes, the stiffness matrix in global CS is obtained in the supplement to the chapter in the form (26S.5). For element 2, we have direction cosines $\lambda_x = 0$ and $\lambda_y = -1$. According to (26S.5), we obtain stiffness matrix:

$$\mathbf{K}_2 = \begin{pmatrix} \frac{12EI}{l^3} & 0 & \frac{6EI}{l^2} & -\frac{12EI}{l^3} & 0 & \frac{6EI}{l^2} \\ & \frac{EA}{l} & 0 & 0 & -\frac{EA}{l} & 0 \\ & & \frac{4EI}{l} & -\frac{6EI}{l^2} & 0 & \frac{2EI}{l} \\ & & & \frac{12EI}{l^3} & 0 & -\frac{6EI}{l^2} \\ & & & & \frac{EA}{l} & 0 \\ & & & & & \frac{4EI}{l} \end{pmatrix}.$$

At the high dimension of displacement vector $\mathbf{u}_f$, it is convenient to use the numbers of components selected into the DOF vector of FE as the codes of columns and rows in the FE's stiffness matrix. In the obtained matrix, we reduce axial rigidities to flexural rigidities and title rows and columns of stiffness matrix $\mathbf{K}_2$ by the components of vector $\mathbf{S}_2$:

$$
\mathbf{K}_2=\frac{EI}{l^3}
\begin{array}{|c|c|c|c|c|c|c}
1 & 2 & 3 & 6 & 7 & 8 & \mathbf{S}_2^{\mathrm{T}}/\mathbf{S}_2 \\
\hline
12 & 0 & 6l & -12 & 0 & 6l & 1 \\
 & 3000 & 0 & 0 & -3000 & 0 & 2 \\
 & & 4l^2 & -6l & 0 & 2l^2 & 3 \\
 & & & 12 & 0 & -6l & 6 \\
 & & & & 3000 & 0 & 7 \\
 & & & & & 4l^2 & 8
\end{array}
$$

The stiffness matrix of the structure removed from bearings is obtained as the sum of expanded matrices:

$$
\mathbf{K}=\frac{EI}{l^3}
\begin{pmatrix}
3000+12 & 0+0 & 0+6l & -3000+0 & 0+0 & 0-12 & 0+0 & 0+6l \\
 & 3+3000 & 3l+0 & 0+0 & -3+0 & 0+0 & 0-3000 & 0+0 \\
 & & 3l^2+4l^2 & 0+0 & -3l+0 & 0-6l & 0+0 & 0+2l^2 \\
 & & & 3000+0 & 0+0 & 0 & 0 & 0 \\
 & & & & 3+0 & 0 & 0 & 0 \\
 & & & & & 0+12 & 0+0 & 0-6l \\
 & & & & & & 0+3000 & 0+0 \\
 & & & & & & & 0+4l^2
\end{pmatrix}
$$

$$
=\frac{EI}{l^3}
\begin{pmatrix}
3012 & 0 & 6l & -3000 & 0 & -12 & 0 & 6l \\
 & 3003 & 3l & 0 & -3 & 0 & -3000 & 0 \\
 & & 7l^2 & 0 & -3l & -6l & 0 & 2l^2 \\
 & & & 3000 & 0 & 0 & 0 & 0 \\
 & & & & 3 & 0 & 0 & 0 \\
 & & & & & 12 & 0 & -6l \\
 & & & & & & 3000 & 0 \\
 & & & & & & & 4l^2
\end{pmatrix}.
$$

The system of equilibrium equations (26.22) has the following form:

$$
\begin{cases}
\dfrac{EI}{l^3}\begin{pmatrix} 3012 & 0 & 6l \\ & 3003 & 3l \\ & & 7l^2 \end{pmatrix}\begin{pmatrix} u_1 \\ v_1 \\ \varphi_1 \end{pmatrix}-\begin{pmatrix} 0 \\ -\dfrac{5}{8}wl \\ -\dfrac{wl^2}{8} \end{pmatrix}=0; \\[2ex]
\dfrac{EI}{l^3}\begin{pmatrix} -3000 & 0 & 0 \\ 0 & -3 & -3l \\ -12 & 0 & -6l \\ 0 & -3000 & 0 \\ 6l & 0 & 2l^2 \end{pmatrix}\begin{pmatrix} u_1 \\ v_1 \\ \varphi_1 \end{pmatrix}-\begin{pmatrix} -R_{2h} \\ R_{2v}-\dfrac{3}{8}wl \\ R_{3h} \\ R_{3v} \\ M_3 \end{pmatrix}=0.
\end{cases}
$$

The displacements are found by applying Cramer's rule to the first system of equations. We get:

$$u_1 = \frac{2241}{63180036}\frac{wl^4}{EI}; \quad v_1 = -\frac{12025.5}{63180036}\frac{wl^4}{EI}; \quad \varphi_1 = -\frac{1124982}{63180036}\frac{wl^3}{EI}.$$

After substitution of the obtained displacements into the second system, we establish support reactions:

$$R_{2h} = 0.106wl; \quad R_{2v} = 0.429wl; \quad R_{3h} = 0.106wl;$$

$$R_{3v} = 0.571wl; \quad M_3 = \ 0.0354wl^2.$$

By checking the equilibrium of external forces and moments, the correctness of calculations is confirmed. Note that the found reactions satisfy both the checking with loading 2 and checking with loading in the problem being solved.

The final stage of analysis lies in construction of internal force diagrams. In order to construct diagram M^2 of the bending moments in state 2, we have to establish the moments* M_{12}^2 and M_{13}^2 in the neighborhood of node 1. These moments are calculated as the moments at the nodes of FE by using formula (26.10) for vector $\mathbf{F}_e$ of nodal forces (moments) in the element CS. We shall denote vectors $\mathbf{F}_e$ and $\mathbf{u}_e$ appertaining to the i-th element as $\mathbf{F}_{ei}$ and $\mathbf{u}_{ei}$, respectively; the j-th row of stiffness matrix $\mathbf{K}_e$ of the i-th element we denote $\mathbf{K}_{eij*}$. We obtain:

$$M_{12}^2 = F_{e13} = \mathbf{K}_{e1\,3*}\mathbf{u}_{e1} = \frac{EI}{l^2}\begin{pmatrix} 0 & 3 & 3l & 0 & -3 \end{pmatrix}\begin{pmatrix} u_1 \\ v_1 \\ \varphi_1 \\ u_2 \\ v_2 \end{pmatrix} = -0.0540wl^2;$$

$$M_{13}^2 = -F_{e23} = -\mathbf{K}_{e2\,3*}\mathbf{u}_{e2} = -\frac{EI}{l^2}\begin{pmatrix} 0 & 6 & 4l & 0 & -6 & 2l \end{pmatrix}\begin{pmatrix} -v_1 \\ u_1 \\ \varphi_1 \\ -v_3 \\ u_3 \\ \varphi_3 \end{pmatrix}$$

$$= 0.0710wl^2.$$

* In denotations of internal forces, we follow the rule stated in Chapter 14: in subscript of the form "ij," the first symbol points to a node beside which the force acts at the member section; the second symbol points to the opposite node of a member wherein the force is determined.

In the same manner, we calculate internal forces specifying the diagrams V^2 and N^2:

$$V_{1-2}^2 = F_{e12} = \mathbf{K}_{e1\,2*}\mathbf{u}_{e1} = -0.0540wl; \quad V_{1-3}^2 = F_{e22} = \mathbf{K}_{e2\,2*}\mathbf{u}_{e2} = -0.106wl;$$

$$N_{1-2}^2 = F_{e14} = \mathbf{K}_{e1\,4*}\mathbf{u}_{e1} = -0.106wl; \quad N_{1-3}^2 = F_{e24} = \mathbf{K}_{e2\,4*}\mathbf{u}_{e2} = -0.571wl.$$

Here, the subscript "*i–j*" points to the member with nodes *i* and *j*, and every force in the boundaries of the member is constant. The presented formulas illustrate the possibility of finding internal forces through established DOFs, though in the given case these forces are easily found by using the support reactions calculated before. The bending moment and shear force diagrams in state 2 are shown in Figure 26.5. The axial force diagram in state 2 coincides with the final diagram of this force, which is shown in Figure 26.6. While constructing diagram M^2, we employed reactive moment M_3 established before. (In a computer algorithm, the bending moment at a support node is calculated, as other internal forces, by formula (26.10).)

The final operation is to add together the diagrams in states 1 and 2. When constructing the final diagram of bending moments, one has to determine the minimum bending moment and the corresponding minimum point of the function $M(x_e)$ that determines this moment in member 1. Function $M^1(x_e)$, represented by diagram M^1 in Figure 26.4, is easily established from the known location of the parabola vertex through completing the square of the polynomial of degree 2. For the total bending moment, we have:

$$M(x_e) = M^1 + M^2 = \frac{w}{2}\left(x_e - \frac{5l}{8}\right)^2 - \frac{9}{128}wl^2 - \frac{M_{12}^2}{l}x_e + M_{12}^2.$$

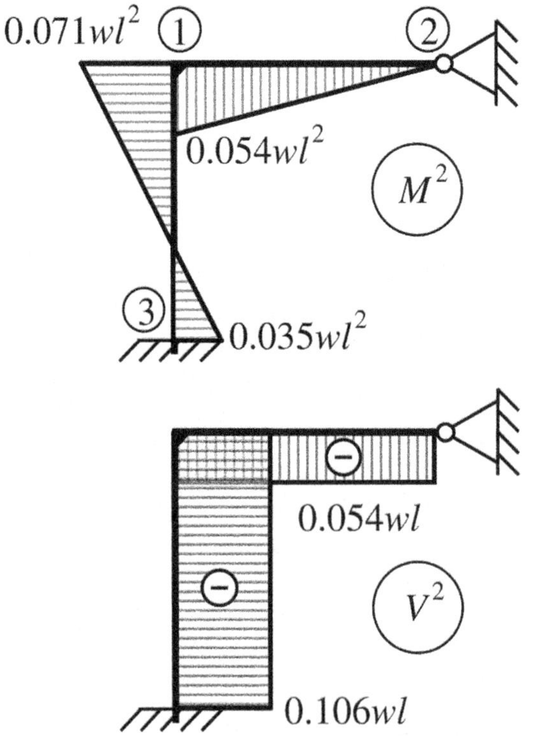

FIGURE 26.5 Diagrams of forces in state 2.

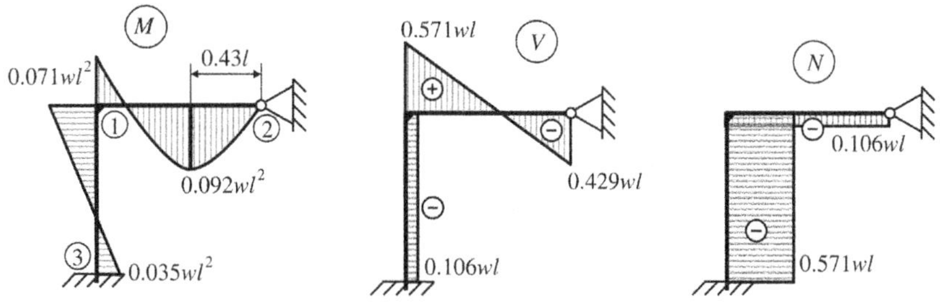

FIGURE 26.6 Final diagrams of forces in the frame.

The minimum point $x_{e\,\min}$ of this function is sought by using the shear force diagram $V = V^1 + V^2$ from the condition $V(x_e) = 0$. It holds:

$$x_{e\min} = 0.571l; \qquad M(x_{e\min}) = -0.0920wl^2.$$

Final diagrams are shown in Figure 26.6.

The considered problem is easily solved by the displacement method, and the results of the calculation will be close to those obtained here. Note, however, that the stiffness method is more exact, because it takes into account the axial strains of members.

SUPPLEMENTS TO CHAPTER

Stiffness Matrices of Bending Members in Global CS

The stiffness matrix of bending member AB with hinged node B is obtained in global CS by means of conversion as follows:

$$\mathbf{K} = \Lambda^{\mathrm{T}}\mathbf{K}_e\Lambda, \tag{26S.1}$$

where $\mathbf{K}_e$ is matrix (26.11), and Λ is the matrix of element CS's rotation, which has the form:

$$\Lambda = \begin{pmatrix} \Lambda_A & \mathbf{O}_{3\times2} \\ \mathbf{O}_{2\times3} & \Lambda_B \end{pmatrix}.$$

Here $\mathbf{O}_{\bullet}$ are null rectangular matrices with their dimensions indicated in subscript;

$$\Lambda_A \equiv \begin{pmatrix} \lambda_x & \lambda_y & 0 \\ -\lambda_y & \lambda_x & 0 \\ 0 & 0 & 1 \end{pmatrix}; \quad \Lambda_B \equiv \begin{pmatrix} \lambda_x & \lambda_y \\ -\lambda_y & \lambda_x \end{pmatrix}.$$

Decompose the stiffness matrix in the element CS into blocks:

$$\mathbf{K}_e = \begin{pmatrix} \mathbf{K}_{e11} & \mathbf{K}_{e12} \\ \mathbf{K}_{e21} & \mathbf{K}_{e22} \end{pmatrix}, \tag{26S.2}$$

where the order of submatrix $\mathbf{K}_{e11}$ is equal to 3, and the order of submatrix $\mathbf{K}_{e22}$ is equal to 2. By substitution of this decomposition into (26S.1) and using the block matrix multiplication, we obtain:

$$\mathbf{K} = \begin{pmatrix} \Lambda_A^{\mathrm{T}}\mathbf{K}_{e11}\Lambda_A & \Lambda_A^{\mathrm{T}}\mathbf{K}_{e12}\Lambda_B \\ \Lambda_B^{\mathrm{T}}\mathbf{K}_{e21}\Lambda_A & \Lambda_B^{\mathrm{T}}\mathbf{K}_{e22}\Lambda_B \end{pmatrix}. \tag{26S.3}$$

Calculating every obtained block separately, we easily establish all elements of the sought matrix:

$$\mathbf{K}=\begin{pmatrix} \frac{EA}{l}\lambda_x^2+\frac{3EI}{l^3}\lambda_y^2 & \left(\frac{EA}{l}-\frac{3EI}{l^3}\right)\lambda_x\lambda_y & -\frac{3EI}{l^2}\lambda_y & -\frac{EA}{l}\lambda_x^2-\frac{3EI}{l^3}\lambda_y^2 & \left(\frac{3EI}{l^3}-\frac{EA}{l}\right)\lambda_x\lambda_y \\ & \frac{EA}{l}\lambda_y^2+\frac{3EI}{l^3}\lambda_x^2 & \frac{3EI}{l^2}\lambda_x & \left(\frac{3EI}{l^3}-\frac{EA}{l}\right)\lambda_x\lambda_y & -\frac{EA}{l}\lambda_y^2-\frac{3EI}{l^3}\lambda_x^2 \\ & & \frac{3EI}{l} & \frac{3EI}{l^2}\lambda_y & -\frac{3EI}{l^2}\lambda_x \\ & & & \frac{EA}{l}\lambda_x^2+\frac{3EI}{l^3}\lambda_y^2 & \left(\frac{EA}{l}-\frac{3EI}{l^3}\right)\lambda_x\lambda_y \\ & & & & \frac{EA}{l}\lambda_y^2+\frac{3EI}{l^3}\lambda_x^2 \end{pmatrix}. \tag{26S.4}$$

The stiffness matrix of a bending member with two rigid nodes in global CS is also obtained by conversion (26S.1), but now the stiffness matrix in element CS has the form (26.3) and the rotation matrix is determined according to (26.6). Similarly to the previous case, we represent matrix $\mathbf{K}_e$ in the form (26S.2) with the distinction that now all blocks have the order 3. Again, reduce the wanted matrix to the block form (26S.3) and calculate all blocks separately. It holds:

$$\mathbf{K}=\begin{pmatrix} \frac{EA}{l}\lambda_x^2+\frac{12EI}{l^3}\lambda_y^2 & \left(\frac{EA}{l}-\frac{12EI}{l^3}\right)\lambda_x\lambda_y & -\frac{6EI}{l^2}\lambda_y & -\frac{EA}{l}\lambda_x^2-\frac{12EI}{l^3}\lambda_y^2 & \left(\frac{12EI}{l^3}-\frac{EA}{l}\right)\lambda_x\lambda_y & -\frac{6EI}{l^2}\lambda_y \\ & \frac{EA}{l}\lambda_y^2+\frac{12EI}{l^3}\lambda_x^2 & \frac{6EI}{l^2}\lambda_x & \left(\frac{12EI}{l^3}-\frac{EA}{l}\right)\lambda_x\lambda_y & -\frac{EA}{l}\lambda_y^2-\frac{12EI}{l^3}\lambda_x^2 & \frac{6EI}{l^2}\lambda_x \\ & & \frac{4EI}{l} & \frac{6EI}{l^2}\lambda_y & -\frac{6EI}{l^2}\lambda_x & \frac{2EI}{l} \\ & & & \frac{EA}{l}\lambda_x^2+\frac{12EI}{l^3}\lambda_y^2 & \left(\frac{EA}{l}-\frac{12EI}{l^3}\right)\lambda_x\lambda_y & \frac{6EI}{l^2}\lambda_y \\ & & & & \frac{EA}{l}\lambda_y^2+\frac{12EI}{l^3}\lambda_x^2 & -\frac{6EI}{l^2}\lambda_x \\ & & & & & \frac{4EI}{l} \end{pmatrix} \tag{26S.5}$$

Note that formula (26S.3) can be also used for the conversion of a stiffness matrix to nodal CSs in which the direction cosines in submatrices $\mathbf{\Lambda}_A$ and $\mathbf{\Lambda}_B$ are different.

FE Model of Reinforced-Concrete Skeleton of Residential House (Brief Description)

The models constructed from elastic FEs are used for analysis of reinforced-concrete skeletons of multi-story buildings. Figure 26S.1 shows a ten-story residential house with a reinforced-concrete skeleton and its FE model. The walls of reinforced concrete, the floor slabs, and the base plate are simulated by shell FEs mostly of rectangular shape; the columns are simulated by beam FEs (above, the planar analogs of the latter were called "bending members with rigid nodes"). Photo a shows the house during its operation; diagram b shows the model of a skeleton through displaying middle surfaces of shells and axial lines of bars; and diagram *c* shows the FEs volumetrically and with no border lines between elements. Exterior walls are multilayered: the layer of foamed concrete is faced with a brick outside and plasterboard inside. Exterior walls are erected upon floor slabs and are considered as the loads upon the slabs. The model includes 28691 elements and 29061 nodes; the stiffness matrix of the structure is of the order 174275.

The model comprises horizontal connections with the earth, which are positioned along two adjacent sides of the base plate at all the nodes of every side. Instead of vertical connections with

FIGURE 26S.1 Residential house in the city of Bataysk, Rostov Region, RF (built in 2011): (a) Photo of an inhabited house; (b) FE model of reinforced-concrete skeleton of the house; (c) Plot of the skeleton obtained by volumetric display of FEs.

the earth, the model involves special FEs of the base plate, which simulate the reaction of the base proportional to the local settlement.

All connections between FEs of the model are rigid.

The calculation of SSS is done separately for several loadings as follows: dead loads of skeleton members, exterior walls, and coverings on slabs; loads of partitions (internal walls); live loads (potential inhabitants, furniture, interior fittings); snow loads; two wind loadings. At the final stage of analysis, for every FE, the most hazardous combinations of loads are established, which serve as an input for constructing the reinforcement of the skeleton.

Concrete subjected to operation loads does not work elastically, yet for analysis of reinforced-concrete skeletons, elastic FE models are in use. The results of such analysis enable us to design reliable structures under observance of some rules of simulation. It is important to select an appropriate size of shell FEs, which should be – in a building with concrete columns – approximately the same as cross-sections of columns, i.e., close in shape to a square with sides of 40–50 cm. In this case, the sites of connections between slabs and columns are adequately described by an elastic model. If the size of a slab's FE decreases, internal forces near a column, evaluated through analysis, increase, and the model loses adequacy.

Together with the elastic model of skeleton, in the design of buildings of the skeleton type, the FE models of structural members are used wherein plastic properties of concrete are taken into account. For example, it is not permitted to establish the deflections of floor slabs by applying generalized Hooke's law for a slab's material. Taking into account the plastic properties of concrete makes the simulation problem complicated, because the results depend not only on the concrete characteristics, but on the arrangement and plastic properties of reinforcement. On the contrary, in the model presented above, there is no need to take into account the effect of reinforcement upon the elastic properties of skeleton members.

Section XI

Stability of Elastic Systems

This section presents the theoretical basics and the methods of structures' stability analysis. The explanation is given to static and energy approaches in analysis of stability. As an example of energy approach, the section sets out the Rayleigh method of stability calculation for structures with a single DOF. Static approach is realized in the section as the Bubnov-Galerkin method and the displacement method. The Bubnov-Galerkin method is exposed in terms of linear operators acting in a Hilbert space. Special attention is paid to the displacement method as being most precise among the methods which allow manual calculations of stability of multi-DOF structures. The assumptions of this method are formulated and the limits of its applicability are established. The basic concepts are defined, among them primary system and primary unknowns of the method. It is shown that stability analysis can be done by setting up a canonical equation set and transitioning from this set to the equation of stability. While developing the technique of composing canonical equations, the section presents the initial parameters method for calculation of the beam-column's SSS and clarifies its usage to this effect. Namely, it is shown that unit reactions (coefficients of canonical equations) are found from equations of statics for separate parts of a primary system.

27 Theoretical Basics of Structure Stability Analysis

27.1 PROBLEM OF ELASTIC STRUCTURE STABILITY AND STATIC APPROACH TO ANALYSIS OF STABILITY

So far, in this course, the problem of determining the SSS of an elastic structure was addressed under assumption of small strains produced by loads. Due to these assumption, as well as the property of strains' elasticity (linear relation of strains and stresses), the SSS is unambiguously determined by loads, and displacements, stresses, and strains are linear dependent on the loads. The linearity here is understood as a proportional relation, when the increase of all loads by k times causes the increase of an SSS characteristic by k times as well.

In reality, the behavior of elastic systems is more complicated. Under significant load, there may arise the ambiguity of SSS, when, besides the state with small displacements, the equilibrium states with large displacements are possible, such that the SSS characteristics do not have the property of linearity in loads. The multiplicity of equilibrium states under given loads is linked with the phenomenon called buckling.

We refer to the small disturbing loads and (or) support displacements (the points of application of disturbing loads and their directions may be arbitrary) as the disturbing impacts (or factors of impact) upon a structure. The equilibrium of an elastic structure is referred to as stable if arbitrary sufficiently small disturbing factors produce as much as necessary small displacements in a system. Otherwise, if as much as necessary small disturbance produces finite displacement in a system (i.e., displacement limited from below by a fixed value), then the equilibrium state is referred to as unstable.

It is known that a system in stable equilibrium after termination of disturbing impact returns to the initial equilibrium state. It is also known that a system being in unstable equilibrium does not return to its initial state after termination of some disturbing factors' action, however small they were. These assertions are confirmed by experience and can be theoretically substantiated through considering the variation of a system's total energy under disturbing impacts. From these assertions, we conclude that the ability to restore initial equilibrium state after termination of impact is a sign of stability.

*The phenomenon of a structure's transition from unstable equilibrium to a new equilibrium state after termination of disturbances is referred to as buckling or stability loss.** Buckling implies the multiplicity of equilibrium states. On the contrary, uniqueness of an equilibrium state means its stability.

In mechanics of materials, we studied the buckling of an elastic column subjected to the compression load. Due to this phenomenon, a column may lose capacity of supporting the load, notwithstanding the fact that the stresses in it haven't reached the material's strength characteristics (yield strength or ultimate compressive strength). The buckling occurs all of a sudden and sets an example of stability loss due to increase of load when a structure comes from unstable equilibrium with small displacements to stable equilibrium with large displacements (spanning an order of the structure's dimensions). In practice, such a transition means the collapse of a structure. The cause of

* The term "stability loss" is also used in another sense: it may designate the phenomenon of a system's pass from stable equilibrium to unstable equilibrium due to increase of load.

stability loss is the puny disturbing loads. Stability loss might occur at relatively small stresses in a structure's members; hence, investigation of this effect is important.

In the 'mechanics of materials' course, the investigation of the bar's stability under a compressive load is reduced to finding the critical force of compression. Despite the diversity of objects in structural analysis, the investigation of their stability has the common methodology: the one or several loads are selected as the cause of the unstable state, and their proportional increase is allowed. (One may say that the step of the loading process wherein selected loads rise from zero and other loads are fixed is under consideration.) One of the variable loads $P > 0$ is taken, and we determine its critical value P_{cr} (i.e., such the quantity the exceeding of which causes the instability of a system's equilibrium). This critical load defines other critical values of variable loads. By finding out critical force P_{cr}, the analysis of stability is usually accomplished. On rare occasions, the strained state after buckling is also to be established.

The underlying statements for the analysis of stability of externally stable system are as follows: **(1) Under sufficiently small loads, equilibrium state is uniquely possible and, along with it, stable; (2) If, under some load, the multiplicity of equilibrium states has arisen, then the state with the small (zeroth as idealized) displacements is unstable.** These affirmations are based not only on experience but can be theoretically substantiated. In particular, the second statement is proved below for such a system with a finite number of DOFs that its potential energy is a second-order polynomial function of generalized coordinates.

For analysis of the stability of plane structures, a new type of connections is introduced: an *elastic hinge*, which may be used together with principal types of connections. A simple elastic hinge is a connection of the second type complemented with an elastic element (spring) which counteracts the mutual rotation of connected bodies and develops a reaction moment proportional to an angle of this rotation. Every simple elastic hinge is characterized by the stiffness of its spring. Figure 27.1 shows two simple elastic hinges, connecting three members in pairs and equivalent to one multiple elastic hinge. In analysis of the system's geometrical stability, elastic hinges are taken for the rigid connections.

When analyzing stability, a system's position in space is specified by the set of generalized coordinates. The generalized coordinate is a synonym of the degree of freedom (see Chapter 26). *A parameter in a set of necessary independent parameters which determine the location of an elastic system in the space at known requirements to the totality of external forces acting upon the system is referred to as a generalized coordinate.* The location of a strained system in the space specifies the field of displacements for every body of the system, i.e., the strained state of its bodies. Before the FEM was developed, the term "degree of freedom" did not replace the notion of a generalized coordinate but was used only for pointing out the total number of generalized coordinates (for instance, by saying "the object with n DOFs"). With the expansion of finite element analysis, these

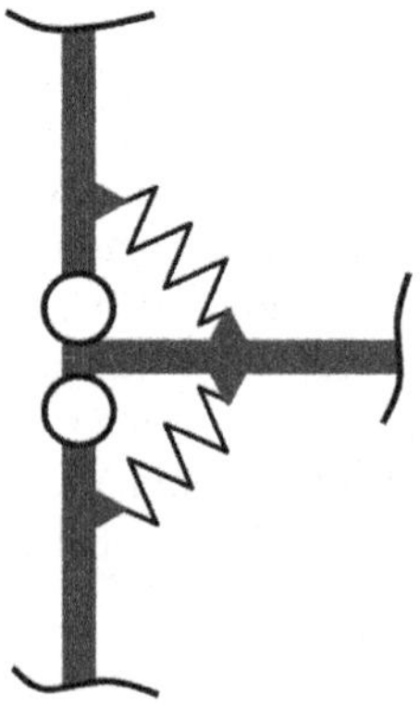

FIGURE 27.1 Multiple elastic hinge.

two terms became the synonyms, yet in stability analysis, the term "generalized coordinate" is still preferable. Thus, we emphasize that generalized coordinates in use are not necessarily the displacements of characteristic points of a structure ("nodes").

Note that in finite element analysis of elastic systems, all bodies of a system are assumed elastic, and the system's DOFs unambiguously determine its SSS. The theory of stability often considers elastic systems which contain perfectly rigid bodies (or even consist of them). One can speak of the strained state of a rigid body, meaning its displacement as an entire whole, but this displacement doesn't specify the stress state of such a body. Therefore, generalized coordinates in the theory of stability do not necessarily determine the stress state of a system.

Figure 27.2 shows examples of a structure's stability loss; the contour of a structure before the loss of stability is depicted by dashed lines. In diagram a, the three-hinged arch buckles under the action of a distributed load. Diagram b (on the left) can be used for stability calculation of a building frame. Load P presents the action of floors; members are assumed to be rigid; and the connections between members are elastic hinges. The possible form of the loss of structure stability is shown in diagram b (on the right). The case c emerges due to the compression of the column being a compound bar with a planar form of buckling. The diagram on the left shows the structure in the state with small displacements. On the right – the same structure after the loss of stability.

Diagrams in Figure 27.2 show the cases of stability loss due to the compression of members. In Figure 27.3, there is another example when the buckling of a bar occurs in the symmetric bending, i.e., while the axial internal force is absent. In this case, both compressed and tensed parts of the member create the conditions for buckling.

There are three kinds of approaches to investigation of stability: static, energy, and dynamic.

The static approach was firstly employed by L. Euler, who in the 18th century obtained the known formula for the critical force of a column's compression. This approach lies in analysis of statics equations containing unknown displacements in order to establish the loads when there emerges the multiplicity of equilibrium states, and thus the displacements of buckling can occur. Let us make static analysis of stability of the rigid cantilever column having length l, fixed in the basis by elastic hinge (Figure 27.4). The system has single DOF; as the generalized coordinate, we choose an angle θ of the member's bias from the vertical position. The column is subjected by vertical load P; the reaction moment is defined by stiffness k of support connection and equals

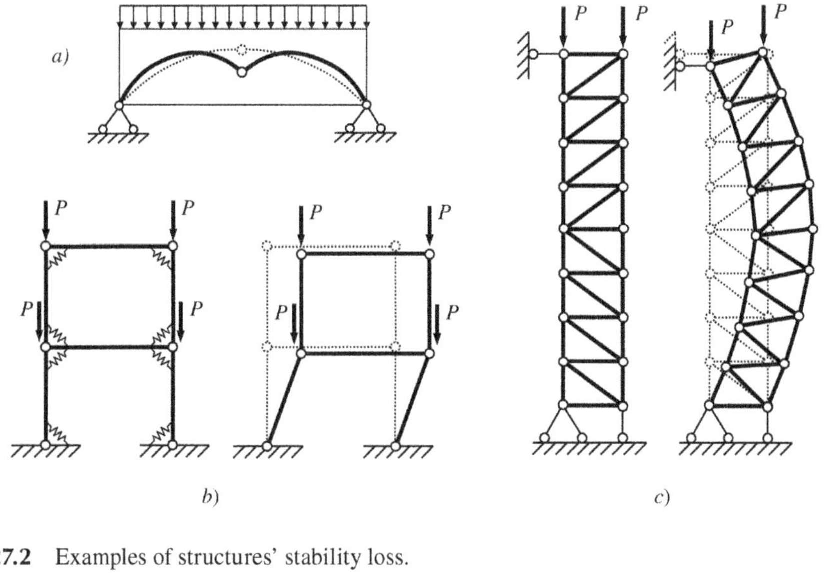

FIGURE 27.2 Examples of structures' stability loss.

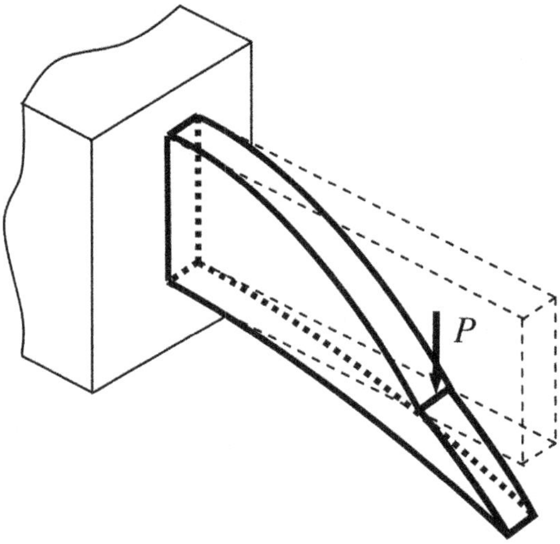

FIGURE 27.3 Buckling of bar in bending.

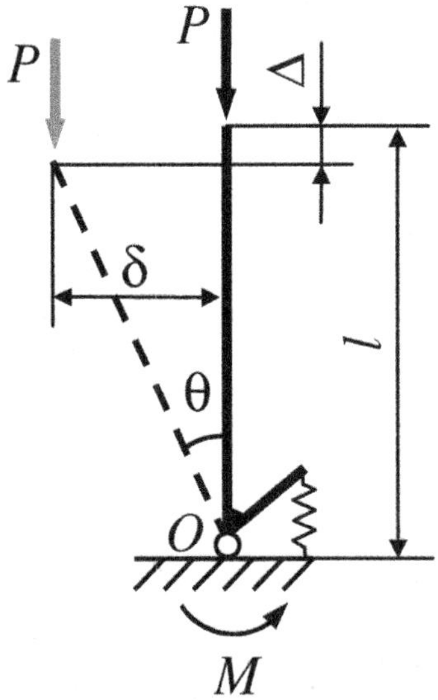

FIGURE 27.4 Cantilever column on elastic hinge before (solid line) and after (dashed line) stability loss.

$M = -k\theta$. To obtain the equation of statics for such a system, we write down the condition of moment equilibrium about the support hinge:

$$\sum M_O = P\delta - k\theta = 0.$$

After replacement $\delta = l \sin\theta$, we come to the equation:

$$Pl\sin\theta - k\theta = 0. \quad (27.1)$$

The equation happened to be nonlinear; we linearize it in the neighborhood of the zero coordinate by substitution $\sin\theta \cong \theta$. We get:

$$\theta(Pl - k) = 0. \quad (27.2)$$

The state $\theta = 0$ is possible at an arbitrary load. The instability of this state is linked to the possibility of the state $\theta \neq 0$. A nonzero (nontrivial) state is possible only if $Pl - k = 0$, wherefrom we obtain the critical force:

$$P_{cr} = \frac{k}{l}. \tag{27.3}$$

The analysis of nonlinear equation (27.1) enables us to verify the correctness of the critical force calculation and to determine displacement θ after stability loss. The principle of analysis remains the same: the state with zero displacements is accepted as unstable if another state is possible. We rewrite equation (27.1) in the form:

$$\frac{k}{Pl}\theta = \sin\theta, \tag{27.4}$$

and use the graphical method in order to establish the existence of a nontrivial solution. Figure 27.5 shows the dependences of the left-hand and right-hand sides of the equation from θ in the range of angles $0 \leq \theta \leq \frac{\pi}{2}$. At $\frac{k}{Pl} \geq 1$, the graphs of the left and right sides intersect at the coordinate origin only, and thus states other than $\theta = 0$ are impossible. On the contrary, at $\frac{2}{\pi} < \frac{k}{Pl} < 1$, there exist nonzero states which comply with equation of statics (27.4). The emergence condition for an unstable state has the form:

$$\frac{k}{Pl} < 1,$$

i.e., formula (27.3) for critical force is confirmed by analysis of the exact equation.

The dependence of displacements on loads in possible equilibrium states of a system is represented by the graph of equilibrium states. In the problem under consideration, this graph is shown in

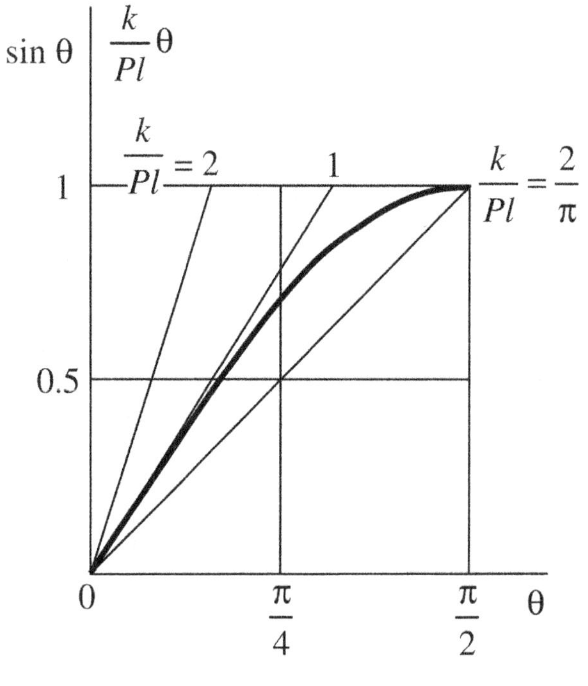

FIGURE 27.5 Graphs of left-hand and right-hand sides of the equation (27.4).

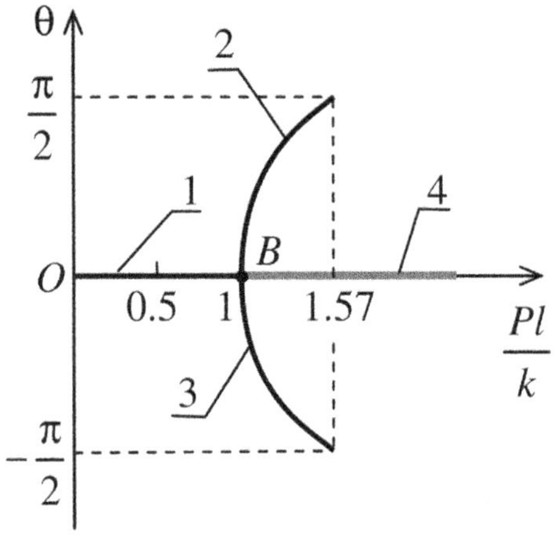

FIGURE 27.6 Graph of equilibrium states: 1, 2, 3 are lines of stable states; 4 is the line of unstable states; *B* is the bifurcation point.

Figure 27.6. At the point of critical load, the graph forks, and afterwards, several states correspond to a single load magnitude. The point of branching is referred to as the bifurcation point. The static approach does not have a criterion to distinguish stable and unstable states under the same loads. States with large displacements a priori are considered stable.

27.2 ENERGY APPROACH TO ANALYSIS OF STABILITY

The energy approach to the analysis of structure stability was thoroughly substantiated by English mathematician George Bryan in his works published in 1888 and the following years. He employed this approach when solving engineering problems.

In the energy approach, the possibility of buckling under the action of some loading is investigated by means of the total potential energy of a system represented as a function of generalized coordinates. The conclusion about stability of equilibrium states is made on the basis of the principle of minimum total potential energy. The complete formulation of this principle was given in Chapter 23 and enabled us to set up equations in unknown generalized coordinates of an elastic system in equilibrium. For analysis of stability, this principle is worded briefly: **the equilibrium of a conservative system is stable if, at the equilibrium point, a rigorous minimum of total potential energy is attained. The absence of a rigorous minimum at the equilibrium point permits us to consider this equilibrium as unstable**. The first assertion (sufficient condition of stability) is known as the Lagrange-Dirichlet theorem and has been proved in the middle of the 19th century by German mathematician Peter Dirichlet. Cases when rigorous minimum is not attained at the equilibrium point can be analyzed by means of two theorems about stability of a conservative system's motion, which was proved by Russian mathematician A.M. Lyapunov at the end of the 19th century (Liapounoff 1897). From these theorems, it follows that if at the equilibrium point a rigorous minimum of total potential energy is not attained, and, besides, the matrix of the second partial derivatives of this energy with respect to generalized coordinates is not positive semi-definite, then equilibrium is unstable. In engineering problems, the assumption about sign indefiniteness of this matrix can be accepted,* and we may take for granted that equilibrium at the stationary point out of rigorous minimum is unstable.

* The sign indefiniteness of a matrix is understood as indefiniteness of the sign of its quadratic form. It is known that an arbitrary positive semi-definite singular symmetric matrix can be transformed into a symmetric matrix indefinite in sign by means of infinitesimal perturbation of its elements. Since input data of engineering problems are subject to margins of error, we can accept the hypothesis about the sign indefiniteness of a matrix formed by the second partial derivatives of the potential energy at a stationary point where there is no rigorous minimum of this energy.

In practice, the energy approach assumes the setting up the equations of statics from conditions of reaching an extremum by potential energy (more precisely, of reaching a stationary point), and then it is reduced to the static approach. But the energy approach has the advantage that it enables us to distinguish stable and unstable equilibrium states.

Next, we employ the energy approach in calculation of the critical force wanted for displacement of the column on an elastic hinge (Figure 27.4). In this case, the total potential energy of the system is determined by the expression:

$$U = \frac{k\theta^2}{2} - P\Delta. \tag{27.5}$$

Here, the first term is the strain energy of the support connection; the second term is the energy of the external force field. For displacement Δ, we can obtain:

$$\Delta = l(1-\cos\theta) = 2l\sin^2\frac{\theta}{2} \cong \frac{l\theta^2}{2}.$$

Therefore, with an accuracy up to the terms of the second order of smallness, we can write the function of potential energy as follows:

$$U = \frac{k\theta^2}{2} - P\frac{l\theta^2}{2}. \tag{27.6}$$

The equilibrium is attained at the stationary point of the system's state space. In the given case, this point is determined from the requirement:

$$\frac{dU}{d\theta} = 0, \tag{27.7}$$

which for function (27.6) brings about the equilibrium equation (27.2). In order to check the stationary point for a rigorous minimum, one should use the criterion:

$$\frac{d^2U}{d\theta^2} = k - Pl > 0. \tag{27.8}$$

It should not be hard to ensure that the critical force partitioning the loads' magnitudes of stable and unstable equilibrium states is determined by formula (27.3). Really, at $P < P_{cr}$, the equation of statics (27.2) allows only the nil equilibrium state, and stability condition (27.8) is fulfilled. Under condition $P \geq P_{cr}$, the nil equilibrium state is unstable according to criterion (27.8).

The exact function of potential energy in the problem under consideration has the form:

$$U = \frac{k\theta^2}{2} - 2Pl\sin^2\frac{\theta}{2}. \tag{27.9}$$

Using this function, we can obtain the exact statics equation (27.1), the analysis of which has been done above. Calculating the second derivative of function (27.9), we can establish that curves 1, 2, and 3 on the graph in Figure 27.5 really represent stable states, whereas line 4 represents an unstable state.

In the energy approach, quadratic approximation of potential energy is usually sufficient for solving engineering problems. We shall consider the peculiarities of this approach in the case of a system with multiple degrees of freedom. Let vector $\mathbf{q}$ of the system's state be determined by n coordinates: $\mathbf{q}^T = (q_1, q_2, \ldots, q_n)$, and there is a given function $U = U(\mathbf{q})$ of total potential energy of a system. We investigate the stability of an equilibrium state in which the dependence of the system's deformation on loads is negligible. We select the generalized coordinates in such a manner that in this state, there would be $\mathbf{q} = 0$.

Remark

The diagram in Figure 27.2b, may serve as an example of a system with two DOFs; here, we take the angles of the columns' deviation from the vertical for the first and second story, respectively, for generalized coordinates.

We assume that the function of potential energy $U = U(\mathbf{q})$ is smooth* in the neighborhood of zero and, therefore, can be expanded into a Taylor series centered at the zero point. In this expansion, we limit ourselves by the terms of the second order:

$$U = U_0 + \left(\frac{\partial U}{\partial \mathbf{q}}\right)^{\mathrm{T}}_{|\mathbf{q}=0} \mathbf{q} + \frac{1}{2}\mathbf{q}^{\mathrm{T}}\mathbf{B}\mathbf{q} \tag{27.10}$$

Here, $\mathbf{B}$ is the symmetric matrix of the second partial derivatives of the potential energy function taken at the zero point:

$$B_{ij} \equiv \frac{\partial^2 U}{\partial q_i \partial q_j}_{|\mathbf{q}=0}.$$

In the general case, matrix $\mathbf{B}$ is dependent on the load parameter P.

Formula (27.10) has been written in matrix form. In the notations we are accustomed to, the same expression looks like the following:

$$U = U_0 + \sum_{i=1}^{n} \frac{\partial U}{\partial q_i}_{|\mathbf{q}=0} q_i + \frac{1}{2}\sum_{i,j=1}^{n} B_{ij} q_i q_j.$$

The stationary point of the function $U = U(\mathbf{q})$ is determined from the vector equation:

$$\frac{\partial U}{\partial \mathbf{q}} \equiv \begin{pmatrix} \dfrac{\partial U}{\partial q_1} \\ \vdots \\ \dfrac{\partial U}{\partial q_n} \end{pmatrix} = 0. \tag{27.11}$$

The state $\mathbf{q} = 0$ satisfies the stationary condition; thus, function (27.10) has the simple form:

$$U = U_0 + \frac{1}{2}\mathbf{q}^{\mathrm{T}}\mathbf{B}\mathbf{q} \tag{27.12}$$

Any nontrivial equilibrium state must also satisfy requirement (27.11), wherefrom, by differentiating function (27.12), we obtain the equation set of equilibrium:

$$\mathbf{B}\mathbf{q} = 0 \tag{27.13}$$

According to the provisions of stability analysis formulated in the previous item, instability of the nil equilibrium state of a system arises on existence of a nontrivial solution of the equilibrium

* A smooth function has partial derivatives of an arbitrary order on a given domain.

equation set. In the given case, the indication of existence of such a solution is the singularity of the coefficient matrix of equation set (27.13), i.e., the condition:

$$\det \mathrm{B}(P) = 0. \tag{27.14}$$

Here, the dependence of matrix **B** on the load is represented in explicit form. Minimum load P_{cr}, for which the condition (27.14) is satisfied, is the critical load. Conclusion: **if the total potential energy of a system is specified by a quadratic function with a zero point of equilibrium, the loss of stability of the nil state occurs when the matrix of second partial derivatives of this function becomes singular at point zero. Condition (27.14) can serve as the equation for finding critical force.**

Provision 2 of structure's stability analysis in the given case is proved rigorously. It is fairly easy to establish that quadratic form $\varphi = \mathbf{q}^T\mathbf{B}\mathbf{q}$ has a rigorous minimum at the point $\mathbf{q} = 0$ if and only if matrix **B** is positive definite. By using this property of quadratic form, we show that if at some load the multiplicity of equilibrium states emerges, then the nil state is unstable. Indeed, on this condition, there appears a nontrivial solution of equilibrium equations (27.13), i.e., matrix **B** degenerates and thus can't be positive definite. Therefore, a zero solution cannot correspond to a rigorous minimum of potential energy and cannot be stable.

Besides the static and energy approaches, there is the dynamic approach to analysis of stability. This one is based on the consideration of a system's oscillations in order to find out such a load when external perturbation causes unlimited increase of oscillation amplitude in time. In structural analysis, this approach is not often applied and is not considered here.

27.3 ANALYSIS OF FRAME'S STABILITY BY RAYLEIGH METHOD

The simplest implementation of the energy approach is known as the Rayleigh method (Jones 2006). The method is named after Lord Rayleigh, who developed the energy approach in the 1870s for analysis of structural vibration, which enabled Russian scientist S. Timoshenko to work out this method in 1910. According to Rayleigh method, we approximately represent the strain state of an elastic system by using a single DOF and determine the load under which second derivative of the system's total potential energy with respect to the generalized coordinate turns into zero. The obtained load will be the wanted critical load. Sometimes, these provisions permit us to investigate the stability of a truss, but they do not explain how the dependence of potential energy versus the DOF could be established. The correct substantiation of the method is known for frames only.

We consider the displacements of cross-sections of an arbitrary system's member in CS yz with the origin at one endpoint of the member and axis z pointed to another endpoint. (CS may get displacement together with end sections of member and may be left-handed.) In the Rayleigh method, it is necessary to specify the elastic curve of every member of a frame by means of some function $y = y(z)$ of the axial coordinate, which depends on two parameters: DOF q and load parameter P. The choosing of the function is done according to the expected shape of the elastic curve after the system's buckling and under compliance with boundary conditions. The latter are determined, mainly, by position and type of frame supports, and are presented as equalities of translational and (or) rotational displacements to zero at the characteristic points of members: $y = 0$ and $y' = 0$, respectively. These conditions are referred to as *geometric*, in contrast to natural boundary conditions which lie in the equality of bending moments and (or) shear forces to zero at some connections and endpoints of members. Natural boundary conditions are not taken into account in the Rayleigh method. In geometric boundary conditions, the displacements of the second-order infinitesimal of DOF q may be neglected (assuming $q = 0$ in the absence of strains). The precision of evaluation is essentially dependent on the error of the elastic curve's approximation by the function chosen for every member. We call these functions elastic curve functions, or functions of transverse displacements of a member's sections.

After choosing functions of displacements, the strain energy $U_1(q)$ of a system is analytically found as a function of DOF q; the potential energy of loads $U_{2P}(q)$ is established as the function of the DOF, dependent on load parameter P; after that, the second derivative of total energy $U(q) = U_1 + U_{2P}$ with respect to the DOF (the load is considered as a parameter of this function) should be developed. The zero condition for an obtained derivative approximately determines the critical load.

In order to calculate the strain energy of a frame, the next formula is used:

$$U_1 = \sum \int_0^l \frac{M^2 dz}{2EI}, \tag{27.15}$$

which follows from general formula (11.6) if strains due to axial force and shear strains in every member are neglected.

For calculation of potential energy U_{2P} of an external force field, the Rayleigh method employs an approximate formula for the length of the curve representing the graph of a function. We give the short development of this formula. The length of the curve of continuously differentiable function $y = y(z)$ at the segment $[0, l_1]$ can be obtained by integration of the lengths $dl = \sqrt{dz^2 + dy^2}$ of elementary portions forming the curve. The exact formula has the form:

$$l = \int_0^{l_1} \sqrt{1 + \left(\frac{dy}{dz}\right)^2}\, dz. \tag{27.16}$$

Let function $y = y(z)$ depend on parameter p and is represented as the product $y = py_0(z)$ where $y_0(z)$ is some continuously differentiable function. By using power series expansion of the integrand in formula (27.16) in the neighborhood of the point $y' = 0$, the integral is converted to the form:

$$l = l_1 + \frac{1}{2} p^2 \int_0^{l_1} \left(\frac{dy_0}{dz}\right)^2 dz + O(p^4) \quad \text{as} \quad p \to 0. \tag{27.17}$$

Here, $O(\cdot)$ is a function of argument p, which is an infinitesimal under $p \longrightarrow 0$ of the same order as the function in parentheses. Hence, we obtain the required approximate formula:

$$l \approx l_1 + \frac{1}{2} \int_0^{l_1} y'(z)^2 dz, \tag{27.18}$$

the error of which is small under a sufficiently small value p, i.e., when the curve is flat-shaped.

Example

Find the buckling force for the frame in Figure 27.7a. Axial rigidities of members are infinitely great; flexural rigidities are identical and equal to EI.

We specify the form of elastic curves according to the following hypothesis: the dependence $y = y(z)$ for transverse displacements of member AC is a polynomial of the second order; the elastic curve of member AB is determined by angular displacement of joint A, ignoring axial force in member AB. This hypothesis permits us to specify the system's deformation by a single DOF q, for which we take, for instance, the angle of joint A's pivot (Figure 27.7b). We introduce CS for member AC by positioning its origin at joint A and directing axis z towards joint C. The distance between joints of deformed member AC will be denoted as l_1. Before buckling, this distance was equal to the length

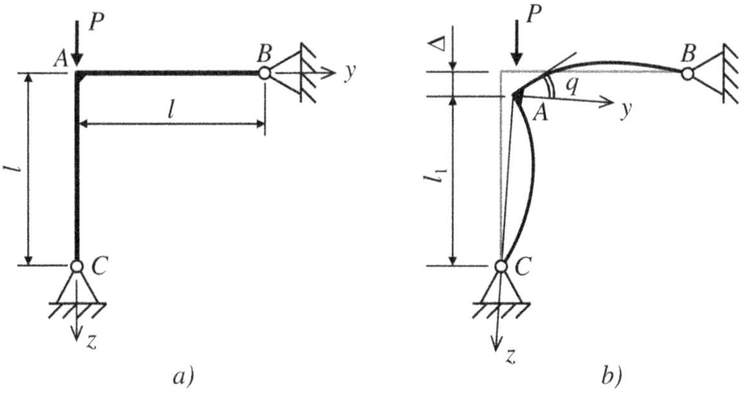

FIGURE 27.7 Buckling of frame: (a) Design diagram of frame; (b) Mode of buckling.

of member l; due to the loss of stability, this distance decreases by the magnitude of displacement Δ of joint A as shown in diagram b. Boundary conditions for member AC have the form:

$$y(0) = y(l_1) = 0. \tag{27.19}$$

The DOF is determined by the derivative of the elastic curve function:

$$q = y'(0). \tag{27.20}$$

The form of the elastic curve function is determined by conditions (27.19)–(27.20) at the given DOF unambiguously:

$$y = qz\left(1 - \frac{z}{l_1}\right). \tag{27.21}$$

The derivatives of this function have the form:

$$y' = q\left(1 - 2\frac{z}{l_1}\right); \quad y'' = -\frac{2q}{l_1}. \tag{27.22}$$

The strain energy of the system is calculated by formula (27.15). The elastic curve equation for a beam enables us to calculate the bending moment through an elastic curve function; for member AC we have:

$$M(z) = EIy''. \tag{27.23}$$

Limiting ourselves by the second-order infinitesimal of q, we may accept that $l_1 = l$ and get strain energy of member AC in the form:

$$U_{1\,AC} = \int_0^l \frac{M^2 dz}{2EI} = 2\frac{EIq^2}{l}.$$

The bending moment in member AB under unit DOF q is determined by the diagram in Figure 18.1b. Integrate function $M^2(z)$ along the length of this member; by taking into account the triangular shape of the diagram, we obtain the strain energy of member AB as follows:

$$U_{1\,AB} = 1.5\frac{EIq^2}{l}.$$

The strain energy of the system is the sum: $U_1 = U_{1\,AB} + U_{1\,AC}$.

In order to obtain the total potential energy of the system, it is required to determine the component of this energy related to the field of loads. This component in the given case is the work done by load P through displacement of joint A taken with the opposite sign: $U_{2P} = -P\Delta$.

Let us express displacement Δ through DOF q. This displacement is the difference between the axial line length of deformed member AC and distance l_1. In accordance with formula (27.18), it holds:

$$\Delta = l - l_1 = \frac{1}{2}\int_0^{l_1} y'^2 dz$$

By substitution of derivative (27.22) we obtain:

$$\Delta = \frac{q^2 l}{6} \Rightarrow U_{2P} = -P\Delta = -\frac{q^2 l}{6}P.$$

Thus we come to the formula for the potential energy of the system:

$$U(q) = U_1 + U_{2P} = \left(3.5\frac{EI}{l} - \frac{l}{6}P\right)q^2.$$

At the right-hand side, the doubled expression in parentheses is the second derivative of function $U(q)$. The condition of buckling and the corresponding critical force are obtained in the form:

$$3.5\frac{EI}{l} - \frac{l}{6}P = 0 \Rightarrow P_{cr} = 21\frac{EI}{l^2}.$$

The exact value of critical force in this problem can be established by means of the displacement method, which will be studied in the following chapters. This exact value is equal to:

$$P_{cr}^{+} = 13.9\frac{EI}{l^2}. \tag{27.24}$$

So, the error of calculation through the Rayleigh method in the given simple problem is 51%. Note that inequality $P_{cr} > P_{cr}^{+}$ is not by chance: the Rayleigh method always results in overestimations. We can give the following **intuitive** explanation for this. We proceed from the idea that any additional restrictions of displacements in a system shouldn't diminish critical force. A subjectively specified shape of a member's elastic curve will almost certainly differ from its true shape after the loss of stability. Therefore, in the Rayleigh method, we impose additional restrictions upon deformations of a system, and, through this means, increase calculated critical force.

The precision of calculation by the Rayleigh method depends on choosing the members' elastic curve functions, and the result of analysis usually renders only an order of magnitude of the sought critical force. Taking into account natural boundary conditions (if they take place) not only complicates calculations but may even increase the error of analysis.

The principal shortcoming of the Rayleigh method is that it doesn't enable us to estimate the precision of the obtained result. Nevertheless, this method is sometimes used owing to its simplicity.

28 Methods of Structure Stability Analysis

28.1 BUBNOV-GALERKIN METHOD AND BUCKLING ANALYSIS

An elastic system, in the general case, has an infinite number of DOFs because it is described by the displacement vector taken at all possible points of the system's bodies. In other words, the state of a system is specified by the displacement vector in the form of the vector-function of position. The theory of elasticity establishes that the equilibrium states of an elastic body are determined by the system of differential equations in unknown displacements of the body's points. Meanwhile, there is the philosophy of idealization of a real mechanical system, which suggests to consider the system as having a finite number of DOFs, and, instead of differential equations of equilibrium, to employ algebraic equations for its analysis; but such philosophy means idealization of a real system.

There is a simple method for stability analysis of systems with an infinite number of DOFs, which was developed during the period of 1913–1915 by Russian scientists: naval engineer professor I.G. Bubnov and professor of mathematics B.G. Galerkin. Initially, this method was employed for approximate calculation of the SSS in the theory of elasticity, but later, it appeared to be an effective instrument of structure stability analysis.

The **Bubnov-Galerkin method gives us a way to approximately specify the strained state of a structure by using a finite number of DOFs. The essence of the method is to replace the differential equations of equilibrium by algebraic linear equations for generalized coordinates.**

We shall provide substantiation of the Bubnov-Galerkin method for structures constituting a statically determinate column of a variable section with a single rigid support or two hinged supports at the ends. After setting forth the method for structures of this type, we generalize it for structures consisting of several members. The examples of statically determinate columns under consideration are shown in Figure 28.1. Each presented column is shown before and after the loss of stability. The centroidal principal axes of cross-sections are parallel to axes x and y of structure-bound CS, the ratio of inertia moments of the cross-sections is such that buckling occurs in the plane $O'yz$.

In the "mechanics of materials" course, we obtained the equation of elastic curve due to buckling for a column of constant cross-section, see (Hibbeler 2011, Sec. 13.4) and also: (Beer et al. 2012, Sec. 10.5). The same equation can be derived for the buckling of a column with a variable cross-section. We position the origin of bound CS O' at the upper end of the bar (at the point of load action) and direct axis $O'z$ vertically downwards (Figure 28.1). Then, for the buckling in the plane $O'yz$, the equation of the deflection curve has the form:

$$\frac{d^2y}{dz^2} + \frac{P}{EI(z)}y = 0, \tag{28.1}$$

where $EI(z)$ is the flexural rigidity; at that, the moment of inertia depends on coordinate z.

The boundary conditions for an elastic curve depend on the type of supports. For scheme a, it holds:

$$y(0) = y(l) = 0. \tag{28.2}$$

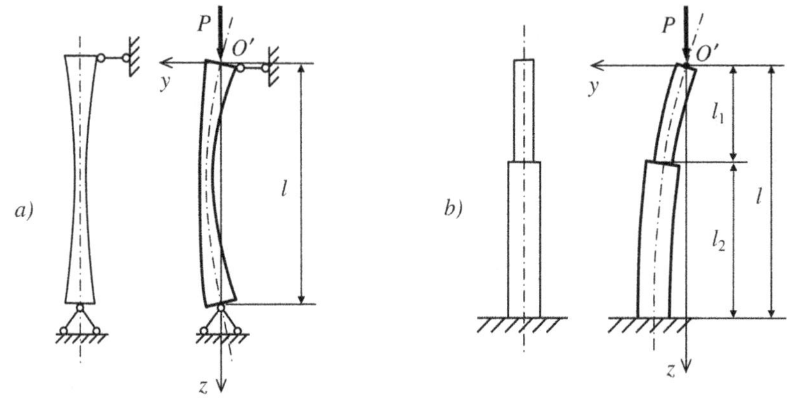

FIGURE 28.1 Statically determinate columns and their buckling.

For scheme b it holds:

$$y(0) = 0; \quad \left.\frac{dy}{dz}\right|_{z=l} = 0. \tag{28.3}$$

In the case under consideration, the strain state of a bar is described by continuously differentiable deflection function $y = y(z)$, which represents dependence "lateral displacement versus coordinate" and satisfies the linear differential equation and linear boundary conditions. The linearity of boundary conditions means that linear combinations of functions which comply with these conditions comply with them as well.

We refer to a function which is defined on some segment $[0, l]$, except a finite number of points, and has continuous derivatives of arbitrary order on the definitional domain as a *piecewise smooth function*. The dependence $I = I(z)$ is a piecewise smooth function; its indetermination points separate the stages of the member, i.e., are positioned at the places of cross-section step changes.

The Bubnov-Galerkin method is used for stability analysis of systems, the equilibrium states of which are specified by linear differential equations and linear boundary conditions. The type of structures under consideration is convenient for substantiation of the method owing to the simple form of both the equation of state (28.1) and boundary conditions.

The separately taken left side of differential equation (28.1) defines the transformation of arbitrary function $y(z)$, smooth within every stage of a bar, into another function:

$$u(z) = y''(z) + \frac{P}{EI(z)} y(z). \tag{28.4}$$

We shall consider the rule of transformation (28.4) as a linear operator in vector space* of piecewise smooth functions with fixed indetermination points. This space we denote $\mathbf{L}_2$. The indetermination points of functions forming space $\mathbf{L}_2$ depend on the problem being solved. In the example in Figure 28.1a, the functions are smooth at the whole segment $[0, l]$; in the example on diagram b, the uncertainty takes place at $z = l_1$. (In a column consisting of several stages, the indetermination points of deflection function are positioned between stages.) We denote operator (28.4) as $\hat{L}$ and refer to it as a differential operator of the equation of state. Expression (28.4) is shortly written in terms of operators as follows:

$$u = \hat{L}y \text{ or } u(z) = (\hat{L}y)(z).$$

* The set of elements wherein the operations of element summation and element-number multiplication are defined is referred to as vector space. The elements of vector space are referred to as vectors.

The later expression clarifies that the result of operator action upon some function is also a function. This result is referred to as an image of vector y. Further, equation (28.1) will be written in terms of operators like this:

$$\hat{L}y = 0. \tag{28.5}$$

This equation is to be solved not on the whole space $\mathbf{L}_2$, but on the subspace $\mathbf{L}' \subset \mathbf{L}_2$ formed by functions, which satisfy the boundary conditions of the problem. It is important, in so doing, that while allowing uncertainty of function $y(z)$ between stages of the bar, we must supplement the conditions at the endpoints (28.2) and (28.3) by the conditions binding together smooth branches of this function.

In the Bubnov-Galerkin method, the SSS of a structure is investigated by means of piecewise smooth deflection function $y = y(z)$, for which two types of boundary conditions are possible:

- Zero boundary conditions of the form:

$$y^{(m)}(z_{ind} + 0) = 0 \;\text{ or }\; y^{(m)}(z_{ind} - 0) = 0,$$

  where z_{ind} is a boundary point of a definitional domain or indetermination point; m marks the derivative's order (in particular, $y^{(0)} \equiv y$); and notations $f(x + 0)$ and $f(x - 0)$ are used for one-sided limits of function f at point x.
- Conditions of stitching at the indetermination points, which link left and right limit values of a function's derivatives (and of the function itself) at the indetermination points. These conditions usually have the form:

$$y^{(m)}(z_{ind} - 0) = k\, y^{(m)}(z_{ind} + 0),$$

  where k is the proportionality factor established for every condition.

Conditions of stitching for a compressed bar of variable cross-section imply that function $y(z) \in \mathbf{L}_2$ can be supplemented by the values at indetermination points in such manner that the function becomes continuous. Besides, it is necessary to secure the continuity of the first derivative for the supplemented function at the indetermination points. In the case of the bar on diagram a, there are no conditions of stitching because all the functions in $\mathbf{L}_2$ are smooth on segment $[0, l]$. In the case of the cantilever column on diagram b, the derivative $y'(z)$ must be continuous at the point $z = l_1$ in order to ensure the absence of kink of the elastic curve. (Here and further on, we do not stipulate that function $y(z) \in \mathbf{L}_2$ was supplemented at indetermination points by the left or right limit value.)

In order to move to a system with a finite number of DOFs according to the Bubnov-Galerkin method, it is required to put into consideration the set of linearly independent coordinate functions $y_1(z), y_2(z), \ldots, y_n(z)$ which belong to space $\mathbf{L}_2$ and comply with two requirements of accordance with the expected mode of buckling:

- Every coordinate function satisfies zero boundary conditions and the conditions of stitching;
- The equilibrium state of a system is represented, at least approximately, by a linear combination of coordinate functions.

By means of such a set of functions, the equilibrium strain state of a structure can be approximately represented by following deflection function:

$$y(z) = q_1 y_1(z) + q_2 y_2(z) + \cdots + q_n y_n(z). \tag{28.6}$$

So, the equilibrium states of the structure are specified through a finite set of coefficients q_i. We call these coefficients generalized coordinates, keeping in mind, however, that they determine the SSS of a structure approximately, and the error of approximation is not known beforehand.

In order to pass from the differential equation of the state to algebraic equations for generalized coordinates, we substitute expression (28.6) into the left side of operator equation (28.5). We get the equation:

$$\sum_{i=1}^{n} q_i \hat{L} y_i = 0. \tag{28.7}$$

We multiply both sides of equation (28.7) by any coordinate function y_k and integrate the left and right sides along all the length of the bar. The left side takes the form:

$$\int_0^l y_k(z)\left(\sum_{i=1}^{n} q_i \hat{L} y_i\right)(z)\,dz = \sum_{i=1}^{n} q_i \int_0^l y_k(z)(\hat{L} y_i)(z)dz. \tag{28.8}$$

The right side remains zero after integration. By making this operation for all coordinate functions y_k, we get the system of equations for generalized coordinates:

$$\sum_{i=1}^{n} L_{ki} q_i = 0, \;\; k = \overline{1,n}, \tag{28.9}$$

where it is denoted:

$$L_{ki} = \int_0^l y_k(z)(\hat{L} y_i)(z)dz. \tag{28.10}$$

The Bubnov-Galerkin method prescribes investigation of the stability of a system's equilibrium states by means of equations (28.9). In order to formulate the substance of the Bubnov-Galerkin method briefly, it is convenient to introduce the notions of linear span and inner product in vector space $\mathbf{L}_2$ of piecewise smooth functions. The set of functions of the form (28.6) at all the possible values of factors q_i is referred to as the linear span of the system of functions y_i, $i = \overline{1,n}$. The integral

$$(u,v) \equiv \int_0^l uv dz \tag{28.11}$$

is referred to as the inner product of functions $u(z)$ and $v(z)$ (of the vectors belonging to the named vector space). The expression at the left side of (28.8) is the inner product $\left(y_k, \hat{L} y\right)$, where y is the linear combination of coordinate functions of the form (28.6).

The Bubnov-Galerkin method is based on constructing the set of coordinate functions which form the linear span containing an approximation of the required solution. In order to find the coordinates of the approximate solution, one constructs the system of linear equations through vector multiplication of both sides of the operator equation by the coordinate functions.

We have been studying the Bubnov-Galerkin method as an instrument for stability analysis of statically determinate columns with the equation of elastic curve given in the form (28.1). In order to employ this method for other types of structures, we should operate in the following manner:

1) To represent the deformed state of a structure through function $y = y(z)$ piecewise smooth at a segment $[0, l]$, which determines the dependence of deflections (transverse displacements) of members' cross-sections on their position in the space. To establish the indetermination points within segment $[0, l]$, and by this to specify vector space $\mathbf{L}_2$ containing function $y(z)$;

2) To obtain linear operator $\hat{L} = \hat{L}(P)$, dependent on load parameter P, which acts in space $\mathbf{L}_2$ and enables us to represent the equation of the structure's equilibrium state in the form (28.5). Also, to establish boundary conditions which define the subspace for solving this equation;
3) To construct in $\mathbf{L}_2$ a system of linearly independent coordinate functions $y_i(z)$, $i = \overline{1, n}$, which satisfy boundary conditions and enable us to reproduce the expected shape of the structure's buckling through linear combination (28.6);
4) To calculate the n-th order matrix of inner products $L_{ki} = \left(y_k, \hat{L}y_i\right)$.

Under these four conditions, the operator equation of a structure's equilibrium state is represented in the form of linear equation set (28.9), and the problem of stability analysis lies in investigation of this equation system about the solution's uniqueness at different loads. Note that the requirement of stitching branches together and of accordance to zero boundary conditions, imposed above upon coordinate functions of a compressed structure with the equation of deflection curve (28.1), is specified according to the character of a structure and is intended to secure the membership of coordinate functions to subspace $\mathbf{L}'$, wherein solutions of equation (28.5) represent equilibrium states.

A possible application area for this method is stability analysis of rectilinear structures consisting of sequentially connected prismatic members positioned in a straight line. (An example of such a structure is shown in Figure 6.2a. In contrast to the multi-support beams, these structures are subjected only to longitudinal loads.) The position of every member's cross-section in structures like these can be specified by length z of the axial line portion between the section and structure endpoint. Respectively, the right bound of definitional domain [0, l] for deflection function $y(z)$ is the sum of member lengths: $l_\Sigma = \sum l_i$. For the structures of this kind, the conditions of stitching prescribe continuity of the deflection function, yet the discontinuities of derivative $y'(z)$ are allowed at the points of hinged connections between members.

We shall distinguish geometric zero boundary conditions for member deflection curves, which have the form $y = 0$; $y' = 0$, and natural zero boundary conditions, which may be fulfilled at the members' endpoints and have the form: $M(z) = 0$; $V(z) = 0$. Similarly, we refer to conditions of stitching, imposed upon limit values of function $y(z)$ and derivative $y'(z)$ at the indetermination points, as geometric, in distinction from stitching conditions for derivatives $y''(z)$, $y'''(z)$, which we shall call natural. For a statically determinate column with the deflection curve equation (28.1), considered above, natural boundary conditions are fulfilled for an exact solution as the effect of geometric conditions. While choosing coordinate functions for rectilinear structures, it is sufficient to take into account geometric conditions only.

Despite the fact that equations (28.9) underlying the method determine the state of a system approximately, this method is broadly employed because of its simplicity and usually correct conclusions.

In the Bubnov-Galerkin method, there is the possibility to estimate the error of the obtained critical structure buckling load. The evaluation of error is done through comparison of results of two-three calculations which should be accomplished in such a way that the sequence of calculated critical loads converges to an exact value. For analysis of method error, the coordinate functions are selected in special manner: one introduces an infinite sequence of these functions so that an exact function of deflections $y(z)$ can be obtained as the limit of linear form (28.6) when $n \to \infty$. In other words, the sequence of coordinate functions must secure the convergence of the functional series constructed from these functions to the exact solution of equation (28.5). To this end, the power series and Fourier series are in use. While making the analysis of the error, we assume that if an employed series converges to an exact solution of the problem, then the calculated critical force converges to an exact value as well. The technique of calculation of a critical force through the Bubnov-Galerkin method with estimation of error is described in the book *Buckling of Bars, Plates, and Shells* (Jones 2006). The accuracy analysis of this method should be considered as a relevant scientific problem.

Example

For the two-stage column in Figure 28.1b, specify the shape of buckling by a single generalized coordinate and determine critical force by the Bubnov-Galerkin method. The result should be obtained in general form and also under the next input data: $l_1 = 0.4l$; inertia moment of upper stage $I_1 = 0.4I_2$, where I_2 is the known inertia moment of the lower stage (both stages are made of the same material with Young modulus E).

The shape of buckling on diagram b resembles the sine curve; thus, we choose following coordinate function of the Bubnov-Galerkin method:

$$y_1 = \sin\frac{\pi}{2l}z. \tag{28.12}$$

The equation of the equilibrium state in the given case is single:

$$L_{11}(P)q_1 = 0.$$

Here, we indicate the dependence of coefficient L_{11} on load. According to the static approach, the critical load is the load under which there is a nontrivial solution of the equilibrium equation, i.e., when

$$L_{11}(P) = 0. \tag{28.13}$$

In the given case, we have for coefficient L_{11}:

$$L_{11} = \int_0^l y_1(z)(\hat{L}y_1)(z)dz = \int_0^l \sin\frac{\pi}{2l}z\left(-\frac{\pi^2}{4l^2}\sin\frac{\pi}{2l}z + \frac{P}{EI(z)}\sin\frac{\pi}{2l}z\right)dz$$
$$= \int_0^l \left(-\frac{\pi^2}{4l^2} + \frac{P}{EI(z)}\right)\sin^2\left(\frac{\pi}{2l}z\right)dz. \tag{28.14}$$

In the integrand, we make the substitution:

$$\sin^2\frac{\pi}{2l}z = \frac{1}{2}\left(1 - \cos\frac{\pi}{l}z\right)$$

and integrate it separately on both portions:

$$\int_0^l = \int_0^{l_1} + \int_{l_1}^l.$$

For the first portion, we take $I(z) = I_1$, for the second, $I(z) = I_2$. The result of integration has the form:

$$L_{11} = \frac{1}{2}\left[-\frac{\pi^2}{4l} + \frac{P}{E}\left(l_1 I_1^{-1} + l_2 I_2^{-1} + (I_2^{-1} - I_1^{-1})\frac{l}{\pi}\sin\frac{\pi l_1}{l}\right)\right].$$

Equating the latter coefficient to zero, we come to the general relationship for critical force:

$$P_{cr} = \frac{\pi^2 E}{\left(l_1 I_1^{-1} + l_2 I_2^{-1} + (I_2^{-1} - I_1^{-1})\frac{l}{\pi}\sin\frac{\pi l_1}{l}\right)4l}. \tag{28.15}$$

After substitution of the data of the example, we get:

$$P_{cr} = 8.61\frac{EI_2}{(2l)^2}. \tag{28.16}$$

In the book *Theory of Elastic Stability* (Timoshenko 1936), there is an exact solution of the stability analysis problem for a column consisting of two prismatic members. In the case considered above, we can get the exact result as follows:

$$P_{cr} = 8.51\frac{EI_2}{(2l)^2}.$$

One can see that there is a small error in the approximate magnitude of critical force (28.16).

28.2 INTRODUCTION TO DISPLACEMENT METHOD

During the first half of the 20th century, an Austro-German school of mechanics developed the displacement method for analysis of stability of frames, which is the generalization of the displacement method for determining the SSS of bar structures, studied in Chapters 17–18. The basic clauses of the displacement method for investigation of stability were set out by Austrian scientist Friedrich Bleich in 1919. In the 1940s, this new method acquired the modern form, owing, first and foremost, to the works of Soviet scientist N.V. Kornoukhov, and also Austrian scientist Ernst Chwalla. Among all the methods of analysis of a bar system's stability, the displacement method is most applicable for manual calculations.

Let us consider any elastic plane frame, wherein the members have a prismatic shape and are jointed by connections of the second and third types, whereas connections of the first type are possible only with the earth (there are no elastic hinges).* In terms of the displacement method, the joints of elastic members to each other, to simple rods, or together with the earth, as well as the cantilever ends, are referred to as nodes, and a member has nodes only at the ends. We assume that loads are exerted on the nodes of a frame, and the equilibrium state of a frame is possible when members are subjected only to axial compression and, maybe, tension. The frames in this state are referred to as *purely compressed*; they can lose stability under large enough loads. Among these frames, we shall consider only the frames wherein the internal axial forces are statically determinate. This means that for a frame in the pure compression state, the equation set of the nodes' equilibrium and the members' equilibrium enables us to establish unambiguously the axial forces in members. We call a hinged truss with minimum number of support connections, which is made from a frame by the substitution of the hinges for all rigid joints and the installation of additional support links at some nodes as *a truss conjugate with respect to the given frame.* In order that in a purely compressed frame, the axial forces are statically determinate, it is necessary and sufficient that the conjugate truss is statically determinate.†

An example of a purely compressed frame with statically determinate axial forces and a corresponding conjugate truss are shown in Figure 28.2, diagrams a and b. An example of a purely compressed frame in which the axial forces are statically indeterminate is shown in Figure 28.3a. While passing to a conjugate truss, mixed node A is transformed into multiple hinge A' (Figure 28.3b). The example of a compressed-bent frame, containing purely compressed members, is shown in

* We do not refer the support simple rods to members of the frame: the support rods are the sources of support reactions and in this sense are referred to the earth.

† The necessity: in statically indeterminate truss internal forces are possible at zero loads and zero reactions of required supports, but such a state contradicts the assumption of uniqueness of internal forces determined from statics equations at loads of pure compression. The sufficiency: in statically determinate truss internal forces are determined from equations of statics. In conjugate truss the loads and internal forces of frame's pure compression comply with these equations. Thus, axial forces of pure compression are statically determinate.

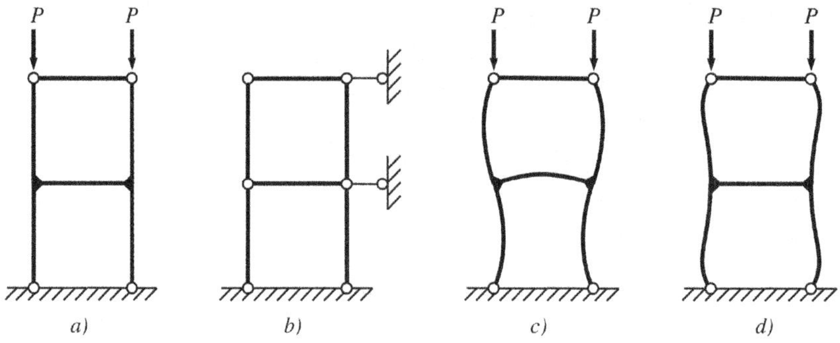

FIGURE 28.2 Purely compressed frame (a), conjugate-to-it statically determinate truss (b), and diagrams of stability loss for this frame (c and d).

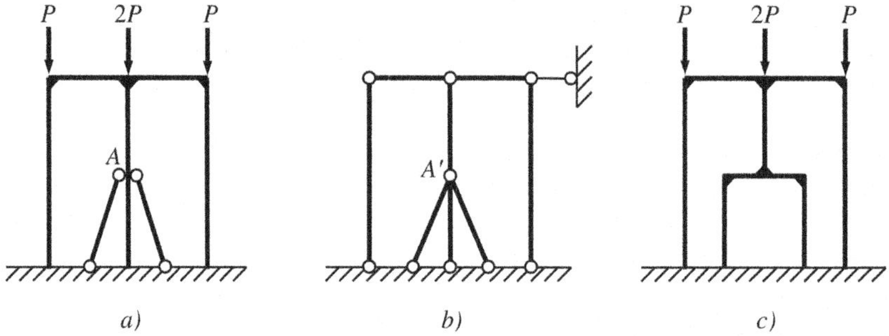

FIGURE 28.3 Purely compressed frame (a), conjugate-to-it statically indeterminate truss (b), and compressed-bent frame (c).

Figure 28.3c. Though such a frame also may lose stability, there are always bent members in it. The frames of these two types are not considered further on.

The displacement method for stability investigation is intended for calculation of purely compressed frames with statically determinate axial forces. In its main points, this method reproduces the displacement method for calculation of the frames in bending.

According to the method under consideration, the SSS of a frame being under given loads is calculated through the totality of linear and angular displacements of nodes by means of calculating these displacements. The *full set of displacements of structure nodes* is specified in the following manner: for each node with no supports, we introduce two independent linear displacements; for a node with hinged immovable support, we introduce one linear displacement which the support permits; and besides, the position of every rigid node not fixed by the earth* is specified by angular displacement. But not all the displacements of the full set are used for analysis by the displacement method; at the first stage of analysis, one constructs the set of displacements employed in the method, which includes a relatively small number of displacements from the full set, in view that problem could be solved by the graphic-analytical method.

In the displacement method, we take as given that, besides loads specified in the problem, the external forces or force couples can be exerted upon those nodes, the displacements of which determine the SSS of a structure. These external forces act in the directions of named nodal displacements (linear or angular, respectively). In a frame assembled of elastic members, owing to the action of these additional external forces, the displacements of the set employed in the method are independent, i.e., these displacements can take arbitrary values. The set of nodal displacements is constructed in such a

* We say that a node is fixed by the earth if it is directly fastened to the earth. Example: the frame in Figure 28.3a includes five nodes fixed by the earth, among them three rigid support nodes and two support hinges.

manner that they define the SSS of a frame. These displacements are called degrees of freedom, and they correspond to the definition of DOFs given in previous chapters. We call independent linear and angular displacements of nodes translational and rotational DOFs, respectively.

The displacement method for investigation of structural stability is based on two assumptions. The first assumption states that **the set of DOFs and the SSS of a structure can be established by neglecting the variation of distance between nodes of every member due to loads**. While investigating the bending of frames by the displacement method, the same assumption was made. This reduced the number of independent nodal displacements and enabled us to specify the strain of members within the bending-moment theory. In analysis of stability, the independent displacements of nodes are established in the same way as in analysis of bending. The number of rotational DOFs is the number of the system's rigid nodes not fixed by the earth. In order to determine the number of translational DOFs, we introduce the *mechanism conjugate with the frame* through substitution of hinges for all rigid nodes and make kinematic analysis of the obtained mechanism. The joints and free ends of members in the conjugate mechanism are still called nodes. The totality of independent nodal displacements in the conjugate mechanism, which uniquely determine its position, is the totality of translational DOFs of the given system. The total number of DOFs is equal to the degree of kinematic indeterminacy (17.17). These DOFs determine the position of a system in the space (accurate to within errors produced by negligence of axial deformations of members). We shortly refer to the set of DOFs obtained using the first assumption as a *reduced set*. The DOFs of this set we refer to as *reduced DOFs*. A reduced set is composed of a full set of rotational and translational nodal displacements by deletion of some (maybe all) translational displacements. The reduced totality of independent nodal displacements is denoted as $Z_i,\ i=\overline{1,n}$.

While accepting the first assumption of the method, we assume that the stress state of a structure can be caused by two assemblages of external forces:

- Given assemblage of loads producing pure compression of frame;
- External nodal forces and moments acting in the directions of reduced nodal displacements upon the same nodes.

We call the second assemblage of forces, which is matched to reduced displacements, the reduced assemblage of forces.

By virtue of the superposition principle, at given loads of pure compression, the reduced assemblage of forces and reduced totality of DOFs are determined in a one-to-one manner by linear equations (see Chapter 17). Besides, the increments of internal forces in members caused by reduced external forces at constant loads of pure compression are linear homogeneous functions of reduced DOFs. But these assertions are correct only at sufficiently small loads. Under large loads of compression, the superposition principle is violated, and these assertions may be unacceptable. Nevertheless, **while accepting the first assumption, we take for granted that the functional dependence of internal forces on DOFs is secured, and this dependence is differentiable in variables $Z_i,\ i=\overline{1,n}$, at the zero point**.

The main limitation of applicability of the displacement method for investigation of stability is linked to the possibility of *local buckling of the frame's members, which lies in deformation of individual members under increasing loads according to the scheme of buckling for a compressed member stable relative to the earth (the scheme studied by L. Euler)*. Let us consider the example of a structure in Figure 28.2a. This structure may lose stability as shown on diagram c of this figure, when the elastic curve of buckled members is defined by displacements of nodes, but the shape of buckling is also possible as shown on diagram d, when angular nodal displacements do not occur. In the latter case, the same position of nodes may correspond to the state of unstable compression of members (diagram a),* i.e., the position of nodes does not uniquely determine the SSS of a system.

* In the case shown in Figure 28.3d, there are small horizontal displacements of nodes caused by axial forces in horizontal members, but we neglect these displacements.

In the case in Figure 28.2d, all vertical members have lost local stability. Local buckling occurs without noticeable displacements of nodes. Further on, we exclude from consideration the possibility of local buckling.

When we have accepted the first assumption, we also have excluded from consideration the structures in which global buckling may arise on pure compression as, for example, in the case of the truss in Figure 27.2c. Note as well that assumption of unchangeable length of members might be unacceptable at the preliminary stage of analysis, when the type of frame compression is being established. For example, during investigation of the stability of multi-story building frameworks, one may conclude that a framework is purely compressed, whereas there is compression with bending due to nonuniform shortening of columns, and then the employment of the displacement method leads to incorrect inferences. The limitation of applicability of the displacement method for stability analysis of multi-story frameworks was established by Soviet scientist A.I. Segal' (1949).

Having defined the reduced set of DOFs, we can introduce the primary system of the displacement method in the same way as was done in analysis of bending: The nodes of a system are attached to the earth by additional support connections which eliminate all independent displacements of nodes and produce the reduced assemblage of external forces. These connections may be either of the moment type (floating supports) or the force type (support links). In the primary system, the displacements of additional supports $Z_i,\ i=\overline{1,n}$, are permitted. These displacements have the same physical meaning as the DOFs of the given system. They define the state of the primary system at specified loads and are referred to as *primary unknowns of the displacement method.* The loads in combination with the displacements of additional supports produce the reactions of these supports $R_i,\ i=\overline{1,n}$. In states with zero reactions, the primary system becomes equivalent to the given one.

We write down the equivalence conditions in the form:

$$R_i(\mathbf{Z},P)=0, \quad i=\overline{1,n}. \tag{28.17}$$

Here, we point out the reaction R_i's dependence on two arguments: $\mathbf{Z}=(Z_1,\dots,Z_n)^{\mathrm{T}}$, which is the vector of displacements of additional supports, and P, which is the parameter of loads exerted upon a given system. Further on, we consider parameter P as the sought proportionality factor for the given assemblage of loads. The conditions (28.17) constitute the *canonical set of equations in primary unknowns of the displacement method.* Under these conditions, the equilibrium of the given system is ensured.

In analysis of bending, the canonical equation set served the determination of internal forces in frame members. In analysis of stability, equation set (28.17) serves another purpose: by its use, the critical value of load parameter P is determined. In the given case, we have a mechanical system described by a finite number of DOFs $Z_i,\ i=\overline{1,n}$, in which the states with small displacements might be unstable. Besides, we have the system of equations (28.17), which determine the DOFs in the equilibrium states of a structure. So, we have all preconditions of the static approach to investigation of stability. The critical parameter of loads in the given case is minimal parameter P at which there arises the multiplicity of solutions of the canonical equation set.

In order to represent the system of equations (28.17) in a form suitable for analysis of stability, we note at the beginning that in the case of purely compressed frames, all the DOFs and loaded reactions are zeroth:

$$R_{iP}\equiv R_i(0,P)=0. \tag{28.18}$$

Indeed, **the first assumption of the method prescribes neglecting the displacements of nodes in a purely compressed frame.*** For example, in the frame shown in Figure 28.4, the angular dis-

* In a purely compressed frame, a constant length of members means that there are no deformations at all. Consequently, there are no displacements either.

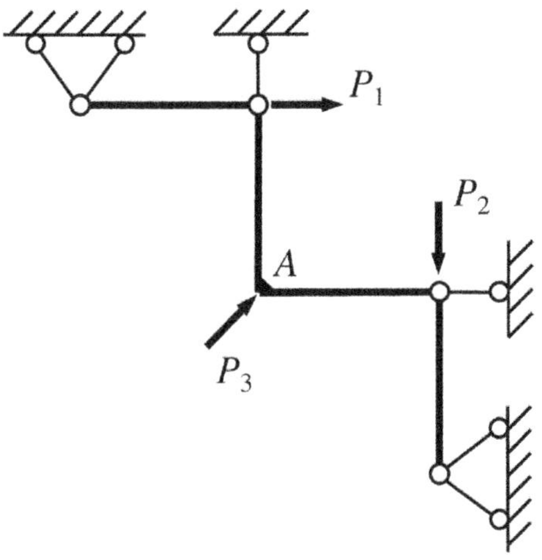

FIGURE 28.4 Purely compressed frame with small pivot of node A under loading.

placement of rigid node A is considered to be zero because the pivot of this node occurs only due to tension-compression of members. So, in the state of pure compression, all DOFs of a frame are zeroth. If one installs additional support upon a node which is not displaced under a load, then such a support will not develop a reaction due to load. Therefore, in a primary system not subjected to displacements of additional supports, the reactions of these supports are zeroth.

By taking into account equality (28.18), equation system (28.17) can be written by means of finite increments of reactions developed by additional supports, taken at zero point $\mathbf{Z} = 0$:

$$\Delta R_i(\mathbf{Z}, P) = 0, \quad i = \overline{1, n}, \tag{28.19}$$

where $\Delta R_i(\mathbf{Z}, P) \equiv R_i(\mathbf{Z}, P) - R_i(0, P)$.

Next, we introduce the second assumption of the displacement method: **while finding the critical load, the finite increments of reactions in canonical equations could be replaced by their differentials**. In other words, it is allowed to specify the increments of reactions of additional supports by the functions linear in primary unknowns, which we write down in the form:

$$\Delta R_i(\mathbf{Z}, P) \cong \sum_{j=1}^{n} r_{ij}(P) Z_j. \tag{28.20}$$

The ground for the second assumption is an arbitrarily small error of approximation (28.20) owing to sufficiently small displacements of supports.

By substitution (28.20), the canonical equation set (28.19) is represented in linear form:

$$\sum_{j=1}^{n} r_{ij}(P) Z_j = 0, \quad i = \overline{1, n}. \tag{28.21}$$

The displacement method is to set up and analyze the system of canonical equations (28.21), which determine the equilibrium state of the given mechanical system. The coefficient matrix of this equation set is made up by the unit reactions of additional connections (the definition of which was given in Chapter 18):

$$\mathbf{r}(P) = \left(r_{ij}(P)\right).$$

The setting up of the system of canonical equations in the given case is understood as finding out the relationship of unit reactions versus load parameter P.

The critical force is determined as the minimal force under which the canonical equation set has a nontrivial solution, i.e., this force can be established from the equation:

$$\det \mathbf{r}(P) = 0. \tag{28.22}$$

If the system of equations of equilibrium states is a linear homogeneous system of algebraic equations, then the determinant of the coefficient matrix of the system is referred to as the *determinant of stability*. The determinant det **r** on the left-hand side of equation (28.22) is the determinant of stability, and the equation itself is sometimes called an equation of stability.

It is known that the matrix of unit reactions is symmetric. This fact is proved in the supplement to the present chapter.

At zero loads, the unit reaction matrix **r**(P) becomes a stiffness matrix of the displacement method for calculation of a structure's SSS. In this latter method, a stiffness matrix shouldn't depend on loads, for such dependence means nonlinear relation between loads and displacements, i.e., contradicts the superposition principle.

In conclusion, we note that the full set of displacements of structure nodes is used for analysis of stability in the stiffness method, which has much in common with the displacement method. In particular, in the stiffness method, the critical loads are also determined from the equation of stability of the form (28.22). Just like the stiffness method in the case of deformation of a frame under small loads is the generalization of the displacement method for analysis of bending, the stiffness method in the case of large loads of pure compression is the generalization of the displacement method for analysis of stability.

SUPPLEMENT TO CHAPTER

Symmetry of Unit Reaction Matrix

The proof of reciprocity of unit reactions in the case of a purely compressed frame is done by following the same pattern as used for proving Rayleigh's first theorem (see Chapter 12). Having considered two states of a primary system caused by displacements of additional supports m and n, respectively, and by the action of nodal loads P_i in both states, we come to the equality of virtual works, resembling relationship (12.1) but including the components of virtual work related to loads:

$$R_{mn}\Delta_m + \sum P_i\Delta_{i(1)} = R_{nm}\Delta_n + \sum P_i\Delta_{i(2)}. \tag{28S.1}$$

Here, $\Delta_{i(1)}$ is the displacement of the node subjected to load P_i in the direction of this load in the first state (with displaced support m); $\Delta_{i(2)}$ is the same displacement in the second state (with displaced support n); and the summation is done over all nodal loads.

Each of the additional sums in equality (28S.1) happens to be of zero value, wherefrom follows the relationship (12.1) and required equality (12.2). We show that, indeed, under displacement of any additional support of the primary system, it holds:

$$\sum P_i\Delta_i = 0, \tag{28S.2}$$

where Δ_i is the displacement of the node subjected to load. If the support is of the floating type, then translational displacements of nodes do not arise, and equality (28S.2) is evident. If the additional support is of the hinged-movable type, then displacements of nodes in the primary system due to the bias of this support are virtual displacements of the mechanism made up by replacement of rigid nodes by hinges in the given system. But this mechanism keeps equilibrium under nodal loads P_i. Indeed, the equilibrium of this mechanism is secured by pure compression of the frame: every frame node keeps equilibrium owing only to axial forces in the members, i.e., the replacement of rigid nodes by hinges does not affect this equilibrium. So, displacements Δ_i form the totality of virtual displacements of nodes in the mechanism in equilibrium under loads P_i. Using the principle of virtual work, we get equality (28S.2).

29 Investigation of Frames' Buckling by Means of Displacement Method

29.1 DETERMINATION OF SSS OF BEAM-COLUMN USING INITIAL PARAMETERS METHOD

We shall consider a beam-column bent in the plane $O'yz$ due to the action of forces exerted on the end-sections, among them the compression force P (Figure 29.1). We briefly call this structural member a bar. We assume that axis $O'z$ of the earth-bound CS coincides with the initial position of the unstrained bar, and the origin of this CS is selected in such a way that the projection of the bar upon axis z is some segment $[0, l]$. **We select the directions of axial and shear forces with respect to the undeformed axial line of a bar**; the moments and shear forces are determined under the beam sign convention (as shown in Figure 29.1). An angle of slope of the elastic curve is read in a clockwise direction, and this angle is assumed to be small; thus, it can be determined as the derivative of the deflection function:

$$\theta = y'(z).$$

We set the problem of finding the deflection function $y(z)$ and internal forces at arbitrary section $M(z)$, $V(z)$ as the functions of initial parameters $y(0)$, $\theta(0)$, $M(0)$, and $V(0)$. The equilibrium condition for a bar's portion on the left of the selected cross-section gives the relationship:

$$M(z) = M(0) + V(0)z + P\left[y - y(0)\right]. \tag{29.1}$$

In the mechanics of materials, we set up the equation of the beam's elastic curve, which for the bar in Figure 29.1 takes the form:

$$\frac{d^2y}{dz^2} = -\frac{M(z)}{EI}.$$

Substitute the moment (29.1) into this equation. We get:

$$\frac{d^2y}{dz^2} = -\frac{M(0) + V(0)z + P[y - y(0)]}{EI}. \tag{29.2}$$

We convert this equation to the standard form of a linear equation with a right side:

$$y'' + k^2 y = -\frac{M(0) + V(0)z - Py(0)}{EI}; \tag{29.3}$$

$$k \equiv \sqrt{\frac{P}{EI}}.$$

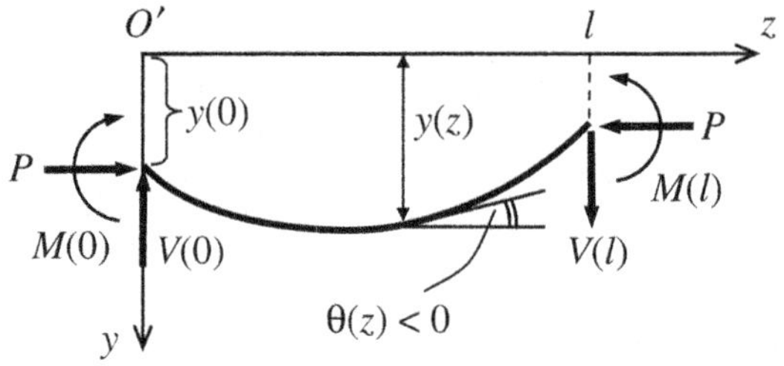

FIGURE 29.1 Beam-column under action of forces exerted on the end-sections.

The general integral of this equation, as it is easy to see, has the form:

$$y(z) = C_1 \sin kz + C_2 \cos kz - \frac{M(0) + V(0)z - Py(0)}{k^2 EI}. \tag{29.4}$$

Coefficients C_1 and C_2 are determined from boundary conditions:

$$y(0) = C_2 - \frac{M(0) - Py(0)}{k^2 EI};$$

$$y'(0) = C_1 k - \frac{V(0)}{k^2 EI}.$$

We have:

$$C_2 = \frac{M(0)}{k^2 EI}; \quad 4C_1 = \frac{y'(0)}{k} + \frac{V(0)}{k^3 EI}.$$

After substitution of these relationships into (29.4), we come to the final expression:

$$y(z) = y(0) + \frac{\theta(0)}{k} \sin kz - \frac{M(0)}{P}(1 - \cos kz) - \frac{V(0)}{P} \cdot \frac{kz - \sin kz}{k} \tag{29.5}$$

(here, the replacement $\theta(0) = y'(0)$ has been done). The angle of the slope of the elastic curve is easily obtained by derivation of the latter formula:

$$\theta(z) = \theta(0)\cos kz - \frac{M(0)k}{P}\sin kz - \frac{V(0)}{P}(1 - \cos kz). \tag{29.6}$$

The bending moment as a function of the cross-section's coordinate can be obtained by repeated differentiation using the formula $M(z) = -EIy''$:

$$M(z) = \theta(0)kEI \sin kz + M(0)\cos kz + \frac{V(0)}{k}\sin kz. \tag{29.7}$$

For shear force, we have, from equilibrium conditions:

$$V(z) = V(0). \tag{29.8}$$

Formulas (29.5)–(29.8) are known as the basic formulas of the initial parameters method for a beam-column. This method is aimed to determine the SSS of a beam-column through boundary conditions at its ends. For calculation of internal forces in a bar, basic formulas are used twice; the first time they enable us to set up the equations in unknown initial parameters by using boundary conditions, and the second time, they allow us to determine displacements y, θ, and internal forces M, V at an arbitrary cross-section through established initial parameters. Note that these formulas were obtained for nonzero compression force: $P > 0$. We may pass to special case $P = 0$ by evaluation of indeterminate form 0:0 in the obtained formulas. The result will be the known formulas of the initial parameters method for bending (Karnovsky and Lebed 2010, Ch. 6).

By using basic formulas, one establishes internal forces in a beam-column in its standard unit states, and obtained internal force distributions are employed in the displacement method. The standard unit states are considered in the next items of the chapter, while here, we use these formulas in order to show that internal forces in a beam-column are linear forms of end-sections' displacements with coefficients that depend on compression force. More precisely, this fact is established for four types of bars which differ in the method of fixing the ends. These types are assembled in Table 29.1. For each type, the table shows what displacements of end-sections are given and what additional requirements determine the deformation of bars. The shown displacements are determined by primary unknowns of displacement method Z_j. We establish the linear relation of internal forces with end-section displacements for a beam-column of first type (from the first row of the table). For other types of bars, such a relation is established in similar way.

It is convenient to specify the stress state of beam-columns by *parameter of compression* $\nu = kl$. For every type of bar under consideration, one can determine the critical magnitude of this parameter ν_{cr}, which indicates the instability of a bar with zero displacements of its end-sections. The shapes of buckling and critical magnitudes of the compression parameter are shown in Table 29.1. The characteristic ν_{cr} is important due to the fact that the linear relation of internal forces and displacements of end-sections of a bar is ensured under condition:

$$0 < \nu < \nu_{cr}. \tag{29.9}$$

On the contrary, if $\nu = \nu_{cr}$, then the forces are determined ambiguously by position of the end-sections, or an equilibrium of a bar under given boundary conditions is impossible.

While investigating the stability of columns, the fixing of end-sections is characterized by effective-length factor K_e (Hibbeler 2011). Using this factor, the critical magnitudes of compression parameter are represented in the form:

$$\nu_{cr} = \frac{\pi}{K_e}.$$

TABLE 29.1
Types of Members Making up a Frame

Types of Nodes at the Endpoints; Coordinates of Nodes	Boundary Conditions: Displacements of End-Sections	Boundary Conditions: Forces at the End-Sections	Shape of Buckling at Zero Specified Displacements of End-Sections	Critical Compression Parameter ν_{cr}
Rigid joints at both endpoints	$y(0)$, $\theta(0)$, $y(l)$, $\theta(l)$			2π
Rigid joint ($z=0$) and hinge ($z=l$)	$y(0)$, $\theta(0)$, $y(l)$	$M(l)=0$		$\frac{\pi}{0.6992}$
Hinge and hinge	$y(0)$, $y(l)$	$M(0)=0$, $M(l)=0$		π
Rigid joint of cantilever ($z=0$)	$y(0)$, $\theta(0)$	$M(l)=0$, $V(l)=0$		0.5π

Let us consider the first type of beam-column, presented by Table 29.1. We show that under condition:

$$0 < v < 2\pi \tag{29.10}$$

the forces $M(0)$ and $V(0)$ are linearly expressed through translational and rotational displacements of ends $y(0)$, $\theta(0)$, $y(l)$, and $\theta(l)$.

By using formulas (29.5) and (29.6), we write down boundary conditions at the right end as follows:

$$\begin{cases} y(l) = y(0) + \dfrac{\theta(0)}{k}\sin v - \dfrac{M(0)}{P}(1-\cos v) - \dfrac{V(0)}{P}\cdot\dfrac{v-\sin v}{k}; \\ \theta(l) = \theta(0)\cos v - \dfrac{M(0)k}{P}\sin v - \dfrac{V(0)}{P}(1-\cos v). \end{cases}$$

We convert these conditions to the standard form of a system of linear equations in unknown initial parameters $M(0)$ and $V(0)$:

$$\begin{cases} M(0)\dfrac{1-\cos v}{P} + V(0)\dfrac{v-\sin v}{kP} = -y(l) + y(0) + \dfrac{\theta(0)}{k}\sin v; \\ M(0)\dfrac{k\sin v}{P} + V(0)\dfrac{1-\cos v}{P} = -\theta(l) + \theta(0)\cos v. \end{cases} \tag{29.11}$$

Linearity and uniqueness of relation between unknown internal forces and specified displacements is ensured at a nonzero determinant of the coefficient matrix of this equation set, the latter having the form:

$$\Delta = \frac{(1-\cos v)^2}{P^2} - \frac{\sin v(v-\sin v)}{P^2}.$$

Conversely, the forces at the end-sections of a bar are not uniquely determined through displacements of the ends when this determinant is zero. We easily obtain:

$$P^2\Delta = (1-\cos v)^2 - \sin v(v-\sin v) = 2 - 2\cos v - v\sin v$$

$$= 4\sin^2\frac{v}{2} - 2v\sin\frac{v}{2}\cos\frac{v}{2} = 2\sin\frac{v}{2}\left(2\sin\frac{v}{2} - v\cos\frac{v}{2}\right).$$

As a result, the parameter v, which indicates the violation of linearity of the relation between forces and displacements of end-sections, is obtained as a solution of one of the equations:

$$\sin\frac{v}{2} = 0; \quad \tan\frac{v}{2} = \frac{v}{2}.$$

It can be shown that positive solutions of the second equation satisfy the inequality: $\frac{v}{2} > \pi$. For the smallest positive solution of the first equation, we have: $\frac{v}{2} = \pi$. Hence, we come to requirement (29.10), which secures the representation of forces through linear forms of displacements.

If the compression parameter has reached magnitude $v = 2\pi$, then on clamping of end-sections when:

$$y(l) = y(0); \quad \theta(0) = \theta(l) = 0, \tag{29.12}$$

there arises the multiplicity of bar's equilibrium states, while at the violation of restraint conditions (29.12), an equilibrium can be impossible.

It is sufficiently obvious that linearity of the relation between forces at the left end of a bar $M(0),V(0)$ and displacements of end-sections means a similar linear relation regarding the forces at arbitrary section $M(z),V(z)$.

29.2 THEORETICAL PROVISIONS OF CALCULATION OF REACTIONS DEVELOPED BY ADDITIONAL SUPPORTS IN A PRIMARY SYSTEM

The first stage of calculation by the displacement method is to determine the degree of kinematic indeterminacy and to construct a primary system. The next stage is to set up the set of canonical equations in primary unknowns, and it is important that these equations are linear and homogeneous. In the previous chapter, the reactions of additional supports were considered as functions $R_i = R_i(\mathbf{Z}, P)$ of DOF vector and load. These functions are equal to their finite increments in the neighborhood of zero DOFs at fixed parameter P, because $R_i(0,P)=0$. It was pointed out that the composing of canonical equations assumes the replacement of finite increments of reactions developed by additional supports with differentials of reactions at the point $\mathbf{Z}=0$. We now show how the differentials of functions $R_i(\mathbf{Z}, P)$ should be calculated.

Reaction R_i must equilibrate the total internal forces (force couples) at the ends of members forming some node, together with the nodal load. In order to obtain functions $R_i(\mathbf{Z}, P)$, one has to establish the dependences of internal forces acting at the end-sections of members on the displacements of these ends. To this purpose, we shall use known internal forces in the unit stress states of the beam-column, assembled in the Table 29.2 (presented at the end of the chapter).

Every unit state is the stress state caused by unit displacement of a member's end-section while compression force P is exerted. The stress state is specified by distributions of internal forces M and V in the member subjected to given impacts. Table 29.2 shows a typical shape of bending moment diagrams and gives the relationships for internal forces at the end-sections of the member as functions of force P. The unit states have been established by means of the initial parameters method substantiated in the previous item. By using the tabled formulas for forces at the ends of the member in unit states 1–5, one can obtain the dependences of these forces on small independent nodal displacements for all the types of members presented in Table 29.1.

The second column of Table 29.1 contains independent end-section displacements of a compressed bar, which define its stress state depending on the types of joints at the ends. Since forces M and V are linear forms of independent displacements of end-sections, any internal force $M(z)$ or $V(z)$ can be obtained as a sum in which every component has been caused by only a single end-point displacement. The sum is taken over all independent displacements. For example, any force M or V produced in the beam-column of the first type under given nodal displacements $y(0)$, $\theta(0)$, $y(l)$, $\theta(l)$ is established by summing up four components of this force; for each component, one of the boundary displacements is given and remaining three are zero. Every component is obtained by multiplying the characteristics of unit state 3 or 4 by a given boundary displacement (in the cases of given displacement $y(0)$ or $\theta(l)$, it is required, besides, to permute boundary conditions of unit states at the left and right ends of the bar). The distributions of forces M and V together with compression force P determine the stress state of the selected bar. The stress states of bars of the second and fourth types from Table 29.1 under the given nodal displacements are obtained likewise, through transformation and subsequent superimposing of forces M and V in unit states. For the bar of the third type, the stress state is trivial: $M=0$, and $V=0$. Linear displacements were considered here in the transverse direction only, but we may allow the identical displacements of both ends in the direction of the member's axial line because these displacements do not affect internal forces.

Note that superimposing forces M or V caused by different displacements of the endpoints of the compressed bar (in order to consolidate them into a single force caused by the set of displacements) doesn't mean the superposition of the stress states: while superimposing these forces, the compression force remains the same as opposed to being summed up over all the states.

Denoting the force at any end-section of the bar (whether it is shear force or bending moment) as Q, we get the general formula:

$$Q = \mathbf{K}_e(N)\mathbf{Z}_e, \tag{29.13}$$

where $\mathbf{Z}_e$ is the vector of independent nodal displacements of a bar, according to Table 29.1; $\mathbf{K}_e(N)$ is a row of factors composed of boundary forces determined at unit states by the formulas in Table 29.2 (the member with two hinged nodes is not presented by the table; for this one it holds $\mathbf{K}_e(N) = 0$). These factors are determined for compression forces P which are not greater than Euler critical force, and, as it is seen from the table, they are smooth functions of axial force $N = -P$ acting in a bar. We note that shear force does not change along a bar subjected to nodal loads.

Now, we return to analysis of a bar system, the state of which is determined by loads of pure compression with parameter P and vector $\mathbf{Z}$ of independent nodal displacements. For clarity's sake, it is allowed to assume that the cause of displacements Z_i, $i = \overline{1,n}$, is the totality of reactions of additional supports installed in the directions of these displacements, i.e., one may consider a primary system of the displacement method.

Introduce the notations:

m is the number of members in a system;
$\mathbf{V}$ is the vector of shear forces in system's members, dim $\mathbf{V} = m$;
$\mathbf{M}$ is the vector of bending moments acting at the end-sections of all the members of the system, dim $\mathbf{M} = 2m$;
$\mathbf{N}$ is the vector of axial forces in the system's members as a function of DOFs, dim $\mathbf{N} = m$; $\mathbf{N} = \mathbf{N}(\mathbf{Z}, P)$;
$\mathbf{N}_0$ is the vector of axial forces in a purely compressed frame, i.e., $\mathbf{N}_0 = \mathbf{N}(0, P)$.
$\mathbf{\Delta N} = \mathbf{N}-\mathbf{N}_0$ is the vector of increments of axial forces in members, arisen due to nodal displacements forming vector $\mathbf{Z}$.

Theorem 1

The increment vector of axial forces in members of the primary system is obtained by linear transformation of the vector of shear forces as follows:

$$\Delta\mathbf{N} = \mathbf{L}\mathbf{V}, \tag{29.14}$$

where matrix $\mathbf{L}$ is defined by the geometric characteristics of the primary system (location of members and positioning of all given supports and additional support links).

Proof

When proving the theorem, we shall consider the primary system in which all support nodes are attached to the earth indirectly, through floating supports (moment connections) and simple rods (force connections). If, in the given structure, some support node is a rigid support, then we represent it in the primary system as the subsystem of three supports: two support links and a floating support. A hinged immovable support is represented by two links. Thus, the nodes of the primary system do not have direct contact with the earth, but the total force of the earth's reaction upon the support node is developed by one or two support links. Below, we distinguish the members of the primary system, of m altogether, and support simple rods producing reaction of the earth.

The theorem is proved by analysis of internal forces in the conjugate truss which is obtained from such a primary system by transformation of all floating supports into hinges. The truss is statically determinate; therefore: (1) internal forces and support reactions in it comply with the superposition principle; (2) the equation set of statics is solved unambiguously at given loads. The complete equation set of statics is formed by equations of equilibrium of nodes and members, but the equations of member equilibrium have trivial form (7.6), and we exclude these equations from consideration by substitution of the axial forces into the equations of nodes' equilibrium. We did the same in Chapter 7, when we represented the equation set of statics as the equation set of nodes' equilibrium (7.7). The unknowns in these equations are axial forces and support reactions; the latter are axial forces in support links which are present in equations (7.7) together with forces in the members of a system.

Below, we use the same numbering of nodes, members, and support links both in the primary system and conjugate truss. For each i-th node of the primary system, we calculate the vector sum of transverse reactions of members forming the node (these reactions are some components of shear force vector **V**). This sum will be referred to as the *fictitious load upon node*. Every i-th node of the primary system keeps equilibrium owing to the balance of the next forces:

- Axial forces in members of the i-th node, being among the components of vector **N**;
- Reactions of support links, including additional supports;
- Fictitious load upon the node;
- Actual load of pure compression.

Let the conjugate truss be subjected to loads of pure compression and, additionally, fictitious loads upon nodes. We write down the equation set of statics for the truss in a form similar to (7.7):

$$\begin{cases} \sum_{j=1}^{m} I_{ij} N_j \cos\alpha_j + \sum_{j=m+1}^{m+m_s} I_{ij} \tilde{R}_j \cos\alpha_j = -V_{\Sigma ix} - P_{ix}; \\ \sum_{j=1}^{m} I_{ij} N_j \sin\alpha_j + \sum_{j=m+1}^{m+m_s} I_{ij} \tilde{R}_j \sin\alpha_j = -V_{\Sigma iy} - P_{iy}; \\ \qquad i = \overline{1, n_n}. \end{cases} \tag{29.15}$$

Here, $V_{\Sigma ix}$ and $V_{\Sigma iy}$ are the components of the vector of the fictitious load upon the i-th node, which have the form:

$$V_{\Sigma ix} = \sum_{j=1}^{m} I_{ij} V_j \sin\alpha_j; \qquad V_{\Sigma iy} = -\sum_{j=1}^{m} I_{ij} V_j \cos\alpha_j. \tag{29.16}$$

The set (29.15) consists of $2n_n$ equations, where n_n is the total number of the structure's nodes. Support links are numbered next to the members of the primary system and have the numbers $m+1, \ldots, m+m_s$. $\tilde{R}_j$ is the reaction of the j-th support connection; positive reactions of support links are directed to the earth (additional links of the primary system are included in the total number of m_s support links). I_{ij} and α_j are geometric characteristics of the conjugate truss (an explanation of them is given in Chapter 7). Unknown forces are presented at the left-hand side of equations and the loads at the right-hand side. The relationships (29.16) for fictitious loads have been written with account for the sign convention for shear forces V_j in members.

It is easy to see that **the system of statics equations for nodes of the conjugate truss coincides with the system of equations of forces' equilibrium for nodes belonging to the primary system**. So, the equations of forces' equilibrium for nodes in a conjugate truss and primary system have the same form (29.15), and this equation system is solved unambiguously at a given right-hand side.

On zero DOFs, a given frame is in the state of pure compression and fictitious loads are equal to zero. Consequently, equation set (29.15) under given loads $\boldsymbol{P}_i$ and in the absence of fictitious loads determines the vector of axial forces $\mathbf{N}_0$. While DOFs have arbitrary small magnitudes, the right-hand side of (29.15) may include, besides loads $\boldsymbol{P}_i$, nonzero fictitious loads, and now this system of equations determines the vector of axial forces $\mathbf{N} = \mathbf{N}_0 + \boldsymbol{\Delta}\mathbf{N}$. Therefore, keeping only fictitious loads on the right side, we get on the left side only the increment vector $\boldsymbol{\Delta}\mathbf{N}$ of axial forces. Because the unknowns ΔN_j in the latter case are defined by homogeneous functions of absolute terms $V_{\Sigma ix}$ and $V_{\Sigma iy}$, through substitution of relations (29.16) into these functions, we can represent the increment vector of axial forces in the form (29.14).

The theorem is proved.

Theorem 2

The shear force (bending moment) at a cross-section of a frame member, represented as a function of DOF vector $\mathbf{Z}$ under constant load P, has the differential at the point $\mathbf{Z} = 0$ equal to shear force (bending moment) at this section under displacements of the member's nodes determined by the vector of differentials $d\mathbf{Z}$, and internal axial force corresponding to the pure compression of the frame.

Remark to Theorem 2

The idea of the theorem is that named differentials of internal forces can be calculated neglecting the dependence of axial forces on the DOF vector. Its proof is based on the postulate about differentiability of vector-function $\mathbf{N} = \mathbf{N}(\mathbf{Z}, P)$ at the point $\mathbf{Z} = 0$ under fixed P.

Proof

Formulas (29.13) written for shear forces acting in all the members of a frame can be assembled into the next general formula:

$$\mathbf{V} = \mathbf{K}_V(\mathbf{N})\mathbf{Z}, \tag{29.17}$$

where $\mathbf{K}_V$ is the matrix of order $m \times n$, dependent on axial forces acting in members, such that every constituent of this matrix K_{Vij} is a smooth function of vector $\mathbf{N}$ (the definitional domain for this function is limited by the Euler critical compression forces obtained for every member). Because the components of vector $\mathbf{N}$ are the functions of DOFs differentiable at point $\mathbf{Z} = 0$, there exists the total differential of the matrix $\mathbf{K}_V(\mathbf{N}(\mathbf{Z}))$ (dependence of $\mathbf{N}$ on P is not denoted). Hence, the total differential of function $\mathbf{V} = \mathbf{K}_V(\mathbf{N}(\mathbf{Z}))\mathbf{Z}$ at the zero point can be calculated in the form:

$$d\mathbf{V} = d\mathbf{K}_V(\mathbf{N}(\mathbf{Z}))_{|\mathbf{Z}=0}\mathbf{0} + \mathbf{K}_V(\mathbf{N}(\mathbf{Z}))_{|\mathbf{Z}=0}d\mathbf{Z} = \mathbf{K}_V(\mathbf{N}_0)d\mathbf{Z}, \tag{29.18}$$

where $\mathbf{0}$ is the zero vector. So, the theorem's assertion is correct for shear forces.

Formula (29.13) is correct for the end bending moments as well, and thus, the corresponding vector can be represented in a form similar to (29.17):

$$\mathbf{M} = \mathbf{K}_M(\mathbf{N})\mathbf{Z}. \tag{29.19}$$

The constituents of matrix $\mathbf{K}_M(\mathbf{N})$ are smooth functions of vector $\mathbf{N}$. Through differentiation of equality (29.19) we prove the assertion of the theorem for bending moments at the end-sections of members.

In order to prove the required for a bending moment at an arbitrary cross-section, we notice that members with the hinges at both endpoints are not supposed to be considered (we exclude the possibility of local buckling, and therefore, hinged members are not subjected to bending). For any member with a rigid node, we introduce the earth-bound CS $O'yz$ with an origin at the rigid node (before displacement) and employ formula (29.7) for bending moment $M(z)$ at a point $z \in [0, l]$. In this formula, initial parameters $M(0)$ and $V(0)$ can be represented through vector $\mathbf{Z}$ by linear forms according to equalities (29.19) and (29.17), respectively. After that, formula (29.7) is represented in the generalized form as follows:

$$M(z) = \mathbf{K}_{Mz}(N)\mathbf{Z},$$

where $\mathbf{K}_{Mz}(N)$ is a row of factors, each of which is a smooth function of axial force N in a bar. By differentiation of the latter equality like it was done in (29.18), we come to the assertion of the theorem, which was required.

The differentials of internal forces with respect to the DOF vector $\mathbf{Z}$ should be calculated in order to establish the reactions of additional supports in the primary system. Having calculated the differentials of shear forces in members, one has to calculate the differentials of axial forces by means of linear dependence (29.14). We remind readers that the reaction of every additional moment or force connection in a primary system is calculated as the differential of the reaction represented as a function of vector $\mathbf{Z}$ formed up by support displacements; this differential is taken at the zero point under constant load P.

In order to calculate additional support reactions by the displacement method, we may follow the next rule: **The reaction of every additional moment or force connection in a primary system is determined from the equations of statics, wherein internal forces at members' sections determined beforehand are substituted. Bending moments and shear forces wanted for calculation of reactions are determined as the differentials with respect to the vector of support displacements Z under axial forces of the frame's pure compression. The increments of axial forces produced by displacements of additional supports are determined from statics equations through established shear forces.**

We clarify this rule.

The necessary equations of statics can be set up as equations of force and moment equilibrium for nodes of a primary system. In the case of additional moment connection, the reaction of this support is the sum of bending moments at abutting cross-sections of members. The reaction of additional force connection is statically determinate if forces acting at boundary sections are known, and this one is established either from the force equilibrium equations composed for the support node or from the principle of virtual work. The example of calculation of the force connection's reaction is shown in Figure 29.2. In the general case, the reaction of an additional support link is calculated in linear form:

$$R = \sum_i a_i N_i + \sum_i b_i V_i + cP_n, \tag{29.20}$$

where N_i and V_i are the forces exerted upon the node through sections of adjacent members of the structure, and P_n is the nodal load of specified direction under pure compression of the frame (if there is no load, then $P_n = 0$).

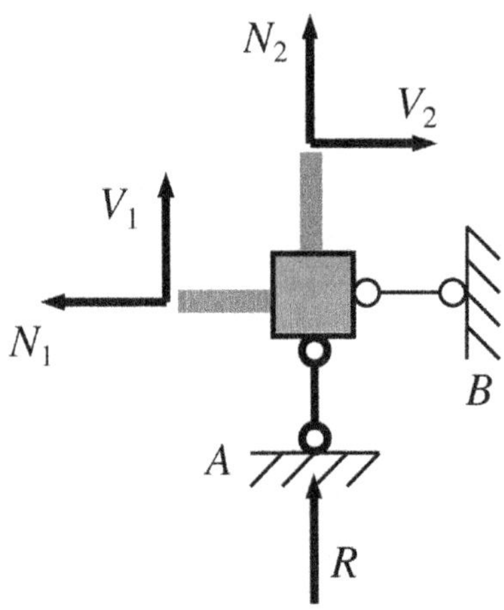

FIGURE 29.2 Example of support node: A is additional support; B is own support; $R = -V_1 - N_2$.

The reaction of additional support can be calculated as the differential at the point $\mathbf{Z} = 0$, thus in linear form (29.20), we may pass to differentials of acting forces and represent the reaction of additional force connection in the form:

$$R = dR = \sum_i a_i dN_i + \sum_i b_i dV_i.$$

This relationship means that the equations of a node's equilibrium can be set up under the assumption that the node is subjected only to increments of internal forces at abutting sections.

Sometimes, while calculating the axial forces' increments and reactions of additional supports, it is convenient to set up the equations of statics for the system's parts including more than one node. But at any rate, we use in these equations the differentials of shear forces and bending moments with respect to vector variable $\mathbf{Z}$, calculated beforehand.

29.3 PROCEDURE OF CALCULATION OF REACTIONS DEVELOPED BY ADDITIONAL SUPPORTS AND EXAMPLE OF ANALYSIS BY DISPLACEMENT METHOD

According to the second assumption of the method, when calculating the reactions of additional supports, we limit ourselves by the first-order terms with respect to primary unknowns, i.e., we use the relations of the form:

$$R_i(Z_1, \ldots, Z_n, P) = \sum_{j=1}^{n} r_{ij}(P) Z_j, \tag{29.21}$$

where $r_{ij}(P)$ are unit reactions. The assertions proved above enable us to establish functions $r_{ij}(P)$ in the following manner. Initially, we are to establish the axial forces of pure compression in all members as the functions of load P. Next, we consider all n unit deformed states of the system produced by displacement of single support $Z_j = 1$ ($j = 1, 2, \ldots, n$). We establish the displacements of all the nodes at every state No j; while doing so, translational displacements are determined by means of a mechanism conjugate with a given system. Having established displacements, we are to determine shear forces and bending moments at the end-sections of members as the functions of load parameter, to which end we accept that **axial forces in members correspond to pure compression**

of the given frame and employ Table 29.2.* Then, we calculate increments of axial forces caused by support displacement $Z_j = 1$. To do that, we calculate the fictitious loads upon nodes of the primary system by vector summation of shear forces in members abutting every node and consider the conjugate truss with nodes subjected to fictitious loads. The axial forces in the conjugate truss are sought increments of axial forces in the primary system. Having established internal forces at members' sections adjacent to the nodes with additional supports, we determine the reactions of these supports $r_{ij}(P),\ i = \overline{1,n}$, from the conditions of nodes' equilibrium. The reactions of additional force connections can be found from the equilibrium equations written for nodes of the conjugate truss subjected to fictitious loads.

The transition to a conjugate truss with a view of seeking axial forces is a calculating trick which is not compulsory in the displacement method. According to the theorems proved above, **shear forces and bending moments at the sections of a primary system are calculated under assumption of axial forces corresponding to the frame pure compression.** By using this inference, we may calculate the reactions of additional supports by solving the equations of statics for individual parts of the primary system.

Every unit state is usually displayed by conventional bending moment diagram obtained by selection of typical diagrams from Table 29.2 for all members of a system. The diagram is constructed as conventional, for the load parameter P which would determine its ordinates is not known. The conventional diagram approximately reproduces the shape of the true diagram and includes math expressions for ordinates at the endpoints of members as functions of the compression parameter. The purpose of drawing the conventional unit bending moment diagram is to obtain bending moments and shear forces at the endpoints as functions of the load. In fact, there is no need of a diagram to do that, but it is enough to present the list of formulas for calculation of internal forces at the end-sections of members in the primary system. But these formulas are customarily written for bending moments on the scheme of a primary system beside corresponding ordinates of the moment profile. Formulas for shear forces are selected from Table 29.2, as necessary.

The forces at the ends of bars are presented in Table 29.2 by using four special functions $\varphi_1(\nu)$, $\varphi_2(\nu)$, $\varphi_3(\nu)$, and $\eta_1(\nu)$. The normalizing of special functions was selected in such fashion that $\varphi_i(\nu) \to 1$ as $\nu \to +0$. It is convenient for calculation of end-section forces when compression force is absent, because the diagrams in Figure 18.1 are particular cases of diagrams for states 1–4 in Table 29.2.

It was noted above that the formulas for internal forces in the unit states of bars were obtained by the initial parameters method. For example, the formulas for state 4 are obtained from equation set (29.11) and relationships (29.7) and (29.8) after next substitutions into (29.11):

$$y(l) - y(0) = 1; \quad \theta(0) = \theta(l) = 0.$$

The conventional bending moment diagrams are presented in Table 29.2 for small magnitudes of compression parameter ($\nu < 0.5\pi$). Figure 29.3 shows the change of the bending moment profile for state 2 while compression force increases. In this state, the moment at the left end-section of a member changes its sense on a compression parameter sufficiently great ($\nu > \pi$).

Example of Analysis by the Displacement Method

For the square frame in Figure 29.4a, determine the critical value of compression force. Take as given that flexural rigidity is the same for all members.

In Chapters 17 and 18, by taking the given frame as an example, we studied the displacement method for analysis of a frame in bending. The only difference of the primary system for

---

* Parameter P of loads upon a structure is not the compression force P on diagrams of Table 29.2. The compression force is an internal axial force with opposite sign, established for a member of the frame.

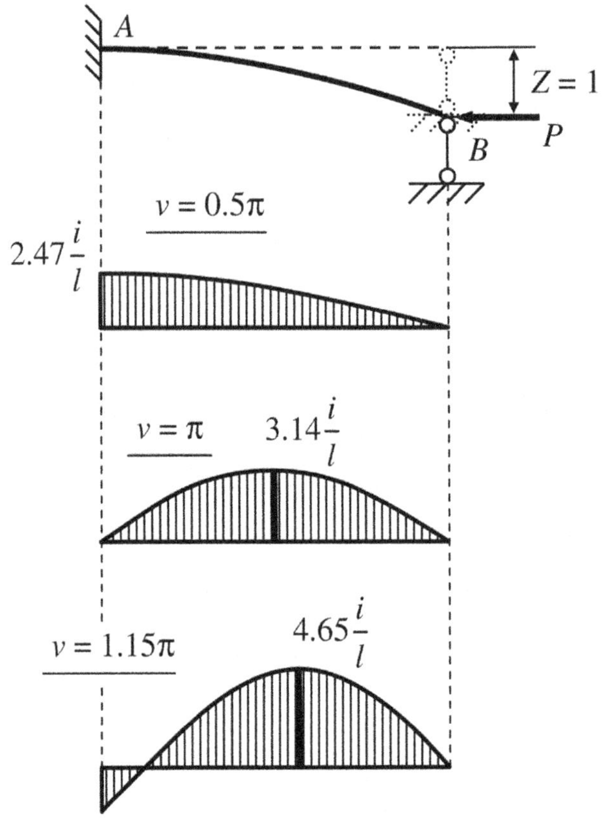

FIGURE 29.3 Bending moment diagrams when compression load increases (in given case $v_{cr} = 1.43\pi$).

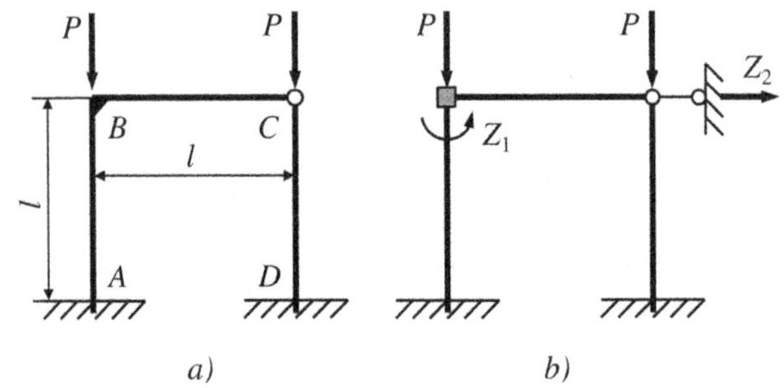

FIGURE 29.4 Given frame (a) and corresponding primary system (b).

investigation of stability lies in loading (Figure 29.4b). In order to construct unit diagrams for the given problem, we first establish axial forces in members of the frame before buckling. It is easy to see that both columns are compressed by force P, while there is no axial force in the girder. Unit diagrams have been constructed in Figure 29.5. They are close in shape to unit diagrams for analysis of bending (see Figure 18.2).

The unit reactions are established in the same manner as in Chapter 18. While the unit reactions are being calculated, we employ the free-body diagrams shown in Figure 29.6 (the superscript about V shows the number of the unit state of primary system). The reactions as functions of load P are obtained in the form:

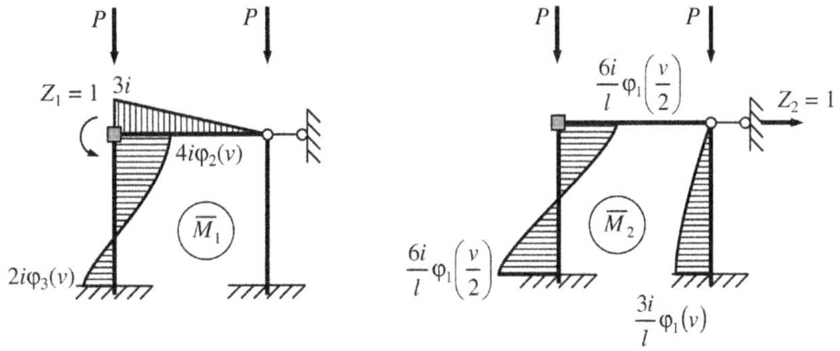

FIGURE 29.5 Unit bending moment diagrams of primary system in Figure 29.4b.

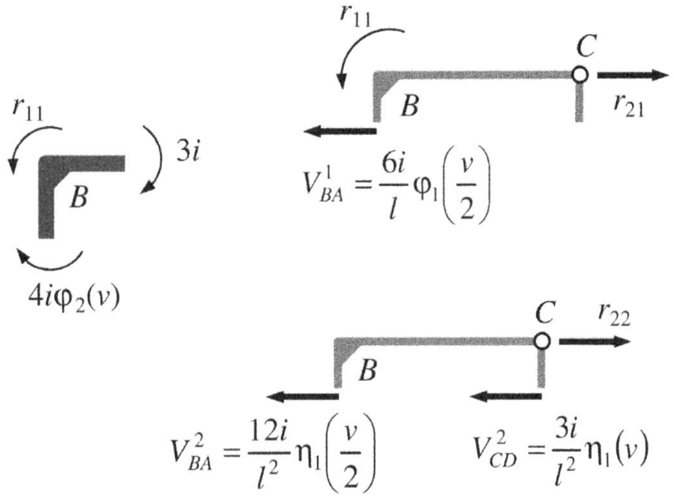

FIGURE 29.6 Free-body diagrams for the states of primary system shown in Figure 29.5.

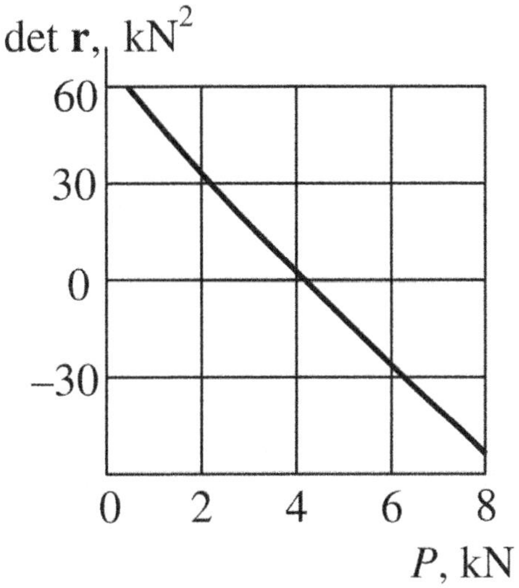

FIGURE 29.7 The curve of stability determinant versus load.

$$r_{11} = 3i + 4i\varphi_2(v); \quad r_{21} = V_{BA}^1 = \frac{6i}{l}\varphi_1\left(\frac{v}{2}\right);$$

$$r_{22} = V_{BA}^2 + V_{CD}^2 = \frac{12i}{l^2}\eta_1\left(\frac{v}{2}\right) + \frac{3i}{l^2}\eta_1(v).$$

Out of this, the determinant of stability is obtained as follows:

$$\det \mathbf{r}(P) = \left(3i + 4i\varphi_2(v)\right)\left(\frac{12i}{l^2}\eta_1\left(\frac{v}{2}\right) + \frac{3i}{l^2}\eta_1(v)\right) - \left[\frac{6i}{l}\varphi_1\left(\frac{v}{2}\right)\right]^2; \quad v = \sqrt{\frac{Pl}{i}}.$$

Usually, the equation of stability (28.22) is solved numerically. We employ the graphical method of solving. For example, if it is given that $l = 1$ m and $i = 1$ kN·m, then the curve of the stability determinant versus load is shown in Figure 29.7. In this case, we obtain critical force $P_{cr} = 4.2$ kN.

TABLE 29.2
Unit States of Beam-Columns

No. of State | **Diagram of Loading, Bending Moment Diagram, Internal Force Functions**

1

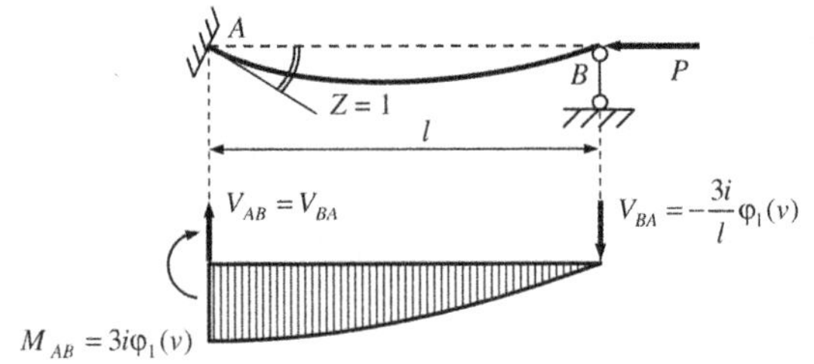

2

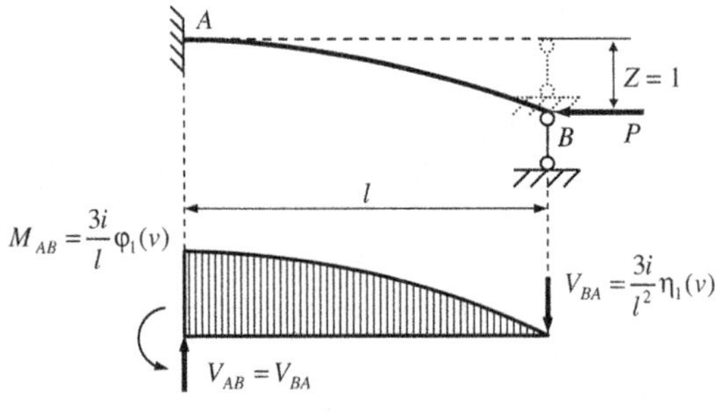

3

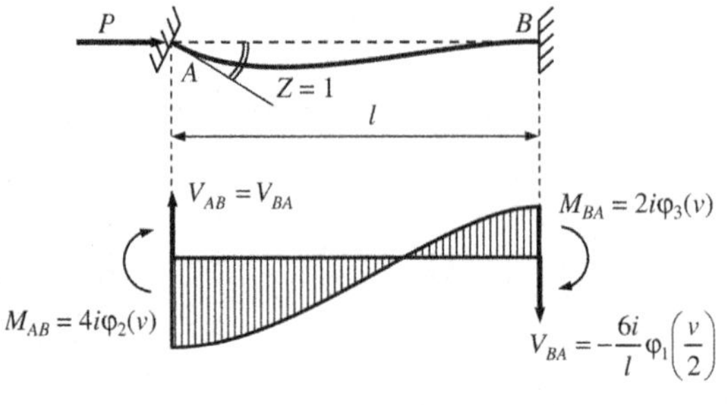

(Continued)

TABLE 29.2 (CONTINUED)
Unit States of Beam-Columns

No. of State	Diagram of Loading, Bending Moment Diagram, Internal Force Functions
4	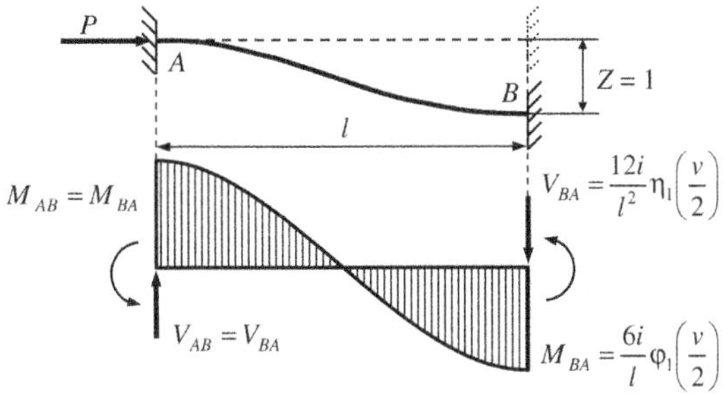
5	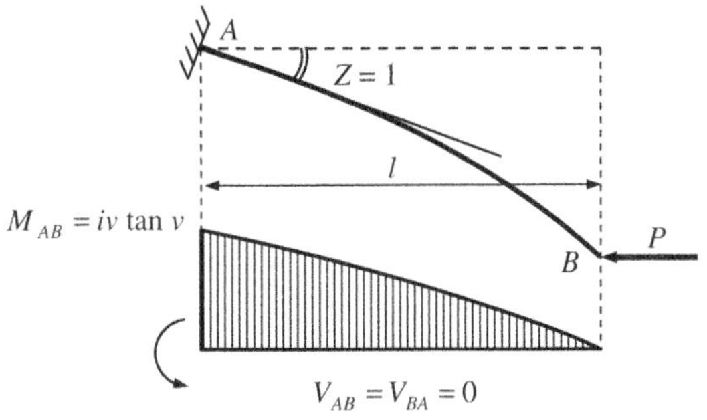

Remarks to Table. Notations of variables:

$v = kl = l\sqrt{\frac{P}{EI}}$ is the parameter of compression; $i = \frac{EI}{l}$ is rigidity per unit length.

Notations of special functions:

$$\varphi_1(v) = \frac{v^2 \sin v}{3(\sin v - v\cos v)}; \quad \varphi_2(v) = \frac{v(\sin v - v\cos v)}{4(2 - 2\cos v - v\sin v)};$$

$$\varphi_3(v) = \frac{v(v - \sin v)}{2(2 - 2\cos v - v\sin v)}; \quad \eta_1(v) = \varphi_1(v) - \frac{v^2}{3}.$$

Section XII

Dynamics of Elastic Systems

In this section inertia force is defined and the D'Alembert principle is stated as the basis for investigation of structural vibration. The sense of static and energy approaches in structural dynamics is disclosed. The section reveals the essence of dynamic versions of the force method and displacement method as the instruments for setting up the equations of an elastic system's motion. Based on the energy approach, the dynamics of a single-DOF structure is studied: it is shown how the energy conservation law enables the equation of motion to be obtained, and the peculiarities of free motion of a SDOF system are deduced from this equation. While investigating forced vibrations of a SDOF system, the relationships are obtained for dynamic magnification factor and phase lag, and the conditions for when the energy dissipation can be disregarded have been established. Next the section considers vibration of multi-DOF systems: it is shown that free motion of an elastic system is the superposition of harmonic vibrations, wherein each is characterized by natural mode and frequency. The dynamic force method is set out for analysis of steady-state vibration of MDOF systems. Also the section presents the modal expansion method for dynamic analysis of elastic system excited by arbitrary load with no energy dissipation. The model of motion has been constructed for an MDOF system taking into account the Rayleigh damping, and for this model the character of forced vibration is established and the general formulas are obtained for system response due to arbitrary load.

30 Free Vibration of Systems with a Single Degree of Freedom

30.1 BASIC CONCEPTS OF STRUCTURAL DYNAMICS; D'ALEMBERT PRINCIPLE AND STATIC APPROACH TO SOLVING THE PROBLEMS OF STRUCTURAL DYNAMICS

Structural dynamics is the branch of structural analysis which develops the methods of structural calculations concerned with the action of dynamic loads. Dynamic analysis may supplement the analysis of static loading action, and for some types of structures, this analysis may be a principal one.

Loads are called dynamic if they vary in time and give acceleration to the parts of the structure, which essentially affects its stress state. This might be periodic loading produced by machines installed in a structure, or by wind blasts acting upon a structure from without, or by other causes. It is known, for instance, that traffic flow near a structure can produce vibrations unacceptable for habitability; in that case, the traffic effect upon the ground can be considered as the source of periodic loads. Another type of dynamic loading is momentary load, when a structure is subjected to the impulse of force caused, for example, by explosion or seismic shock. It is these two types of loading – periodic and momentary – that will be under consideration further on, though impact load and movable load are also classified as dynamic. An impact is a short-time action upon a structure done by another body with initial parameters of motion being specified. The action of a movable load was previously considered in the present course (Chapter 2 and further). In contrast to the studied approaches of statics, in structural dynamics, it is assumed that a structure under movable load action also starts motion which affects its SSS.

Dynamic loading excites the vibration of a structure. The vibration of a mechanical system which goes on in the absence of external forces varying in time is referred to as free vibration. The vibration which occurs due to periodic loading is referred to as forced vibration. Free vibration might occur due to a single action of force impulse. Characteristics of free vibrations – mainly frequencies – enable us to determine internal forces at forced vibrations.

When the frequency of free vibration coincides with the frequency of applied force, resonance may appear, which is characterized by a great (multiple) magnification of oscillation amplitude in comparison to the amplitude under differing frequencies. In structural design, it may be necessary to discover and eliminate the resonance of a structure, which might be inadmissible due not only to the strength requirement, but to the operation conditions. For instance, the amplitude of vibration of a residential building is limited according to conditions of habitability; the vibrations of antenna towers may be limited due to conditions of transmitting-receiving.

There are static and energy approaches to solving the problems of structural dynamics. Both approaches aim to set up the equations of system's motion in order to bring the problem to dynamic calculation, i.e., to a math problem of solving differential equations. We shall consider the static approach. To substantiate this approach, we introduce the notion of inertia force and formulate the d'Alembert principle.

The inertia force is defined for any particle having mass m and moving with acceleration $\boldsymbol{a}$ as a vector quantity

$$X = -m\boldsymbol{a}.$$

In accordance with Newton's second law, one can say that *inertia force is the resultant of forces exerted on a particle, taken with the opposite sign.*

The d'Alembert principle is as follows: *the stresses, strains, and displacements of an elastic mechanical system can be determined by taking the inertia forces as the external body forces and assuming the equilibrium of the system.* This principle was substantiated by French mathematician Jean d'Alembert in 1743. The idea of the principle is that inertia forces acting in a system can be considered as additional forces exerted upon a system in equilibrium in order to determine the SSS at an arbitrary point in time of the system's motion. Fictitious equilibrium of a structure or its part, which is ensured by inertia forces, is referred to as *dynamic equilibrium.* For example, the equation of dynamic equilibrium of forces for any part of a structure can be written in the form:

$$\boldsymbol{P} + \boldsymbol{R} + \boldsymbol{X} = 0, \tag{30.1}$$

where $\boldsymbol{P}$ is the total load vector for the part under consideration; $\boldsymbol{R}$ is the vector of the sum of internal forces at boundary sections and reactions of constraints; and $\boldsymbol{X}$ is the total inertia force vector exerted upon the part of a structure.

The static approach lies in setting up the equations of a system's motion in the form of dynamic equilibrium equations or as the corollary of these equations. The equations of dynamic equilibrium differ from common equilibrium equations in taking into account the inertia forces among external forces. Note that an equilibrium equation, in a strict sense, is an equation of equilibrium of forces or moments, but this term can be understood as the math condition of the system's equilibrium being corollary from equations of statics. For example, canonical equations of the force method also may be named equilibrium equations.

Next, we consider the examples of setting up the equations of motion of an elastic structure using the static approach. In the presented examples, the structure includes a single lumped mass and has a single DOF (in the sense that the position of the structure under a known load is determined by a single generalized coordinate).

Example 1

Figure 30.1 shows a stage on a spring which is subjected to vertical varying load $P(t)$ and gravity. The mass of the stage is m, the spring is weightless and has rate k_{rig}; the resistance of the medium against the stage's motion is absent; and the state of the system is specified by a q-coordinate on the

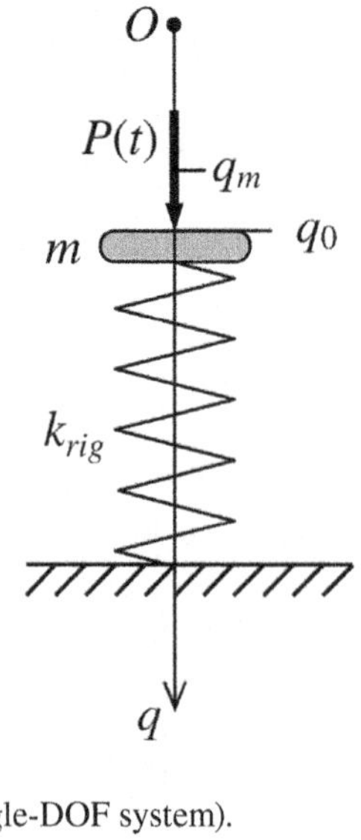

FIGURE 30.1 A stage on a spring (single-DOF system).

axis bound to the earth. We shall obtain the equation of motion as an equation of dynamic equilibrium (30.1) in the projection upon axis Oq. In the given case, the total load is $P_\Sigma = mg + P(t)$, inertia force $X = -m\ddot{q}$. The reaction force of the elastic connection (spring) is determined by the formula:

$$R = -k_{rig}\left(q - q_m\right),$$

where q_m is the coordinate of the stage in equilibrium with no gravity. Equation (30.1) in this case is elementary converted to the form:

$$m\ddot{q} + k_{rig}\left(q - q_0\right) = P(t), \tag{30.2}$$

where q_0 is the generalized coordinate of the system in equilibrium under the action of gravity, which has the form:

$$q_0 = q_m + \frac{mg}{k_{rig}}. \tag{30.3}$$

Example 2a

There is a weightless elastic beam which supports lumped mass m and gets excitation from dynamic load $P(t)$ exerted upon another point of the beam (Figure 30.2a). The beam's deflection y at the point mass is taken for generalized coordinate.

The equation of dynamic equilibrium (30.2) in the given case has the form:

$$X = R, \tag{30.4}$$

where R is the reaction of the beam exerted upon the point mass. The problem will be solved if reaction R is expressed through load P, generalized coordinate y, and characteristics of the system. The use of equation (30.4) in the given case is only the one that reminds us about the static action of external forces X and P upon the beam. We consider the strained equilibrium state of the beam produced by these forces (Figure 30.2, the top of diagram b). The inertia force X is determined functionally through displacement y of the lumped mass and load P (see Chapter 17); therefore, this force can be calculated as the reaction of additional support having settled by distance y (Figure 30.2, the top of diagram c). We enter an additional support link at the place of inertia force's action and write the corresponding reaction down in terms of the displacement method (Figure 30.2c):

$$X = r_{11}y + R_{1P}. \tag{30.5}$$

Here $r_{11} > 0$, and the loaded reaction is proportional to exerted load:

$$R_{1P} = r_{1P}P(t).$$

Out of this, we come to the equation of motion:

$$m\ddot{y} + r_{11}y = -r_{1P}P(t). \tag{30.6}$$

The method of setting up the equations of a system's motion by means of introducing additional support links at locations of lumped masses and composing the relationships for the reactions of these links versus their displacements and loads is referred to as the displacement method. An equilibrated system with additional supports, substituting for a given system in motion, is referred to as a primary system of displacement method.

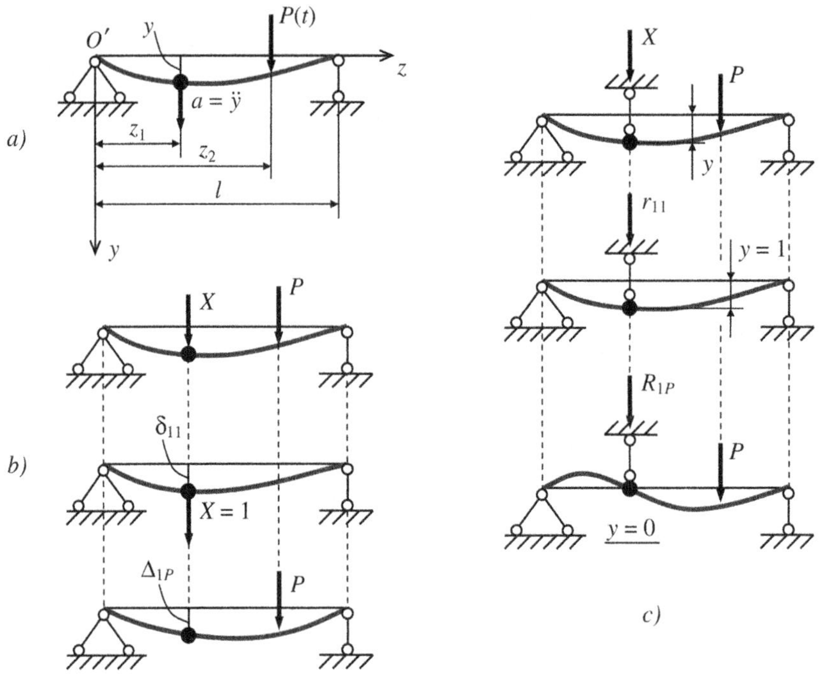

FIGURE 30.2 Vibration of a beam with lumped mass: (a) Given system; (b) Primary system of force method; (c) Primary system of displacement method.

Example 2b

The motion equation for the beam in Figure 30.2a can be set up in another way. One can calculate displacement y as the result of the combined action of inertia force and load. By using denotations of the force method, we have (Figure 30.2b):

$$y = \delta_{11}X + \Delta_{1P}. \tag{30.7}$$

Here $\delta_{11} > 0$; the loaded displacement is proportional to exerted load:

$$\Delta_{1P} = \delta_{1P}P(t). \tag{30.8}$$

Relationship (30.7) is converted into the equation of motion after substitution of the following expression for inertia force: $X = -m\ddot{y}$. We obtain the wanted equation in the form:

$$m\ddot{y} + \frac{1}{\delta_{11}}y = \frac{\delta_{1P}}{\delta_{11}}P(t). \tag{30.9}$$

The method of setting up the equations of a system's motion by means of composing the relationships for the displacements of lumped masses versus inertia forces and loads is referred to as the force method. An equilibrated system in which the motion of lumped masses has been replaced by unknown inertia forces is referred to as a primary system of the force method.

Both the force method and displacement method for setting up the motion equations are based on the d'Alembert principle: a structure is considered in equilibrium under the action of loads and inertia forces.

30.2 ENERGY APPROACH IN SOLVING THE PROBLEMS OF DYNAMICS; MOTION EQUATION FOR SINGLE DOF SYSTEM WITH ENERGY DISSIPATION

The energy approach lies in setting up the equations of a system's motion on the basis of the energy conservation law.

We shall employ the energy approach to set up the motion equation for a stable system with a single DOF. We consider an elastic structure subjected to, besides gravity, concentrated load $P(t)$ changing in time. The positive direction of load P is defined by earth-bound axis Oq (Figure 30.3a). We accept that **all possible strain states of a structure due to dynamic load P are defined by static action of the load exerted in the same manner.** In other words, the displacements in a structure at each moment in time are determined through static action of fictitious load P_{fict} exerted at the same point and direction as real load P. By virtue of the superposition principle, displacement at an arbitrary point of a structure is the sum of initial displacement due to the weight of the structure and additional displacement due to load P_{fict}, and this additional displacement is proportional to the load. Therefore, the field of displacements is determined unambiguously by coordinate q of the point where load P is exerted, and we take this coordinate for the DOF of a structure.

The assumption about admissibility of replacing real load P by fictitious load P_{fict} does not cause an error in determination of displacements in two ideal cases: when all the mass of a structure is concentrated at the point of the load's exertion with a single possible direction of motion, and when the process of loading is quasistatic. It is natural, therefore, to consider this assumption acceptable if the basic mass of a structure is located in a neighborhood of load P, or the load changes slowly enough. **Usually, the structures with distributed mass cannot be described by a single DOF**, and the practical significance of the approach presented below is little. Nevertheless, the conclusions about motion of a system with a single DOF, obtained on the basis of the energy approach, are used in structural dynamics for construction of more complicated models of systems with multiple DOFs.

We denote the value of generalized coordinate q at $P_{fict} = 0$ as q_0, i.e., under the action of only the gravity force in the state of the structure's equilibrium. Strain energy density at a point of an elastic body is the quadratic form of a strain vector, whereas the strain at an arbitrary point of a structure caused by load P_{fict} is proportional to displacement $q - q_0$ produced by this load. From this, we conclude that the strain energy of whole structure U_s is a quadratic function of generalized coordinate q (see the supplement to the chapter). We represent this dependence by completing the square of the polynomial:

$$U_s = \frac{k_{rig}}{2}\left(q - q_m\right)^2 + U_{sm}, \tag{30.10}$$

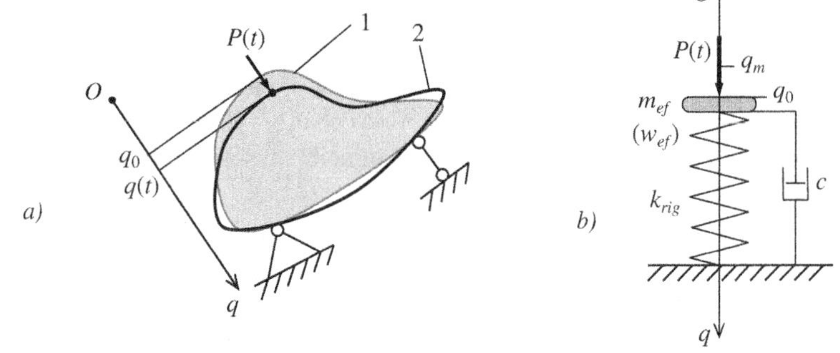

FIGURE 30.3 Example of elastic structure in motion (a) and its mechanical model (b): 1 – outline of body deformed only by weight; 2 – the same body under the action of gravity and load P.

where the constants k_{rig}, q_m, U_{sm} are determined by the design of structure (including elastic constants of materials) together with the application point and direction of load P_{fict} (but not its magnitude). The constant k_{rig} is referred to as the *rigidity factor of the structure.*

The potential energy of any elementary mass in a gravity field is linearly dependent on the displacement of this mass. The gravitational potential energy of a whole structure can be obtained by summing up energies of all elementary masses, and zero reference for potential energy is selected at will. Because additional displacement at any point due to load P_{fict} is proportional to displacement $q - q_0$, the gravitational energy of a structure is represented in the form:

$$U_g = -w_{ef}q, \tag{30.11}$$

where characteristic w_{ef} is determined by the same factors as the constants in formula (30.10), i.e., constructive factors and conditions of load exertion.

We introduce the notation for potential energy of conservative forces in a system as follows: $U \equiv U_s + U_g$. The sum of functions (30.10) and (30.11) can be written with a completed square. We have:

$$U = \frac{k_{rig}}{2}\left(q - q_0\right)^2 + U_0. \tag{30.12}$$

It is easy to see that the parameter of the complete square in this function is really a generalized coordinate q_0 of immovable structure under zero load P_{fict}. At this point, the minimum total potential energy is attained; thus, the point $q = q_0$ is the point of stable equilibrium. Besides, it is the only stationary point. Energy U_0 is the total potential energy of a system in the state of equilibrium.

Given that additional displacement of an arbitrary point of a structure produced by load P_{fict} is proportional to displacement $q - q_0$, we can represent kinetic energy of a structure through the time derivative of the generalized coordinate:

$$T_{kin} = \frac{m_{ef}\dot{q}^2}{2}, \tag{30.13}$$

where constant m_{ef} is called *effective mass of structure* and determined, as other introduced parameters, by constructive factors and conditions of load exertion. (The detailed derivation of formulas (30.11) and (30.13) is presented in the supplement to the chapter.)

Let the motion of a system's particles be of ordered character, i.e., is determined by the time change of the displacement field, and there are only conservative forces of elastic deformation and gravity acting in the system. Then the total energy of the system is the sum:

$$E \equiv T_{kin} + U, \tag{30.14}$$

and the motion of the system obeys the law of conservation of energy. In the case of action of variable load P upon such a system, the total energy (30.14) changes in time, and the variation of total energy through displacement dq is equal to the work done by the load upon the system:

$$dE = P(t)dq. \tag{30.15}$$

The latter equation can be transformed into a differential equation in generalized coordinate q in order to investigate the excitation of a conservative system under the action of a variable load. The relationships (30.10)–(30.15) define the simplified energy model of a structure.

The peculiarity of the simplified model is that the motion produced by load $P(t)$ goes on eternally after termination of load, whereas in an adequate model, it should cease over time. To make the simplified energy model more exact, one has to take into account the processes of energy loss in a system. Denoting the total loss of energy in a system in motion as E_{dis}, we can refine the equation for total energy (30.15) as follows:

$$dE = P(t)dq - dE_{dis}. \tag{30.16}$$

In a system which can be considered elastic, the total loss of energy is determined by the constituents as follows:

- Heat losses, which occur due to heat generation in the bulk of material while it is deformed, with consequent dissipation of heat into the environment;
- Acoustic losses, i.e., the energy of acoustic waves in the environment, produced by the vibration of the system;
- Energy losses because of friction at the joints of constructive elements.

Experience shows that in a system with a single DOF, the energy losses during a short interval of time are determined by expression:

$$dE_{dis} = c\dot{q}dq, \tag{30.17}$$

where c is a so-called viscous damping coefficient dependent on constructive factors and conditions of load exertion. The relationships (30.10)–(30.14) and (30.16)–(30.17) define the energy model of a structure with account for energy dissipation.

Note that the equation of energy balance (30.16) represents, as before, the law of conservation of energy, despite the fact that there is the term dE_{dis} of total energy's losses at the right-hand side. This is because the "lost" energy includes energy released into the environment and the thermal component of the kinetic energy of system's particles, which is not accounted for in the total energy (30.14).

Together with the suggested *energy model of a structure*, we can put the following mechanical model (Figure 30.3b): load P is supported by a stage of finite mass installed on the spring having rate (rigidity) k_{rig}, and the coordinate q_0 of the platform corresponds to the equilibrium of the system in the absence of this load. The mass of the stage is taken as equal to m_{ef} for calculation of its kinetic energy; the weight of the stage is accepted as equal to w_{ef} for calculation of its potential energy in the gravity field; and the mass of the spring is assumed to be zero. The motion of the platform goes on in the viscous medium, and this produces the loss of the system's energy according to formula (30.17) (the resistance of the medium is shown on the diagram by the damper). The mechanical model implements all relationships of the energy model for a given structure. Moreover, this model enables us to obtain parameter q_m of function (30.10) as the coordinate of the platform in the equilibrium state with no external forces, although in the given structure (the model's prototype) under the same conditions, the q-coordinate of the load P's application point is not necessarily equal to q_m.

Among the parameters introduced during construction of the energy model, we will further need four parameters: effective mass m_{ef}, rigidity factor k_{rig}, viscous damping coefficient c, and generalized coordinate q_0 corresponding to zero load P_{fict}. Coordinate q_0 is known at the beginning of dynamic calculations. Coefficient c is established on the basis of experimental data. Factors m_{ef} and k_{rig} are obtained by means of the technique presented in the supplement to the present chapter. Figure 30.4 shows two examples of elastic systems with distributed mass for which these factors are established easily enough.

By means of the model described above, we set up the equation of a structure's motion. We write down the basic relationship of energy approach (30.16) in the form:

$$dT_{kin} + dU = P(t)dq - dE_{dis}, \tag{30.18}$$

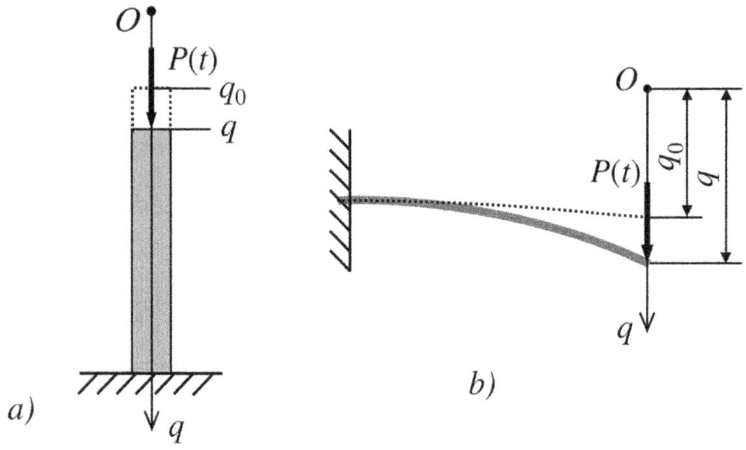

FIGURE 30.4 Representation of displacements in elastic systems by single DOF: (a) Case of a compressed bar; (b) Case of a cantilever beam.

and derive the terms of the left-hand side:

$$dT_{kin} = d\left(\frac{m_{ef}\dot{q}^2}{2}\right) = \frac{m_{ef}}{2}2\dot{q}\ddot{q}dt = m_{ef}\ddot{q}dq;$$

$$dU = k_{rig}\left(q - q_0\right)dq.$$

By substituting obtained relations for differentials into the equation of energy balance (30.18) and taking into account expression (30.17) for the dissipative constituent, we get equation (30.18) with the terms written through the increment of generalized coordinate:

$$m_{ef}\ddot{q}dq + k_{rig}\left(q - q_0\right)dq = P(t)dq - c\dot{q}dq.$$

After cancellation of dq, we come to the required equation of dynamics of a structure:

$$m_{ef}\ddot{q} + k_{rig}\left(q - q_0\right) + c\dot{q} = P(t). \tag{30.19}$$

The energy approach and its guidelines were developed by Lord Rayleigh in the 1850s. In particular, he proposed the use of effective mass in order to describe the dynamics of a structure with a single DOF. In comparison with the static approach, the construction of an energy model may not be a simple problem, for it is required to determine generalized characteristics of structure m_{ef}, k_{rig}, c. But this approach enables one to investigate some structures with distributed mass if the variation of the load is sufficiently slow and does not excite wave processes in construction. The structures in Figure 30.4 are examples of this in the case of impact action of load P. Besides, the energy approach gives us a way to take into account the energy dissipation due to the motion of the structure.

30.3 FREE VIBRATION OF SYSTEM; LOGARITHMIC DECREMENT

We rewrite the obtained equation of dynamics (30.19) for the displacement of a structure from equilibrium position $\Delta q = q - q_0$ and convert it to the standard form handy for analysis (Chopra 2012, p. 48):

$$\Delta\ddot{q} + \omega_0^2\Delta q + 2\zeta\omega_0\Delta\dot{q} = \frac{P(t)}{m_{ef}}, \tag{30.20}$$

where:

$$\omega_0 = \sqrt{\frac{k_{rig}}{m_{ef}}}; \quad \zeta = \frac{c}{2\sqrt{k_{rig} m_{ef}}}. \tag{30.21}$$

Further, we establish the motion of the system described by the equation of dynamics (30.20). In the present chapter, we limit ourselves to the motion of a structure disturbed from its equilibrium position and not subjected to loads. For definiteness, we shall consider the undisturbed motion of a structure from the instant in time $t = 0$. Let $\zeta < 1$, i.e., the energy dissipation in a system is possible but sufficiently low. For this case, we will write down the general solution of equation (30.20) taken with the zero right-hand term. It is known from the theory of differential equations that the general solution of a linear homogeneous differential equation of n-th order is the linear combination of n linearly independent partial solutions. Without dwelling on finding partial solutions of equation (30.20) with a zero right-hand side, we give the chance for the reader to verify that the wanted linear combination has the form:

$$\Delta q = C_1 e^{-\zeta\omega_0 t} \sin \omega t + C_2 e^{-\zeta\omega_0 t} \cos \omega t, \tag{30.22}$$

where:

$$\omega = \omega_0 \sqrt{1 - \zeta^2}. \tag{30.23}$$

The procedure of finding a general solution in the form (30.22) is presented in the supplement to the chapter. Hereinafter, it is convenient to represent this solution in the form:

$$\Delta q = C e^{-\zeta\omega_0 t} \sin(\omega t + \varphi), \tag{30.24}$$

where $C = \sqrt{C_1^2 + C_2^2}$ is the quantity called the initial amplitude of oscillations, and φ is the initial phase of oscillations, also called a phase angle, which is determined from conditions:

$$\tan \varphi = \frac{C_2}{C_1}; \quad \cos \varphi = \frac{C_1}{C}.$$

One can see that the motion of a system is an oscillation process which dies out in time at $0 < \zeta < 1$ (Figure 30.5a). This process is referred to as free vibration, because it is observed in the absence of disturbing force $P(t)$. This process can be considered as the sequence of cycles,* of which the

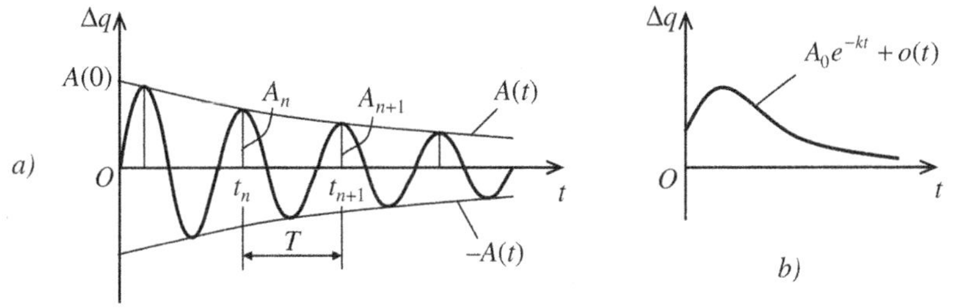

FIGURE 30.5 Motion of system without dynamic loading: (a) Case $0 < \zeta < 1$ (damped oscillations); (b) Case $\zeta \geq 1$ (motion with no oscillations); $o(t)$ is the function such that $o(t)e^{kt} \to 0$ as $t \to +\infty$.

* A cycle is any complete series of occurrences that repeats in time.

boundaries are either the points of the maximum of the function describing oscillations, or the points of the minimum of this function, or zero points taken every other one. The duration of an oscillation cycle is called a period and determined as:

$$T = \frac{2\pi}{\omega}; \tag{30.25}$$

quantity ω is called *damped circular frequency* and has the sense of a number of cycles in a system during 2π units of time. Sometimes, instead of this quantity, it is handy to use cyclic frequency f which is determined as:

$$f = \frac{\omega}{2\pi}$$

and has the sense of a number of cycles executed by a system during a unit of time.

Local maxima and minima (in modulus) of oscillating coordinate Δq are referred to as amplitudes. The decreasing function of time having the form:

$$A(t) = Ce^{-\zeta\omega_0 t} \tag{30.26}$$

is referred to as the *upper envelope function* with respect to the function of oscillations $\Delta q(t)$; the function $-A(t)$ is referred to, respectively, as the *lower envelope function*. The envelope curves $A(t)$ and $-A(t)$ limit the graph of dependence $\Delta q(t)$ from both sides and touch it near the peaks (Figure 30.5a). At the moments of time t_n when coordinate Δq reaches their amplitudes A_n, the magnitudes of function (30.26) are insignificantly different from these amplitudes.

The damping of vibration in a system is characterized by the natural logarithm of the ratio of two successive peaks having the same sign, which is referred to as *logarithmic decrement*. This quantity is represented through coefficient ζ:

$$\delta \equiv \ln\frac{A_n}{A_{n+1}} = \frac{2\pi\zeta}{\sqrt{1-\zeta^2}}. \tag{30.27}$$

The latter formula is easily verified:

$$\ln\frac{A_n}{A_{n+1}} = \ln\frac{Ce^{-\zeta\omega_0 t_n}\sin(\omega t_n + \varphi)}{Ce^{-\zeta\omega_0 (t_n+T)}\sin\left(\omega(t_n+T)+\varphi\right)} = \ln e^{\zeta\omega_0 T}$$

$$= \zeta\omega_0 T = \zeta\omega_0\frac{2\pi}{\omega} = \frac{2\pi\zeta}{\sqrt{1-\zeta^2}}.$$

If $\zeta << 1$, then we have the equality: $\delta \approx 2\pi\zeta$.

At $\zeta = 0$, i.e., when there is no energy dissipation in a system, we have harmonic oscillations with constant amplitude. In this case, $\omega = \omega_0$, and therefore, **quantity ω_0 is the circular frequency of free vibration without damping**.

In the example in Figure 30.1, the frequency of free harmonic oscillations ω_0 is determined by the expression:

$$\omega_0 = \sqrt{\frac{k_{rig}g}{mg}} = \sqrt{\frac{g}{\Delta}}, \tag{30.28}$$

where $\Delta = q_0 - q_m$ is the settlement of the stage due to its own weight.

In the case of significant energy losses in a system, when $\zeta \geq 1$, formula (30.22) can also be used to construct a general solution. The fact is that this formula determines the general solution of a homogeneous equation of vibration on the complex plane under arbitrary parameters $\omega_0 > 0$ and $\zeta \geq 0$. If $\zeta \geq 1$, then parameter ω happens to be imaginary:

$$\omega = i\omega_0\sqrt{\zeta^2 - 1}.$$

In order to obtain the real solution from the complex one, in this case, we may represent trigonometric functions through the exponentials:

$$\sin \omega t = \frac{1}{2i}\left(e^{i\omega t} - e^{-i\omega t}\right); \quad \cos \omega t = \frac{1}{2}\left(e^{i\omega t} + e^{-i\omega t}\right),$$

and take the real part of solution (30.22). As result, we discover that displacement Δq has a single maximum (or minimum) and, if $t \to +\infty$, decreases in modulus according to the exponential law, i.e., without oscillations (Figure 30.5b). The motion of the structure resembles the motion in viscous liquid damping the vibration.

Parameter ζ enables us to establish the character of free motion of a system:

- At $\zeta = 0$, a system does harmonic oscillations with frequency ω_0. Such a system is called *undamped*;
- At $0 < \zeta < 1$, a system does fading oscillations with frequency $\omega < \omega_0$ and is called *underdamped*;
- At $\zeta \geq 1$, the velocity of a system's motion $\dot{q}$ can change the sense only once, and then displacement Δq decreases (in absolute value) to zero with time. If $\zeta = 1$, then the system is called *critically damped*; if $\zeta > 1$, then the system is called *overdamped.*

Damping coefficient c, on which $\zeta = 1$, is referred to as the critical damping coefficient. It is determined by the relationship:

$$c_{cr} = 2\sqrt{k_{rig} m_{ef}}. \tag{30.29}$$

Under the given characteristics m_{ef}, k_{rig}, this is the least viscous damping coefficient when the motion of the system goes without oscillations. Because of relation:

$$\zeta = \frac{c}{c_{cr}}, \tag{30.30}$$

parameter ζ is referred to as a damping ratio or fraction of critical damping. It can be also called a viscous damping factor (Craig 1981, p. 49).

The physical meaning of logarithmic decrement is determined by the loss of a total system's energy during the cycle of free vibration. It is known that potential energy is determined up to addition by a constant. We shall determine the potential energy of a system, assuming this energy to be zero in the state of equilibrium under gravity force, i.e., in formula (30.12), we set $U_0 = 0$. Owing to this, formula (30.12) assumes the form:

$$U = \frac{k_{rig}}{2}\Delta q^2.$$

The ratio of the total loss of energy during a cycle to this energy at the beginning of the cycle, while the origin of potential energy is bound to the equilibrium of the system, is referred to as the *coefficient of energy dissipation* and denoted as ψ. According to the definition we have:

$$\psi = \frac{E(t_n) - E(t_{n+1})}{E(t_n)}, \tag{30.31}$$

where t_n is the instant of the n-th cycle's start. We shall consider an instant of the system's stop to be the cycle's start (Figure 30.5). Then it holds:

$$E(t_n) = U(t_n) = \frac{k_{rig}}{2} A_n^2,$$

and the coefficient of energy dissipation can be expressed first through amplitudes and then through logarithmic decrement:

$$\psi = \frac{A_n^2 - A_{n+1}^2}{A_n^2} = 1 - \frac{A_{n+1}^2}{A_n^2} = 1 - e^{-2\delta} \cong 2\delta. \tag{30.32}$$

One can see that the coefficient of energy dissipation is approximately equal to doubled logarithmic decrement, and the error of approximation can be neglected under sufficiently small decrement.

SUPPLEMENTS TO CHAPTER

Characteristics of Energy Model of Elastic Structure

Further on, we consider the case of a uniform elastic body stable with respect to the earth (Figure 30.3a). This allows us to somewhat simplify the math manipulations, because there will be no need to sum the potential or kinetic energy over all bodies of a system.

We obtain the strain potential energy of a system with a single DOF in the form (30.10). In arbitrary CS $O'xyz$ referenced to the earth, we introduce the strain vector at a body's point by the expression: $\boldsymbol{\varepsilon}^{\mathrm{T}} = (\varepsilon_x, \varepsilon_y, \varepsilon_z, \gamma_{xy}, \gamma_{xz}, \gamma_{yz})$, and consider the deformation field $\boldsymbol{\varepsilon} = \boldsymbol{\varepsilon}(\boldsymbol{r})$ at any time moment of motion. According to the assumption made in the chapter, this field is created by the gravity force and load P_{fict} in the equilibrium state of a structure. Further, such a state of equilibrium is assumed unless the motion of the system is specifically stipulated. Strain energy per unit volume is the quadratic form:

$$U_{s0} = \boldsymbol{\varepsilon}^{\mathrm{T}} \mathbf{C} \boldsymbol{\varepsilon}, \tag{30S.1}$$

where coefficient matrix $\mathbf{C}$ is symmetric and each of its elements is the function of Lame's constant and shear modulus. The representation of strain energy in such a form can be found, for example, in the book *Deformation Theory of Plasticity* (Jones 2009, p. 122). The relationships for the elements of matrix $\mathbf{C}$ will not be further required. Let us denote the displacement of the load application point in the projection on axis Oq as u_q. Displacement u_q is the sum of two components: the displacement produced by the weight of the structure and displacement u_{qP} caused only by load P_{fict}. Using the notations introduced in this chapter, one can write down:

$$u_{qP} = \Delta q = q - q_0, \tag{30S.2}$$

where the terms on the right side are the generalized coordinates of a structure.

Likewise, the strain at an arbitrary point of the structure is the sum of components:

$$\boldsymbol{\varepsilon}(\boldsymbol{r}) = \boldsymbol{\varepsilon}_0(\boldsymbol{r}) + \boldsymbol{\varepsilon}_P(\boldsymbol{r}),$$

where $\boldsymbol{\varepsilon}_0(\boldsymbol{r})$ is the vector of strains under the action of gravity force while $P_{fict} = 0$; $\boldsymbol{\varepsilon}_P(\boldsymbol{r})$ is the vector of strains in the absence of gravity but under given load P_{fict}. Both displacement u_{qP} as well as strain

ε_P are proportional to load P_{fict}, which allows us to express this strain through the corresponding displacement:

$$\varepsilon_P(r) = u_{qP}\,\varepsilon_1(r).$$

Here, $\varepsilon_1(\boldsymbol{r})$ is the strain vector in the absence of gravity and under static load P_1, which produces displacement $u_{qP} = 1$. This vector consists of coefficients which do not depend on the load's magnitude but are determined for the given structure only by the point of application and direction of this load. The manipulation made above enables us to represent the dependence of the deformation field versus the generalized coordinate in the form:

$$\varepsilon(r) = \varepsilon_0(r) + (q - q_0)\,\varepsilon_1(r).$$

After substitution of this relation into formula (30S.1) and integration over the volume of the body, we come to the following expression for the strain energy of a structure:

$$U_s = \int_V U_{s0}dV = (q - q_0)^2 \int_V \varepsilon_1^{\mathrm{T}} \mathrm{C} \varepsilon_1 dV + 2(q - q_0)\int_V \varepsilon_0^{\mathrm{T}} \mathrm{C} \varepsilon_1 dV + \int_V \varepsilon_0^{\mathrm{T}} \mathrm{C} \varepsilon_0 dV. \qquad (30S.3)$$

At the right-hand side, we have the function of the generalized coordinate, which is the polynomial of degree 2 and can be elementarily represented in the form (30.10).

Formula (30S.3) gives us a simple way to calculate the rigidity factor of structure k_{rig}. The polynomial's factor at q^2 in this formula equals $k_{rig}/2$, thus:

$$k_{rig} = 2\int_V \varepsilon_1^{\mathrm{T}} \mathrm{C} \varepsilon_1 dV = 2U_{sP_1}, \qquad (30S.4)$$

where U_{sP_1} is the strain potential energy of the structure in the absence of gravity force and under load $P_{fict} = P_1$ which produces displacement $u_{qP} = 1$. Let us represent energy U_{sP_1} through load P_1. According to Clapeyron's theorem, the strain energy of a structure at given load P_{fict} and absence of gravity is equal to half the work done by this load acting through displacement u_{qP}. Upon unit displacement, this work is equal to $P_{fict}u_{qP} = P_1$; thus, $U_{sP_1} = {}^{P_1}\!/_2$ and finally we obtain:

$$k_{rig} = P_1. \qquad (30S.5)$$

So, **the rigidity factor of a structure is equal to the load P which, being exerted upon a structure in equilibrium, causes the displacement $\Delta q = 1$.**

Now, we shall obtain the gravity potential energy of a structure in the form (30.11). We denote the variation of this energy due to displacement from the deformed state of the system under the action of only a gravity force to the state of equilibrium under the action of load P_{fict} along with gravity as ΔU_g. Introduce denotation $u_v = u_v(\boldsymbol{r})$ for the field of vertical displacements with positive direction downward, caused by static action of separately taken load P_{fict}. The potential energy of elementary mass contained in volume dV varies due to displacement u_v by the quantity of $-\gamma dV \cdot u_v$, where γ is the unit weight of the material. We obtain the variation of the gravity potential energy for the whole body by summing over all elementary masses:

$$\Delta U_g = -\gamma \int_V u_v dV.$$

Displacement $u_v(\boldsymbol{r})$ at an arbitrary point of a body is proportional to load P_{fict}, that allows us to represent this displacement in the form:

$$u_v(r) = u_{qP} u_{v1}(r),$$

where $u_{v1}(r)$ is vertical displacement in the absence of gravity force and under load P_1 which causes displacement $u_{qP} = 1$. From this, there follows:

$$\Delta U_g = -(q - q_0)\gamma \int_V u_{v1} dV = -(q - q_0) w_{ef},$$

where the denotation is used:

$$w_{ef} = \gamma \int_V u_{v1} dV. \tag{30S.6}$$

Accepting the gravitational potential energy of a structure being in equilibrium only under the action of its own weight to be equal to $-w_{ef} q_0$, we come to formula (30.11). The coefficient (30S.6) can be of positive or negative sign, according to the point of application and direction of load P_{fict}.

Further, we derive the kinetic energy of a structure in the form (30.13). We introduce the displacement vector at a point of the structure's body as $\mathbf{u}^T = (u_x, u_y, u_z)$, and consider the field of displacements $\mathbf{u} = \mathbf{u}(\boldsymbol{r})$ under the action of gravity and load P_{fict}. The vector-function $\mathbf{u}(\boldsymbol{r})$ can be represented in the form:

$$\mathbf{u}(\boldsymbol{r}) = \mathbf{u}_0(\boldsymbol{r}) + (q - q_0)\mathbf{u}_1(\boldsymbol{r}), \tag{30S.7}$$

where $\mathbf{u}_0(\boldsymbol{r})$ is the displacement vector at $P_{fict} = 0$ and action of gravity; $\mathbf{u}_1(\boldsymbol{r})$ is the displacement vector in the absence of gravity but under load P_1 which produces displacement $u_{qP} = 1$. Now let a structure move under the action of variable load P. According to the assumption made in the chapter, **displacements in a structure are determined by a generalized coordinate in the same way both under static and dynamic action of load**. Hence, the relationship (30S.7) is correct at an arbitrary moment of motion, and the velocity vector of the structure's elementary mass can be obtained by derivation:

$$\dot{\mathbf{u}}(\boldsymbol{r}) = \dot{q}\mathbf{u}_1(\boldsymbol{r}). \tag{30S.8}$$

For the square of velocity, we have:

$$v^2 = |\dot{\mathbf{u}}(\boldsymbol{r})|^2 = \dot{q}^2 |\mathbf{u}_1(\boldsymbol{r})|^2. \tag{30S.9}$$

Out of this, the kinetic energy of a structure is obtained in the form:

$$T_{kin} = \frac{\rho \dot{q}^2}{2} \int_V |\mathbf{u}_1(\boldsymbol{r})|^2 dV,$$

where ρ is the density of the body's material. The latter formula can be written in the form (30.13) by introduction of effective mass of a structure as follows:

$$m_{ef} = \rho \int_V |\mathbf{u}_1(\boldsymbol{r})|^2 dV. \tag{30S.10}$$

Example 1

Determine the effective mass of the bar shown in Figure 30.4a, given the weight of bar m and its length l.

We place the origin of earth-bound CS $O'xyz$ at the bottom end of the member and direct axis $O'z$ upward (Figure 30S.1). Vertical displacement u_v of the member's arbitrary cross-section arisen under load P_{fict} is proportional to distance z from this section to the bottom end of the member. The corresponding dependence can be represented in the form:

$$u_v(z) = u_{qP}\frac{z}{l}, \tag{30S.11}$$

where u_{qP} is the displacement of the top end of member (30S.2). While the displacement of the load application point is a unit, formula (30S.11) takes the form:

$$u_v(z) = \frac{z}{l}.$$

The mass of the member's elementary portion having length dz is equal to $m\frac{dz}{l}$. Neglecting transverse displacements of the member's points, we replace in formula (30S.10) the integration variable $\boldsymbol{r}$ with scalar variable z by using the substitutions:

$$\rho dV = m\frac{dz}{l}, \tag{30S.12}$$

$$|\mathbf{u}_1(\boldsymbol{r})| = \frac{z}{l}.$$

Finally, we have the integral along the length of the member:

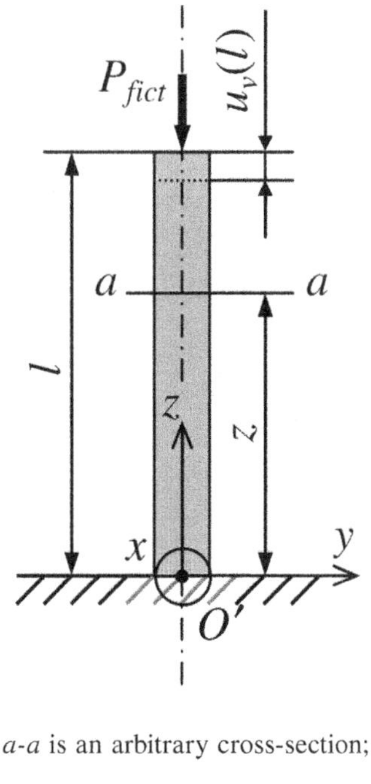

FIGURE 30S.1 Bar under loading: a-a is an arbitrary cross-section; $u_v(l) = u_{qP}$ is displacement of the top butt-end of the bar.

$$m_{ef} = \int_0^l \left(\frac{z}{l}\right)^2 m\frac{dz}{l} = \frac{m}{l^3}\int_0^l z^2 dz = \frac{m}{3}. \tag{30S.13}$$

Example 2

Determine the effective mass of the cantilever shown in Figure 30.4b, given the weight of beam m and its length l.

We determine the deflections of cross-sections of the beam in earth-referenced CS $O'xyz$, in which the z-axis coincides with the axial line of the beam before deformation. We denote the deflection of the section with coordinate z as y (Figure 30S.2). The elastic curve of the beam on the scheme of loading in Figure 30S.2 was obtained in mechanics of materials (Hibbeler 2011, p. 578) and is determined by the function:

$$y(z) = \frac{P_{fict}z^2}{2EI}\left(l - \frac{z}{3}\right),$$

where EI is flexural rigidity of beam. At the point of load action, the displacement is equal to:

$$y(l) = \frac{P_{fict}l^3}{3EI},$$

whence:

$$y(z) = \frac{3y(l)}{2l^3}z^2\left(l - \frac{z}{3}\right).$$

On $y(l) = 1$, we obtain the displacement of the elementary portion of the beam in the form:

$$|\mathbf{u}_1(\boldsymbol{r})| = \frac{3}{2l^3}z^2\left(l - \frac{z}{3}\right),$$

where radius-vector $\boldsymbol{r}$ corresponds to coordinate z. Equality (30S.12) is correct for the mass of the elementary portion, as before. After representation of the integral over the volume (30S.10) by the integral along the length of the member, we obtain:

$$\begin{aligned} m_{ef} &= \int_0^l \left(\frac{3}{2l^3}z^2\left(l - \frac{z}{3}\right)\right)^2 m\frac{dz}{l} = \frac{9m}{4l^7}\int_0^l z^4\left(l - \frac{z}{3}\right)^2 dz \\ &= \frac{9m}{4l^7}\frac{33}{315}l^7 = \frac{33}{140}m. \end{aligned} \tag{30S.14}$$

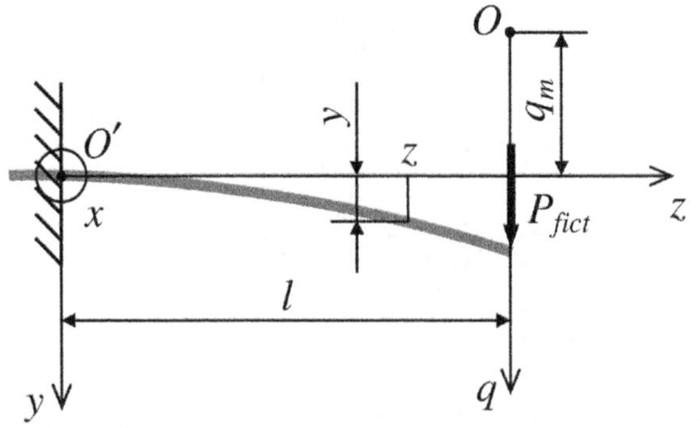

FIGURE 30S.2 Deflection of a cantilever beam.

Solving Equations of a Single-DOF System's Free Vibration

We will get a general integral of the equation of motion, given in the form:

$$\Delta\ddot{q} + \omega_0^2\Delta q + 2\zeta\omega_0\Delta\dot{q} = 0. \tag{30S.15}$$

We assume the solution in the form $\Delta q = e^{st}$, where s is a constant. Substitution of this function gives the equation:

$$\frac{d^2}{dt^2}e^{st} + \omega_0^2 e^{st} + 2\zeta\omega_0\frac{d}{dt}e^{st} = 0.$$

After deriving the left-hand side and cancellation of e^{st}, we obtain the characteristic equation:

$$s^2 + 2\zeta\omega_0 s + \omega_0^2 = 0,$$

which has two roots in the field of complex numbers:

$$s_{1,2} = -\zeta\omega_0 \pm i\omega_0\sqrt{1-\zeta^2}.$$

According to these roots, there is the following system of linearly independent solutions (complex in general case):

$$\Delta q_1 = e^{-\zeta\omega_0 t}e^{i\omega t}; \quad \Delta q_2 = e^{-\zeta\omega_0 t}e^{-i\omega t}, \tag{30S.16}$$

where the denotation is used: $\omega = \omega_0\sqrt{1-\zeta^2}$. By using linear transformations of the obtained solutions, we construct two independent solutions which are real under $\zeta \le 1$:

$$\frac{1}{2}(\Delta q_1 + \Delta q_2) = e^{-\zeta\omega_0 t}\frac{e^{i\omega t} + e^{-i\omega t}}{2} = e^{-\zeta\omega_0 t}\cos\omega t;$$

$$\frac{1}{2i}(\Delta q_1 - \Delta q_2) = e^{-\zeta\omega_0 t}\frac{e^{i\omega t} - e^{-i\omega t}}{2i} = e^{-\zeta\omega_0 t}\sin\omega t.$$

As the result, we obtain the general integral of the equation of motion in the form (30.22).

The general integral of equation (30S.15), written in the form

$$\Delta q = C_1\Delta q_1 + C_2\Delta q_2, \tag{30S.17}$$

should be employed for analysis of the case $\zeta \ge 1$. In this case, the parameter ω is a complex number. By using substitution $\omega = i\omega_0\sqrt{\zeta^2 - 1}$ into formulas (30S.16), we convert the general solution (30S.17) to the form:

$$\Delta q = C_1\exp\left[-\omega_0 t\left(\zeta + \sqrt{\zeta^2-1}\right)\right] + C_2\exp\left[-\omega_0 t\left(\zeta - \sqrt{\zeta^2-1}\right)\right]. \tag{30S.18}$$

This function in the case of $\zeta \ge 1$ and real coefficients C_1 and C_2 describes the motion without oscillations represented by the graph in Figure 30.5b.

31 A Dynamic Factor and Problem of Finding Natural Frequencies

31.1 FORCED VIBRATION AND DYNAMIC MAGNIFICATION FACTOR

The problems of structural dynamics can often be replaced by similar static problems for the same structure. After having solved such a static problem, one can come to the dynamic characteristics of the SSS through multiplying the obtained static characteristics by the magnification factor. A *dynamic magnification factor* shows the number of times the maximum stresses and strains caused by dynamic load in the absence of gravity is greater than the corresponding stresses and strains due to static load equal to the maximal dynamic one. Indicating the parameter of SSS under dynamic loading by subscript "*d*" and the same parameter under static loading by subscript "*st*," we can define the magnification factor by the ratios:

$$k_d = \frac{\sigma_d}{\sigma_{st}} = \frac{\varepsilon_d}{\varepsilon_{st}}. \tag{31.1}$$

We emphasize that the stresses and strains in formula (31.1) do not include the component produced by the gravity force. In particular, characteristics σ_d and ε_d are determined only by dynamic loads and inertia forces.

In a recent chapter, we constructed the energy model for a system with single DOF and obtained the equation of motion (30.20); the latter we write down again:

$$\Delta\ddot{q} + \omega_0^2\Delta q + 2\zeta\omega_0\Delta\dot{q} = \frac{P(t)}{m_{ef}}. \tag{31.2}$$

Now, we determine the dynamic magnification factor for a system with the equation of motion (31.2) at the action of harmonic excitation, i.e., under the load as follows:

$$P = P_0 \sin\theta t, \tag{31.3}$$

where θ is the circular frequency of load oscillations. For this purpose, we establish the relationship for coordinate Δq versus the time.

It is natural to expect that the structure under loading (31.3) vibrates with the same frequency, i.e., the particular solution of vibration equation (31.2) should be found in the form:

$$\Delta q = C_1 \sin\theta t + C_2 \cos\theta t. \tag{31.4}$$

After substitution of solution (31.4) into the vibration equation, we have to obtain the identity:

$$\begin{aligned}&-C_1\theta^2 \sin\theta t + C_1\omega_0^2 \sin\theta t + 2\zeta\omega_0 C_1\theta\cos\theta t\\&-C_2\theta^2 \cos\theta t + C_2\omega_0^2 \cos\theta t - 2\zeta\omega_0 C_2\theta\sin\theta t = \frac{P_0}{m_{ef}}\sin\theta t.\end{aligned}$$

Coefficients C_1 and C_2 are established from linear independence of time functions sin θt and cos θt. The sum of terms containing the function sin θt at the left-hand side of the identity must be equal to the right-hand side; the sum of terms containing the function cos θt at the left-hand side must be equal to zero. We have the system of equations:

$$\begin{cases} \left(\omega_0^2 - \theta^2\right) C_1 - 2\zeta\omega_0\theta C_2 = \dfrac{P_0}{m_{ef}}; \\ 2\zeta\omega_0\theta C_1 + \left(\omega_0^2 - \theta^2\right) C_2 = 0. \end{cases}$$

After solving it, we proceed to representation of the particular solution (31.4) by the single trigonometric function:

$$\Delta q = \Delta q_d \sin(\theta t - \varepsilon), \tag{31.5}$$

where:

$$\Delta q_d = \sqrt{C_1^2 + C_2^2}; \ \tan\varepsilon = -\frac{C_2}{C_1}; \ \cos\varepsilon = \frac{C_1}{\sqrt{C_1^2 + C_2^2}}.$$

We obtain the final result for the amplitude of forced harmonic vibration in the form:

$$\Delta q_d = \frac{P_0}{m_{ef}} \frac{1}{\sqrt{\left(\omega_0^2 - \theta^2\right)^2 + 4\zeta^2\omega_0^2\theta^2}}. \tag{31.6}$$

The general solution of the equation of motion (31.2) is obtained through adding the general solution of the same equation with a zero right-hand side to particular solution (31.5). But the general solution of the homogeneous equation of the system's motion with energy dissipation turns to zero at $t = +\infty$, as shown in the previous chapter. Thus, the motion of such a system under a sufficiently long action of harmonic loading also becomes harmonic with the amplitude (31.6). The vibration (31.5), which a system with energy dissipation executes under harmonic load until long after the excitation has begun, is referred to as *steady-state vibration*.

For most structures, steady-state vibration is considered as the possible cause of a system's failure. The reasons for this are the following. If the operability is understood as keeping the strength of a structure, then the loss of strength induced by vibrations would mainly occur as a result of fatigue phenomena which evolve due to steady-state vibration. If the criterion of operability is the serviceability characteristic, then the vibration amplitude should be limited, and in this case, the amplitude is estimated for a sufficiently long period of operation, i.e., again for the stage of steady-state vibration. Still, in some cases, the analysis under assumption of a steady-state vibration is not permitted. It should be borne in mind that even if at the starting point in time $t = 0$ a structure was at rest, then under action of periodic load (31.3) the displacement of the structure at the initial phase of motion may be noticeably greater (sometimes double) than the amplitude of steady-state vibration.

We assume that a dangerous state of a structure occurs under steady-state vibration. The magnification factor we shall calculate as the ratio:

$$k_d = \frac{\Delta q_d}{\Delta q_{st}}, \tag{31.7}$$

where Δq_d is the amplitude of steady-state vibration, and Δq_{st} is the displacement under action of static amplitude load $P = P_0$. The equivalence of relation (31.7) to the definition of magnification factor (31.1) follows from the superposition principle.

In the case of static loading, the equation of motion (31.2) takes the trivial form:

$$\omega_0^2 \Delta q = \frac{P}{m_{ef}},$$

whence:

$$\Delta q_{st} = \frac{P_0}{\omega_0^2 m_{ef}}. \tag{31.8}$$

Through substitutions (31.6) and (31.8) into (31.7), we obtain the formula for magnification factor under harmonic excitation:

$$k_d = \frac{1}{\sqrt{\left(1-\frac{\theta^2}{\omega_0^2}\right)^2 + \left(2\zeta\frac{\theta}{\omega_0}\right)^2}}. \tag{31.9}$$

So, the problem of the magnification factor calculation has been solved for the steady-state vibration of a system with a single DOF.

We clarify the notion of resonance for a system with a single DOF. *The motion of a vibrating system at exciting frequencies such that the ratio of steady-state vibration amplitude to exciting load amplitude is close to the maximum is referred to as resonance.* Formula (31.9) enables us to study the resonance phenomenon, which manifests itself in a surge of magnification factor at the load frequencies $\theta \approx \omega_0$. For analysis of resonance, we have to represent the expression (31.9) as the following function of magnification factor versus relative frequency of load θ/ω_0:

$$k_d(\beta) = \frac{1}{\sqrt{(1-\beta^2)^2 + (2\zeta\beta)^2}}, \quad \beta \equiv \theta/\omega_0, \tag{31.10}$$

and establish the maximum point of this function from equating its first derivative to zero.

Figure 31.1 shows the graphs of dependence of the magnification factor on relative load frequency θ/ω_0 at different magnitudes of damping ratio ζ. The maximum of function (31.10) appears to be equal to:

$$k_{d,\max} = \frac{1}{2\zeta\sqrt{1-\zeta^2}} \tag{31.11}$$

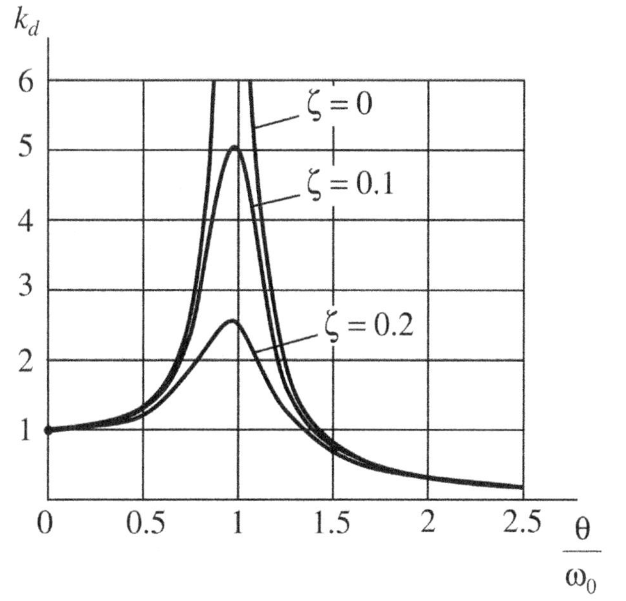

FIGURE 31.1 The curves of magnification factor versus relative load frequency θ/ω_0.

and is attained under:

$$\frac{\theta}{\omega_0} = \sqrt{1-2\zeta^2}. \tag{31.12}$$

For real structures, the damping ratio is small enough to neglect the difference between the maximum point (31.12) and the value $\theta/\omega_0 = 1$. Assuming $\theta/\omega_0 = 1$, we have for the dynamic factor:

$$k_d = \frac{1}{2\zeta}. \tag{31.13}$$

The difference of this coefficient from greatest magnitude (31.11) is usually no more than 1%. Therefore, formula (31.13) can be used in practice for calculation of the maximum value of the dynamic magnification factor.

From formula (31.13), one can obtain the damping ratio for different types of structures by conducting the series of experiments for a selected type in order to determine the maximum factor $k_{d,\max}$. The damping ratio is calculated by the formula:

$$\zeta \cong \frac{1}{2k_{d,\max}}. \tag{31.14}$$

The method of experimental determination of damping ratio in the form (31.14) is called the resonant amplification method. Researches show that the damping ratio of reinforced-concrete and brick structures can reach the magnitude $\zeta = 0.1$, but this value is considered large and is seldom the case. For steel structures, the factor ζ usually does not exceed 0.025 (Chopra 2012, p. 454).

Under constant ratio θ/ω_0, the magnification factor increases as quantity ζ decreases (Figure 31.1). When $\zeta = 0$, there is no energy dissipation in a system, and formula (31.9) takes the simple form:

$$k_d = \frac{1}{\left|1-\dfrac{\theta^2}{\omega_0^2}\right|}. \tag{31.15}$$

Under condition $\zeta = 0$, function $k_d(\theta/\omega_0)$ becomes discontinuous at the point $\theta/\omega_0 = 1$. This means the unlimited increase of forced vibration amplitude due to resonance in a system with no dissipation of energy (Figure 31.2).

It is useful to compare the relationships (31.9) and (31.15) in order to establish when the dissipation of energy can be neglected in calculations. We shall point out the value of ratio ζ in the

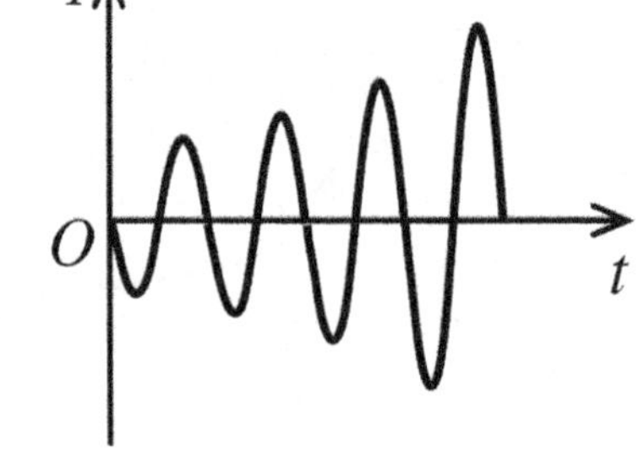

FIGURE 31.2 Increase of vibration amplitude due to resonance.

subscript at magnification factor, i.e., the factor (31.9) will be denoted $k_{d,\zeta}$, and the factor (31.15) is denoted $k_{d,0}$. It is easy to establish that for $\zeta \le 0.05$ under frequencies satisfying the requirement:

$$\theta/\omega_0 \le 0.7 \quad \text{or} \quad \theta/\omega_0 \ge 1.3,$$

the next inequality holds:

$$\frac{k_{d,0} - k_{d,\zeta}}{k_{d,\zeta}} < 0.02,$$

i.e., the difference between dynamic factors (31.9) and (31.15) is negligibly small. This property is formulated as the requirement to the ratio of frequencies θ and ω_0: in order to neglect the energy dissipation phenomenon in a vibrating system, it is sufficient to ensure the deviation from the resonance frequency by 30%.

Now, we will consider quantity ε in the expression (31.5) as a function of relative frequency θ/ω_0. This quantity is the phase shift of steady-state oscillations with respect to oscillations of load. More accurately, quantity ε is referred to as phase lag of system's oscillations with respect to exciting force. In explicit form, the relationship $\varepsilon = \varepsilon(\theta/\omega_0)$ can be obtained through calculation of coefficients C_1 and C_2 in formula (31.4). The reader can verify the correctness of relationships:

$$\tan\varepsilon = \frac{\dfrac{2\zeta\theta}{\omega_0}}{1 - \dfrac{\theta^2}{\omega_0^2}}; \quad \cos\varepsilon = \frac{1 - \dfrac{\theta^2}{\omega_0^2}}{\sqrt{\left(1 - \dfrac{\theta^2}{\omega_0^2}\right)^2 + \left(2\zeta\dfrac{\theta}{\omega_0}\right)^2}}. \tag{31.16}$$

It is easy to see from these formulas that in the absence of energy dissipation in a system (when $\zeta = 0$), sine load (31.3) causes steady-state vibration of the form:

$$\Delta q = \pm\Delta q_d \sin\theta t. \tag{31.17}$$

At $\theta < \omega_0$, the displacements of a system are in phase with excitation (the sign "+" is taken in the latter formula); at $\theta > \omega_0$, the displacements of a system are out of phase with excitation (the sign "–" is taken in formula (31.17)).

At $\zeta > 0$, the function (31.17) taken with appropriate sense can often approximately represent the steady-state oscillations. The error of approximation can be characterized by phase shift ε_1 in exact time dependence:

$$\Delta q = \pm\Delta q_d \sin\left(\theta t + \varepsilon_1\right).$$

For the values of damping ratio $\zeta \le 0.05$ and the deviation from a resonance frequency no less than 30%, parameter ε_1 appears to be small: $|\varepsilon_1| < 11^\circ$.

31.2 FREE VIBRATION OF SYSTEMS WITH A FINITE NUMBER OF DOFS

Now, we proceed to studying the motion of bar systems which may have a number of DOFs greater than 1. We consider a plane elastic system with weightless members and several lumped masses fastened upon the members. We allow the system to include members with infinitely large axial rigidity (the length of such members doesn't vary under loading).

The displacement of every mass is allowed either along the axis specified for this mass or in an arbitrary direction, or displacement of the mass is impossible. In the first case, the position of mass is determined by one displacement in a specified direction. If displacement is allowed in an arbitrary direction, then we specify the position of the mass by two displacements obtained as orthogonal projections of the displacement vector upon two non-parallel directions. One or two displacements which specify the location of lumped mass we refer to as *possible displacements* of this mass. Further on, we assume that **the totality of possible displacements of a system's masses makes up an independent set of displacements**. We call this statement the independence requirement upon displacements of lumped masses.

The totality of independent displacements of concentrated masses of a system will be represented by displacement vector $\mathbf{u} = (u_1, u_2, \dots, u_n)^{\mathrm{T}}$.

Figure 31.3 shows three systems, for which we neglect the deformations of members' tension-compression, i.e., assume that the distance between the endpoints of each member doesn't change under loading. Diagram a presents the frame wherein there are four independent displacements of masses; the masses at the columns can perform only horizontal displacements, and the mass on the girder is displaced both in horizontal and vertical directions. Directions of possible displacements are designated on the diagram. In the case b, there is but one possible displacement of lumped mass. Diagram c shows an example of a system which does not satisfy the independence requirement upon displacements of lumped masses. In this system, there are two independent displacements – such, for instance, as the ones presented on the diagram. Horizontal displacements of masses m_1 and m_2 are possible but are determined by displacement of mass m_3, i.e., they are not to be taken as independent along with the shown displacements. Such structures are not in the scope below.

Note that if all members of a system have finite axial rigidity and the connections of the first type can be only supporting, then at arbitrary positioning of lumped masses at different points, the independence requirement upon displacements of these masses is fulfilled. If finite axial rigidity of members is allowed in the examples in Figure 31.3, then two mutually orthogonal displacements are possible for every concentrated mass, and these displacements constitute the independent totality for each of the shown systems.

The state of a system with lumped masses at any moment of motion can be considered as the equilibrium if the inertia forces are exerted upon these masses. We decompose the vector of inertia force for a mass with two possible displacements into the components along the directions of possible displacements. With every possible displacement u_i at any $i = \overline{1,n}$, we put together the inertia force X_i acting in the same direction as the displacement is determined. These inertia forces make up vector $\mathbf{X} = (X_1, X_2, \dots, X_n)^{\mathrm{T}}$.

Note that while two possible displacements of a mass are the orthogonal projections on the directions chosen for this mass, the according projections of inertia force are entered as oblique projections of the force vector upon corresponding axes (Figure 31.4). Further on, in order to avoid treating oblique projections, two possible displacements of the same mass are determined in mutually perpendicular directions (as in the case of displacements u_3, u_4 in Figure 31.3a).

Now, we use the results of Chapter 17. The vector of external forces $\mathbf{X}$ (with no regard to the nature of these forces) determines corresponding displacement vector $\mathbf{u}$ for a system in equilibrium.

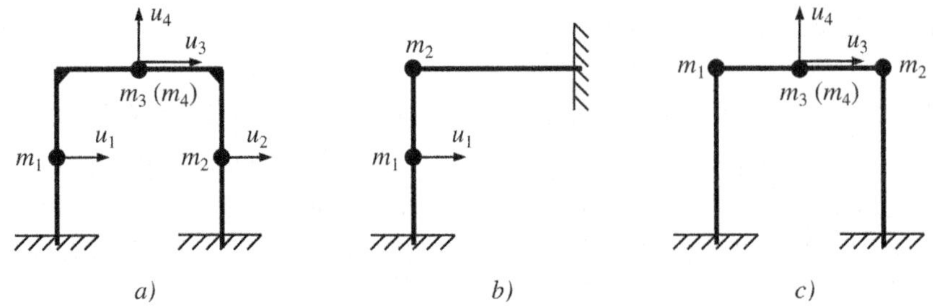

FIGURE 31.3 Systems with concentrated masses (mass m_3 has additional notation m_4).

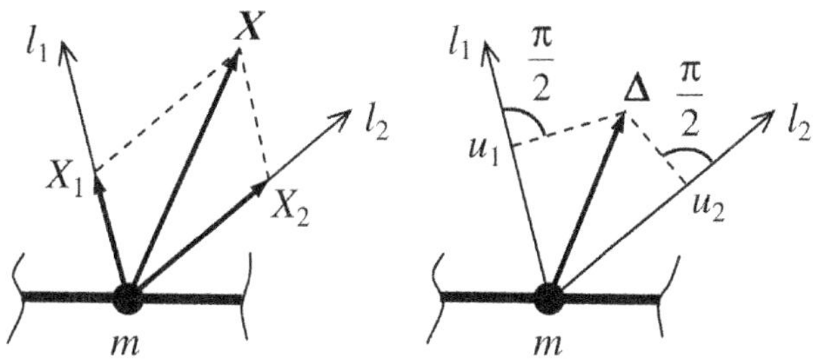

FIGURE 31.4 Representation of external force X and displacement vector $\mathbf{\Delta}$ for point mass m by means of pairs of generalized forces and displacements: l_1, l_2 are axes of oblique CS; (X_1, u_1), (X_2, u_2) are pairs of generalized forces and displacements.

In an absence of loads, the relation between these vectors is represented by using matrix $\boldsymbol{\delta}$ of order n in the form:

$$\mathbf{u} = \delta\mathbf{X}. \tag{31.18}$$

Every pair X_i, u_i makes up generalized force and displacement; thus, matrix $\boldsymbol{\delta}$ is a flexibility matrix. Because the components of displacement vector $\mathbf{u}$ are independent, matrix $\boldsymbol{\delta}$ is nonsingular. Therefore, vector of generalized forces $\mathbf{X}$ is unambiguously determined by vector of generalized displacements $\mathbf{u}$. The latter affirmation is correct both in the absence of loads and when the loading has been **specified**. So, the possible displacements of concentrated masses under given loads determine the SSS of a structure and can be accepted as generalized coordinates.

Below, we show that elastic systems with several DOFs can execute a free motion in which all generalized coordinates change according to the same sine law. Such free vibration is referred to as *natural vibration* of a system, and its frequency is referred to as *natural frequency*. A nonzero vector of generalized coordinates at an arbitrary time instant of natural vibration is referred to as a *modal vector*, and the corresponding shape of the deformed structure is referred to as *natural vibration mode*.

The goal of studying free vibration is to establish the motion of a system under action of external forces. In particular, the analysis of free vibration enables us to establish forced vibration when a system is subjected to in-phase harmonic exciting forces. If the exciting frequency coincides with one of the natural frequencies, *resonance* might occur, *i.e., the vibration process in which the amplitude of forced vibration is close to one of the local maxima at the given amplitudes of exciting loads*. The resonance state of a structure should be avoided. In the case of a system with a single DOF, the dynamic calculation begins from the verification if natural vibration frequency is close to exiting frequency. If system appears to be in the resonance state, then measures may be required to change the natural frequency. These measures are not only desired from the point of view of structure operation, but ensure the simplicity of consequent analysis; because, if there is a noticeable difference between frequencies θ and ω_0, then the dissipation of energy in a system can be neglected. In the case of a system with multiple DOFs, resonance usually arises on low frequencies of natural vibration and manifests itself in great magnification of forced vibration amplitude due to the exiting frequency approaching the natural frequency. In the case of a single DOF system, it was established that the deviation from resonance frequency about 30% allows the negligence of energy dissipation. The same pattern is manifested in systems with multiple DOFs. The main difference of the calculation procedure for such systems from the simple case of single-DOF system analysis is that, at the beginning, one has to determine the natural frequencies of vibration.

So, we consider an elastic bar system with several lumped masses, and we take independent displacements u_i, $i = \overline{1,n}$, of lumped masses from position of equilibrium as the generalized coordinates.

Each mass can execute displacements either in one or two directions or does not have the DOFs in the frame of assumptions about the rigidity of members. We set up the system of equations of free motion using the force method.

We pass to the equilibrated primary system of the force method by replacing accelerated motions of concentrated masses with inertia forces X_i, $i=\overline{1,n}$, exerted in the directions of corresponding displacements u_i. In the case of a system's motion with no loading, the basic relationships of this method are determined by vector equality (31.18) and have the following scalar form:

$$u_i = \sum_{j=1}^{n} \delta_{ij} X_j, \quad i = \overline{1,n}. \tag{31.19}$$

As noted above, the flexibility matrix $\boldsymbol{\delta}$ is nonsingular. Further, we need a stronger property of this matrix: its positive definiteness, which follows from its nonsingularity (see Chapter 17). Also, the concepts of eigenvector and eigenvalue of a matrix, which have been studied in the linear algebra, will be required. We remind readers of them as follows:

The eigenvector of matrix $\mathbf{A}$ *is column vector* $\mathbf{z}$ *such that pre-multiplication of it by the matrix is reduced to vector-number multiplication:*

$$\mathbf{Az} = \lambda \mathbf{z}. \tag{31.20}$$

Factor λ*, which determines the result of matrix-eigenvector multiplication is referred to as the eigenvalue of the matrix.*

Some properties of matrices pertaining to the problem of seeking the eigenvectors and eigenvalues will be introduced as necessary.

In relationships (31.19), we pass from inertia forces to accelerations according to the formulas:

$$X_j = -m_j \ddot{u}_j. \tag{31.21}$$

Here, m_j is the mass, the displacement of which is specified by coordinate u_j. The same mass can have two denotations as, for example, in Figure 31.3a, where notations m_3 and m_4 are used for the mass at the girder.

After substitution of (31.21) into relationships (31.19), we get the system of differential equations for generalized coordinates:

$$u_i = -\sum_{j=1}^{n} \delta_{ij} m_j \ddot{u}_j, \quad i = \overline{1,n}. \tag{31.22}$$

In matrix form, this system looks like the next:

$$\mathbf{u} = -\boldsymbol{\delta}\,\mathbf{m}\ddot{\mathbf{u}}, \tag{31.23}$$

where $\mathbf{m} = \mathrm{diag}\{m_1, m_2, \ldots, m_n\}$ is diagonal matrix of concentrated masses (all enumerated masses are positioned at the main diagonal).

Now, we proceed to solving the equation system (31.23) that describes free motion of a structure. We resolve this system with respect to derivatives:

$$\ddot{\mathbf{u}} = -\mathbf{m}^{-1}\boldsymbol{\delta}^{-1}\mathbf{u}. \tag{31.24}$$

This form of the system of the second order differential equations is called normalized. It is known that such systems have exactly $2n$ linearly independent solutions (the so-called fundamental set of solutions), and general solution of a system is the linear combination of arbitrary $2n$ linearly independent solutions. This assertion is proved in the supplement to the chapter as the particular inference in the theory of ordinary differential equations. Based on this affirmation, **we shall obtain the fundamental set of solutions as the set of natural vibrations of a structure**.

We show that the fundamental set can be constructed from the vector functions of time having the form:

$$\mathbf{u}_i = \mathbf{a}_i \sin \omega_i t; \quad \mathbf{u}_{i+n} = \mathbf{a}_i \cos \omega_i t, \quad i = \overline{1, n}, \tag{31.25}$$

where $\mathbf{a}_i = (a_{i1}, a_{i2}, \ldots, a_{in})^{\mathrm{T}}$ are the column vectors of amplitude values of coordinates (these values may be negative).

Firstly, we make sure that on linear independence of vectors $\mathbf{a}_i$, $i = \overline{1, n}$, the presented set of functions is independent too. Linear combination of the pair of functions taken from set (31.25) at any i is represented by a single harmonic with some initial phase:

$$\begin{aligned} \mathbf{u} &= \mathbf{a}_i \left(C_{i1} \sin \omega_i t + C_{i2} \cos \omega_i t \right) = \mathbf{a}_i C_i \sin(\omega_i t + \varphi_i); \\ C_i &= \sqrt{C_{i1}^2 + C_{i2}^2}. \end{aligned} \tag{31.26}$$

Linear combination of the set of $2n$ functions (31.25) is represented by linear combination of n harmonics:

$$\mathbf{u} = \sum_{i=1}^{n} \mathbf{a}_i C_i \sin(\omega_i t + \varphi_i). \tag{31.27}$$

Owing to linear independence of vectors $\mathbf{a}_i$, the time function $\mathbf{u}$ turns identically into zero only if all coefficients C_i equal zero; this, with account of (31.26), means linear independence of the set of $2n$ functions under consideration.

Now, it is sufficient to establish that under some linearly independent amplitude vectors $\mathbf{a}_i$, $i = \overline{1, n}$, the functions of time (31.25) will be the solutions of equation of motion (31.23). By differentiating assumed solution $\mathbf{u} = \mathbf{a} \sin \omega t$ with respect to time, we get:

$$\ddot{\mathbf{u}} = -\mathbf{a}\omega^2 \sin \omega t.$$

The substitution of this derivative into equation (31.23) will give, after deletion of $\sin \omega t$, the following equation in unknown vector $\mathbf{a}$ and frequency ω:

$$\mathbf{a} = \omega^2 \boldsymbol{\delta}\, \mathbf{ma},$$

or, which is the same:

$$\boldsymbol{\delta}\, \mathbf{ma} = \frac{1}{\omega^2} \mathbf{a}. \tag{31.28}$$

If we set up the equation in the amplitude vector and frequency by differentiation of assumed solution $\mathbf{u} = \mathbf{a} \cos \omega t$ in the same manner, then we again come to equation (31.28). Now, it remains to prove that there exist n linearly independent vectors $\mathbf{a}_i$ which are the solutions of the obtained equation under appropriate frequencies ω_i. But here, we want some results from the theory of matrices.

The square matrix $\mathbf{A}$ is referred to as diagonalizable if there exists nonsingular matrix $\mathbf{S}$ which converts matrix $\mathbf{A}$ to a diagonal form by means of relationship: $\Lambda = \mathbf{SAS}^{-1}$. This definition is applicable both for matrices over the field of complex numbers and for real matrices. Here, we consider only real matrices and real vector spaces. It is known that a matrix of order n is diagonalizable if and only if it has n linearly independent eigenvectors. In matrix analysis, it is proved that **the product of two symmetric positive definite matrices is a diagonalizable matrix, and eigenvalues of this product are positive.***

Let us compare equation (31.28) with basic equation of the eigenvalue problem (31.20). We can see that vector **a** is the eigenvector of the matrix **δm**, whereas the quantity ω^{-2} is the eigenvalue for this vector. Matrix **δm** is formed by two multiplicands; each is symmetric positive definite matrix. Thus, matrix **δm** is diagonalizable and there exist n linearly independent vectors $\mathbf{a}_i$ which, together with frequencies ω_i, make up solutions of equation (31.28). Therefore, the fundamental set of solutions of the motion equation can be constructed as the set of functions (31.25). The general solution of this equation is represented as linear combination of n harmonics (31.27). If the vector of generalized coordinates representing a system's motion is the sum of vectors of generalized coordinates at other possible motions, then we refer to the total motion as superposition of these possible motions (or superposition of dynamic responses of a system). Note that summing of vectors of generalized coordinates means the summing of the displacement fields determined by vector terms of the sum.

Inferences. Free motion of an elastic system with n DOFs is the superposition of n harmonic vibrations. Every vibration is characterized by the vector of natural mode $\mathbf{a}_i$ and frequency ω_i, which are found through solving matrix equation (31.28). The natural modes constitute the system of n linearly independent vectors.

The fundamental problem of free vibration analysis is to find the natural modes and frequencies. These characteristics are necessary for solving the problem of dynamics of a structure under loading, including the detection of resonance. It was shown above that natural frequencies are linked with eigenvalues λ_i of matrix **δm** by relation:

$$\omega_i^2 = \frac{1}{\lambda_i}, \quad i = \overline{1, n}. \tag{31.29}$$

There are effective computer algorithms of seeking the eigenvalues, but manual calculations are also possible for the matrices of small order ($n \leq 3$). Manual calculations usually lie in solving the characteristic equation. Every eigenvalue λ of any matrix **A** of order n is obtained from condition (31.20) which can be rewritten in the form:

$$(\mathbf{A} - \lambda\mathbf{I})\mathbf{z} = 0, \tag{31.30}$$

where **I** is the identity matrix. In order that nonzero vector **z** comply with this condition, it is required that:

$$\det(\mathbf{A} - \lambda\mathbf{I}) = 0. \tag{31.31}$$

The latter condition is the polynomial equation of degree n in unknown λ. This equation is referred to as the characteristic equation of **A**. It has precisely n roots (including all repeated roots) and is complex in the general case. In the case **A** = **δm**, all the roots of the characteristic equation are real positive numbers and determine n natural frequencies.

The characteristic equation for determining the natural frequencies is called the *frequency equation*. When $n = 3$, the frequency equation takes the form:

* The product of two symmetric positive definite matrices is a matrix diagonalizable with positive eigenvalues over the field of complex numbers (Horn and Johnson 2013, Corollary 7.6.2*a*, p. 486). If any matrix is diagonalizable over the complex numbers and has real eigenvalues only, then it is diagonalizable over the real numbers (Horn and Johnson 2013, Theorem 1.3.29, p. 67). Out of this follows the feasibility to diagonalize the matrix product under consideration over the real numbers.

$$\det\left(\delta\mathbf{m}-\lambda\mathbf{I}\right)=\begin{vmatrix}\delta_{11}m_1-\lambda & \delta_{12}m_2 & \delta_{13}m_3\\ \delta_{21}m_1 & \delta_{22}m_2-\lambda & \delta_{23}m_3\\ \delta_{31}m_1 & \delta_{32}m_2 & \delta_{33}m_3-\lambda\end{vmatrix}=0, \tag{31.32}$$

where unknown λ determines the frequencies according to relation (31.29).

All natural frequencies are numbered in ascending order, as follows: $\omega_1 \le \omega_2 \le \ldots \le \omega_n$; the vibration modes $\mathbf{a}_i$ are numbered accordingly. The lowest natural frequency and according mode are referred to as the fundamental ones. Under a large number of DOFs, it is not required to determine all natural frequencies and modes. The fundamental frequency is determined mandatorily, because the resonance on this frequency goes on with greatest amplitude and is most dangerous for a structure. Sometimes, several higher natural frequencies are wanted in ascending order. Together with frequencies, it is required to determine the vibration modes.

Although the study has been focused on the problem of motion of plane elastic bar systems, the inferences of the theory remain in force for a spatial system of bodies. The peculiarity of spatial problems is that every concentrated mass of a structure has up to three possible displacements (whereas in planar systems, there are no more than two such displacements).

The possibility of representing free motion of an elastic system in the form of superposition of harmonic vibrations was established by Italian mathematician and astronomer J.L. Lagrange. In the book *Analytical Mechanics* published in 1811, he showed that the motion of point masses near the position of stable equilibrium is the superposition of harmonic vibrations, and the number of vibration modes is equal to the number of the system's DOFs (Lagrange 1997).

The FEM software complexes enable us to obtain the natural modes and frequencies of a structure's vibration by modeling a structure as a system of point masses positioned at the nodes of the FE model. Figure 31.5 shows a 17-story residential building with a reinforced-concrete skeleton (photo a), the FE model of the skeleton constructed from the shell and beam finite elements (diagram b), and also two modes of vibrations obtained by modeling. The fundamental frequency is implemented by flexural mode of vibration (diagram c); the first higher frequency corresponds to the torsion mode of vibration (diagram d). Multi-story buildings should be designed with taking into account the possible dynamic loadings. For example, the wind excitation upon high-rise buildings is always taken into account both in the strength analysis and in the verification of habitability. In the next chapter, it will be shown how the natural frequencies and modes are used in analysis of the structure response to dynamic loads.

SUPPLEMENT TO CHAPTER

Constructing a Fundamental Set of Solutions for System of Second Order Linear Differential Equations

We shall represent the solution of a system of m differential equations in unknown functions $y_1(t)$, $y_2(t)$, …, $y_m(t)$ as m-dimensional vector-function:

$$\mathbf{y}(t)=\left(y_1(t), y_2(t),\ldots, y_m(t)\right)^{\mathrm{T}}$$

and call it the solution of the equation system.

While considering a linear homogeneous system of differential equations, we call the set of its linearly independent solutions such that an arbitrary solution of the equation system is a linear combination of solutions from this set *the fundamental set of solutions*. It can be said succinctly that the fundamental set of solutions of a linear homogeneous system of differential equations is the basis of the space of solutions of this system.

FIGURE 31.5 Building with reinforced-concrete skeleton and natural modes of its vibration: (a) Photo of an inhabited house (city of Bataysk, Rostov Region, RF, yr 2019); (b) Undeformed skeleton (shown by axial lines of columns and median surfaces of plates); (c) Fundamental vibration mode, frequency $f_1 = 0.30$ Hz; (d) Second natural mode, frequency $f_2 = 0.38$ Hz.

In the course of differential equations, the theory of systems of first-order linear differential equations is set out. Among these systems, the important place is taken by homogeneous equation systems, which are represented in the normalized form:

$$\mathbf{y}'(t) = \mathbf{A}(t)\mathbf{y}(t), \tag{31S.1}$$

where $\mathbf{A}(t)$ is the matrix of order m, the elements of which are continuous functions of argument t, $t \in (a, b)$. The fundamental set of solutions of the vector equation (31S.1) consists of m solutions, and there are the standard methods to find out this set (Boyce and DiPrima 2012).

The system of n second-order linear homogeneous equations of the form:

$$\mathbf{u}''(t) = \mathbf{B}(t)\mathbf{u}(t) \tag{31S.2}$$

with matrix $\mathbf{B}$, the elements of which are continuous functions of variable t, can be transformed into the system of $2n$ first-order equations (31S.1). Indeed, by introducing vector-function $\mathbf{v}(t) = \mathbf{u}'(t)$, we represent equation (31S.2) in the form of the equation system:

$$\begin{cases} \mathbf{v}'(t) = \mathbf{B}(t)\mathbf{u}(t); \\ \mathbf{u}'(t) = \mathbf{v}(t). \end{cases} \tag{31S.3}$$

The latter equation system is easily represented in the form (31S.1) by means of vector-function:

$$\mathbf{y}(t) = \begin{pmatrix} \mathbf{u}(t) \\ \mathbf{v}(t) \end{pmatrix} \tag{31S.4}$$

and matrix:

$$\mathbf{A}(t) = \begin{pmatrix} \mathbf{O} & \mathbf{I} \\ \mathbf{B}(t) & \mathbf{O} \end{pmatrix} \tag{31S.5}$$

We show that the fundamental set of solutions for obtained vector equation (31S.1), which consists of $m = 2n$ scalar equations, determines the fundamental set of $2n$ solutions of system (31S.2). Really, let the system (31S.1) have fundamental set:

$$\mathbf{y}_1(t) = \begin{pmatrix} \mathbf{u}_1(t) \\ \mathbf{v}_1(t) \end{pmatrix}, \quad \mathbf{y}_2(t) = \begin{pmatrix} \mathbf{u}_2(t) \\ \mathbf{v}_2(t) \end{pmatrix}, \quad \mathbf{y}_{2n}(t) = \begin{pmatrix} \mathbf{u}_{2n}(t) \\ \mathbf{v}_{2n}(t) \end{pmatrix}. \tag{31S.6}$$

The general solution of system (31S.3), represented by vector (31S.4), is the linear combination of functions (31S.6). Accordingly, the general solution of equation (31S.2) is the linear combination of vector functions $\mathbf{u}_i(t), i = \overline{1, 2n}$. Besides, the subvectors of every vector in the set (31S.6) are linked by relationship: $\mathbf{v}_i(t) = \mathbf{u}_i'(t)$. This means that from the linear independence of set (31S.6), there follows linear independence of the set of vector-functions $\mathbf{u}_i(t), i = \overline{1, 2n}$. So, the functions $\mathbf{u}_i(t), i = \overline{1, 2n}$, make up the fundamental set of solutions of equation (31S.2).

We have established that the vector space of solutions of equation (31S.2) has a dimension of $2n$. From this, it follows that the arbitrary set of $2n$ linearly independent solutions of this equation is the fundamental set of solutions.

32 Forced Vibration of Systems with a Finite Number of DOFs

32.1 DYNAMIC FORCE METHOD FOR VIBRATION ANALYSIS OF FRAMES

We continue studying elastic bar systems with several concentrated masses. As before, we choose the independent displacements of lumped masses from the equilibrium position as general coordinates and assume that two independent displacements of the same mass are mutually perpendicular. We consider the case of the frame's excitation by harmonic in-phase loads:

$$P_k = P_k^A \sin\theta t, \;\; k = 1, 2, \ldots, \tag{32.1}$$

where amplitude value is marked by superscript "*A*." Below, we allow that amplitude values of harmonic parameters may be negative; for example, it may be that $P_k^A < 0$. Thus, we take out-of-phase oscillations for in-phase oscillations with the amplitude opposite in sign. An example of a frame under harmonic loads is shown in Figure 32.1a. Here, the left column is subjected to distributed load:

$$w = w^A \sin\theta t,$$

which can be represented in the form of a series of small concentrated loads (32.1).

We shall establish the motion of lumped masses and show how the distribution of bending moments must be determined in a frame at any instant of the vibration's time. While solving this problem, we assume that the response of a structure is harmonic and has the frequency of loading, i.e., the time dependence of the displacement vector has the form:

$$\mathbf{u} = \mathbf{u}^A \sin(\theta t + \varphi). \tag{32.2}$$

The problem, therefore, is reduced to seeking the vector of vibration amplitude $\mathbf{u}^A$ and phase angle φ.

Given the harmonic law of displacements' change in time, the inertia forces will be proportional to displacements:

$$X_i = -m_i\ddot{u}_i = m_i u_i^A \theta^2 \sin(\theta t + \varphi) = \theta^2 m_i u_i. \tag{32.3}$$

Further, we solve the problem of finding the amplitude values of these forces and their initial phase.

We pass to a primary system of the dynamic force method, which differs from the dynamic system under consideration in that it keeps equilibrium under the action of loads together with unknown forces X_i, $i = \overline{1,n}$, exerted in the directions of corresponding displacements u_i (Figure 32.1b). The additional forces X_i are called primary unknowns of the force method. The loads are taken at an arbitrary moment of time and assumed as static. If the primary system is subjected to external forces X_i only, then displacements of lumped masses are determined by formula (31.18). The action of loads makes it necessary to supplement this formula by the vector of loaded displacements $\mathbf{\Delta}_P$, made up of displacements only due to loads at a given instant of time (under zero additional forces X_i). We get:

$$\mathbf{u} = \delta\mathbf{X} + \Delta_P. \tag{32.4}$$

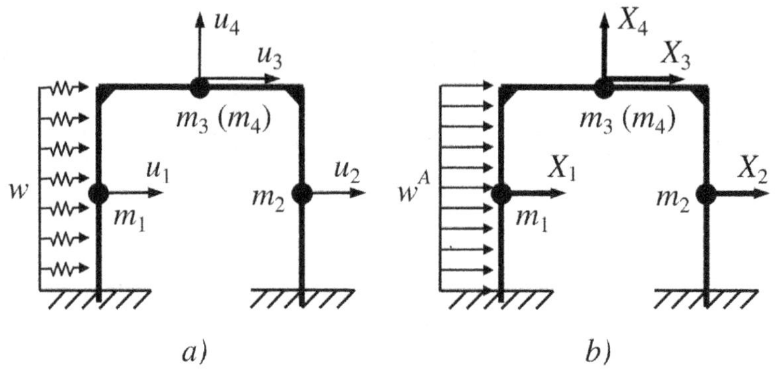

FIGURE 32.1 Given system with lumped masses (a) and primary system of force method (b).

If vector **X** represents inertia forces, then, owing to d'Alembert principle, the system can be considered as static (keeping equilibrium) at each point in time under loading, and relationship (32.4) remains correct for displacements at an arbitrary instant of time. In order to turn this relationship into the equation in the unknown vector of inertia forces, we represent the displacement vector through the inertia forces vector. To do that, we rewrite expressions (32.3) in the vector form:

$$\mathbf{X} = \theta^2 \mathbf{mu}, \tag{32.5}$$

where **m** is the matrix of lumped masses introduced in the previous chapter. From this, it holds:

$$\mathbf{u} = \frac{1}{\theta^2}\mathbf{m}^{-1}\mathbf{X}, \tag{32.6}$$

and the substitution of the latter expression into formula (32.4) for displacements gives:

$$\frac{1}{\theta^2}\mathbf{m}^{-1}\mathbf{X} = \delta\mathbf{X} + \Delta_P. \tag{32.7}$$

The equation for inertia forces, correct at any instant of time of load action, has been obtained. We rewrite it in a convenient form:

$$\left(\delta - \frac{1}{\theta^2}\mathbf{m}^{-1}\right)\mathbf{X} + \Delta_P = 0. \tag{32.8}$$

This vector equation represents the *system of the force method of canonical equations for a structure with a finite number of DOFs.* We denote the coefficient matrix of this system as:

$$\delta^* = \delta - \frac{1}{\theta^2}\mathbf{m}^{-1}. \tag{32.9}$$

If the coefficient matrix is nonsingular, then the canonical equation set has a unique solution for the arbitrary vector of loaded displacements $\mathbf{\Delta}_P$. But if the coefficient matrix is singular, the solution of the given system of equations exists only in particular cases when vector $\mathbf{\Delta}_P$ has a specific form. It can be proved that singularity of the coefficient matrix arises when the excitation frequency θ coincides with one of the system's natural frequencies ω_i. The absence of a solution of the canonical equation set means unlimited increase of the amplitude of forced oscillations over time, as shown in Figure 31.2. The case of singularity of matrix (32.9) is further excluded from consideration.

Vector $\boldsymbol{\Delta}_P$ is calculated on given loads under the assumption of the equilibrium state of a system. Obviously, its variation over time is in phase with loads, i.e., it holds:

$$\Delta_P = \Delta_P^A \sin\theta t,$$

where superscript "A" denotes the vector of amplitudes. It follows that the sought inertia forces are also in phase:

$$\mathbf{X} = \mathbf{X}^A \sin\theta t. \tag{32.10}$$

So, we may accept $\varphi = 0$ in relationship (32.2). The amplitude vector $\mathbf{X} = \mathbf{X}^A$ in equation (32.8) corresponds to the amplitude vector $\Delta_P = \Delta_P^A$. Below, the primary system and corresponding canonical equation (32.8) are considered in the time instant when the loads reach their amplitudes.

Equation (32.8) of the force method is written in scalar form as follows:

$$\left\{\begin{array}{l} \delta_{11}^* X_1 + \delta_{12} X_2 + \ldots + \delta_{1n} X_n + \Delta_{1P} = 0; \\ \delta_{21} X_1 + \delta_{22}^* X_2 + \ldots + \delta_{2n} X_n + \Delta_{2P} = 0; \\ \ldots\ldots\ldots\ldots\ldots\ldots\ldots\ldots\ldots\ldots\ldots\ldots \\ \delta_{n1} X_1 + \delta_{n2} X_2 + \ldots + \delta_{nn}^* X_n + \Delta_{nP} = 0. \end{array}\right. \tag{32.11}$$

Here, δ_{ii}^* are the diagonal constituents of matrix (32.9):

$$\delta_{ii}^* = \delta_{ii} - \frac{1}{\theta^2 m_i}. \tag{32.12}$$

While setting up canonical equations, we have used the proportionality of every displacement u_i to inertia force X_i, which is represented by relationship (32.6). This proportionality requirement could be taken for the condition of the equivalence of primary and given systems.

When canonical equation set (32.11) has been set up and solved under the amplitude values of the loaded displacements, one can construct the diagram of amplitude bending moments. The bending moment diagram changes in time under vibration, but the bending moment at the arbitrary cross-section changes according to the same law as the totality of external forces P_k, X_i. In other words, bending moments oscillate in phase with loads and attain their greatest values on amplitude loads. M-profile corresponding to amplitude loads is referred to as the *dynamic banding moment diagram* and enables us to evaluate the strength of a structure. Its construction can be implemented by using formula (13.9) derived from the force method, which we rewrite:

$$M = \sum_{i=1}^{n} \bar{M}_i X_i + M_P. \tag{32.13}$$

Here, $\bar{M}_i$ is the bending moment diagram corresponding to the action of the only force $X_i = 1$ in the primary system; M_P is the bending moment diagram due to amplitude loads exerted upon the given system in equilibrium. Note that a primary system in vibration analysis is usually statically indeterminate (as, for example, in Figure 32.1), and thus, the construction of named diagrams may be a separate and complicated problem.

The procedure of vibration analysis of a frame by means of the force method is explained next. First, the unit displacements δ_{ij} and loaded displacements Δ_{iP} should be calculated. In order to calculate the unit displacements, one has to construct unit diagrams $\bar{M}_i$ and employ Mohr's formula:

$$\delta_{ij} = \sum \int_0^l \frac{\bar{M}_i \bar{M}_j dz}{EI}. \tag{32.14}$$

To calculate the loaded displacements, one has to construct loaded diagram M_P for amplitude loads. Mohr's formula in this case has the form:

$$\Delta_{iP} = \sum \int_0^l \frac{\bar{M}_i M_P dz}{EI}. \tag{32.15}$$

In order to construct diagrams $\bar{M}_i$, $i = \overline{1,n}$, M_P in the case of a statically indeterminate primary system of dynamic force method, it will be necessary to resolve static indeterminacy, and to do that, we would use the force method repeatedly – but now, for the system at rest. This method should be employed as many times as the number of diagrams to be constructed. For instance, in the case of the system in Figure 32.1b, one must construct bending moment diagrams for unit forces X_1, X_2, X_3, X_4, exerted in the directions of corresponding displacements, and also for distributed load w^A. Thus, we have to consider five different loadings of the primary system, and every time, we have to pass to the primary system of static force method, which is obtained by removing three redundant constraints from the primary system of the dynamic force method. Only after construction of the required diagrams can one return to calculation of the parameters appertaining to the dynamic force method and determined by formulas (32.14) and (32.15).

Having established the unit and loaded displacements, we set up and solve the system of canonical equations (32.8) in order to determine the amplitude inertia forces and then to construct the dynamic bending moment diagram by means of formula (32.13).

In real bar systems, mass is not concentrated at separate points but is distributed. The usual practice in dynamic calculations is to construct a design scheme in which the mass of every member is concentrated at three points: 50% of the mass is assumed to be located in the middle of the member and 25% at each endpoint. As a result of such positioning of masses, even a relatively simple frame is represented by a design scheme with a number of DOFs in the order of tens, and the problem of dynamic calculation appears to be cumbersome.

For calculating the SSS of a vibrating elastic system, besides the force method, the dynamic displacement method may be used. In the latter method, the relation between inertia forces and displacements of concentrated masses is established by using the stiffness matrix $\mathbf{r} \equiv \boldsymbol{\delta}^{-1}$. Formerly by this method, we have obtained the motion equation (30.6) for a single-DOF system. Both these methods usually require a large amount of calculations.

32.2 MOTION OF ELASTIC SYSTEMS UNDER ARBITRARY LOADING AND WITHOUT ENERGY LOSSES

Now, we somewhat complicate the problem. Let us take as given that loads are exerted upon lumped masses in the directions of displacements and vary with time according to arbitrary law. Therefore, mass m_i is subjected to the load which is exerted in the direction of u_i and specified by continuous function:

$$P_i = P_i(t). \tag{32.16}$$

We will establish the displacements of a system under the assumption that the system was at rest at the initial instant $t = 0$.

We shall set up the equation of motion by means of the principal relationship of the force method (32.4), which we supplement with the next expression for loaded displacements:

$$\Delta_P = \delta \mathbf{P}. \tag{32.17}$$

Here, **P** is the vector of loads, the components of which are functions (32.16). The relationship (32.4) is represented in the form:

$$\mathbf{u} = -\delta \mathbf{m}\ddot{\mathbf{u}} + \delta \mathbf{P}. \tag{32.18}$$

In the previous chapter, the similar differential equation (31.23) was represented in normalized form (31.24) in order to establish the existence of a fundamental set of $2n$ solutions. If we resolve equation (32.18) in the same way, with respect to acceleration vector $\ddot{\mathbf{u}}$, then it provides grounds to state that the solution of the given equation is unambiguously determined by $2n$ initial parameters $\mathbf{u}(0)$ and $\dot{\mathbf{u}}(0)$. In the problem addressed here, the initial conditions have the form:

$$\mathbf{u}(0) = 0; \quad \dot{\mathbf{u}}(0) = 0. \tag{32.19}$$

Linear independence of natural modes $\mathbf{a}_i$, established in Chapter 31, enables us to represent the arbitrary load vector as the sum:

$$\mathbf{P}(t) = \sum_{i=1}^{n} H_i(t)\mathbf{m}\mathbf{a}_i. \tag{32.20}$$

This sum is referred to as the decomposition of a load over natural modes. Each separate addend of this sum can be called a projection of load onto the corresponding natural mode of vibration. The possibility of such representation of load follows from the fact that vector set $\{\mathbf{a}_i, i = \overline{1,n}\}$ is the basis of n-dimensional vector space, which can be transformed to a new basis $\{\mathbf{m}\mathbf{a}_i, i = \overline{1,n}\}$ if matrix **m** is nonsingular. Note also that from the continuity of vector-function $\mathbf{P}(t)$, there follows continuity of functions $H_i(t)$. The technique of decomposition (32.20) will be described below.

We seek the solution of motion equation (32.18) in the form of linear combination of natural modes:

$$\mathbf{u} = \sum_{i=1}^{n} q_i(t)\mathbf{a}_i. \tag{32.21}$$

We call coefficients q_i of this decomposition modal coordinates of the system; these coefficients constitute a special type of generalized coordinates useful for determining the response of a system to arbitrary perturbation. Expression (32.21) can be called decomposition of motion over natural modes of vibration. Further math manipulations are aimed at obtaining the equations for unknown functions of time $q_i(t)$ and to solve these equations under specified initial conditions (32.19).

In the beginning, we show that, under zero initial conditions, for every load vector's component $H_i(t)\mathbf{m}\mathbf{a}_i$ acting separately, the solution of the motion equation is matched as follows:

$$\mathbf{u} = q_i(t)\mathbf{a}_i. \tag{32.22}$$

That is, the components of decompositions (32.20) and (32.21) with the same number could, in couple, specify the motion of a system from its rest. The substitution of these components into the equation of motion (32.18) gives the equation in unknown function $q_i(t)$:

$$q_i\mathbf{a}_i = -\ddot{q}_i\delta\mathbf{m}\mathbf{a}_i + H_i(t)\delta\mathbf{m}\mathbf{a}_i. \tag{32.23}$$

Recall that the natural mode is an eigenvector of matrix $\boldsymbol{\delta m}$, or, more precisely, the natural mode satisfies equation (31.28). This allows us to simplify equation (32.23) as follows:

$$q_i\mathbf{a}_i = -\ddot{q}_i\frac{1}{\omega_i^2}\mathbf{a}_i + H_i(t)\frac{1}{\omega_i^2}\mathbf{a}_i,$$

where ω_i is the natural frequency corresponding to mode $\mathbf{a}_i$. After deletion of vector $\mathbf{a}_i$, we come to the equation with respect to modal coordinate q_i:

$$\ddot{q}_i + \omega_i^2 q_i = H_i(t). \tag{32.24}$$

The conditions (32.19) for solution $\mathbf{u} = q_i(t)\mathbf{a}_i$ take the form:

$$q_i(0) = 0; \quad \dot{q}_i(0) = 0. \tag{32.25}$$

The wanted solution $q_i(t)$ exists at the arbitrary continuous right-hand side of equation (32.24). According to (32.22), this solution determines the system's motion proportional to the natural mode $\mathbf{a}_i$ (motion over the natural mode).

One can see that the obtained equation (32.24) for modal coordinate q_i appears to be identical to equation (30.20) of motion of a single-DOF system with no damping (when $\zeta = 0$). Thus, we make the inference: **A system's motion started from rest under action of a separate load's component in decomposition over natural modes goes on proportionally to the chosen vibration mode according to the law of motion of an elastic single-DOF system at the frequency of natural vibration corresponding to this mode**.

The sum (32.21) is the solution of the original equation of motion (32.18) with load (32.20) if every modal coordinate satisfies equation (32.24). To make sure of that, it is sufficient to sum up identities (32.23) over all i-s. If, besides, every function $q_i(t)$ satisfies zero initial conditions (32.25), then total solution (32.21) satisfies zero conditions (32.19). Thus, we come to the inference: **The displacement of a system from its initial state of rest due to load is the sum of displacements produced by the components of load in its decomposition over natural modes**.

In order to determine the system's motion due to arbitrary loads, we must solve differential equation (32.24) for every $i = 1, 2, \dots, n$ under zero initial conditions. Such a solution is known; it is called Duhamel's integral and has the form (Chopra 2012, p. 129):

$$q_i(t) = \int_0^t \frac{H_i(\tau)}{\omega_i}\sin\omega_i(t-\tau)\,d\tau. \tag{32.26}$$

The reader can verify that function (32.26) really turns equation (32.24) into an identity and satisfies conditions (32.25).

The time-dependent multipliers $H_i(t)$ in decomposition (32.20) are easily determined by means of the *orthogonality property of natural modes*. Firstly, we phrase and prove the theorem representing this property.

Theorem about Orthogonality of Natural Modes

For two arbitrary natural modes $\mathbf{a}_i$ and $\mathbf{a}_j$ which correspond to different natural frequencies, the next equality holds:

$$\mathbf{a}_i^{\mathrm{T}}\mathbf{m}\mathbf{a}_j = \sum_{k=1}^{n} a_{ik}m_k a_{jk} = 0, \tag{32.27}$$

where $\mathbf{a}_i^{\mathrm{T}} = (a_{i1}, a_{i2}, \dots, a_{in})$ is the transpose of the modal vector.

Proof

We shall consider two equilibrium states of a system produced by static forces exerted upon concentrated masses and equal to the amplitude inertia forces during natural vibrations of system. In state 1, we exert amplitude inertia forces which arise due to natural vibration having mode $\mathbf{a}_i$, i.e., with oscillations $\mathbf{u} = \mathbf{a}_i \sin \omega_i t$. These forces constitute the vector

$$\mathbf{X}_i = -\mathbf{m}\ddot{\mathbf{u}} = \mathbf{m}\omega_i^2 \mathbf{a}_i$$

and are attained on the displacements $\mathbf{u} = \mathbf{a}_i$. In state 2, we exert the forces which arise due to vibration with mode $\mathbf{a}_j$. These forces constitute the vector:

$$\mathbf{X}_j = \mathbf{m}\omega_j^2 \mathbf{a}_j$$

and are attained on the amplitude displacements $\mathbf{u} = \mathbf{a}_j$. By applying the theorem of reciprocal works to these two states, we obtain equality:

$$\mathbf{a}_j^{\mathrm{T}} \mathbf{X}_i = \mathbf{a}_i^{\mathrm{T}} \mathbf{X}_j,$$

or:

$$\omega_i^2 \mathbf{a}_j^{\mathrm{T}} \mathbf{m} \mathbf{a}_i = \omega_j^2 \mathbf{a}_i^{\mathrm{T}} \mathbf{m} \mathbf{a}_j.$$

Without matrix notations, this condition can be rewritten as follows:

$$(\omega_i^2 - \omega_j^2) \sum_{k=1}^{n} a_{ik} m_k a_{jk} = 0.$$

The latter equality, taking into account that $\omega_i \neq \omega_j$, proves the theorem.

Note that coincidence of natural frequencies for different vibration modes is a possible but very rare occurrence. In the latter case, natural modes might be non-orthogonal in the sense of equality (32.27), but in this case, too, we can pass to a set of orthogonal natural modes. Below, we presume that all natural modes are mutually orthogonal.

Now multiply the left-hand and right-hand sides of equality (32.20) by any vector $\mathbf{a}_j^{\mathrm{T}}$. The identity is obtained:

$$\mathbf{a}_j^{\mathrm{T}} \mathbf{P}(t) = \sum_{i=1}^{n} H_i(t) \mathbf{a}_j^{\mathrm{T}} \mathbf{m} \mathbf{a}_i.$$

Due to the orthogonality of modes, on the right side, all addends with the numbers $i \neq j$ turn into zero, which gives:

$$\mathbf{a}_j^{\mathrm{T}} \mathbf{P}(t) = H_j(t) \mathbf{a}_j^{\mathrm{T}} \mathbf{m} \mathbf{a}_j.$$

Finally, we get the required expression, which determines decomposition of load over natural modes:

$$H_j(t) = \frac{\mathbf{a}_j^{\mathrm{T}} \mathbf{P}(t)}{\mathbf{a}_j^{\mathrm{T}} \mathbf{m} \mathbf{a}_j}. \tag{32.28}$$

The scalar form of this expression is cumbersome but convenient for calculations:

$$H_j(t) = \frac{\sum_{k=1}^{n} a_{jk} P_k(t)}{\sum_{k=1}^{n} a_{jk} m_k a_{jk}}. \tag{32.29}$$

The described method of determining the motion of a system is referred to as the *modal expansion method*. The assumption about the at-rest condition of a system at the initial point in time was entered for simplicity of substantiation of the method. The generalization of the method for the case of arbitrary starting conditions of motion is given in the supplement to the present chapter.

The modal expansion method enables us to solve the important problem of calculation of a structure's response to seismic shock. The simplest formalization of the problem of finding the earthquake response is as follows. A structure is represented by a simple design diagram of an elastic system with concentrated masses. For example, towers can be represented by the model of a column with concentrated masses (Figure 32.2). The same model is applicable for multi-story buildings if their size in plan is much less than their height and their walls are bearing. For the period of excitation, the translational movement of the earth's crust is specified; for this purpose, the reference system $O'xyz$ is affixed to the bulk of soil in the neighborhood of a structure, and the acceleration vector $\boldsymbol{W} = \boldsymbol{W}(t)$ of this CS is specified under the assumption that movement occurs with no rotation. The movement of the crust is considered in another CS $Ox_0y_0z_0$ which can be taken as inertial (Figure 32.2). The dynamics of a structure are considered in the CS bound to the structure, i.e., in CS $O'xyz$, which moves together with the soil basis. It is known that in moving CS, the inertia forces upon an object arise additionally to the forces of the object's interaction with an environment. The inertia forces are applied to every mass m_i of the object and, in the case of translational movement of the CS, are specified by the vectors:

$$\boldsymbol{P}_i = -m_i \boldsymbol{W}. \tag{32.30}$$

While forming the design scheme, we replace the movement of soil by variable-over-time seismic loads (32.30) exerted upon all masses of a structure.

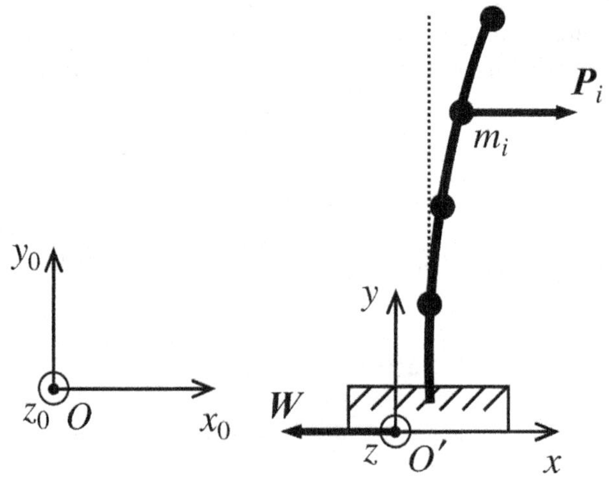

FIGURE 32.2 Design diagram of a tower subjected to seismic shock.

The dynamics of structures with multiple DOFs was studied here with disregard for energy dissipation. Such neglect is not always allowable. In the next item of the chapter, we show how the solutions of problems in structural dynamics can be made more precise by taking into account the energy losses in a system.

32.3 ANALYSIS OF AN ELASTIC SYSTEM'S MOTION TAKING INTO ACCOUNT ENERGY DISSIPATION

In Chapter 30, we studied the motion of an elastic system with a single DOF, the defining characteristic of which is proportionality of the system's displacements, i.e., one can take any nonzero displacement in a system being deformed, and consider **arbitrary** displacement in the system as proportional to the selected displacement. In Chapter 30, for such "basic" displacement, we took the displacement of the load application point and accepted it as the DOF. This property of the displacements' proportionality allowed us to construct an energy model of a single-DOF system and to set up the equation of its motion, taking into account the energy dissipation. In the present chapter, we have established that the motion of a multi-DOF elastic system is represented as the superposition of the motions, each of which has the property of proportionality to one of the natural vibration modes. One could say that response $\mathbf{u}(t)$ to an impact $\mathbf{P}(t)$ is the superposition of responses (32.22), each of which was produced by the impact constituent in the expansion (32.20). It is allowed to determine the energy dissipation caused by a motion proportional to the vibration mode, according to the law established for single-DOF systems, i.e., by the relation (30.17) where q is the proportionality factor. We shall accept the hypothesis that **every response in superposition of mode-proportional responses is independent of the rest**. In accordance with this hypothesis, we supplement equation (32.24) in modal coordinate q_i by a dissipative term and shall consider the system of uncoupled equations:

$$\ddot{q}_i + \omega_i^2 q_i + 2\zeta_i\omega_i\dot{q}_i = H_i(t), \quad i = \overline{1,n}. \tag{32.31}$$

Here, ζ_i is the damping ratio which depends on natural vibration frequency, i.e., is determined by some function $\zeta = \zeta(\omega)$ with possible argument's values $\omega = \omega_1, \omega_2, \dots, \omega_n$. Further on, we employ the equation set (32.31) in order to take into account energy dissipation in multi-DOF systems. We execute the transition from modal coordinates to displacement vector $\mathbf{u}(t)$ by means of the superposition (32.21) of natural vibrations. The construction of the math model of system motion with account for energy dissipation requires the following problems to be solved:

1. The method must be developed to determine dependence $\zeta = \zeta(\omega)$ which would ensure the math modeling consistent with empirical data.
2. The equation of motion should be composed in unknown displacement vector (32.21) from the condition of equivalence to the equation system (32.31) in modal coordinates.

The solution of the second problem, besides its theoretical significance, allows us to take into account the energy dissipation in the dynamic force and displacement methods.

The equation for displacement vector $\mathbf{u}(t)$ is customarily written in a form similar to equation (30.19) of a single-DOF structure's motion. In the absence of energy dissipation, the traditional form of a dynamic equation for a multi-DOF system has the form:

$$\mathbf{m}\ddot{\mathbf{u}} + \mathbf{r}\mathbf{u} = \mathbf{P}; \quad \mathbf{r} \equiv \delta^{-1},$$

which is equivalent to equation (32.18). To take into account the energy dissipation, we supplement this equation as follows:

$$\mathbf{m}\ddot{\mathbf{u}} + \mathbf{r}\mathbf{u} + \mathbf{c}\dot{\mathbf{u}} = \mathbf{P}. \tag{32.32}$$

The term $\mathbf{c}\dot{\mathbf{u}}$ represents damping forces which counteract the loads P_i. These forces, in the given case, can be called the forces of viscous friction, for they are proportional to the velocity of displacements. Matrix $\mathbf{c}$ is referred to as a damping matrix, and we are required to obtain it from the condition of equivalence of vector equation (32.32) to the scalar equations system (32.31).

The stated problems of the math model's construction are related to one another. The point is, that the damping matrix in the vector equation of the structure's motion can be matched to arbitrary positive function $\zeta = \zeta(\omega)$ with definitional domain $\omega \in (0, +\infty)$, but such a matrix is not always easily calculated. On another hand, there is no need to aim for high precision of the motion representation while finding function $\zeta = \zeta(\omega)$, and the reasons of that are as follows: (1) formula (30.17) determines energy dissipation with some error, sometimes large, and thus original equations (32.31) do not describe the system motion exactly at any damping ratio; (2) for practical purposes, it is important to ensure the generality in the selection of function $\zeta = \zeta(\omega)$, i.e. the possibility to determine this function without any difficulties for different types of structures, even to the prejudice of calculation accuracy. The most known and convenient model of motion with energy dissipation is the model proposed by Lord Rayleigh (1877, p. 98). The damping of oscillations compliant with this model is referred to as *Rayleigh damping*, and corresponding matrix $\mathbf{c}$ is referred to as a *classical damping matrix*.

While choosing a damping function, we proceed from its indicative curve in Figure 32.3a. At low natural frequencies, the damping factors are approximately the same and the curve is close to the horizontal; at higher natural frequencies, the damping increases sharply, and owing to this, the high frequencies are not contained in the vibration spectrum and do not manifest themselves at all. To such shape of the curve, the next function corresponds:

$$\zeta = \frac{1}{2}\left(\frac{b_1}{\omega} + b_2\omega\right). \tag{32.33}$$

Coefficients b_1 and b_2 are determined by using damping ratios known from experience for two natural frequencies (so-called underlying frequencies); multiplier ½ will ensure the simple form of the damping matrix. Usually, the damping ratio is taken as the same for both underlying frequencies by using empirical data. We denote this ratio as ζ_{emp}. The underlying frequencies are selected in such manner that the vibrations at these frequencies make the main contribution in the motion of a system; fundamental frequency ω_1 is always selected and, if there are no reasonable assumptions about system vibration, the first higher natural frequency ω_2 is selected as a second underlying frequency ω_s (i.e., index $s = 2$ is accepted). If the system is subjected to harmonic loads (32.1) with a frequency that is not too large ($\theta < 3\omega_1$), then one might select a second underlying frequency ω_s

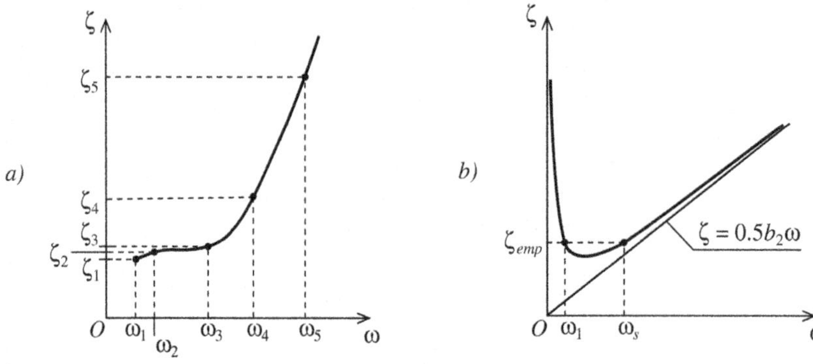

FIGURE 32.3 Typical dependence $\zeta = \zeta(\omega)$ constructed by interpolation of empirical points (a) and Rayleigh approximation of similar dependence (b). On diagram b: $\zeta = 0.5b_2\omega$ – right asymptotical line; axis $O\zeta$ – left asymptotical line.

$\approx \theta$. The curve of the dumping function with specified damping ratio ζ_{emp} and selected frequencies ω_1, ω_s is presented in Figure 32.3b, and coefficients of this function are easily obtained in the form:

$$b_1 = \zeta_{emp}\frac{2\omega_1\omega_s}{\omega_1+\omega_s}; \quad b_2 = \zeta_{emp}\frac{2}{\omega_1+\omega_s}. \tag{32.34}$$

We shall obtain the damping matrix **c** for the damping ratio (32.33). There is the following set of motion equations in modal coordinates:

$$\ddot{q}_i + \omega_i^2 q_i + \left(b_1 + b_2\omega_i^2\right)\dot{q}_i = H_i(t), \quad i = \overline{1,n}. \tag{32.35}$$

Let us repeat the math manipulations which enabled the transition from equation (32.18) to uncoupled equations (32.24), but take the damping term in (32.31) into account and do the manipulations in inverse order. Multiply the left-hand and right-hand sides of every equation (32.35) by the vector $\omega_i^{-2}\mathbf{a}_i$ and make the replacement $\omega_i^{-2}\mathbf{a}_i = \boldsymbol{\delta}\mathbf{m}\mathbf{a}_i$ on both sides. We obtain (with no permutation of terms):

$$\ddot{q}_i\boldsymbol{\delta}\,\mathbf{m}\mathbf{a}_i + q_i\mathbf{a}_i + b_1\dot{q}_i\boldsymbol{\delta}\,\mathbf{m}\mathbf{a}_i + b_2\dot{q}_i\mathbf{a}_i = H_i(t)\boldsymbol{\delta}\mathbf{m}\mathbf{a}_i. \tag{32.36}$$

Sum up the left-hand sides of all n equations, and do the same with the right-hand sides. By equating both sums and making replacement (32.20), we come to the following equation for displacement vector (32.21):

$$\boldsymbol{\delta}\,\mathbf{m}\ddot{\mathbf{u}} + \mathbf{u} + \left(b_1\boldsymbol{\delta}\,\mathbf{m} + b_2\mathbf{I}\right)\dot{\mathbf{u}} = \boldsymbol{\delta}\mathbf{P}, \tag{32.37}$$

where **I** is the identity matrix. Pre-multiplication by matrix $\boldsymbol{\delta}^{-1}$ gives equation (32.32) in which the damping matrix has the form:

$$\mathbf{c} = b_1\mathbf{m} + b_2\mathbf{r}. \tag{32.38}$$

So, the problem of transforming the motion equation to the standard form (32.32) has been solved, and the obtained classical damping matrix is easily calculated.

Lord Rayleigh, in the book *The Theory of Sound* (1877), pointed out that the form of damping matrix (32.38) is convenient for analysis of the multi-DOF system's response, because it enables us to obtain the set of uncoupled differential equations governing modal coordinates. However, practical applicability of such a model of motion wasn't confirmed at that time, and it required more than 80 years to establish that the model of structure motion consisting of equation (32.32) and matrix (32.38) is applicable for a wide range of structures. During this period, the limitations of the model with classical damping matrix were established as well, and the methods of damping matrix refinement appeared, also based on empirical data (Caughey 1960).

Now, we come back to the problem of forced vibration analysis, which was considered in i.1. While substantiating the dynamic force method, we stuck to the assumption (32.2) of harmonic vibration of a system. The motion equation in the given problem is obtained by substitution of Newton's second law into equation (32.4) of dynamic equilibrium and has the form:

$$\boldsymbol{\delta}\,\mathbf{m}\ddot{\mathbf{u}} + \mathbf{u} = \Delta_P. \tag{32.39}$$

The general solution of this equation is the sum of its general solution under zero right-hand side (so-called complementary solution) and any particular solution, for instance, (32.2). We write down this general solution in the form:

$$\mathbf{u} = \mathbf{u}^{(1)} + \mathbf{u}^{(2)}, \tag{32.40}$$

where $\mathbf{u}^{(1)}$ is the complementary solution obtained earlier in the form (31.27), and $\mathbf{u}^{(2)}$ is the particular solution (32.2). The first constituent of the general solution is the superposition of natural system vibrations, each of which is harmonic vibration proportional to some natural mode. Every displacement $u_j^{(1)}(t)$ is represented, according to expression (31.27), as the time function:

$$u_j^{(1)}(t) = \sum_{i=1}^{n} a_{ij} C_i \sin(\omega_i t + \varphi_i). \tag{32.41}$$

Constants C_i may take arbitrary values corresponding to initial conditions of motion. Let the initial state be such that the k-th harmonic of displacement $u_j^{(1)}(t)$ dominates both over all other harmonics of the sum (32.41) and over all harmonics of particular solution (32.2). In other words, we assume that $a_{kj} \neq 0$ at some k and j, and initial conditions are such that $|a_{kj}C_k| >> |a_{ij}C_i|$ for all $i \neq k$, and $|a_{kj}C_k| >> |u_i^A|$ at arbitrary i. Then, we have at some moment of time within an **arbitrary** period of the k-th harmonic:

$$u_j^{(1)}(t) \approx |a_{kj}C_k| >> |u_i^A|,$$

i.e., component $u_j^{(1)}$ of vector $\mathbf{u}^{(1)}$ periodically becomes much greater than amplitude values of all the components of vector $\mathbf{u}^{(2)}$. Moreover, it can be shown that even if the structure was at rest at the starting time point of observation, then inequality $u_j^{(1)}(t) >> |u_i^A|$ also can take place with a certain periodicity (Chopra 2012, p. 67). So, neglecting the constituent $\mathbf{u}^{(1)}$ in the general solution of equation (32.39) is not allowed, and assumption (32.2), made for substantiation of the force method, generally speaking, is inapplicable.

But now we shall take into account the energy losses in a system by means of the model with Rayleigh damping: assume that the loads are exerted upon lumped masses and the motion equation has the form (32.37). In this case, the general solution of the homogeneous motion equation will represent the dying-out process of motion and looks like the following:

$$\mathbf{u}^{(1)} = \sum_{i=1}^{n} q_i^{(1)}(t)\mathbf{a}_i, \tag{32.42}$$

$$q_i^{(1)}(t) = \begin{cases} C_{1i} e^{-\zeta_i \omega_i t} \sin(\omega_{Di} t) + C_{2i} e^{-\zeta_i \omega_i t} \cos(\omega_{Di} t), & \text{at } \zeta_i < 1; \\ C_{1i} \exp\left[-\omega_i t\left(\zeta_i + \sqrt{\zeta_i^2 - 1}\right)\right] + C_{2i} \exp\left[-\omega_i t\left(\zeta_i - \sqrt{\zeta_i^2 - 1}\right)\right], & \text{at } \zeta_i \geq 1. \end{cases} \tag{32.43}$$

In the latter formula, we used general solutions (30.22) and (30S.18) of the equation describing free motion of a single-DOF structure; for the frequency of damped vibration, the next denotation is used:

$$\omega_{Di} = \omega_i \sqrt{1 - \zeta_i^2}; \tag{32.44}$$

constants C_{1i} and C_{2i} are real and determined by initial conditions. On a sufficiently long vibration process, the constituent of free vibration $\mathbf{u}^{(1)}$ will be as small as necessary, and it will be sufficient to take into account the only constituent of forced vibration $\mathbf{u}^{(2)}$ in the total displacement vector (32.40). This constituent – under condition of deviation from resonant frequencies – can be determined with no account for energy losses, i.e., according to equations (32.11). So, **the dynamic force method is applicable for seeking the forced steady-state vibration of a structure while preventing the possibility of resonance.**

In the case of Rayleigh damping of a system subjected to arbitrary loading, usually, we can simplify the problem of determining the displacement vector by seeking an approximate solution which is obtained by taking into consideration only underdamped modal coordinates in solution (32.21). In other words, we find the solution in the form:

$$\mathbf{u} \approx \sum_{i=1}^{n_1} q_i(t)\mathbf{a}_i, \tag{32.45}$$

where n_1 is the greatest index i such that $\zeta_i < 1$. If a system is at rest at the initial instant of time, then the modal coordinates of the approximate solution are found by means of Duhamel's integral, which now has the somewhat more complicated form:

$$q_i(t) = \int_0^t \frac{H_i(\tau)}{\omega_{Di}} e^{-\zeta_i\omega_i(t-\tau)} \sin\omega_{Di}(t-\tau)\,d\tau, \quad i = \overline{1, n_1}, \tag{32.46}$$

where frequency ω_{Di} is determined by formula (32.44). The solution (32.46) is easily verified by substitution into (32.31).

In structural design, an engineer almost always limits himself by approximate solutions obtained through taking into account several modal coordinates corresponding to lower natural frequencies. Theoretically, the situation is possible when n_1 solutions (32.46) do not secure precision of calculations, but in practice, this possibility is ignored. The issue of the necessary number of modal coordinates is considered in the special course of dynamics. Here, we note that the models of structures which a designer would deal with might have several hundreds of thousands of DOFs, but in dynamic analysis based on such models, it is sufficient to take into account only three-five natural modes.

SUPPLEMENT TO CHAPTER

Taking into Account Initial Motion of System in the Modal Expansion Method

In the present chapter, we have established that the vector equation of a structure's motion (32.18) is equivalent to the system of equations (32.24) under $i = \overline{1, n}$, governing modal coordinates. The equivalence was established through representation of sought displacement vector in the form of a linear combination of natural modes (32.21) and decomposition of load over natural modes in the form (32.20). The equivalence takes place with no regard for the initial conditions of system motion. Here, we shall obtain the solution for a system composed of uncoupled equations (32.24) at arbitrary given vectors of initial displacements $\mathbf{u}(0)$ and displacement velocities $\dot{\mathbf{u}}(0)$.

Initial conditions for coordinates $q_i(t)$ are obtained by solving vector equations:

$$\sum_{i=1}^{n} q_i(0)\mathbf{a}_i = \mathbf{u}(0); \quad \sum_{i=1}^{n} \dot{q}_i(0)\mathbf{a}_i = \dot{\mathbf{u}}(0). \tag{32S.1}$$

These equations have unique solutions $q_i(0)$, $\dot{q}_i(0)$, $i = \overline{1, n}$, owing to linear independence of natural modes.

The solution of every equation (32.24) can be found as the sum of two terms:

- The solution of the same homogeneous equation (assuming $H_i(t) = 0$) under established initial conditions $q_i(0)$, $\dot{q}_i(0)$;
- The solution of the given equation with the given right-hand side and under zero initial conditions.

It is obvious enough that a such sum satisfies both the equation itself and the initial conditions. The second constituent of the sought solution was obtained in this chapter in the form (32.26). We shall dwell upon finding the first constituent; the latter we denote $q_i^{(1)}(t)$.

In Chapter 30, we obtained the solution of the differential equation governing free vibration of a single-DOF system. This solution has the form (30.22) and is applicable to the problem under consideration if $\zeta = 0$. Changing denotations in formula (30.22) according to the problem being considered, we write down the wanted constituent as follows:

$$q_i^{(1)} = C_{i1}\sin\omega_i t + C_{i2}\cos\omega_i t.$$

We determine the factors at trigonometric functions from the initial conditions for the coordinate $q_i(t)$ and obtain:

$$q_i^{(1)} = \frac{\dot{q}_i(0)}{\omega_i}\sin\omega_i t + q_i(0)\cos\omega_i t. \tag{32S.2}$$

The wanted complete solution has the form:

$$q_i(t) = \frac{\dot{q}_i(0)}{\omega_i}\sin\omega_i t + q_i(0)\cos\omega_i t + \int_0^t \frac{H_i(\tau)}{\omega_i}\sin\omega_i(t-\tau)\,d\tau. \tag{32S.3}$$

So, the generalization of the method for the case of arbitrary initial conditions of motion lies in supplementing this method with equations (32S.1) in unknown initial modal coordinates and replacement of solution (32.26) by more general solution (32S.3).

In this method, we also can take into account the forces of viscous damping acting in a system. It was established in the chapter that motion equation (32.32) with damping matrix (32.38) is equivalent to equation system (32.31) with damping ratios which depend on natural frequencies according to the function (32.33). The method of modal expansion considered in the chapter enables us to determine the underdamped modal coordinates and obtain an approximate solution of the problem in the form (32.45). In general case, the initial conditions for modal coordinates are determined from equations (32S.1). The technique of finding damped modal coordinates remains the same: we seek the solution of equation (32.31) as the sum of two solutions, the sense of which was defined above. As a result, we come to following cumbersome relationship (Chopra 2012, pp. 49, 129):

$$q_i(t) = e^{-\zeta_i\omega_i t}\left(\frac{\dot{q}_i(0)+\zeta_i\omega_i q_i(0)}{\omega_{Di}}\sin\omega_{Di}t + q_i(0)\cos\omega_{Di}t\right) + \int_0^t \frac{H_i(\tau)}{\omega_{Di}}e^{-\zeta_i\omega_i(t-\tau)}\sin\omega_{Di}(t-\tau)\,d\tau. \tag{32S.4}$$

Here, ω_{Di} is the frequency of damped vibration (32.44). The obtained expression is reduced to formula (32S.3) at $\zeta_i = 0$.

Bibliography

AASHTO LRFD. 2012. *Bridge Design Specifications* LRFDUS-6. Washington, D.C.: American Association of State Highway and Transportation Officials.

Beer, F.P., E.R. Johnston Jr., J.T. Dewolf and D.F. Mazurek. 2012. *Mechanics of Materials.* New York: McGraw-Hill.

Boyce, W.E. and R.C. DiPrima. 2012. *Elementary Differential Equations and Boundary Value Problems.* Hoboken, NJ: John Wiley & Sons.

Building Code Requirements for Structural Concrete (ACI318-11) and Commentary. 2011. American Concrete Institute.

Case, J., L. Chilver and C.T.F. Ross. 1999. *Strength of Materials and Structures.* Oxford: Butterworth-Heinemann.

Caughey, T.K. 1960. Classical Normal Modes in Damped Linear Dynamic Systems. *Journal of Applied Mechanics.* Vol. 27, Issue 2 (June): 269–271.

Chopra, A.K. 2012. *Dynamics of Structures: Theory and Applications to Earthquake Engineering.* Boston, MA: Prentice Hall.

Craig, R.R. Jr. 1981. *Structural Dynamics (An Introduction to Computer Methods).* New York: John Wiley & Sons.

Davies, N. and E. Jokiniemi. 2008. *Dictionary of Architecture and Building Construction.* Oxford: Architectural Press (Elsevier Ltd.).

Eurocode 1: Actions on Structures – Part 2: Traffic Loads on Bridges. 2003. EN 1991-2:2003 E. Brussels: European Committee For Standardization.

Gvozdev, A.A. 1949. *Analysis of Structures' Bearing Capacity by Method of Ultimate Equilibrium. I. The Essence of Method and Its Grounding. (Raschet Nesushchei Sposobnosti Sooruzhenii po Metodu Predel'nogo Ravnovesiya. I. Sushchnost' Metoda i ego Obosnovanie.)* Moscow: Stroiizdat. (In Russian).

Hibbeler, R.C. 2011. *Mechanics of Materials.* Upper Saddle River, NJ: Pearson Prentice Hall.

Hibbeler, R.C. 2012. *Structural Analysis.* Upper Saddle River, NJ: Pearson Prentice Hall.

Horn, R.A. and Ch.R. Johnson. 2013. *Matrix Analysis.* New York: Cambridge University Press.

International Building Code. 2015. Country Club Hills, IL: International Code Council, Inc.

Jones, R.M. 2006. *Buckling of Bars, Plates, and Shells.* Blacksburg, VA: Bull Ridge Publishing.

Jones, R.M. 2009. *Deformation Theory of Plasticity.* Blacksburg, VA: Bull Ridge Publishing.

Karnovsky, I.A. 2012. *Theory of Arched Structures: Strength, Stability, Vibration.* New York: Springer Science + Business Media.

Karnovsky, I.A. and O. Lebed. 2010. *Advanced Methods of Structural Analysis.* New York: Springer Science + Business Media.

Kohnke, Peter (ed.). 2001. *ANSYS, Inc. Theory. Release 5.7.* Canonsburg, PA: ANSYS, Inc.

Lagrange, J.L. 1997. *Analytical Mechanics (Translated from the "Mécanique analytique, novelle edition" of 1811).* Dordrecht: Springer Science + Business Media.

Liapounoff, A.M. 1897. Sur l'instabilité de l'équilibre dans certains cas où la fonction de forces n'est pas un maximum. *Journal de Mathématiques Pures et Appliquées.* Ser. 5. Vol. 3: 81–94.

Loitsianskii, L.G. and A.I. Lurie. 1940. *Course in Theoretical Mechanics. Part I. (Kurs teoreticheskoi mehaniki. I.)* Leningrad, Moscow: GITTL. (In Russian)

Lord Rayleigh. 1874. A Statical Theorem. *The London, Edinburgh, and Dublin Philosophical Magazine and Journal of Science.* Vol. 48, No. 320: 452–456.

Lord Rayleigh. 1877. *The Theory of Sound. Vol. I.* London: Macmillan and Co.

Love, A.E.H. 1906. *A Treatise on the Mathematical Theory of Elasticity.* Cambridge: University Press.

Lu, L.W., B.G. Chapman and G.C. Driscoll Jr. 1958. Plastic strength and deflection of a gabled frame. Fritz Engineering Laboratory Report No. 205D.8. Bethlehem, PA: Lehigh University.

Manual for Railway Engineering. Vol. 2. Structures. 2015. Lanham, MD: American Railway Engineering and Maintenance of Way Association.

Mertz, D. 2012. *Steel Bridge Design Handbook: Loads and Load Combinations.* FHWA-IF-12-052 – Vol. 7. Washington, D.C.: U.S. Department of Transportation. Federal Highway Administration. Pittsburgh, PA: HDR Engineering, Inc.

Minimum Design Loads for Buildings and Other Structures. 2016. ASCE/SEI 7-16. Reston, VA: American Society of Civil Engineers.

Neal, B.G. 1977. *The Plastic Methods of Structural Analysis*. London: Chapman and Hall.

Pak, Igor. 2006. A Short Proof of Rigidity of Convex Polytopes. *Siberian Mathematical Journal*. Vol. 47, No. 4 (July–August): 710–713. (Translated from *Sibirskii Matematicheskii Zhurnal*, Cambridge: Springer Science + Business Media, Inc.)

Segal', A.I. 1949. *High-Rise Structures: Calculation of Strength, Rigidity, and Stability*. (*Vysotnye sooruzhenia: Raschet na Prochnost', Zhestkost' i Ustoychivost'*.) Moscow: Stroiizdat. (In Russian).

Smirnov, A.F., A.V. Aleksandrov, B.Ya. Lashchenikov and N.N. Shaposhnikov. 1981. *Structural Mechanics: Bar Systems. (Stroiteljnaya Mehanika: Sterzhnevye Sistemy.)* Moscow: Stroiizdat. (In Russian).

Timoshenko, S. 1936. *Theory of Elastic Stability*. New York: McGraw-Hill Book Company, Inc. (Second edition is coauthored with J.M. Gere in 1961.)

Timoshenko, S. 1948. *Strength of Materials. Part 1. Elementary Theory and Problems*. New York: D. Van Nostrand Company, Inc.

Timoshenko, S. and J.N. Goodier. 1951. *Theory of Elasticity*. New York: McGraw Hill Book Company, Inc.

Trussed Rafter: Technical Manual. 2004. Shilton, Coventry: Wolf Systems Ltd.

White, D. 2012. *Steel Bridge Design Handbook: Structural Behavior of Steel*. FHWA-IF-12-052 – Vol. 4. Washington, D.C.: U.S. Department of Transportation. Federal Highway Administration. Pittsburgh, PA: HDR Engineering, Inc.

Index

O

P

Q

R

S

T